T0362454

Contemporary Ergonomics 2004

Contemporary Ergonomics 2004

Edited by

Paul T. McCabe

Atkins, UK

THE **E**rgonomics society

CRC Press
Taylor & Francis Group
Boca Raton London New York

CRC Press is an imprint of the
Taylor & Francis Group, an **informa** business

On page 496, Figure 1a is reprinted courtesy of Ed McFadden, IBM Global Services, and Don Tepas, the University of Connecticut. The authors' data represented in the figure were collected on 18 nonconsecutive days.

Library of Congress Cataloging-in-Publication Data

Catalog record is available from the Library of Congress

Visit the CRC Press Web site at www.crcpress.com

CRC Press
Taylor & Francis Group
6000 Broken Sound Parkway NW, Suite 300
Boca Raton, FL 33487-2742

First issued in hardback 2018

© 2004 by Taylor & Francis Group, LLC
CRC Press is an imprint of Taylor & Francis Group, an Informa business

No claim to original U.S. Government works

ISBN 13: 978-1-138-43305-2 (hbk)
ISBN 13: 978-0-8493-2342-3 (pbk)

CONTENTS

INCLUSIVE DESIGN

MODELLING & USABILITY

INTERFACE DESIGN

BIOMECHANICS

MUSCULOSKELETAL DISORDERS

OCCUPATIONAL HEALTH AND SAFETY

ROAD TRANSPORT

RAIL

AIR TRAFFIC CONTROL

DEFENCE

HUMAN RELIABILITY

PATIENT SAFETY

WARNINGS

METHODS

GENERAL ERGONOMICS

PREFACE

Contemporary Ergonomics 2004 is the proceedings of the Annual Conference of the Ergonomics Society, held in April 2004 at Swansea University, Swansea, UK. The conference is a major international event for ergonomists and human factors specialists and attracts contributions from around the world.

Papers are chosen by a selection panel from abstracts submitted in the autumn of the previous year and the selected papers are published in *Contemporary Ergonomics*. Papers are submitted as camera ready copy prior to the conference. Authors are responsible for the presentation of their papers. Details of the submission procedure may be obtained from the Ergonomics Society.

The Ergonomics Society is the professional body for ergonomists and human factors specialists based in the United Kingdom. It also attracts members throughout the world and is affiliated with the International Ergonomics Association. It provides recognition of competence of its members through its Professional Register. For further details contact:

The Ergonomics Society
Devonshire House
Devonshire Square
Loughborough
Leicestershire
LE11 3DW
UK

Tel: (+44) 1509 234 904
Fax: (+44) 1509 235 666

Email: ergsoc@ergonomics.org.uk
Web page: http://www.ergonomics.org.uk

SLIPS, TRIPS & FALLS

Predicting Losses of Balance on Moving Platforms

RS Bridger[1] and P Crossland[2]

[1]Institute of Naval Medicine,Crescent Rd, Alverstoke, Hants, PO12 2DL,
hhfd@inm.mod.uk.
[2]QinetiQ,Haslar Marine Technology Park,
Haslar Rd, Gosport, Hants, PO12
2AG, pcrossland@qinetiq.com

Two groups of 5 subjects carried out 5 tasks on a ship at sea while deck motions were measured. Losses of balance (MIIs) and instantaneous deck accelerations were recorded. Significant effects on MIIs due to the direction of the accelerations and the nature of the task were observed. Some tentative guidelines for workspace design on ships are offered.

Introduction

About 30% of injuries on Royal Navy ships are caused by slips, trips and falls and another 7% by manual handling. About two thirds of falls are from 1 level to another. The other third involves tripping over things or slipping on decks. Ship motion can be seen as a biomechanical stressor that places personnel at risk of injury due to losses of balance and it is likely that it plays a role in the incidence of injury. The present paper describes an investigation of losses of balance during the performance of naval tasks aboard a ship at sea.

The simplest model of postural stability, widely used in both the naval and commercial ship design community, is a rigid body similar in size and shape to a human body. In general terms, the model predicts that a person will lose balance during a simple gross motor task when the accelerations experienced by that person exceed a threshold. The thresholds of accelerations can be defined by the tipping coefficient - a function of body COG height and half the stance or shoe width (i.e. half the length of the base of support). For example, an individual facing athwartships, adopting a natural stance with feet apart, but in-line, would be less able to resist the lateral accelerations caused by roll motions in their saggital plane than if facing longitudinally and experiencing the same accelerations in the transverse plane. For equal stance width, a short individual would be more stable than a taller one.

Human subjects are not rigid and can, in some cases, anticipate the motion and utilise both ankle and knee strategies to resist de-stabilising forces. In real shipboard tasks, COG varies with posture. In dynamic tasks COG height varies and simple mechanical models overestimate postural instability.

The purpose of the present investigation was to investigate losses of balance (stumbles, falls or shifts in posture to remain balanced - hereafter "motion-induced interruptions" or MIIs) during the performance of generic activities ("tasks"). By observing the MII rate in different tasks under known conditions of deck acceleration on a ship at sea, empirical values of tipping coefficients could be derived (see Crossland and Rich, 1998 and Crossland et al., 2003, for a mathematical discussion of these matters).

Method

Ten Royal Naval subjects took part in the study, in 2 groups of 5 on separate voyages, each volunteering and giving their informed, written consent. The study complied at all times with the Declaration of Helsinki, as adopted at the 52nd WMA General Assembly, Edinburgh, October 2000. The trial took place on board RV Triton in January 2001 in slight to moderate seas and in March 2001 in rough to severe seas. The trial was carried out in the Western Approaches to the English Channel. Due to the limited accommodation, there were only 5 subjects per voyage.

Tasks
Subjects performed a series of tasks each of 4 minutes duration:

1. Standing facing aft (SFA).
The subject was required to stand along the roll axis of rotation of the ship with the feet spaced approximately shoulder-width apart and not offset.

2. Walking a designated track (WA).
This task was chosen to investigate the effects of ship motions on walking. Each subject was required to walk a track marked out in the trial laboratory along the roll axis of the ship. The task was 'self paced' and the subject instructed to walk as smoothly as possible, at their usual walking speed.

3. Carrying out a simulated weapons loading task (WL).
This task represented the chaff missile-loading task undertaken at sea. Each subject was required to shift two 11.1kg cylinders from the bottom of the missile rack to the top and then return them to the bottom again. Subjects worked at their own pace. The task was designed to force gross positive repositioning on the subject, in order to stimulate MIIs.

4. Standing facing athwartships (SFX).
This task was chosen to validate the longitudinal or body fore-and-aft MIIs with the subject facing the pitch axis of rotation. The feet were placed slightly apart (approximately 480 mm, varying slightly according to the subject's preference within a range of 450 – 510 mm) and not offset.

5. Simulated fire-fighting task (FF).
Each subject was required to stand holding a fire hose and simulate boundary cooling. The simulation required the subject to move the hose nozzle in a figure of eight pattern, as if covering the bulkhead with a spray of water. The experimenter gave instructions for the method of holding the hose and manoeuvring it in the figure of eight. The weight to be supported was approximately 10 kg. There is negligible load to be resisted when operating a hose used for boundary cooling; consequently the trial did not incorporate a method of simulating the force of expelled water.

Procedure
All subjects performed task 1, one after another. When this task was completed by all, task 2 was carried out in the same order of subjects. This design has the advantage of maximising the chance that all the subjects carried out the same task in the same motion

condition. About 30 minutes was needed for all subjects to complete each task. Interaction effects between tasks were avoided because of the 30 minutes separation between tasks for each subject. The routine was conducted twice a day on a daily basis for the duration of the trial except on day 1, when subjects were given a practice session in calm water. Each subject entered the lab at the start of the session and completed the appropriate questionnaires. This was repeated for all tasks, twice a day. MIIs were recorded by an experimenter seated at a dedicated monitor, curtained from the main lab. A camera in the lab displayed subject movements on the monitor

Results

Ship motions
Sea conditions differed in Jan and March 01. Generally, conditions were much calmer in January with the result that there were fewer MIIs in all tasks. Furthermore, the range of conditions to which subjects were exposed was more limited. The acceleration data from accelerometers positioned closest to the Trials Laboratory on RV Triton were used to estimate the acceleration at each subject's COG as described in Evans et al. (2002). Table 1 summarises the root mean square (RMS) accelerations at the COG of the subject.

Table 1. Mean (and range) RMS Accelerations for Jan 01 (Group 1) and March 01 (Group 2) Voyages on RV Triton

	RMSLateral	RMSLongitudinal	RMSVertical
Group 1*	0.072 (0.057-0.082)	0.0023 (0.0016-0.0033)	0.029 (0.025-0.034)
Group 2**	0.092 (0.041-0.138)	0.0076 (0.0035-0.0187)	0.064 (0.023-0.104)

* Recorded over 4 experimental sessions
** Recorded over 7 experimental sessions

MIIs
Falls to the side and losses of balance in the frontal plane were categorised as MII1. Falls forward of backward and losses of balance in the saggital plane as MII2. Table 2 summarises the MII rates for both voyages in each of the tasks.

The MII rates were greater for group 2 than in group 1. For group 1, MII1s tended to occur when standing facing aft and when walking athwartships. MII2s (falling forwards) tended not to occur when standing facing aft but did occur when standing facing athwartships. For group 2, higher frequencies of MII1s were observed in walking athwartships, firefighting and weapons loading. MII1s were not observed when standing facing athwartships. MII2s were more frequently observed in walking athwartships, standing facing athwartships and the fire-fighting task.

Tipping Coefficients
Tipping coefficients were calculated from the MII data. As can be seen from table 3, the coefficients varied across the tasks and for type of MII.

Table 2. Mean MII Rate (MIIs/min) in 2 Groups of Subjects in 5 Tasks

Task	Group 1 MII1	Group 1 MII2	Group 2 MII1	Group 2 MII2
1.	0.125	0.000	0.300	0.250
2.	0.125	0.825	0.550	2.130
3.	0.000	0.250	0.750	0.525
4.	0.000	1.812	0.000	3.125
5.	0.000	0.315	0.700	1.780

Table 3. Mean Tipping Coefficients for 2 Groups of 5 Subjects Carrying Out 5 Different Tasks

	Task Number 1	2	3	4	5
Group 1	0.223	0.157	0.113	0.145	0.172
Group 2	0.294	0.172	0.168	0.138	0.175

The differences in the magnitude of the coefficients indicate a difference in susceptibility to MIIs when performing the different tasks.

Discussion

For the standing tasks, higher values of RMS accelerations in the direction of least body stability were associated with a higher incidence of MIIs. When walking athwartships, the frequency of MII2 increased as roll accelerations increased and the frequency of MII1 increased as pitch accelerations increased. In the weapons loading task, the frequency of MII1 increased with pitch acceleration and the frequency of MII2 increased with roll acceleration. When standing facing athwartships, the frequency of MII1s increased with pitch accelerations and the frequency of MII2 increased with roll acceleration. In the fire-fighting task, the frequency of MII1 increased with pitch accelerations. The frequency of MII2 increased as roll accelerations increased.

The findings indicate that, not only do MII rates rise with increases in deck acceleration but that this occurs in a way that would be predicted by a simple mechanical model. The size and shape of the base of support described by the position of the feet in relation to the greatest lateral accelerations determines the likelihood that an MII will occur in the direction of the acceleration. This has been discussed elsewhere, for the purposes of ship design and refinement of MII estimates (Crossland et al., 2003). For present purposes, only the ergonomic implications will be discussed.

Comparing the MII results for standing facing aft and standing facing athwartships, it can be seen that the incidence of MII2s (falling forwards or backwards) was much greater in the latter than the former. This is because, when subjects are facing athwartships, the lateral accelerations due to the roll of the vessel act along the length of the feet, that is, in the direction of the narrowest part of the base of support – the part that offers the least resistance to destabilisation. In other words, subjects were better able to resist vessel roll when they were perpendicular to it (i.e. facing forward or aft).

Conversely, MIIs (falling to the side) did not occur when subjects stood athwartships during either of the voyages. When subjects are facing athwartships, the base of support of the feet is large in relation to longitudinal accelerations caused by pitching of the vessel and these accelerations are easily resisted. When standing aft, a small number of MIIs was observed in both voyages. Taken together, these findings indicate that the likelihood of an MII depends on:

- the direction of the greatest deck accelerations (due to the interaction between the vessel and the sea state)
- The direction a person is facing and the nature of the task being performed
- The position of the feet

The last two of these factors are under the control of designers or can be influenced by design. By extension some tentative ergonomic guidelines for safeguarding postural stability on ships can be proposed. These are of most relevance to small vessels or vessels designed for use in heavy seas:

- Align workstations for standing personnel perpendicular to the largest motions in the plane of the deck
- Allow people to move their feet to maintain balance. Provide plenty of free space for the feet around workstations and in areas such as storerooms and gunbays, where manual handling takes place. Items should be stored on raised shelves or pallets to provide foot space underneath the load
- Design workstations so that personnel can work facing aft or facing athwartships, depending on the conditions
- Provide foot and legroom underneath the surfaces of map reading tables and galley worktables. Consider including footrests or foot straps to aid postural stability
- Provide knee space below work surfaces to allow postural adaptation to motion in the legs
- Design manual handling tasks so that lifting is in the same direction as the largest lateral accelerations in the plane of the deck
- Manual handling tasks must not raise the centre of gravity of user and load. Lift no higher than unloaded body COG height
- Place grab rails on the deck head around work surfaces or around bridge consoles.
- Place rails or poles at the tops of hatches and stairwells and along bannisters of stairs to prevent falling, minimise under-shooting and over-shooting when walking along or between levels.
- Maximise the opportunities for bracing

References

Crossland, P., Evans, M.J., Grist, D., Lowten, M., Jones, H., Bridger, R.S. 2003. Motion-induced interruptions aboard ship: Model development and application to ship design. Submitted to *Ergonomics*.
Crossland, P., Rich, KNJC. 1998. Validating a model of the effects of ship motion on postural stability. *The 8th International Conference on Environmental Ergonomics, San Diego, USA*.

MINIMUM GOING REQUIREMENTS AND FRICTION DEMANDS DURING STAIR DESCENT

Marianne Loo-Morrey, Kevin A. Hallas and Steve C. Thorpe

Health & Safety Laboratory,
Broad Lane, Sheffield, S3 7HQ, U.K.
www.hsl.gov.uk

Stair descent is inherently risky, approximately 500 fatalities a year occur as a result of falls on stairs. A number of factors such as going size, the type of flooring material (e.g. carpet, vinyl tile etc.) used on stairs, and friction demands during stair descent all contribute the likelihood of a fall. The main aim of this work was to determine the effect of going size and to investigate the effect of using different flooring materials on stair safety. Findings suggest that minimum permissible stair goings should be increased as follows: Private stairs minimum going = 250 mm, Public stairs minimum going = 300 mm, Assembly stairs minimum going = 350 mm. Ground reaction force data generated during the study was subsequently analysed to determine the friction required during stair descent.

Introduction

Each year there are approximately 500 fatalities on domestic stairs and 100 fatalities on non-domestic stairs making them one of the most hazardous locations in buildings. Studies have shown that accidents may occur during both stair ascent and descent, but it is accidents during descent that present the greater potential for serious outcomes, 80% of injuries reported occurring during descent. The main aims of the study were to improve the safety of stairs by determining the effect of going size and flooring material.

There are significant differences between the gait used for level walking and that used on stairs. During level walking initial contact between the shoe and ground is made by the heel, with the walking speed and stride length influencing the required coefficient of friction. When walking up or down stairs the first contact between the shoe and the stair tread is made with the toes or sole of the shoe with the forward speed being significantly influenced by the stair geometry. These differences in gait have lead to debate in the literature as to how the required coefficient of friction (RCoF) for stair descent compares with that required for level walking. Some studies suggest that the RCoF for stair descent is lower than RCoF for level walking [Riener et al], others suggest that the RCoF levels are comparable [Christina & Cavanagh]. A secondary aim (of this work) was to gain a better understanding of RCoF requirements during stair descent.

Terminology
The geometry of a stair may be described by three terms, the rise, going and pitch, defined as follows: Rise is the vertical distance between consecutive treads, or between a tread and a landing; Going is the horizontal distance between consecutive nosings. (The nosing is defined as the part of the tread that overlaps the tread below); Pitch is the angle between a line joining consecutive nosings and the horizontal. Building regulations

control the limits for these dimensions. The RCoF is the calculated ratio of the horizontal foot force (FH) to the vertical foot force (FV):

$$RCoF = \frac{F_H}{F_V}$$

This represents the minimum coefficient of friction that must be available at the shoe / floor interface to prevent the initiation of a slip

Experimental Method

Experimental work was carried out using the BRE variable going stair rig. The going was varied between 150mm and 425mm, at 25mm intervals, on an eight step stair with a fixed rise of 175mm. Two Kistler Type 9253 force plates were suspended in cradles below the middle two treads to measure the vertical and horizontal forces applied to the tread during normal use by the subjects.

In the first phase of work 60 subjects were asked to walk up and down the 12 different stair geometries. The youngest subject was 16 and the eldest 84 years of age and 60% were female. They varied in height between 1250mm and 2030mm. Objective and subjective observations were used to evaluate the ease of use and relative safety of the stairs for each going.

In the second phase of testing seven different materials were selected to represent a cross section of the flooring materials typically found in both domestic and non-domestic environments. Two contaminants, water and glycerol, were used. To eliminate cross contamination, runs for all materials were carried out first in dry conditions, then with water and finally with glycerol contamination. As before objective and subjective observations were used to evaluate the ease of use and relative safety of the stairs.

Results and Discussion

Objective observations of the subjects for four key measures (hesitations, missteps, glances at feet and use of the handrail) were recorded to gauge the ease with which the subjects negotiated the stair rig as the going was changed. The higher the frequency of these actions the greater the subjects feeling of insecurity when using the stairs. The frequency with which all of these actions were observed decreased as the going increased. Critical going sizes were observed for each action beyond which further increase in the going size had little effect on the frequency of the action, see Table 1.

Table 1. Summary of the objective observations of the subject's actions on the stairs.

Action	Variation with Size of Going	Critical Going
Number of hesitations	Decreased for goings ≥ 200 mm. Occasional for going = 425 mm among smallest subjects.	200 mm
Number of missteps	Decreased as going increased, dramatically for goings ≥ 250 mm.	250 mm
Number of glances at the feet	Decreased as going increased, approximately constant for goings ≥ 300 mm.	300 mm
Use of the handrail	Decreased as going increased, approximately constant for goings ≥ 275 mm.	275 mm

The going sizes at which the frequency of the actions in Table 1 decreased confirmed the subjective data that the user's perception of the safety and ease of use of the stairs changes around a going size of 250-275 mm. If the minimum going were increased to 250 mm, this would equal or exceed the critical going for both hesitations and missteps, which have the most serious implications for stair safety.

Accurate empirical data of the placement of the subjects' feet was available from pressure pads, which showed that the proportion of the shoe overhanging the nosing decreased linearly as the going increased. Calculations based on population data indicated that over 70% of the foot needs to be placed on the tread to ensure that the initial contact with the tread is made by the metatarsal heads. Comparison of these calculations with the results of the pressure pad data show that a going size of at least 250 mm would be required to enable half the population to accomplish this.

These results suggest that a going of 250 mm represents a critical threshold value below which a significant proportion of the population will experience difficulties using stairs. It would therefore seem intuitively sensible to recommend that the minimum allowable going be increased to 250 mm, and the minimum permissible goings for the different classes of stairs be increased as follows; Private stairs: 250mm, Semi Private stairs: 300 mm, Public stairs: 350 mm.

Analysis of the data captured from the force plates to determine the RCoF was limited to stair descent as this poses the highest risk to users. The stair rig design made force plate data for small goings noisy and difficult to interpret. No attempt has therefore been made to determine RCoF for individual going sizes.

Analysis of the force plate data showed a clear peak in RCoF at $\mu = 0.15$, with a mean value of $\mu = 0.18$. 50% and 95% of the subjects group required $\mu = 0.148$ and $\mu = 0.27$ respectively during stair descent, in line with RCoF values for level walking reported in the literature [Budzeh et al.].

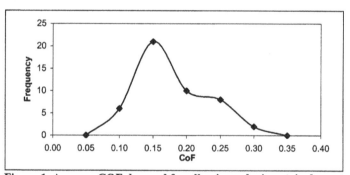

Figure 1. Average COF demand for all goings, during stair descent

Seven different flooring materials were selected to represent a broad cross-section of flooring materials commonly found on both domestic and non-domestic stairs. The slip resistance of the flooring materials was determined in accordance with the Guidelines recommended by the United Kingdom Slip Resistance Group (UKSRG) Issue 2 (2000), using the "Stanley" TRRL Pendulum. Data for dry, water wet and glycerol contaminated conditions was generated and is given in Table 2. Results show that all the selected floors present a low slip risk in the clean and dry condition. Under water contamination only the carpet, resilient safety floor and the quarry tiles present a low slip risk, all the other materials present either a moderate or high slip risk. Under glycerol contamination of the flooring materials, all except the carpet would be considered to present a high slip risk.

Table 2. Slip resistance value (SRV) on the level for the selected flooring materials.

Flooring Material	Dry		Wet		Glycerol	
	SRV	Slip Risk	SRV	Slip Risk	SRV	Slip Risk
Carpet	-		-		-	
Untreated wood	68	Low	27	Moderate	17	High
Resilient safety flooring	64	Low	46	Low	25	High
Quarry tiles	60	Low	50	Low	19	High
Linoleum	55	Low	30	Moderate	19	High
Profiled rubber	47	Low	14	High	20	High
Glazed ceramic tile	86	Low	6	High	15	High

Both subjective and objective data were collected for each ascent and descent of the stair rig. Two volunteers took part in the contamination trials and generally reported that they experienced more difficulties during descent than ascent, and incidences of misstep, slip, or catching either the toe (ascent) or heel (descent) occurred under contaminated conditions and increased as the size of the going decreased.

As predicted from the pendulum tests no slips were recorded in dry conditions for any of the floors or goings investigated on the stair rig. The stair results in contaminated conditions are summarised in Table 3 below.

Table 3. Table summarising stair rig tests under contamination.

Flooring Material	Water Contamination		Glycerol Contamination	
	Min. Going (mm)	Incidence of Difficulty	Min. Going (mm)	Incidence of Difficulty
Carpet	150	Misstep or catching the foot – going <250 mm	225	Catching the foot – going <275 mm
Resilient safety flooring	250	Single instance of caught toe. Going = 250 mm	225	Slight slips, mainly when going < 350 mm. Caught foot going < 300 mm.
Quarry tiles	250	Single misstep. Going = 250 mm	150	Occasional misstep. Significant slips observed going < 275 mm
Linoleum	250	None	225	Slight slips in descent. Foot caught going < 250 mm
Untreated wood	250	Occasional incidence of catching the foot	225	Slight slips, missteps, & catching the foot at all goings. Severity of slip increased as the going decreased.
Profiled rubber	150	Caught foot, going <300 mm. Significant slips occur, going< 225 mm.	225	Caught foot, going <375 mm. Slight slips occur for all goings.
Glazed ceramic tile	200	Missteps, caught foot & slips at all goings. Severity of slips increases as going decreases.	150	Missteps, caught foot & significant slips, going <250 mm.

The results show that both users, despite appreciable differences in their gait, managed to successfully negotiate the stair rig under contaminated conditions when μ > 0.2 with little or no difficulty. When μ was reduced to below 0.2 the users experienced appreciable levels of slipping, the severity of which increased as μ decreased. The study also confirms earlier finding that the size of going dramatically affects the ease with which the user negotiates stairs. This corroborates the force plate data, which indicates that the majority of the population have a CoF requirement of more than 0.15 during stair descent. Contaminated flooring, which the users were able to successfully negotiate at larger going was found to become hazardous, with large slips being recorded when the going was reduced below 275mm. This would appear to indicate that while the required CoF during stairs gait is low, the placement of the user's feet significantly affects the level of friction required.

Conclusions

The study indicates that the minimum recommended goings for the different classes of stairs be revised as follows; Private stairs: 250mm, Semi Private stairs: 300 mm, Public stairs: 350 mm.

The pendulum test did correctly rank the anti-slip performance of the flooring materials used in this study, is therefore a valid test for stair materials.

The results of the current work on contaminated flooring materials indicates that the required coefficient of friction for ascent and descent of stairs may be comparable with that required for walking on a level. The force plate data from phase 1 of the experiment is also in good agreement.

Further work
Further work is required to capture better force plate data for each going, with a much larger sample, and a more robust design of stair rig will be needed both in terms of limiting vibrations and eliminating interactions between subsequent stairs.

References

Buczek F.L. et al 1990, Slip resistance needs of the mobility disabled during level and grade walking, *Gray B.E. (editor) Slips Stumbles and Falls: Pedestrian footwear and surfaces*, ASTM STP 1103, 1990:39-54.

Christina K.A. and Cavanagh P.R. 2001, Ground reaction forces and frictional demands during stair descent: effects of age and illumination, *Gait and Posture*, **15** (2002), 153-158

Riener R., Rabuffetti M. and Frigo C. 2001, Stair ascent and descent at different inclinations, *Gait and Posture*, **15** (2002), 32-44

Roys M.S. 2000, Serious stair injuries can be prevented by improved stair design, *Applied Ergonomics*, **32** (2001), 135-139

Roys M. and Wright M. 2003, Proprietary nosings for non-domestic stairs (BRE Bookshop)

UKSRG, 2000, The measurement of floor slip resistance – guidelines recommended by the United Kingdom Slip Resistance Group, issue 2. UKSRG, RAPRA, Shropshire, June 2000. ISBN: 1-85957-227-8.

SLIP, TRIP AND FALL ACCIDENTS IN A SOUTH AFRICAN AUTOMOTIVE PLANT: PRACTICAL RECOMMENDATIONS

Andrew I. Todd, Jonathan P. James and Pat A. Scott

Department of Human Kinetics and Ergonomics
Rhodes University,
Grahamstown, 6140, South Africa

Slip, trip and fall hazards are commonplace in the South African automotive industry. The awkward postures assumed during the execution of manual tasks within the industry predispose workers to high levels of discomfort and concomitant work-related musculoskeletal disorders (WMSDs).The present study aimed to assess the slip, trip and fall risks associated with push-pull forces required to complete a series of sub-tasks in the paintshop of a South African Automotive industry. *In situ* analyses was conducted to assess the forces required to move the vehicle frames. In addition Laboratory testing was conducted (n=15) in order to analyse the interaction between the task and the human operator. The results from the *in situ* evaluation demonstrated that workers were required to exert push-pull forces of up to 35 kg.f, necessitating workers to adopt poor working postures. The excessive forces and awkward postures indicated the need to automate the conveyor pulling component in order to minimize the risk to the operator, and to increase the efficiency of production.

Introduction

Slip, trip and fall accidents have been shown to be a substantial cause of occupational accidents (Lortie and Rizzo, 1999; Bently and Haslam, 1997; Leamon and Murphy, 1995) accounting for between 16 to 30% of all accidents. The associated compensation costs are high, with slip, trip and fall accidents costing 24% of total compensation in the United States (Leamon and Murphy, 1995). Gauchard *et al* (2001) argued that the frequent occurrence of occupational falls leads to important social and economic consequences both for the employee and the employer. Much of the research relating to slip, trip and fall accidents has focused on falls from some form of elevation, however, as recently as 2002 Chang identified that falls on the same level account for 65% of all claims. The push-pull actions which are frequently required to move vehicle frames, particularly in the paintshop, are of prime concern to the Ergonomist. James and Todd (2003) found that the associated high forces for pushing and pulling tasks require operators to take up working postures where the centre of mass (CM) is located at the extreme limits of their support base (SB) and beyond, which significantly increases the likelihood of slip, trip and fall accidents.

The present study aimed to assess the slip, trip and fall risks associated with the push-pull forces required to complete a series of sub tasks in the paintshop of a local automotive industry. In order to be able to generate sufficient force operators are required to adopt awkward working postures. They are required to lean forwards while pushing, and backwards while pulling in order to optimise the use of body mass in generating the required force output, which simultaneously increases the risk of slip, trip and fall accidents occurring. Lortie and Rizzo (1999) identified the body's centre of gravity falling outside of the supporting foot base as being one of the three major events leading to stability difficulty or the loss of balance. This study investigated the incompatibility between the demanding push and pull forces required to move the motor vehicle frames and the operator capabilities of generating the required force at various angles of lean observed in the paintshop of an automotive industry.

Methodology

In Situ Testing
The *in situ* testing aimed to assess the push and pull forces required to complete a series of sub-tasks in the paintshop of a South African Automotive industry, and to assess the postures adopted by workers in moving the motor vehicle frames. The forces required to move the vehicle frames were assessed using the Chatillon™ Hand-Held Dynamometer (Model CSD 500) following a standardised protocol. Repeated measures were collected in order to ensure the validity of the responses obtained.

Photographic evidence illustrated in Figure 1 demonstrates that workers were required to assume awkward working postures with the centre of mass falling outside of the base of support. Angles as extreme as 50° of forward and backward lean from an upright posture during pushing and pulling respectively were observed in the paintshop.

Figure 1: Picture showing extreme working postures adopted by workers

Laboratory Testing
From the observed angles in industry four pushing and four pulling conditions (a total of eight experimental conditions) were established to investigate a range of pushing and pulling angles under laboratory conditions.

Table 1. The eight test conditions:

	Angle				
Pushing	50°	63°	76°	90°	90 - 50° Forwards (40°)
Pulling	90°	103°	116°	129°	90 - 129° Backwards (39°)

All subjects (n =15) were required to complete maximal isometric efforts using the Chatillon™ Hand-Held Dynamometer (Model CSD 500) as the testing equipment.

Participants were required to complete all eight test conditions, with the testing protocol requiring 3 maximal isometric efforts at each test angle. Pushing and pulling height was standardised for each subject to 80% of acromiale height to ensure optimal testing conditions.

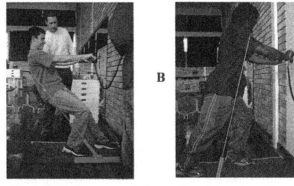

Figure 2: Subjects completing pulling (A) and pushing (B) tests

Informed Consent
Informed consent was administered to all participants (who were free from recent or ongoing injury) who volunteered to partake in this research.

Demographic and Anthropometric Data

Table 2. Participant demographic and anthropometric data

	Age (yr)	Mass (kg)	Stature (mm)	BMI	Grip (kg.f)	Acromiale ht (mm)	Stylion ht (mm)
Mean	20.63	75.43	1800	23.25	54.06	1460	820
SD	2.31	9.97	51.08	2.81	8.55	49	42
CV	11.18	13.22	2.84	12.11	15.82	3.36	5.06

Results and Discussion

Isometric testing revealed significant differences between testing conditions as the angle of push and pull changed.

Table 3. Pushing and Pulling responses (Means with SD in brackets)

Pushing			Pulling		
Angle	*Peak*	*Mean*	*Angle*	*Peak*	*Mean*
90°	12.05* (3.19)	10.08 (3.5)	90°	12.23* (4.33)	10.22 (3.8)
76°	30.64* (7.04)	28.15 (6.46)	103°	23.01* (7.6)	19.2 (5.49)
63°	35.88* (9.15)	33.18 (8.45)	116°	29.29* (10.03)	26.25 (9.59)
50°	44.52* (7.10)	41.55 (6.25)	129°	40.29* (9.77)	37.28 (9.54)

* Significant (P < 0.05) between all angles

The lowest force output was generated at 90° with a mean force of 10.08 kg and 10.22 kg for pushing and pulling respectively (see Table 3). As the angle was reduced for pushing so the force generated increased, due to the increased leverage and greater use of body mass, with a maximum pushing force of 41.55 kg at an angle of 50°. The responses to the pulling conditions followed a similar trend with the greatest force of 37.28 kg being generated at the greatest angle of 129°.

These results indicate that when workers are required to exert forces greater than approximately 10 kg they have to lean forwards or backwards to use their body mass to aid in force generation. The concomitant effect is that the CM is moved towards the edges of the SB, and once an angle greater than approximately 73° for pushing and 107° for pulling has been reached the CM is most likely falling outside of the base of support (assuming a base of support of 300 mm and mean stature of 1800 mm). The result is that workers are having to use solely the upper extremity in order to maintain balance and a lose of grip on the vehicle frame will result in workers falling. Therefore the extreme body lean which was evidenced in a local automotive industry will have the concomitant effect of increasing the risk of slip, trip and fall accidents.

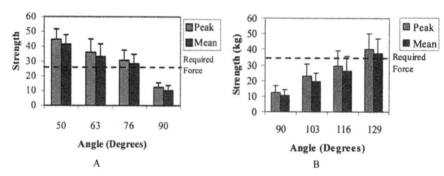

Figure 3: Average and peak force output for pushing (A) and pulling (B) conditions and the force required to move the vehicle frame in industry

The results of the present study illustrated the forces required to move the motor vehicle frames *in situ* necessitated the operators to assume awkward working postures. The forces required to pull the frames required workers to lean backwards to an angle of 129°, below which the forces generated were not sufficient to move the frame (see Figure 1). Likewise under the pushing conditions the operators had to lean as much as 63° forward in order to achieve the required force output. These extreme angles result in the centre of mass falling outside of the base of support, therefore increasing the likelihood of slip, trip and fall accidents. Furthermore the reliance on the upper extremities to maintain the upright body position increases the likelihood of work-related musculoskeletal disorders (WMSDs). Both slip, trip and fall accidents and WMSDs are going to negatively impact on costs of production and therefore hinder the financial bottom line.

Other Considerations

The *in situ* analysis of the working environment identified several other factors which increased the risks of slip, trip and fall accidents. Lortie and Rizzo (1999) recognized interaction problems between the feet and the supporting elements as being a major event in the initialisation of loss of balance. The floor surface evidenced in the paintshop had an uneven and bumpy surface and with the poor visual range whilst pulling the frames this area was classified as being a high risk area for slip, trip and fall accidents.

Conclusions and Recommendations

The slip, trip and fall risks associated with the push-pull tasks evidenced in the automotive industry were high, with workers having to adopt postures where the centre of gravity fell outside the base of support provided by the operator's feet.

It was therefore recommended that the excessive components of push-pull process in the paintshop be automated, reducing the physical strain being placed on the workers in this area, as well as eliminating the need to adopt poor working postures and hence reducing the likelihood of musculoskeletal problems and the potential for operators to slip, trip or fall. It was also recommended that any pushing and pulling tasks which are retained, do not require workers to adopt postures that place the centre of mass outside of, or at the extreme limits of the SB.

Figure 4: Changes made by the automotive company in order to reduce the push-pull component of the operators job

Reference List

Bentley, TA. And Haslam, RA. 2001, Identification of risk factors and countermeasures for slip, trip and fall accidents during the delivery of mail, *Applied Ergonomics,* **32**, 127-134

Chang, W. 2002, From research to reality on slips, trips and falls, *Safety Science,* **40**, 557-558

Gauchard, G., Chau, N., Mur, JM. And Perrin, P. 2001, Falls and working individuals: role of extrinsic and intrinsic factors, *Ergonomics,* **44**, 1330-1339

James, JP. And Todd, AI. 2003, Push-pull force evaluation in an IDC automotive industry, *CD-Rom Proceedings of the XVth Triennial Congress of the International Ergonomics Association and The 7th Joint Conference of the Ergonomics Society of Korea/Japan Ergonomics Society*

Leamon, TB. And Murphy P. 1995, Occupational slips and falls: more than a trivial problem, *Ergonomics,* **38**, 487-498

Lortie, M. and Rizzo, P. 1999, Reporting and classification of loss of balance accidents, *Safety Science,* **33**, 69-85

SLIP, TRIP, FALL RELATED HAZARDS AT
A FOOTBALL STADIUM

Roger A. Haslam[1], S. Y. Zachary Au[2], Jenny Gilroy[3], Andrew D. Livingston[3]

[1]*Health & Safety Ergonomics Unit, Loughborough University, Leicestershire, LE11 3TU*
[2]*Independent Consultant, 13 Meadow Lane, Warrington, Cheshire, WA2 0PR*
[3]*Atkins Consultants Limited, Birchwood Boulevard, Warrington, Cheshire, WA3 7WA*

Slip, trip, fall hazards have been examined in a large capacity football stadium, using a combination of document review, site inspection and crowd observations. Problems were identified with the physical design and layout of seating areas and entrance and egress routes, especially the narrow clearway between seat rows, irregular and excessive step dimensions and unmarked step edges. Spectator behaviour was considered to be a risk factor, in connection with the movement of individuals within seated areas at moments of excitement and due to boisterousness. The frequent entering and leaving of seated areas while matches are in progress was also considered to increase risk. Control measures to reduce risk of falling in football stadia require comprehensive attention to environmental, behavioural and crowd management factors.

Introduction

This study, part of the wider investigation reported in a companion paper (Au *et al*, 2004), sought to identify slip, trip, fall related hazards at a large capacity football stadium. Of particular interest was risk of falling arising from persistent standing in the all-seated stadium. Although there is a wider general literature on the causes of falls and their prevention (eg Leamon, 1992a & 1992b), it is surprising that Pauls (1991 and 2002, personal communication) is one of the few researchers to have looked at fall risk in sports grounds (during commissioning of the Olympic Stadium in Montreal in particular). However, the focus of Pauls' work has tended to be on the design of steps and stairs, rather than seated or standing areas. Pauls' research demonstrated that the dimensions of steps and stairs, their conspicuousness, and the provision of handrails, affected numbers of falls that occurred. Au *et al* (1993) and Smith (1995) recognised falling as a risk in crowd situations, but did not go into detail regarding the circumstances under which falls may occur.

Recently, consideration has been given to the influence of behaviour on risk of falling (eg Hill *et al*, 2000; Bentley and Haslam, 2001; Haslam *et al*, 2001). Using the behavioural classification derived by Hill *et al*, it can be predicted that the risk of falling among spectators in a football ground will be affected by design of the environment (eg presence of slip and trip hazards); individual interaction with the environment (eg hurrying, carrying items); behaviour which modifies the environment (eg dropping food and other litter, which then becomes a slip or trip hazard); and behaviour affecting the individual (eg use of alcohol affecting balance, but also leading to boisterousness and increased visits to the toilet to urinate).

The 'Green Guide' (DCMS, 1997), an authoritative source in the UK, provides recommendations for the physical layout of sports grounds, including provision of spectator seating. The guide does not, however, consider adequately circumstances where spectators are standing in areas with fixed seating. Also, the evidence base from which recommendations for physical dimensions of seating and spacing have been derived is unclear from the Guide.

Methods

The investigation entailed document reviews (eg steward briefing notes, post match reports, annual safety reports), a site inspection, and crowd observations (as per Au *et al*, 2004). The post-match reports and annual safety reports contained only limited information useful in assessing fall risks, although they did demonstrate that falls had occurred.

Findings

Physical design and layout of seating area
In general, the seating met the dimensional requirements of the Green Guide (DCMS, 1997) in the areas of the stadium examined. Stadium lighting and weather protection were all deemed to be good. Dimensions of particular interest with regard to falling are the clearway (distance between the foremost projection of a seat and the back of the seat in front) and the rear seat height (distance between floor level of the row behind to the top of the seat back), Table 1.

Table 1. Clearway and rear seat height dimensions (range measured)

	clearway (mm)	rear seat height (mm)
recommended by Green Guide	minimum 305	*no recommendation*
Stand 1 (upper level)	410-430	450-470
Stand 1 (lower level)	380-430	510-520
Stand 2	380-390	330-340
Stand 3	370-380	320-330

The clearway affects the space available for individuals to stand or for one person to pass another along a row. The seating in the stadium was considered to provide sufficient room for individuals to stand in safety, once people are in and remain in position, providing they are not pushed or jostled. However, despite the clearway in the seating areas exceeding the minimum recommendation of the Green Guide, there is very little room for an individual to pass another person, whether they are standing or seated. By way of illustration, the body depth of a 50th percentile standing British male (approximate average size) is 334mm (PeopleSize software, Open Ergonomics Ltd 1993-9). Even in the seating with the greatest clearway (430mm), this leaves less than 100mm for someone to pass. The only way this can happen is if individuals cooperate, either leaning backwards or forwards, depending on whether the person is attempting to pass behind or in front. Where individuals are seated, with a typical seating row depth of 660mm and a 50th percentile buttock to front of knee dimension for British males of 619mm, space is very restricted, Figure 1.

The problem with limited space is exacerbated where larger individuals are involved, or where someone is carrying bulky items, such as a rucksack. The body depth of a 95th percentile standing British male (only 5% of males exceed this dimension) is 383mm (PeopleSize software, Open Ergonomics Ltd 1993-9). Where space is insufficient for passing, a degree of physical contact will occur, with the potential for loss of balance.

Figure 1. Limited space available to pass a seated spectator

With a rear seat height ranging from 320-520, the seat backs do not present a tripping hazard in the conventional sense. However, in situations where a spectator is pushed or moved forward for some reason (eg at moments of excitement), the seat backs prevent the normal response of taking a step forward in an attempt to maintain balance and act as a fulcrum against the individual's leg. With the rear seat height being much lower than the centre of gravity of the adult human body, the seat backs are a potential source of instability for spectators.

Although the coefficient of friction of the flooring in the seated areas was not measured, the surface is smooth concrete in all stands examined and will become slippery when liquids or contaminants such as food or ice-cream are present. This situation is especially undesirable, because should an individual slip in these areas, the presence of the seating restricts ability to recover balance. Irregularities in the floor surface of as little as 5mm can be sufficient to cause someone to trip (Marletta, 1991). In some seating areas, concrete joints along the seating walkways, sometimes covered with tape, resulted in irregularities approaching 10mm. It is considered that this could be sufficient to cause individuals failing to notice the joint (possible in a crowd situation) to trip.

A number of problems were observed with the radial gangway stairs. As well as being a safety issue in their own right, these interact with persistent standing of spectators. When standing occurs, individuals tend to move onto the gangway due to the limited room for standing in the seated areas. In many instances, the stairway dimensions did not meet the requirements of the current edition of the Green Guide, Table 2. The dimension of most importance for safety is the rise (step height). It can be seen from Table 2 that the steps in Stand 1 (upper level) significantly exceed the maximum height of 190mm specified by the Guide and in practice these steps are cumbersome and awkward to climb and descend. The Green Guide also stipulates that goings and risers in a stairway should be uniform. This is because irregularities of as little as 5mm in riser height between one step and the next are sufficient to cause a user to misstep or trip (Templer, 1992). Irregularities exceeding 5mm between adjacent steps were found in all areas of the stadium examined. In Stand 1 (upper level) and Stand 3, irregularities greater than 10mm were measured.

In most cases it would not be 'reasonably practicable' to modify the stair dimensions to meet the recommendations of the Green Guide. This makes it particularly important that the step edges are easily distinguished and that hand rails are provided. While the conspicuousness of the steps is good when ascending, due to the white lines painted on the step risers, it is more difficult to see the step edges when descending, figure 1. This could be a problem where spectators are hurrying to reach their seats at the start of a match, after the interval or leaving at the end of a game. Barriers, which also serve to act as handrails, were present in some areas of the stadium, but not all.

Table 2. Radial gangway stair dimensions (range measured)

	going (mm)	riser (mm)
recommended by Green Guide	minimum 280	minimum 150
(for existing construction)	preferred 305	maximum 190
Stand 1 (upper level)	660-675	250-300
Stand 1 (lower level)	322-340	117-156
Stand 2	346-356	201-224
Stand 3	323-337	220-235

Spectator behaviour and fall risk
There are various ways in which spectator behaviour might affect risk of falling. It was apparent from match observations that there is a steady movement of spectators to and from seated areas when matches are in progress. This increases as the interval approaches and towards the end of a match. This appears to be motivated by a desire to avoid queues for refreshments, to use the toilets and to avoid congestion when leaving the stadium. The movement causes problems because of the limited space available for spectators to pass each other in the seated areas, with the possibility of loss of balance occurring.

Vigorous movement when standing within the seated areas, as might happen during moments of celebration or in response to incidents on the pitch, was also considered to increase risk of falls. This is due to the limited space available, together with restricted room to move to recover balance, should this be disturbed.

Fall consequences

In most situations, a person falling is likely to make contact with either the floor, plastic seating or other spectators. In some respects, the close proximity of other individuals may be beneficial in reducing severity, as other human bodies can be less injurious objects on which to fall than hard physical surfaces. However, a fall in such circumstances may increase the number of people involved and exposed to risk of injury. The plastic seating has a degree of flexibility which, again, will operate to reduce and dissipate impact forces to an extent. In some areas, however, unprotected metal seat pillars were present, particularly at the end of rows. Serious injury could result if a person's head struck one of these pillars.

Of the various types of fall that might occur in the seating areas, a fall forwards, over the seat in front appears to be the most serious situation. One concern is whether the gradient of the stands could lead to a domino effect in these circumstances. This would be more likely the steeper the gradient. However, if this did occur, it is thought the consequences would be limited to only a small number of rows. Perhaps the most worrying situation would be if a fall occurred among crowded spectators descending the radial gangway stairs. It is conceivable in this situation that, in the absence of any physical barriers or supports, a cascade might occur involving a number of individuals. Risk of injury is increased in falls on stairs, due to the vertical distance through which it is possible to fall and the forces generated by this.

Conclusions

There is scope to reduce the risk of falling in the stadium examined, through addressing the issues identified above. In some instances, hazards are integral to the design and construction of the stadium, so that secondary control measures are necessary. Persistent standing, of itself, is not considered to be a major concern as far as risk of falling is concerned, although from a wider safety perspective there are benefits from encouraging spectators to sit rather than stand. Because of the limited space available to move past other spectators, whether they are seated or standing, spectators should be encouraged to remain at their seats when play is in progress.

References

Au S Y Z, Ryan M C, Carey M S and Whalley S P, 1993, Managing crowd safety in public venues: a study to generate guidance for venue owners and enforcing authority inspectors (HMSO: London), HSE Contract Research Report 53/1993

Au S Y Z, Gilroy J, Livingston A D and Haslam R A, 2004, Assessing spectator safety in seated areas at a football stadium. In: *Contemporary Ergonomics 2004* (edited by P T McCabe) (Taylor & Francis: London)

Bentley T A and Haslam R A, 2001, Identification of risk factors and countermeasures for slip, trip and fall accidents during the delivery of mail. *Applied Ergonomics*, 32, 127-134

Department for Culture, Media and Sport (DCMS), 1997, Guide to Safety at Sports Grounds (Stationary Office: London), 4th edition

Haslam R A, Sloane J, Hill L D, Brooke-Wavell K and Howarth P, 2001, What do older people know about safety on stairs? *Ageing and Society*, 21, 759-776

Hill L D, Haslam R A, Howarth P A, Brooke-Wavell K and Sloane J E, 2000, Safety of older people on stairs: behavioural factors (Department of Trade and Industry: London), DTI ref 00/788

Leamon, T B, 1992a, The reduction of slip and fall injuries: part I - guidelines for the practitioner. *International Journal of Industrial Ergonomics*, 10, 23-27

Leamon, T B, 1992b, The reduction of slip and fall injuries: part II - the scientific base for the guide. *International Journal of Industrial Ergonomics*, 10, 29-34

Marletta W, 1991, Trip, slip and fall prevention. In Hanson D (ed), *The work environment, vol. 1: occupational health fundamentals* (Michigan: Lewis), pp 241–76

Pauls J L, 1991, Safety standards, requirements and litigation in relation to building use and safety, especially safety from falls involving stairs. *Safety Science*, 14, 125-154

Smith R A, 1995, Density, velocity and flow relationship for closely packed crowds. *Safety Science*, 18, 321-327

Templer J A, 1992, *The staircase: studies of hazards, falls and safer design* (MIT Press: Cambridge, Massachusetts)

ASSESSING SPECTATOR SAFETY IN SEATED AREAS AT A FOOTBALL STADIUM

S. Y. Zachary Au[1], Jenny Gilroy[2], Andrew D. Livingston[2], Roger A. Haslam[3]

[1]*Independet Consultant, 13 Meadow Lane, Warrington, Cheshire, UK. WA2 0PR*
[2]*Atkins Consultants Limited, WS Atkins House, Birchwood Boulevard, Warrington, Cheshire, UK. WA3 7WA*
[3]*Health & Safety Ergonomics Unit, Department of Human Sciences, Loughborough University, Loughborough, Leicestershire, UK. LE11 3TU*

All UK Premier League and First Division football clubs are required to provide only seated accommodation at their stadia. However, many spectators do stand up in seated areas, sometimes persistently for long periods of time. This paper looks at a study to assess the hazards and risks of such behaviour and the methodological issues involved, including the risk assessment method used and the collection of risk data and information for the study.

Introduction

Following the Taylor Report into the 1989 Hillsborough Disaster, all UK Premier League and First Division grounds are required to be all-seated. The intention is that spectators should be seated for the duration of the match although it is expected that seated spectators could "move from their seats" at moments of excitement (Lord Justice Taylor, 1990). However, at football grounds up and down the country many spectators do stand up during matches and sometimes they do so persistently for long periods of time. The enforcing authorities have been concerned about the safety implications of such behaviour. Persistent standing in areas designed for sitting has been deemed to be "less safe", but to what extent this is so had not been entirely clear. In order to determine this, and to establish what actions are justifiably required, it is necessary to examine the hazards and the risks involved. A study was therefore commissioned to assess the risks of standing in seated areas at a major football stadium.

Assessing crowd safety risks – the methodological issues

The risk assessment principles are clear; essentially it involves the identification of hazards and the evaluation of the extent of the risks involved (e.g. HSC, 2000). However, how they should be applied to the context of crowds at public venues requires careful consideration. Under the Management of Health and Safety at Works Regulations 1999, it is necessary to ensure that the risk assessment is "suitable and sufficient". This means that it should be appropriate to the nature of the operations concerned and the types of hazards and risks that may arise.

To an extent, crowd management is a human factors issue. It is about dealing with people and their interactions with each other, with the physical environment and with the circumstances. But at the same time, it is also very different to a workplace or other contexts to which human factors and risk assessment are normally applied. For example, in conventional ergonomics the work activities or the manner in which equipment or products are used are restricted by the work procedures, rules, the tasks that people are required to do, the functions of the equipment/ product, etc. These types of restrictions are far less prevalent and stringent for crowd activities and people are much freer to do as they wish. This makes their "tasks" less predictable and far harder to define and analyse in a highly systematic manner - established human factors techniques such as task analysis are, therefore, not applicable in this context. The presence of large crowds can also pose a range of hazards that are unique to the public venue environments. For example, past crowd disasters (e.g. Dickie, 1993) have shown that people could be seriously harmed by crushing, human pile-up, structural failure and other such hazards that are not found in other contexts. Therefore, a suitable and sufficient risk assessment method for crowds should adequately reflect the complexity of the subject matter and facilitate the assessors to identify the types and range of hazards that could arise.

Conventional assessment methods range from a simple approach based on subjective judgement for offices, shops light industrial premises, etc. (e.g. HSE, 1998) to the highly sophisticated quantified risk assessment (QRA) for complex high hazard systems. There is also a range of "semi-quantified" methods. However, recent research (Au, 2001) suggested that none of them are suitable for assessing crowds. The basic approach is not sufficiently robust given the complexity of the subject matter and the severe consequences a crowding accident could have. The QRA method might be desirable in principle but the lack of relevant risk data means that it is not feasible to conduct. Semi-quantified methods, on the other hand, are considered as fundamentally flawed because of the ways in which ratings, which are qualitative in nature, are used as something quantitative for mathematical calculations. Also, all of these methods are intended for the work environments and are not particularly appropriate to the kind of hazards and risks that could arise in a public venue. A survey of the same research has found that by using the conventional methods, some assessors were able to identify the "physical hazards" posed by the buildings but were not able to identify the "behavioural hazards" that arise from the crowding.

For assessing crowd safety, what is required is a method that is qualitative, structured and encourages lateral thinking. A structured approach is necessary where the subject matter is complex, to ensure that all key areas are covered in the assessment. Lateral thinking and being able to think beyond normal conventions is important because crowd safety is not a precise science and past disasters have also shown us that even relatively minor changes to the make-up of the crowd, the physical environment and/or the circumstances could result in a very different outcome. Based on this principle, a methodology was developed (Au, 1998) specifically for assessing crowd safety risks. Designed to encourage a structured and robust assessment, it contains key features such as (a) generic keywords for behavioural related and physical hazards that can arise in the public venue environments, and (b) a risk rating regime that consists of separate ratings for likelihood and severity and a likelihood-severity matrix for determining the extent of risk. The keywords are similar in nature to the HAZOP[1] keywords; they were designed to be thought provoking and to provide a structured approach for hazard

---

[1] HAZOP stands for <u>Haz</u>ard and <u>Op</u>erability Study. It is widely used in the high hazard industries for hazard identification. A human factors version (i.e. Human HAZOP) is also available for human error analysis.

identification. The risk rating regime allows a wide spectrum of risks, ranging from high-magnitude low-probability to low-magnitude high-probability risks, to be accounted for whilst limiting each rating scale to a manageable size. This methodology was therefore adopted, with slight modifications made to cater for the specific needs of this assessment.

The risk assessment of standing in seated areas

The assessment consisted of three key elements: data and information gathering, ergonomics assessment of the physical design, and risk assessment based on the said methodology.

Risk data and information
Risk data and information is essential for identifying the hazards and establishing their causes, consequences and risks. The better the data/information is, the more objective the assessment can become. In those industries where QRA has been used extensively, much effort has been devoted to the collection of risk data (e.g. failure rates of system components). In human factors, human reliability techniques (e.g. Williams, 1988) are available for estimating human error probabilities. Even for a qualitative assessment, data and information is needed to provide evidence, indications or guidance on the types of hazards that may arise and the extent of the risks involved.

There has been no published scientific research or other direct evidence on how much risk people are exposed to when standing in seated areas. Hence, as a key part of the assessment, an extensive information gathering exercise was conducted to obtain information from a variety of sources. They include: (a) review of past records to collect historical data, (b) observations of crowd activities during matches, (c) consultation meetings to gauge the views and experience of the supporter associations and stakeholders (i.e. the football club, the licensing and enforcing authorities, emergency services), and (d) questionnaire survey targeting the spectators and the front line staff (e.g. stewards).

The review of past records involved an examination of the post-match reports produced by the club following each match. The reports contain a combination of numerical data (e.g. injury figures, number of ejections and arrests) and qualitative information (e.g. extent of persistent standing, description of crowd behaviour and the circumstances surrounding the match). A number of interesting findings were revealed. For example, persistent standing tended to occur mostly in certain sections of the stadium and more frequently in certain types of matches (e.g. European matches). But there does not appear to be a relationship between the overall injury figures and persistent standing, which suggests that the risk of persistent standing could be fairly low. Furthermore, probably contrary to popular believe, there were more matches with widespread persistent standing and drunkenness related problems when alcohol was not sold inside the stadium.

The aim of the observations was to look at spectators' behaviour and movements and their interaction with the built environment in various parts of the stadium. It is necessary to observe different parts of the stadium because they were built at different times and have different characteristics. The observations covered a range of matches including European matches, domestic matches, important matches (to the club) and less high profile matches. The findings suggest that although persistent standing is generally deemed to be less safe, people normally stood relatively still without significant movements or crowd surge. But the observations have also identified a number of associated activities that could give rise to some concerns, such as people standing close to the edge and standing on seats. Standing up at moments of excitement (e.g. when a goal is scored), however, is deemed to be more dangerous because of the vigorous movements involved.

The consultation meetings provided a valuable opportunity for open discussions with people of different backgrounds and viewpoints on issues of standing in seated areas. These meetings enabled the assessors to gain an in-depth and more revealing understanding of the nature, the causes (and historical background) and the extent of the problem. The meetings also served to highlight the needs and expectations of all parties and areas of misunderstanding between them. This can be particularly useful for formulating remedial measures to encourage mutual understanding and closer cooperation.

The survey of the wider spectator population and front line staff also generated a wealth of information on the extent of the problem. A total of 1,100 questionnaires were sent out and the response rate was good, at 38% overall. On the whole, the survey results have supported that the likelihood of an accident, such as falling over, due to persistent standing per se is relatively low, although standing on seats is more problematic. But the danger increases when there are significant movements, such as at moments of excitement and during celebrations. Spectators entering and leaving their seats and the seated areas was also found to be rather problematic because of the physical design of the stands.

Site survey and ergonomics assessment

In addition to information gathering, the designs and the built environment of the seated areas were also assessed. A survey was carried out when the stadium was not in use to examine the seated areas and collect measurements and data about the physical design. This was followed by an ergonomics assessment to determine whether and how key design features such as seat dimensions, clearways, rear seat height, flooring, gangway stairs, seating layout and front barriers could affect safety when spectators stand. Ergonomics is an important consideration in this study for various reasons. Firstly, even though the stadium was designed in accordance with relevant guidance (i.e. Guide to Safety at Sports Grounds), it was intended for seated accommodation. Persistent standing represents a deviation from this intention. Hence, it is necessary to examine whether and to what extent the design is still adequate. Secondly, space is often at a premium at many stadia. This, in combination with the movements that could take place in that space, could give rise to a number of hazards such as slips, trips and falls. Finally, the fact that some seated areas are built at a relatively steep angle (to maximise the use of space and ensure unrestricted viewing) could exacerbate the problem. Details of the ergonomics assessment and the types of hazards that cold arise in a stadium environment are discussed in a separate paper by the same authors.

Together with the information generated above, the output of the ergonomics assessment was fed into the risk assessment to assist in the evaluation of risks and establishing how the physical design of the stadium could be improved to control the risks.

Assessment of risks

Based on the methodology described at the beginning of this paper, an assessment was carried out to establish the following: the hazards, the causes (of hazards), the consequences, existing mitigating factors and precautions, the level of risk posed by each hazard, and possible remedial measures.

Although it is a qualitative method, the variation and extent of the data collection methods used has enabled the assessors to identify the hazards and evaluate the risks in a more objective and impartial manner. Using the risk rating regime mentioned above, each hazard was allocated one of the following six risk levels: ranging from A (most serious risk) to E (trivial risk). This was done based on the likelihood of harm as a result of the hazard and its severity. The risk level helps the assessor to determine the extent and priority of the measure(s) required to control the risk.

Conclusions

This paper gives an overview of a study undertaken to assess the risks of standing in seated areas at a football stadium. The methodology for the risk assessment is an important consideration. It must be appropriate to the activities and the environment involved and the types of hazards and risks that may arise. In this study, the nature of the assessment called for the use of a specific methodology that is different to the conventional methods. The gathering of information about the risks is also important to ensure that the assessment is as objective as possible. In this study, information was collected from a range of sources and an ergonomics assessment was also carried out to provide a specialist analysis of the physical design of the stadium. In the event, the study has found that the safety risk of persistent standing per se is relatively low, but it could give rise to other hazards such as people standing on seats. What is of more concern, however, is when there are active movements in the crowd such as at moments of excitement and, to a lesser extent, when people are entering and leaving. The risk assessment has provided an indication of the risk associated with each key hazard. It has highlighted the more important risks and led to the identification of risk reduction measures covering stadium design, crowd safety management and public education/awareness issues.

References

Au, S.Y.Z. 1998, *Research to Develop a Methodology for the Assessment of Risks to Crowd Safety in Public Venues Parts 1 & 2*, HSE Contract Research Report 204/1998, (HMSO, London)

Au, S.Y.Z. 2001, *Assessing Crowd Safety Risks – A Research into the Application of the Risk Assessment Principles to Improve Crowd Safety Management and Planning in Major Public Venues*, PhD Thesis, Loughborough University

Dickie, J.F. 1993, Crowd Disasters. In: R.A. Smith and J.F. Dickie (ed.) *Engineering For Crowd Safety*, (Elsevier Science, Amsterdam)

Guide to Safety at Sports Grounds, 4th edition (HSMO, London)

Health and Safety Commission 2000, *Management of Health and Safety at Work Regulations 1999 Approved Code of practice and Guidance*, (HSE Books)

Lord Justice Taylor 1990, *The Hillsborough Stadium Disaster Inquiry - Final Report* (HMSO, London)

Williams, J.C. 1988, A data-based method for assessing and reducing human error to improve operational performance. *Conference Record for 1988 IEEE Fourth Conference on Human Factors and Power Plants*. Monterey, California, June 5-9 1988.

OPTIMAL CLEANING FOR SAFER FLOORS

François Quirion

QI Recherche et Développement Technologique inc.
1650 boulevard Lionel-Boulet
Varennes, Québec, Canada, J3X 1S2,

Since 1997, QInc has conducted research activities for the **Institut de recherche Robert Sauvé sur la santé et sur la sécurité du travail, IRSST**, aiming at reducing the incidence of slip and fall accidents in the workplace. The hypothesis behind the research is that a clean floor is less slippery than a floor contaminated with slipping agents such as fats and oils. Laboratory experiments showed that floor cleaners can be classified into categories and that the cleaning efficiency of these categories depends on the nature and amount of the fat, the type of flooring, the temperature of the cleaning solution, the concentration of the floor cleaner and , most importantly, the cleaning methodology adopted by the worker. This paper presents an example of the results and conclusions obtained for the cleaning of stripped vinyl flooring exposed to shortening.

Slip and fall accidents in Québec

In the five year period between 1996 and 2000 (Hébert et Massicotte, 2002), slip and fall accidents accounted for 5.5% of all Québec work injuries with an incidence rate of Tx=2,4 per 1000 workers. Every year, there are around 6 500 slip and fall accidents with a gravity of 8.2 weeks and annual compensations over 25 million dollars.

Workers, male and female, from every economic sector are affected but some sectors are more at risk than others such as Restaurant Trade (with 7% of all slip and fall accidents) and Lodging and Catering (with an incidence rate of 6,2).

Addressing the problem through optimal floor cleaning

Since 1997, the **IRSST, Institut de recherche Robert Sauvé sur la santé et sur la sécurité du travail**, has mandated QInc to conduct research activities aiming at the reduction of floor slipperiness through the optimisation of floor cleaning. To this day, our research showed that the cleaning efficiency depends on many parameters among which the most important are : the category and concentration of the floor cleaner, the nature and amount of contaminant, the type of flooring, the temperature of the cleaning solution and the cleaning method.

Floor cleaners

There are hundred of floor cleaners manufactured in the province of Québec. To minimise the number of variables, the floor cleaners were classified into six categories depending on the chemical nature of their main ingredients : surfactants, cosolvents and alkaline salts.

Table 1. Description of floor cleaners categories

Category	Code	Main Surfactant	Main Cosolvent	pH
Neutral Anionic	1 or NA	Anionic	None	7-11
Neutral Non-ionic	2 or NN	Non-ionic	None	7-11
Degreaser Anionic	3 or DA	Anionic	Glycol ether	9-13
Cationic	4 or C	Cationic	Glycol ether or none	7-13
Degreaser based on glycol ether	5 or DG	Non-ionic	Glycol ether	9-13
Degreaser based on limonene	6 or DL	Non-ionic or anionic	Limonene	7-13

Some floor cleaners contain ingredients that are not accounted by this classification. Nevertheless, we were able to classify 94% of 300 floor cleaners by comparing their MSDS with our definitions (Quirion and L'Homme, 1999). Note also that floor cleaner ingredients are often very similar from one manufacturer to another.

Floor contaminants

In the food industry, the most common floor contaminants are fats and oils. They are particularly dangerous because they are not visible to the workers and because very small amounts can make floors slippery. This paper concentrates on vegetable shortening made of partly hydrogenated canola and palm oil.

Floorings

The slipperiness of a given flooring exposed to a given amount of contaminant will depend mostly on its roughness and permeability to the contaminants and those parameters will themselves depend on the degree of wear and clogging. This paper concentrates on stripped vinyl flooring, i.e. a vinyl tile freshly stripped from its protective layer with the stripper recommended by the manufacturer. Such floorings are often found in small restaurants and, although the manufacturer recommends the application of a sealer and acrylic finish, they are often left unfinished.

At QInc, we used light reflection to monitor the surface concentration of fats and oils applied on different floorings (Massicotte et al, 2000). The application of "grease" on a surface usually increases its lustre. When the surface becomes saturated, the reflectivity remains constant at the value of the grease film. **Figure 1** shows how the surface of vinyl tiles becomes saturated as the concentration of shortening increases. The application of an acrylic finish fills the pores of the stripped vinyl and makes it smoother so that the saturation occurs at a lower concentration.

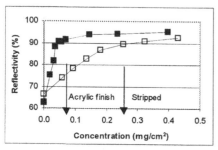 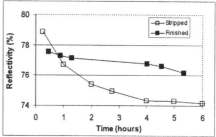

Figure 1 (Left) : Reflectivity curves showing the saturation concentration of shortening on vinyl floorings : □, stripped and ■ protected with an acrylic finish.
Figure 2 (Right) : Reflectivity curves showing the penetration of shortening, initially at 60% saturation, into vinyl floorings : □, stripped and ■ protected with an acrylic finish.

The acrylic finish protects the vinyl flooring and keeps the shortening at the surface where it is easily cleaned away. On the contrary, shortening penetrates within the unprotected vinyl flooring where it becomes very difficult to dislodge. This is shown in **Figure 2** where the continuous drop in the reflectivity is associated with shortening leaving the surface.

Cleaning Methods

We visited many restaurants and institutions and we found that in all cases, the floors are cleaned with a mop, a cleaning solution, a bucket and a wringer. The common procedure is called damp mopping, i.e. to pass a damp mop on the floor and leaving it to dry. With this method, the chemical ingredients of the cleaning solution only have one or two seconds to act on dirt so that most of the cleaning is performed through the mechanical action of the damp mop on the floor contaminants.

For extremely dirty floors, a method called immersion mopping may be used. The method consists of two steps : 1) application of a film of cleaning solution on a given area of the floor and 2) recovering the cleaning solution with a wrung mop. During the time it takes to apply the solution and to wring the mop, the chemical ingredients can act on dirt. At QInc, we believe that immersion mopping should replace damp mopping, at least when the work schedule allows it.

Measuring the cleaning efficiency

To determine the cleaning efficiency, we have reproduced the action of cleaning the floors on a laboratory scale. Small flooring samples of 7,5 cm x 7,5 cm were homogeneously covered with shortening. The samples were then cleaned with the damp or immersion mopping procedure using small mops and the cleaning solutions of the six floor cleaner categories.

Workers slip on what is left on the floorings and not on what has been removed. For that reason, we define the cleaning efficiency in terms of the residual amount of fat left on the flooring, **RM**, relative to the amount of fat required to saturate the surface of that flooring, **sat**. The resulting parameter, **RM/sat**, is expressed in % and the lower it is, the more efficient is the cleaning.

Unless otherwise stated, the concentration of the floor cleaner was 0,15% in active matter for categories 1 and 2 and 0,4% active matter for categories 3, 4, 5 and 6. All cleaning experiments are performed with clean mops and clean solutions. The temperature of the cleaning solution was $23\pm1°C$. The surface concentration of shortening was 0,7 mg·cm^{-2} and the saturation concentration of the stripped vinyl flooring was taken as 0,25 mg·cm^{-2} (~1¼ tablespoon per 10 m^2). All the results reported are the average of at least two independent samples. At QInc, we have defined that an optimal condition is encountered when the cleaning, as performed in our laboratory, leads to a **RM/sat** < 30%.

Results

This paper offers an example of the laboratory work necessary to identify the optimal cleaning conditions of shortening on a stripped vinyl flooring. Recently, the **Commission de la santé et de la sécurité du travail du Québec** has published a guide (Choisir un nettoyant ... , 2003) that summarises the optimal cleaning conditions for vegetable oil, shortening and chicken fat on stripped vinyl, finished vinyl, quarry tiles and glazed ceramic tiles.

Damp versus immersion mopping
Figure 3 compares the values of **RM/sat** for the damp and immersion mopping of stripped vinyl covered with shortening with water and floor cleaner solutions. The addition of a floor cleaner to water improves the efficiency of immersion mopping but it does not affect much that of damp mopping. This confirms a combination of chemical and mechanical actions during immersion mopping while the action of damp mopping is mostly mechanical (for shortening on stripped vinyl at room temperature).

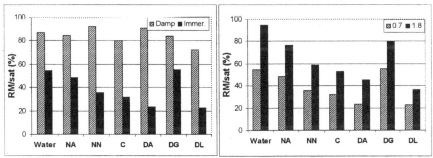

Figure 3 (Left) : Cleaning efficiency of damp and immersion mopping of shortening at 0.6 mg·cm^{-2} on stripped vinyl at room temperature.
Figure 4 (Right) : Cleaning efficiency of immersion mopping of shortening at 0.7 and 1.8 mg·cm^{-2} on stripped vinyl flooring at room temperature.

Impact of fat concentration

Figure 4 compares the efficiency of immersion mopping at two shortening concentrations. As one would expect, the efficiency decreases as the contaminant concentration increases. That confirms the importance of cleaning the floors frequently in order to avoid excessive contaminant accumulations. Note that damp mopping is even more affected by fat accumulation than immersion mopping (results not shown).

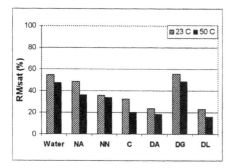

Impact of the temperature of the cleaning solution

Figure 5 compares the efficiency of immersion mopping using temperate ($23\pm1°C$) and warm ($50\pm1°C$) cleaning solutions. The warmer solutions are slightly more effective but the difference is very small so that the cleaning could be performed at either temperatures.

Figure 5 : Cleaning efficiency of immersion mopping of shortening at 0.7 mg·cm^{-2} on stripped vinyl at 23 and 50°C.

Optimal cleaning conditions

From the results presented above, one can state that the optimal cleaning conditions for shortening on stripped vinyl floorings are obtained for **immersion mopping** with a cleaning solution of **DA, C** or **DL** between **23 and 50°C** .

Acknowledgements

This work was supported by the **Institut de recherche Robert Sauvé sur la santé et sur la sécurité du travail du Québec**. We also acknowledge the collaboration of the **Commission de la santé et de la sécurité du travail du Québec** and of the **Canadian sanitary supplier association**.

References

Hébert, F., Massicotte, P., Statistical analysis on slips and falls in Québec from 1993 to 2000, IRSST, 2002 (internal communication).

Quirion, F., L'Homme, P., Répertoire des nettoyants à plancher, rapport de recherche (R-230), 1999, volume 1, 20 p. http://www.irsst.qc.ca/htmfr/pdf_txt/R-230.pdf

Massicotte, A., Boudrias, S., Quirion, F., Conditions optimales d'utilisation des nettoyants à planchers : une approche globale, rapport de recherche (R-258), 2000, 95 p. http://www.irsst.qc.ca/htmfr/pdf_txt/R-258.pdf

Choisir un nettoyant pour plancher : Guide de l'acheteur, CSST (DC 200-16225 03-02), Bibliothèque Nationale du Québec, 2003, ISBN 2-550-40227-8.

COMPARISON OF STANDARD FOOTWEAR FOR THE OIL WET RAMP SLIP RESISTANCE TEST

Richard Bowman, Carl Strautins, My Dieu Do, David Devenish and Geoff Quick

CSIRO Manufacturing & Infrastructure Technology
Graham Road
Highett 3190
Australia.

While the German DIN 51130 oil-wet ramp slip resistance test method has been accepted in Australia and elsewhere as a viable means of assessing the slip resistance potential of floor coverings, it is dependent on the availability of standardised shoes and calibration boards. The specified Bottrop footwear has not been produced for several years and has not been available for purchase since 1998. BIA, in searching for replacement footwear, has found that Lupos Picasso S1 boots give similar results to the Bottrop footwear. CSIRO has compared the Lupos and Bottrop boots on a wide range of floor coverings using both the inclining ramp platform and a sophisticated slip resistance test device. The Lupos shoes give results that were about 0.3 degrees higher than the Bottrop boots. This is sufficiently close to enable their adoption in forthcoming revisions of AS/NZS 4586 and DIN 51130.

Introduction

The selection of slip resistance test devices is highly contentious (Bowman *et al*, 2004). Manufacturers of proprietary test devices have a vested interest in their sale. Flooring manufacturers want rapid test methods, and a preference for automated measurements. They do not want tests that discriminate against products, which have long performed successfully in specific settings. Architects want readily available slip resistance data and design guidance, so that they can specify a suitable level of slip resistance and effortlessly identify appropriate products. Property managers want floors that retain their appearance and perform without an undue maintenance burden. Access auditors need ergonomically friendly devices that enable them to monitor slip resistance, and to advise when remedial action is necessary. Forensic investigators want a test method that is credible in court. Health and safety authorities have to ensure the safety of workers and the general public. They require reliable means of determining slip resistance under water-wet and other types of contamination situations. They also have a responsibility to establish compliance requirements, including a specification of reliable test methods and an identification of their limitations and the limits of their usefulness.

Test methods, which have been developed for measuring the water-wet slip resistance of relatively smooth flat pedestrian surfaces, are often inappropriate for measuring the slip resistance of highly profiled surfaces. AS/NZS 3661.1: 1993, *Slip Resistance of Pedestrian Surfaces: Requirements*, recognised that: 'The inclining ramp test method used in Germany may be more suitable to measure the slip resistance of heavily profiled surfaces under laboratory conditions'. DIN 51130: 1992, *Testing of floor coverings, determination of slip resistance, work rooms and areas of work with an increased risk of slipping, walking method ramp test*, was subsequently incorporated into AS/NZS 4586: 1999, *Slip Resistance Classification of New Pedestrian Surface Materials*.

Jung and co-workers at BIA (Berufsgenossenschaftlichen Institut für Arbeitsticherheit), refined the inclining platform (ramp) test method for determining the slip resistance of floor surfaces (DIN 51130) and footwear. This work on footwear was in conjunction with an international study of sophisticated testing machines, which sought to replicate the critical heel-strike phase of the gait cycle (Jung *et al*, 1989-93).

Jung and Schenk (1990a) found that, when oil was used as a lubricant, Bottrop footwear best discriminated between four reference floor surfaces: steel sheet and the DIN 51130 calibration tiles (manufactured by Ostara in 1987).

In a study involving 98 people walking on the DIN 51130 calibration boards in Bottrop footwear, Jung and Schenk (1989) determined the critical difference in the limit between the R classification groups as a function of the number of walkers, the number of walks each walker made, and whether the results were standardised or not. The critical difference is calculated from the reproducibility and repeatability limits, and represents the value of difference with which two tests can be expected. Where two walkers each made three walks and the tests were not standardised (by applying correction factors), the critical difference decreased from 2.1 degrees at 10 degrees to 0.7 degrees at 35 degrees, i.e., the test method becomes more discriminating the higher the walker is able to walk. This is reflected in the decrease in the acceptable critical differences when walkers are assessed on higher angle calibration boards. When the test was standardised (by means of individual correction factors), the critical difference became 1.0 degree for all angles. This decreased to 0.7 and 0.5 respectively when five and ten walkers each made three walks. Increasing the number of walks from three to six did not improve the separation characteristics. In a study involving Bottrop footwear and a broad range of 61 different floor surfaces, Jung and Rütten (1992) found that the critical difference was 2.2 degrees at a 95% significance level.

BIA have nominated Lupos Picasso S1 boots as the best replacement for Bottrop boots.

Work program

Five ramp walkers compared the Lupos Picasso and Bottrop footwear as an adjunct to the routine commercial testing of floor products to DIN 51130 (and AS/NZS 4586). This study was not funded. Additional tests were made with (initially) a single pair of Lupos shoes only when circumstances permitted. One influential factor was to try to test a wide spectrum of flooring products. The 42 materials tested included hard (ceramic tiles, etched glass, stone) and resilient (vinyl, rubber, cork-rubber, PVC, carpet) specimens. The specimens covered a wide range of surface textures, profiles and hardness.

Further oil-wet testing was performed using the SATRA STM 603 slip resistance tester. This work was principally conducted using 400 N force, a speed of 100 mm/s, and a distance of 75 mm, where the heel of the footwear was tested at a five-degree angle. Some further testing has been performed with the footwear being placed in a flat position.

The Bottrop (form ST, design S1) and Lupos Picasso S1 shoes both have nitrile rubber outsoles of IRHD hardness 72 ± 5 and 72 ± 2 respectively. However, the tread pattern profiles are quite different, as can be seen in Figure 1.

Figure 1. Comparison of footwear tread pattern: Lupos Picasso (left); Bottrop (right)

Experimental results

Some of the experimental results are given in Tables 1 and 2 and Figure 2. The coefficients of friction, obtained during testing with the SATRA STM 603, were converted to degrees by means of the inverse tangent of the value taken 0.3 seconds after movement commenced.

The data has been analysed using a paired-T test at a 95% confidence interval, in conjunction with Anderson-Darling normality tests.

Table 1. Summary of results (in degrees) for five ramp test walkers

Walker	1	2	3	4	5	All
Number of tests	17	11	9	6	6	49
Mean difference*, raw data	-0.68	-0.03	-4.20	-1.08	-0.70	-1.23
Raw standard deviation	0.57	1.59	1.50	2.79	1.36	2.04
Corrected mean difference	-0.32	-0.03	-1.44	-0.15	0.40	-0.35
Corrected standard deviation	0.53	1.67	0.49	2.67	0.96	1.37
P-value	0.024	0.958	0.000	0.898	0.352	0.079

Table 2. Summary of SATRA test data (in °); comparison with used Bottrop footwear

	New Lupos	New Bottrop	New Lupos	New Bottrop
	Heel		Shoe, flat	
Number of tests	21	15	5	5
Mean difference*	-1.27	1.31	-0.12	0.30
Standard deviation	1.87	3.10	0.91	1.45
95% confidence interval	-1.27 ± 0.85	1.31 ± 1.72	-0.12 ± 1.13	0.30 ± 1.80
P-value	0.005	0.124	0.78	0.67
Significant difference	Yes	No	No	No

* Mean difference = Used Bottrop – New Lupos (or new Bottrop)

Discussion of results

Figure 2 indicates that walker 1 obtained a higher angle value when walking in new Lupos boots than in used Bottrop boots. In testing 28 samples, the raw difference varied from -0.5 to 1.8 degrees, averaging 0.57 degrees. Since walker 1 had some missing Lupos calibration board results, only 17 sets of results were included in Table 1. Use of the standard correction criteria (for both sets of shoes) reduced the mean difference to 0.32 degrees.

Figure 2. Plot of raw data for walker 1 on 28 test surfaces in order of slip resistance

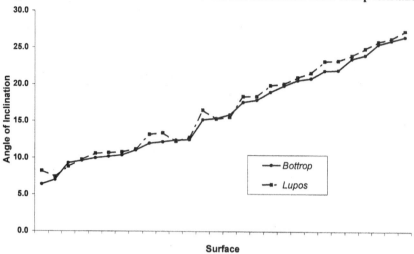

There was a significant statistical difference of 0.32 degrees between the shoes for walker 1, considered to be the most reliable walker, but not for walkers 2 to 5. They had a higher scatter of results, as can be seen from the standard deviation of the raw test data individually and as a group. The high differences that walker 3 recorded are considered to be due to poorly fitting Lupos boots. There was a significant statistical difference between the boots for walker 3, who failed on the calibration boards when wearing Lupos boots. However, when these results were included in the group results, there was no significant statistical difference between the boots. When the data for new Lupos boots was plotted against used Bottrop boots with a zero intercept, the relationships for the raw and corrected data were respectively y = 1.0595 x (R^2 = 0.9094) and y = 1.0203 x (R^2 = 0.9599).

There was some divergence in the results (differences between shoes) that the five ramp walkers obtained on individual test boards. Apart from walker 3, any variability appeared to be due to the sensitivity of the walkers rather than the type of shoes that they were wearing. In order to eliminate any effect of walker variability, some comparative testing of the shoes was conducted using the SATRA STM 603 slip resistance tester. Such SATRA test results are not corrected by means of calibration boards.

SATRA STM 603 testing was principally undertaken using the heel of the shoes, since heel-strike is normally the most critical phase of the gait cycle. There was a significant difference at a 95% confidence interval between the heels of new Lupos and used Bottrop footwear. While it may be purely coincidental, the mean difference between these shoes was -1.27, which is similar to the mean difference of the raw data for all the walkers, -1.23.

While there was a higher scatter (standard deviation) in the difference in results between the heel of the new and used Bottrop footwear, it was not found to be statistically significant. The used Bottrop shoes generally gave higher results than the new Bottrop shoes. From a ramp testing perspective, this difference is not significant, since it should be largely overcome by individual corrections based on performance on the calibration boards.

Since the ramp-walking gait is more typical of a shuffle of half a foot's length, this tends to eliminate the heel-strike phase. When footwear was assessed in a flat position, the mean differences were lower. This testing did not provide conclusive evidence of a significant difference between the used Bottrop boots and the new Lupos shoes. However, a larger sample size would have been preferable. A study of the relative slip resistance of the forepart of the boots would also have been instructive. Despite the small sample size, there was still a fair scatter of results and a moderate confidence interval.

It is interesting to note that the SATRA results were generally higher than the ramp results. If there is any variation of the slip resistance of a product that is being tested on the ramp, one would expect the walker to slip on the area of least slip resistance. The SATRA test method involves a static delay of 0.2 second, when the 400 N force is being established. This delay enables draping of the shoe bottom about the asperities of the floor, and the establishment of true contact between the surfaces.

While the Lupos Picasso shoes are considered to be a suitable replacement for Bottrop boots in forthcoming revisions of AS/NZS 4586 and DIN 51130, a more formal study of the footwear would have been preferable. This could have included more extensive SATRA STM 603 testing of the shoes, including their forepart. While the heels of the Bottrop and Lupos boots are quite dissimilar, the foreparts are more alike, and this may be responsible for the small mean difference in the ramp test results.

In the longer term, one also has to consider the ongoing availability of the existing calibration boards, and whether their characteristics change as a function of wear.

Conclusions

When using corrected values for walkers who qualified on the calibration boards, the corrected mean difference for individual walkers was 0.4 degrees or less between the shoes. Given the inevitable limitations of ramp walkers and the greater variation between them, such a small difference in footwear over a wide range of angles is considered acceptable. Differences in tread patterns would be expected to alter the drainage capability of shoe–floor contact surfaces, and to thus cause differences in results on individual products. This was observed during mechanical slip resistance testing of the footwear. It is likely that only those products, which are close to the classification limits, may change classifications.

References

Bowman, R. 2004, Discrete progress in the development of an international slip resistance standard, 8th World Congress on Ceramic Tile Quality, Castellon, Spain, in press.

Jung, K. and Fischer, A. 1993a, An ISO test method for determining slip resistance of footwear: determination of its precision, *Safety Science*, **16**, 115-127.

Jung, K. and Fischer, A. 1993b, Methods for checking the validity of technical test procedures for assessment of slip resistance of footwear, *Safety Science*, **16**, 189-206.

Jung, K. and Rütten, A. 1992, Development of a method for testing anti-slip properties of floor surfaces for workshops, working-zones and traffic routes, *Zbl. Arbeitsmed.*, **42**, 227-235.

Jung, K. and Schenk, H. 1989, Objectification and accuracy of the walking method for determining the anti-slip properties of floor surfaces, *Zbl. Arbeitsmed.*, **39**, 221-228.

Jung, K. and Schenk, H. 1990a, Objectification and accuracy of the walking method for determining the anti-slip properties of shoes, *Zbl. Arbeitsmed.* **40**, 70-78.

Jung, K. and Schenk, H. 1990b, An international comparison of test methods for determining the slip resistance of shoes, *J. Occup. Accid.*, **13**, 271-290.

USING A STATISTICAL MODEL TO ESTIMATE THE PROBABILITY OF A SLIP ON PORTABLE LADDERS

Wen-Ruey Chang and Chien-Chi Chang

Liberty Mutual Research Institute for Safety, Hopkinton, MA 01748, USA

As a major cause of injuries, a slip at the interface between the ladder bottom and floor happens when the required friction coefficient exceeds the available friction coefficient. A statistical model to estimate the probability of a slip incident on portable ladders based on the comparison of both the available and required friction coefficients with stochastic distributions is presented. The required friction coefficient to support human activities on the ladder and the available friction at the ladder shoe and floor interface were measured with separate setups, and were assumed to have a normal distribution. The results indicate that the slip potential for an oily surface was quite high. Although dry surfaces reduced the slip potential significantly, some dry conditions still could result in some potential risk in ladder climbing, especially when the inclined angle of the ladder is low.

Introduction

Ladder accidents constitute a major problem in safety despite standards and regulations. Ladders are involved in 1-2% of all occupational accidents in industrialized countries and roughly one out of every two thousand employees has a ladder accident annually (Häkkinen et al., 1988; Axelsson and Carter, 1995). Axelsson and Carter (1995) reported that nearly 40% of individuals involved in ladder accidents were absent from work for more than one month, and half of injured individuals experienced continuing, possibly permanent, disability. Björnstig and Johnson (1992) reported that almost half of ladder accidents resulted in moderate or serious injuries. Straight ladders were involved in the majority of ladder accidents (Häkkinen et al., 1988; Björnstig and Johnson, 1992; Axelsson and Carter, 1995). Slipping at the base of the ladder was the most common cause of the accidents involving straight ladders (Dewar, 1977; Häkkinen et al., 1988; Björnstig and Johnson, 1992; Axelsson and Carter, 1995).

Required friction is the minimum friction needed at the ladder base and floor interface to support human activities on the ladder. Available friction is the maximum frictional force that can be supported without a slip at the same interface, constituted by the material combination and surface conditions. When the required friction exceeds the available friction at the interface, a slip may happen. The required friction of ladder climbing was systematically investigated by Chang et al. (2002). The available friction between the ladder bottom and floor surface was measured by Chang et al. (2003) and Pesonen and Häkkinen (1988) by forcing the ladder bottom to move on the floor surface.

Based on the results from a subject climbing on a straight ladder with the ladder angle of 68°, Pesonen and Häkkinen (1988) concluded that the required friction coefficient was $0.17 - 0.28$ for climbing up to the 8th rung. Häkkinen et al. (1988), and

Pesonen and Häkkinen (1988) classified conditions with the available friction coefficient less than 0.3 as dangerous and 0.3 to 0.5 as marginal. No model is available in the literature to estimate the probability of a ladder slip at the base, although models are available to predict slip incidents of human walking. Some of these walking models can potentially be applied to ladder climbing; however, there are shortcomings in these models. Hanson et al. (1999) developed a logistic regression model to predict the probability of a fall incident in which actual fall incidents were compared with the differences between the average available friction coefficient and the average required friction coefficient. However, both required and available friction coefficients are not constants, but have a stochastic distribution. The concept of a distribution in the available friction was discussed by Barnett (2002) and Marpet (2002). Barnett (2002) used a critical friction coefficient above which no slip occurs. The probability of a slip incident could be obtained by calculating the cumulative distribution of having an available friction coefficient that falls below the critical friction coefficient. Marpet (2002) also discussed a similar concept with a minimum friction coefficient for safety. Hsiang and Chang (2002) demonstrated the variation in the normal force at the shoe and floor interface during repeated walking. In the models introduced by Barnett (2002) and Marpet (2002), the probabilities of the available and required friction coefficients above the threshold value for the required friction were completely ignored, leading to an underestimated risk of a slip event. Major simplifications in the distributions of the available and required friction coefficients could have a serious impact on the validity of the prediction model.

In this paper, a statistical model was developed to compare the available and required friction coefficients in predicting the probability of a slip incident at the base of a portable ladder using the data collected in the previous experiments (Chang et al., 2002 and 2003). In this model, a stochastic distribution was assumed for both the available and required friction coefficients. With this assumption, the estimate of the probability of a slip incident may improve.

Proposed Model

In this model, it is assumed that the probability density functions for the required and available friction coefficients, as shown in Figure 1, have a stochastic distribution of $p_r(\mu)$ and $p_a(\mu)$, respectively, where μ is a symbol for the friction coefficient.

For a given available friction coefficient μ, the probability of the required friction coefficient μ_r higher than μ is

$$P(\mu \leq \mu_r) = \int_{\mu}^{+\infty} p_r(y)dy \qquad (3)$$

A slip occurs when the required friction coefficient exceeds the available friction coefficient. Since the probability of having the available friction coefficient at this given μ is $p_a(\mu)$ and the determinations of the available friction and required friction are independent events, the probability of a slip event is

$$p_s(\mu) = p_a(\mu) \int_{\mu}^{+\infty} p_r(y)dy \qquad (4)$$

Therefore, the cumulative probability for $\mu_a \leq \mu_r$, derived from Equation (4), is

$$P(\mu_a \leq \mu_r) = \int_{-\infty}^{+\infty} p_a(x) \int_x^{+\infty} p_r(y) dy dx \qquad (5)$$

Alternatively, the cumulative probability for $\mu_a \leq \mu_r$ can be rewritten as

$$P(\mu_a \leq \mu_r) = \int_{-\infty}^{+\infty} p_r(x) \int_{-\infty}^{x} p_a(y) dy dx \qquad (6)$$

The cumulative probabilities obtained from Equations (5) and (6) should always be identical.

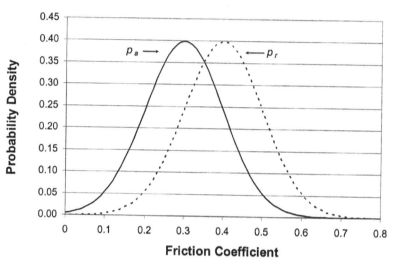

Figure 1 Examples of the probability density function for the available (p_a) and required (p_r) friction coefficients

For measuring the required and available friction coefficients, a force plate was placed under a portable ladder inclined against a wall for measuring the normal and shear ground reaction forces at the bottom. An instantaneous friction coefficient was obtained by dividing the shear force by the corresponding normal force at the same instant. For the required friction, data from one of the three male subjects (weight 69.4 kg; height 183 cm) who participated in the previous experiment (Chang et al., 2002) were used here for illustration. The effects of climbing speed (55, 75 steps/min.), ladder type (aluminum, fiberglass), ladder angle (65°, 75°) and type of support at the top of the ladder (normal, reduced friction) on the friction requirements between the ladder bottom and floor surface were investigated. Instead of the average of the maximum friction coefficient in the last 3 steps calculated in the previous study (Chang et al., 2002), the maximum friction coefficient during ascending for each climb was extracted. Each climbing condition was repeated 5 times. The mean and standard deviation of these 5 data points were used to generate a normal distribution of the required friction coefficient for each climbing condition.

A static weight of 96.2 kg (943 N) was put onto the ladder for the measurements of the available friction to simulate the normal force generated by subjects climbing the ladder (Chang et al., 2003). Among commonly measured friction types, static friction is relevant for the interface between the ladder bottom and floor surface in the presence of liquid contaminants since the squeeze-film effect is not significant at the ladder bottom. Six pairs of rubber ladder shoes, which differed in the shoe materials and tread patterns on the surfaces, were evaluated. Additional independent variables were floor surface (smooth ceramic tile and a stainless steel floor), pulling speed (slow, fast) and surface condition (dry, oily). The available friction measurement was repeated 4 times for each condition. The mean and standard deviation of these 4 repeats were used to generate the normal distribution of the available friction coefficient for this condition.

Results and Discussion

The required friction coefficients are summarized for each independent variable by collapsing over all other independent variables. The average required friction coefficients are 0.22 and 0.21 for fast and slow climbing speeds, respectively, 0.28 and 0.15 for 65° and 75° inclined angles, respectively, 0.22 and 0.21 for aluminum and fiberglass ladders, respectively, and 0.22 and 0.21 for reduced friction and normal top contacts, respectively. Similarly, the available friction coefficients are summarized for each independent variable as shown in Table 1.

Table 1. Average available friction coefficient measured for the oily conditions

Speed	Shoe #	Smooth Tile		Stainless Steel	
		Mean	SD	Mean	SD
Fast	1	0.33	0.023	0.32	0.028
	2	0.28	0.086	0.27	0.033
	3	0.44	0.038	0.44	0.039
	4	0.51	0.043	0.49	0.015
	5	0.34	0.040	0.29	0.051
	6	0.31	0.093	0.30	0.039
Slow	1	0.32	0.014	0.30	0.027
	2	-*	-*	0.23	0.022
	3	0.41	0.040	0.38	0.030
	4	0.47	0.038	0.38	0.049
	5	0.25	0.014	0.28	0.038
	6	0.27	0.023	-*	-*

* missing data

Table 2. Average probability of slips for the oily conditions

Speed	Shoe #	Smooth Tile	Stainless Steel
Fast	1	0.013	0.043
	2	0.29	0.28
	3	1.7×10^{-5}	3.2×10^{-5}
	4	1.2×10^{-7}	1.3×10^{-16}
	5	0.040	0.21
	6	0.21	0.14
Slow	1	0.013	0.14
	2	-*	0.49
	3	4.8×10^{-4}	5.0×10^{-4}
	4	7.5×10^{-7}	0.010
	5	0.48	0.24
	6	0.30	-*

* missing data

The probability of a slip incident was calculated according to Equation 5 for the combinations of each climbing condition and material testing condition. The average probabilities of slips on oily floors are 0.18 and 0.15 for fast and slow climbing speeds, respectively, 0.31 and 0.025 for 65° and 75° inclined angles, respectively, 0.18 and 0.16 for aluminum and fiberglass ladders, respectively, and 0.19 and 0.15 for reduced friction and normal top contacts, respectively. The probabilities for each independent variable for the material testing for the oily conditions are summarized in Table 2. The results indicate that an identical mean value in friction coefficient indeed yielded different probabilities due to very different standard deviations that they had. The examples are shoe number 1 on the stainless steel surface at the fast pulling speed and on the smooth ceramic tile at the slow pulling speed, and shoe numbers 2 on the smooth ceramic tile at the fast pulling speed and shoe number 5 on the stainless steel surface at the slow pulling speed.

Due to a higher friction coefficient on dry surfaces than on the oily conditions, the probability of a slip incident on dry surfaces is much lower than that on the oily conditions. Among six shoe pads evaluated, shoe pad number 1 had the highest average probability of 2.9×10^{-5}. Even on dry surfaces, the average probability of 3.3×10^{-5} at the 65° inclined angle is much higher than that of 1.6×10^{-6} at 75°. Although it might appear to be safe to climb a ladder on dry surfaces, the subject was asked to climb to the 10th rung in the previous experiment. The required friction coefficient increases as the subject climbs higher. One can expect the probability of a slip to reach a dangerous level even on dry surfaces if the subject is asked to climb higher on the ladder, especially when the inclined angle is low.

Conclusions

A statistical model to estimate the probability of a slip incident by comparing the stochastic distributions of the required and available friction coefficients was introduced. The results demonstrated the model's ability to differentiate two conditions of the available or required friction coefficient with nearly identical mean values, but different standard deviation. The estimate of this model provides an alternative perspective in comparing the results of different conditions in addition to the common methods based on the mean values.

References

Axelsson, P.-O. and Carter, N., 1995, Measures to prevent portable ladder accidents in the construction industry, *Ergonomics*, 38 (2), 250-259.

Barnett, R. L., 2002, Slip and fall theory – Extreme order statistics, *International Journal of Occupational Safety and Ergonomics (JOSE)*, 8 (2), 135-159.

Björnstig, U. and Johnson, J., 1992, Ladder injuries: mechanisms, injuries and consequences, *Journal of Safety Research*, 23, 9-18.

Chang, W. R., Chang, C. C. and Son, D. H., 2002, Friction requirements for different climbing conditions on straight ladders, The Proceedings of the XVIth International Annual Occupational Ergonomics and Safety Conference '2002, Toronto, Canada, June 9-12, 2002.

Chang, W. R., Chang C. C. and Son, D. H., 2003, Available friction and slip potential of straight ladders, The Proceedings of the 15th Triennial Congress of the International Ergonomics Association, Seoul, Korea, August 25 – 29.

Dewar, M. E., 1977, Body movements in climbing a ladder, *Ergonomics*, 20 (1), 67-86.

Hanson, J. P., Redfern, M. S. and Mazumdar, M., 1999, Predicting slips and falls considering required and available friction, *Ergonomics*, 42 (12), 1619-1633.

Häkkinen, K. K., Pesonen, J. P. and Rajamäki, E., 1988, Experiments on safety in the use of portable ladders, *Journal of Occupational Accidents*, 10, 1-9.

Hsiang, S. M. and Chang, C. C., 2002, The effect of gait speed and load carrying on the reliability of ground reaction forces, *Safety Science*, 40 (7-8), 639-657.

Marpet, M. I., 2002, Improved characterization of tribometric test results, *Safety Science*, 40 (7-8), 705-714.

Pesonen, J. P. and Häkkinen, K. K., 1988, Evaluation of safety margin against slipping in a straight aluminum ladder, *Hazard Prevention*, 24, 6-13.

A SURVEY OF FLOOR SLIPPERINESS AND EXPERIENCE OF SLIPS AND FALLS IN RESTAURANTS IN TAIWAN

Kai Way Li[1], Wen-Ruey Chang[2], Yueng-Hsiang Huang[2], Theodore K. Courtney[2], Alfred Filiaggi[2], and Kuei-hsiung Hsu[1]

[1]Department of Industrial Management
Chung-Hua University, Hsin-Chu, Taiwan 300, ROC
[2]Liberty Mutual Research Institute for Safety, Hopkinton, MA 01748, USA

The objectives of this study were to conduct employee surveys rating floor slipperiness over major working areas and to investigate their slip and fall experience in western-style fast-food restaurants in Taiwan. Fifty-six employees in ten restaurants participated in the study and rated the floor slipperiness of the seven areas based on their perception of conditions during the lunch period on the survey day. The results of comparing ratings of the areas across all the restaurants using the Kruskal-Wallis test indicated a statistical significance and showed that the back vat and sink areas were the two most slippery areas. The results also showed that 60.7% of the participants experienced a slip without a fall in the kitchen, and that 12.5% of the participants had slip and fall incidents on their jobs without major injuries. It was concluded that slips and falls were common in the fast-food restaurants and floor slipperiness might be differentiated based on employees' perceptions.

Introduction

Slips and falls are a serious safety problem in the workplace (Leamon, 1992; Swensen et al., 1992). In Taiwan, falls accounted for 14.5% of all occupational injuries, second only to traffic accidents in 2001 (Council for Labor Affairs, 2002). Among these reported fall cases, 73.7% were falls on the same level. Statistics show that the majority of falls in the USA and European countries also occur on the same level, and roughly 40-50% of same level falls are attributable to slips (Courtney, et al., 2001).

Contaminants such as grease and water are common on the floors of restaurant kitchens. Hence, slippery floors, which are a critical factor for falls on the same level, are common in restaurants (Chang et al., 2003). Leamon and Murphy (1995) reported that slips and falls resulted in the second most frequent workers' compensation injury claims and were the most costly claims within the restaurant industry in the USA. They reported that the incidence rate of falls on the same level over a 2 year period was 4.1 per 100 full-time equivalent restaurant employees, resulting in an annual per capita cost of US$116 per employee.

Perception of floor slipperiness is essential in assessing slipperiness. Myung et al. (1993) compared the subjective ranking of slipperiness, produced from a paired comparison after walking on surfaces, and the static coefficient of friction (COF) of ceramic, steel, vinyl, plywood, and sandpaper measured with a mechanical device to simulate a foot slip. They found that the higher the measured COF value, the less slippery the subjective ranking, with the exception of vinyl tile. Their results suggested that humans have a promising ability to subjectively differentiate floor slipperiness reliably, even though the measured static COF differences of these floor surfaces might not be prominent. The results from Cohen and Cohen (1994) were, however, somewhat different. Their subjects visually compared 23 tested tiles to a standard tile in a laboratory and reported whether the tile was more slippery than the standard tile. They found a significant number of disagreements between the subjective responses and the COF values of the tiles.

The studies reported in the literature, such as Myung et al. (1993) and Cohen and Cohen (1994), were mainly conducted in laboratories. The conditions of those studies may not represent what most employees encounter daily in the workplaces. A field study can better reflect realistic conditions of floor surfaces. However, field studies of floor slipperiness using employees' ratings of slipperiness are rare. The objectives of this study were to conduct employee surveys of floor slipperiness over major working areas and to investigate their slip and fall experience in western-style fast-food restaurants in Taiwan.

Methods

Restaurants and Participants
Ten western-style fast-food restaurants participated in the study. Forty (40) females (71.4%) and 16 males (28.6%), out of 58 employees working during the lunch period on the day of the visits from all ten restaurants, participated in the survey, yielding a participation rate of 96%. The number of participants per restaurant ranged from 4 to 10 with an average of 5.6. The means (± standard deviations) of the age, length of tenure, and working hours per week of the participants were 22.6 (±5.94) years, 13.1 (±13.3) months, and 37.9 (±9.55) hours, respectively.

Major Working Areas
The general kitchen areas investigated in this study included the cooking, food preparation and front counter/service areas. Seven major working areas of the kitchen were investigated: fryer, back vat, oven, sink, beverage stand, front counter and walk through. These work areas included most of the highly contaminated areas and some less contaminated areas for comparisons. The fryer and back vat are the areas for cooking french fries and fried chicken, respectively. The front counter is the area to take customers' orders and payments, and to deliver food. The beverage stand is typically located next to the front counter inside the kitchen. The sink is mainly used to defrost chicken pieces and to wash cookware. The oven is mainly used to roast chicken. The walk through area is the main path where employees enter and exit the kitchen.

Quarry tile was the typical floor in the kitchens of these restaurants. The type of tiles in seven out of ten restaurants had grit particles imbedded on the surface originally, but most of the grit appeared to be worn. The ages of the tiles were unknown, but believed to be older than the ages of the restaurants since all the restaurant properties were rented and there was no replacement of tiles prior to opening of the current businesses at these locations. The average age of these restaurants at the time of the visits was 32.4 months with a standard deviation of 26.7 months.

Survey of experience of slips and falls and floor slipperiness

A floor slipperiness survey, developed by this research team, was used in the study. Those employees that worked during the lunch period on the day of the visit were invited to participate and were compensated for their time in completing the survey. The surveys were conducted immediately after the peak lunch period, starting at approximately 1 p.m. on weekdays. The protocol was approved by an institutional review committee for the protection of human subjects. All subjects gave written informed consent.

A researcher individually interviewed all the participants in the survey. The participants were asked whether they slipped and/or fell in the past four weeks in the kitchen. The participants also reported the types of shoes they were wearing at the time of the survey. The same researcher also documented shoe pattern wear as either 'not worn,' 'partially worn,' or 'fully worn.'

The subjects rated the slipperiness of the seven working areas according to their recall of experience in the kitchen during that lunch period. A four-point rating scale was used, with 1 as "extremely slippery" to 4 as "not slippery at all." In addition to the seven areas marked on the survey, the subjects were given an opportunity to identify other areas of the restaurant that they felt were slippery.

Results and Discussion

Subjective rating of floor slipperiness

The subjective ratings of floor slipperiness in different areas were tested statistically using the Kruskal-Wallis test. The result was strongly significant ($p < 0.001$). Table 1 shows the means, standard deviations and results of the Kruskal-Wallis tests among the areas. The back vat was rated as the most slippery area with a mean rating of 2.68 which was significantly lower than that of the oven area (3.15) ($p < 0.05$), and those of the beverage area (3.72), walk through (3.74) and front counter (3.74) ($p < 0.001$). The sink area was rated as the second most slippery area with a mean subjective rating of 2.70, also significantly lower than that of the oven area ($p < 0.05$) and those of the beverage, walk through and front counter areas ($p < 0.001$). The fry vat area was rated as the third most slippery area with a mean rating of 2.96. This rating was significantly lower than those of the beverage, walk through and front counter areas ($p < 0.001$). The floor slipperiness of the oven area was rated as the fourth most slippery area in the kitchens. It was also significantly lower than those of the beverage, walk through and front counter areas ($p < 0.001$). The differences among the beverage, walk through and front counter areas were not statistically significant.

Table 1 Means, standard deviations and multiple comparison results of different areas for the subjective rating of floor slipperiness

	Back vat	Sink	Fryer	Oven	Beverage	Walk through	Front counter
	2.68	2.70	2.96	3.15	3.72	3.74	3.74
	(0.97)	(1.04)	(0.85)	(0.88)	(0.49)	(0.48)	(0.44)
Back vat		-	-	*	**	**	**
Sink			-	*	**	**	**
Fryer				-	**	**	**
Oven					**	**	**
Beverage						-	-
Walk through							-

*$p < 0.05$, ** $p < 0.001$; numbers below each area are the mean (S.D.) rating of that area

Results of floor slipperiness ratings indicated that the back vat, sink and fryer were perceived as the most slippery areas in the kitchens. This was not surprising because the floors in these areas are more likely to be contaminated. The front counter, beverage stand and walk through were rated as less slippery areas.

Experience of slips and falls in the kitchen
Among the 56 participants, 41 (73.2%) indicated experiencing a slip and/or a fall in the past four weeks and, among them, seven (12.5%) reported a slip and fall experience. The results from employees in one restaurant indicated that 38% of the employees experienced slips and/or falls. The results from employees in all other restaurants indicated that 60% or more of the employees experienced slips and/or falls. In two of the restaurants, every employee (100%) experienced slip and/or fall incidents in the past four weeks. However, no major injuries were reported during this period in all the participating restaurants.

Types of shoes and degree of shoe tread wear
All the employees are required to wear black shoes to work. Seventeen of them stated that their shoes were safety shoes. The degree of wear of their shoe patterns was determined by one of the researchers. Eleven employees' shoes were classified as 'not worn,' 26 were 'partially worn,' and 19 were 'fully worn.' Table 2 relates the experience of slips and falls to the degree of shoe tread wear. The degree of shoe tread wear was believed to be one of the important factors associated with the slip and fall experience of restaurant employees. Fully and partially worn shoes were believed to be more likely to result in slip and fall incidents. The results in Table 2 showed that among the 34 employees who slipped without a fall in the past four weeks, the number of their shoes classified as 'not worn,' 'partially worn,' and 'fully worn' were 7, 13, and 14, respectively. For the 7 employees who fell, the number of their shoes classified as 'not worn,' 'partially worn,' and 'fully worn' were 1, 5, and 1, respectively. The 'partially worn' or 'fully worn' shoes seemed to be more likely to result in slip and fall incidents. The experience of slips and falls to the degree of shoe tread wear was tested using a Chi-square test for differences. The result was not statistically significant at a $\alpha=0.05$ level which indicated that the degree of shoe tread wear made no statistical difference in the experience of slips and falls.

The results further indicate that shoe factors are far more complicated than just the shoe tread wear. Other factors that might have a significant impact on the slip resistance of the shoes could include floor surface, contaminants, the type of shoe tread pattern, and the shoe bottom profile, material properties and structure. Further investigations are needed to identify critical criteria that could be used to properly select slip resistant shoes for various applications.

Table 2 Experience of slips and falls and the extent of shoe tread wear

| Degree of tread wear | Experience of slips and falls | | | |
	Slip without a fall	Fall	No slip or fall	Total
Not worn	7 (12.5%)	1 (1.8%)	3 (5.4%)	11 (19.6%)
Partially worn	13 (23.2%)	5 (8.9%)	8 (14.3%)	26 (46.4%)
Fully worn	14 (25.0%)	1 (1.8%)	4 (7.1%)	19 (33.9%)
Total	34 (60.7%)	7 (12.5%)	15 (26.8%)	56 (100%)

Other slippery areas

In addition to the seven working areas marked in the questionnaire, the participants were requested to identify locations that a slip and fall incident might occur in the restaurant. The locations and numbers of reports were: walk-in freezer (7), storage room (5), customer dining area (4), lavatory (2), ice machine (2), ramp (2) and stairs (1). Among the 10 participating restaurants, 6 have stairs and 3 have ramps. The walk-in freezer was perceived slippery because of ice buildup on the floor. Slippery conditions were reported at all other locations due to spillage of water and/or other liquids on the floor.

Conclusion

This research investigated the slip and fall experience and perception of floor slipperiness of the employees in ten fast-food restaurants in Taiwan. Restaurant kitchens are normally perceived as slippery work places. The results showed that slips and falls were very common in these restaurants even thought no major injuries were reported. The back vat, sink and fryer were perceived as the most slippery areas in the kitchens according to the results of floor slipperiness ratings. Floor slipperiness might be differentiated based on employees' perception. Further investigations are useful to explore the relationship between employees' perception ratings and friction coefficients measured at these working areas.

References

Chang, W.R., Cotnam, J. P., Matz, S., 2003. Field evaluation of two commonly used slipmeters, *Applied Ergonomics* 34 (1), 51-60.

Cohen, H.H., Cohen, D.M., 1994. Psychophysical assessment of the perceived slipperiness of floor tile surfaces in a laboratory setting, *Journal of Safety Research* 25(1), 19-26.

Council for Labor Affairs, 2002, Labor Inspection Yearbook, Taipei.

Courtney, T.K., Sorock, G., Manning, D.P., Collins, J.W., Holbein-Jenny, M.A., 2001. Occupational slip, trip, and fall-related injuries – can the contribution of slipperiness be isolated? *Ergonomics* 44(13), 1118-1137.

Leamon, T.B., 1992. The reduction of slip and fall injuries: part II- the scientific basis (knowledge base) for the guide, *International Journal of Industrial Ergonomics* 10, 29-34.

Leamon, T.B., Murphy, P.L., 1995. Occupational slips and falls: more than a trivial problem, *Ergonomics* 38(3), 487-498.

Myung, R., Smith, J.L., Leamon, T.B., 1993. Subjective assessment of floor slipperiness, *International Journal of Industrial Ergonomics* 11, 313-319.

Swensen, E.E., Purswell, J.L., Schlegel R.E., Stanevich, R.L., 1992. Coefficient of friction and subjective assessment of slippery work surface, *Human Factors* 34(1), 67-77.

IMPROVING SLIP RESISTANCE WITH OPTIMAL FLOOR CLEANING : A PRELIMINARY FIELD STUDY

François Quirion

QI Recherche et Développement Technologique inc.
1650 boulevard Lionel-Boulet,
Varennes, Québec, Canada, J3X 1S2

In the autumn of 2001, QInc has conducted a field investigation to measure the impact of cleaning on the dynamic friction coefficient of floorings, μ_k. The study took place in the kitchens of 12 restaurants where μ_k of 45 zones, directly or indirectly exposed to fat accumulation, was determined using a horizontal pull method. Data were collected upon arrival on site, after the owner cleaned the floor and after the QInc's personnel cleaned the floor. The aim of this preliminary study was to determine if putting more effort on cleaning the floors could result in significant improvement of μ_k. For this purpose, the absolute values of μ_k are irrelevent and only the change in μ_k relative to $\mu_{k, dirty}$ are presented. Owners use damp mopping to clean the kitchen floors and the average change in friction after cleaning the zones directly exposed to fat accumulation was not significant, i.e. ~ +1%. After the same zones were cleaned using an optimal method, the dynamic friction increased significantly by 24%. These results, although preliminary, confirm that optimal cleaning can reduce significantly the slipperiness of kitchen floors in the "restaurant trade" sector.

Introduction

Since 1997, the **IRSST, Institut de recherche Robert Sauvé sur la santé et sur la sécurité du travail**, has mandated QInc to conduct research activities aiming at the reduction of floor slipperiness through the optimisation of floor cleaning.

The working hypothesis is that a cleaner floor will be less slippery, hence safer. Laboratory experiments led to the identification of optimal cleaning conditions for food contaminants such as fats and oils on different floorings. This paper reports the preliminary results of a field study where the slipperiness of kitchen floors was evaluated before and after cleaning.

Site description

We asked the collaboration of the **"Association des restaurateurs du Québec"** and **"Groupe AST"** (a prevention mutual) to identify volunteers for a preliminary study. The twelve restaurant owners identified were contacted and the purpose of the investigation as well as the test schedule were discussed with them.

The visit on site was planned before the owner cleans the floor. Once on site, we identified test zones and their friction was determined before and after the cleaning by the owner and after the the cleaning by QInc. Two types of zones were tested. Those with direct exposure to fat such as areas close to counters, frying pans or garbage cans. And those with indirect exposure to fat, i.e. the flooring areas exposed to contaminated shoe soles. Typically, the amount of "food contaminants" is higher on direct exposure areas than on indirect exposure areas.

Friction Measurements

At QInc, we have always been interested at increasing the friction of floorings through optimal cleaning. This implies that we are not interested in the absolute values of the friction coefficient but only on its relative change due to the cleaning process. We also assume that the change in the static friction coefficient, μ_S, is correlated with the change in the dynamic friction coefficient, μ_K. Since all friction measurements were made on dry floorings, we do not have to account for capillary effect associated with some test methods. With this in mind, we chose the horizontal pull method because it is simple and it provides a lot of information on the nature of the surface tested.

Like other test methods, the horizontal pull depends on many parameters, the most important being the nature of the shoe sole used to test the floorings. Leather, rubber or neolite are common materials but their condition, sanding and contamination, may affect the reproducibility of the results obtained. At QInc, we concentrate on the contribution of the flooring to friction. To do so, we use a slider mounted on three stainless steel skates. Since stainless steel is smooth and hard (little deformation), we assume that the contribution to friction comes mostly from the deformation and roughness of the flooring tested.

Stainless steel has the advantage of being resistant to wear and easy to clean. Thus the same skates can be used over and over with a good reproducibility. **Figure 1** illustrates the experimental set-up used for the determination of friction on site. The parameters are summarised in **Table 1**.

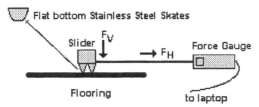

Figure 1 : Illustration of the experimental set-up used to determine the friction of floorings on site.

Table 1 : Horizontal pull parameters.

Parameter :	Value
Length tested :	Tile length, typically 15 cm
Speed :	Manual, 2,5 to 3,5 cm/sec
Slider's weight :	126 g
Number of skates :	3, flat bottom round nuts
Material for the skates :	Stainless steel
Diameter of the flat bottom of skates :	2 mm
Test conditions :	Dry flooring, clean skates

The slider is attached to the force gauge and the data are transmitted to a laptop through RS232 connection. The data is analysed and converted in friction coefficient. **Figure 2** compares typical data obtained on the same zone, before and after optimal cleaning.

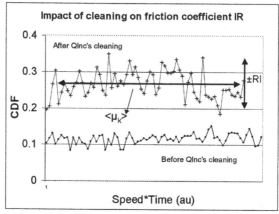

Figure 2 : Comparison of the friction data on a zone (quarry tile, 18 years of age, directly exposed to chicken fat), • not cleaned, and + after optimal cleaning. The average value is used as one determination of the dynamic friction coefficient and the standard deviation is used as a indicator of the roughness of the flooring.

The dynamic friction coefficient is defined as the mean friction coefficient measured as the slider moves on the surface at a constant speed. The ups and downs in the data are caused by the peaks and valleys encountered by the skates as the slider moves on the surface. The rougher the surface and the higher the standard deviation, σ_μ, of the dynamic friction coefficient, μ_K. We use the standard deviation as an indicator of the roughness of the surface , $RI = 100 * \sigma_\mu$.

Cleaning methods

Not surprisingly, the owners favour damp mopping as the cleaning method for their kitchen floors. Damp mopping consists in passing a damp mop on the floor and leaving it to dry. That method is not adopted as a choice but because employees do what they are asked to do, i.e. "pass the mop on the floor".

The cleaning performed by QInc on site is based on immersion mopping, a method that proved more efficient in the laboratory. The procedure was limited to the test zones as follows:

1. Apply a film of cleaning solution (DA, category 3, recommended concentration) over the test surface and let the solution act for 2 minutes.
2. Gently scrub the surface with a red pad and recover the cleaning solution with paper towels (absorption without scrubbing).
3. Rinse the clean surface with water and dry with paper towels (absorption without scrubbing).

Results

The results of the field study are presented in two sections : zones indirectly exposed to fat and zones directly exposed to fat. Most floorings had more than 10 years of age so that they were probably worn with their pores clogged. For instance, the average μ_K and **RI** measured on quarry tiles in the field ($\mu_K = 0,18\pm0,03$ and **RI** $=1,7\pm0,5$) were much lower than the values obtained for new tiles ($\mu_K = 0,49\pm0,02$ and **RI** $= 6,1\pm0,8$) and in very good agreement with tiles worn and clogged using a procedure developed in our laboratory ($\mu_K =0,21\pm0,01$ and **RI** $= 1,4\pm0,2$)

Data representation

As mentioned earlier in the text, we are interested in the impact of floor cleaning on the friction parameters. The data are expressed as the relative change in the dynamic friction coefficient following the cleaning by the owner, $\mu_{K, owner}$, and the cleaning by QInc, $\mu_{K,QInc}$, with respect to the dynamic friction coefficient of the flooring as obtained upon arrival on site, μ_K.

$$\Delta \mu_K = 100 \cdot (\mu_{K, owner} - \mu_K,)/ \mu_K, \ or = 100 \cdot (\mu_{K, QInc} - \mu_K,)/ \mu_K,$$

Typically, the uncertainty on μ_K, is around $\pm3\%$ so that changes of $\pm6\%$ were considered as significant. The data are presented as the number of zones for which the cleaning caused a given change in the dynamic friction coefficient.

Zones indirectly exposed to fat accumulation

Figure 3 compares the impact of the cleaning by the Owner with the impact of the cleaning by QInc. Of the 14 zones cleaned by the Owner, two had a lower μ_K, five showed no significant changes in friction and seven showed a significant increase. Overall, the cleaning by the Owner resulted in an average increase of 6% in μ_K,.

Of the 17 zones cleaned by QInc, two showed no significant changes in friction while fifteen showed a significant increase. Overall, the cleaning by QInc resulted in an average increase of 28% in μ_K,.

Zones directly exposed to fat accumulation

The amount of fat on these zones is generally higher than on the zones indirectly exposed to fat and we expect the cleaning efficiency to be less. **Figure 4** compares the impact of the cleaning by the Owner with the impact of the cleaning by QInc.

Of the 23 zones cleaned by the Owner, seven showed a significant decrease in friction, eleven showed no significant changes and only five showed a significant increase in μ_{K}. Overall, the cleaning by the owner resulted in an average increase of only 1% in μ_{K}.

Of the 28 zones cleaned by QInc, two showed a reduction in friction, one showed no significant changes while twenty five showed a significant increase. Overall, the cleaning by QInc resulted in an average increase of 24% in μ_{K}.

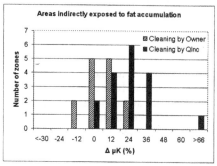 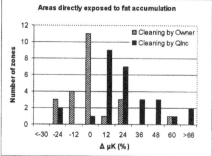

Figure 3 (Left) : Distribution of the zones indirectly exposed to fat according to the impact of the cleaning on their dynamic friction coefficient, Δ μ_K (Full = QInc ; Stripes = Owner).
Figure 4 (Right) : Distribution of the zones directly exposed to fat according to the impact of the cleaning on their dynamic friction coefficient, Δ μ_K (Full = QInc ; Stripes = Owner).

Conclusions

The results obtained suggest that damp mopping, as usually performed by restaurant employees, has little impact on the floor friction. Damp mopping of areas directly exposed to fat accumulation may even reduce floor friction. On the other hand, putting more effort on floor cleaning, for instance cleaning with the QInc method, leads to a significant improvement of the dynamic friction coefficient. This confirms that optimal cleaning can be used as a preventive mean against slippery floors in the workplace.

Acknowledgements

This project was supported by the **Institut de recherche Robert Sauvé sur la santé et sur la sé.curité du travail du Québec**. We acknowledge the collaboration of **Association des restaurateurs du Québec** and **Groupe AST** as well as the kind collaboration of the restaurant owners who volunteered for this preliminary investigation.

References

Massicotte, A. Quirion, F., *Étude préliminaire de la friction des planchers recouverts de matière grasse,* Rapport de recherche de l'IRSST (R-294), 2002, 31 p.
http://www.irsst.qc.ca/htmfr/pdf_txt/R-294.pdf

Investigating Slips, Trips and Falls in Two Major New Zealand Industries

David Tappin[1]
Tim Bentley[2]
Dave Moore[1]
Liz Ashby[3]
Richard Parker[3]
Stephen Legg[4]

[1]*Centre for Human Factors and Ergonomics, Forest Research, Auckland, New Zealand*
[2]*Department of Management and International Business, Massey University at Albany, Auckland, New Zealand*
[3]*Centre for Human Factors and Ergonomics, Forest Research, Rotorua, New Zealand*
[4]*Department of Human Resource Management, Massey University, Palmerston North, New Zealand*

The paper presents findings from a detailed study of 67 slips, trips and falls (STF) in two of New Zealand's high-STF-risk industry sectors: dairy farming and small business residential construction. Key risk factors and their interactions are briefly discussed. The paper also provides commentary on the advantages and drawbacks in the use of this investigation methodology in STF research, concluding that the potential resource, practical and logistical problems associated with conducting detailed STF investigations are outweighed by the opportunity to collect rich data on key risk factors and their interactions. The ability to collect information on latent failure risk factors is a major advantage of this method when compared to the limited detail of data usually available for analysis in injury epidemiological studies.

Introduction

Occupational slips, trips and falls (STF) are receiving increased attention in New Zealand and internationally as the extent of the STF problem and its associated human and economic costs are recognised at organisational, industry and national levels. This growing focus on STF reflects the realisation that such incidents are preventable, rather than inevitable 'everyday accidents', particularly where intervention measures are based on a detailed understanding of the factors that are contributory in STF risk (e.g. Bentley and Haslam, 2001).

The present study was a direct consequence of recognition at national level in New Zealand that STF were a leading cause of injury and compensation cost across all the major industries (ACC, 1997, OSH, 2001). As a consequence, three national bodies, the Health Research Council of New Zealand, the Accident Compensation Corporation (ACC) and the Occupational Safety and Health division of the Department of Labour (OSH), jointly funded a major programme of research concerned with the prevention of STF injuries in two high-STF incidence industry sectors: dairy cattle farming and small business residential construction.

The aims of the research presented in this paper were to identify and analyse key individual, equipment, task, environment, design and work organisation factors and their interactions in STF injury risk, for the purpose of designing research-led interventions for the prevention of STF in dairy farming and small business residential construction.

A review of the published ergonomics and safety literature on STF in these industry sectors revealed a disappointing body of knowledge. Indeed, STF in the area of dairy farming have received little or no attention in the published research, with most studies in this sector concerned with the issues of musculoskeletal problems during milking, other farming activities or farm vehicle safety. Much research in construction is epidemiological or behavioural in nature, and has often focused on the problem of falls from elevations, where injury surveillance has identified ladders, scaffolding and roofs to be the major origins of falls in building and construction (e.g. NIOSH, 2000). No studies concerned specifically with the STF problem for small business residential construction were identified.

The present study looked to address these gaps in knowledge on risk factors for STF in these New Zealand industry sectors. The study had three major research phases: an *exploratory phase* (to determine key areas of risk within each industry sector for detailed research attention in the analysis phase); an *analysis phase* (detailed analysis of STF incidents and tasks identified as key areas of concern in phase 1, together with incident-independent information interviews generating additional general STF information unrelated to the specific event under investigation); and an *intervention phase* (design and initial evaluation of measures targeting the control of risk factors identified in the analysis phase).

This paper outlines the exploratory phase findings, and the methods and findings from the analysis phase. Based on our experiences with the method, a commentary on the experienced benefits and limitations of this approach to STF research (also see Bentley and Haslam, in press) is integrated within the conclusion.

Exploratory Phase Findings

Data on 475 dairy farming and 998 small business residential construction new entitlement claims to the Accident Compensation Corporation were subjected to a descriptive epidemiological analysis. Bentley and Tappin (in press) and Bentley et al (personal communication) provide further details on the findings of the analyses summarised below. For dairy farming, key risk tasks and environments were found to include STF injuries occurring in the milking shed and yard (usually concreted areas), in paddocks and upon mounting or alighting from farm vehicles. Slips were the most frequent fall initiating event (FIE) (45%). In small business residential construction, falls from an elevation occurred in 59% of cases, with falls from ladders (26%), roofs (12%) and scaffolding (10%) the major origins of falls in this sector. Falls from ladders appear to be a specific small business problem, with ladder falls involved in relatively few STF claims in the NZ corporate construction sector. Slips were the leading fall initiating event (38%), with loss of balance also featuring prominently (30%). These findings represented the basis for where to focus research in the analysis phase.

Analysis Phase

Method
Respondents were primarily Accident Compensation Corporation (ACC) claimants during 2002/03, identified from selected incident categories (e.g. 'slipping and skidding'). Dairy

farming (n=39) and small business residential construction (n=28) STF incident follow-up investigations involved a detailed face-to-face semi-structured interview with the injured person, along with injury site observations, photographic evidence and other ergonomics investigation methods where feasible. The site visits and interviews took, on average, around 90 minutes to complete, and wherever possible took place at the site where the injury occurred. The data produced from these methods were triangulated and analysed by the investigators to create case study reports for each incident. Injury events and circumstances were recorded on an 'Events and Contributory Factors Chart' (Haslam and Bentley, 1999), detailing events from 'activity immediately preceding the STF' to 'post injury event'. The major aim of this approach was to identify risk factor data for dairy farming STF injuries that could not be obtained through other methods, such as injury data epidemiology.

Findings and Discussion

Despite the small sample sizes, the demographic, temporal, geographic and location within farm/site distributions for investigated cases were sufficiently similar to those found in the epidemiological analysis of ACC claims data in *Phase 1* to suggest a reasonably representative sample of all STF injury claimants. The largest FIE category for dairy farming STF cases investigated was foot slips (60%), involving a range of underfoot hazards, although the major surface type was concrete, comprising 30% of all cases investigated. The contaminants making concrete surfaces slippery varied, but usually also involved water which is used extensively for cleaning in and around the milking shed. The injured worker's foot tripped or caught on some object in 21% of cases. In construction, 36% involved the underfoot surface giving way, with slips (18%), trips (14%) stepping off surfaces (14%) also prominent. The most common underfoot surfaces were: concrete floor (21%), ladder (21%), scaffolding plank (18%), and uneven/soft ground (18%). Table 1 shows key risk factors identified from detailed investigations to be contributory to STF injuries for each sector.

Table 1. Key risk factors for STF cases investigated

Key risk factor	Factor present in dairy farming cases (n) (%)		Factor present in residential construction cases (n) (%)	
Running, rushing, short-cut	24	60	13	46
Design of plant, equipment, materials	24	60	6	21
Work organisation (esp. time pressure)	24	60	14	50
Water or other surface contamination	20	50	6	21
Inappropriate/worn footwear	20	50	2	7
Concurrent visual task	19	48	14	50
Poor injury assessment	19	48	4	14
Uneven/obstructed underfoot surface	10	25	14	50
Pre-existing injury	10	25	3	11
Unpredictable cattle behaviour	8	20	-	
Housekeeping	7	18	7	25
Maintenance/equipment failure	7	18	5	18
Fatigue	6	15	6	21
Working alone	5	13	1	3
Hazard perception/assessment failure	5	13	12	43
Third party action / inaction	4	10	10	36
Adverse weather (rain, wind gusts)	-		4	14

A major benefit of detailed STF follow-up investigations is the ability to identify the presence and role of latent failure factors in injury events. Indeed, Table 1 shows work organisational factors playing a key role in STF injuries in both industry sectors. In dairy farming, time pressure was common to a large number of incidents, and often motivated unsafe or time-saving behaviours that contributed to STF. Time pressure was reported as an inevitable consequence of variations in workloads over time, minimisation of staff numbers, seasonal fluctuations and the semi-paced nature of the work itself. In residential construction, organisational factors similarly involved time pressures created through highly competitive contract pricing, working in with other contractors involved in the building process, and achieving weather-tightness before the onset of rain. Other organisational factors included building practices impacting negatively on others involved later in the building process, poor housekeeping standards where STF hazards are created by one trade and their effects felt by another, and delays in the provision of appropriate scaffolding resulting in unsafe practices being adopted until the required equipment arrives.

A wide range of design factors were identified as contributory for dairy farming STF, including equipment design errors, such as: absence of safe positioning for feet when climbing onto/off equipment, the necessity to climb up onto or jump down from elevations to view aspects of the task, apparel design and fit issues and aspects of task design that necessitate unsafe operation. Design factors were found to commonly interact with the presence of slippery underfoot conditions in dairy farming. Specifically, many cases involved slippery underfoot conditions, worn footwear tread (or inappropriate footwear design and selection) and rushing. Design factors featured less commonly in residential construction STF, though most of those identified concerned falls from a height due to: sudden failure of the weight bearing surface (e.g. inadequately nailed noggins), narrow slippery surfaces (e.g. top plates), or scaffolding and ladders not fit for the task being undertaken.

Another key risk area highlighted in these investigations was the presence of concurrent visual tasks – to be undertaken at the same time as walking and/or monitoring a hazardous underfoot environment (Marletta, 1991; Bentley and Haslam, 2001). For dairy farming, the major source of divided attention was the need to also watch cattle closely during stock movement or milking-related activities, while in residential construction the primary visual task often required movement while on standing on a small surface (e.g. scaffolding plank) or walking across an uneven surface. This factor was identified as a risk where it contributed to the injured worker not seeing the underfoot hazard involved in the STF (e.g. a sudden change in surface friction characteristics or abrupt vertical protrusion). Aspects of task design may play a role in exaggerating this problem for certain activities undertaken in each sector. This factor occurred in combination with rushing and slippery underfoot conditions in several cases in both dairy farming and residential construction.

The role of behavioural factors, specifically running or moving quickly, was also found to be contributory in a large number of STF cases investigated. Running was often necessary as workers were working alone in dairy farming, but were required to cover a large area (e.g. milking shed and race) in order to control cattle movements and other factors. In construction, constantly moving quickly to keep up with the workload, or enable the next step of the building sequence to occur was common to almost all cases. Inappropriate or worn footwear was a common contributory factor in dairy farming, often being used in the presence of very slippery and/or damaged underfoot conditions (e.g. old concrete steps) and when running in response to cattle who had behaved unpredictably.

Conclusions

Findings from detailed STF follow-up investigations and incident-independent interviews with STF claimants have allowed the identification and analysis of a range of key STF risk factors and their more common interactions. Of particular benefit is information about latent failure factors in STF risk, such as work organisation, task, and physical design shortcomings. The relative paucity of information scope, detail, validity and reliability of nationally collected statistical injury data leads the authors to argue that risk factor information is best obtained through this methodology or equivalent. However, these advantages must be weighed against the many practical and logistic problems follow-up investigations present. As Bentley and Haslam(in press) observe, this method of investigation is time and labour-intensive and therefore costly. Reliance on the cooperation of the STF injured person presents a further challenge. This is exacerbated by ethical concerns, including complex participant contact methods necessitated by the need to protect the claimant's identity, leading to low response rates. Furthermore, in some cases it was not possible to interview the STF injured worker at the injury site, this factor representing an important limitation in the present study. The use of the Events and Contributory Factors Chart (Haslam and Bentley, 1999) and other measures to aid participant recall for events went some way to overcoming this problem. Further research targeting intervention design and evaluation will proceed from a solid base of information, that, arguably, could have only been provided through the methods used in this study.

References

Accident Compensation Corporation., 1997. *Analysis of Work Related Injuries: Construction*. ACC, New Zealand.

Bentley, T.A. and Tappin, D.C. (in press). An exploratory analysis of slips, trips and falls in New Zealand dairy farming. *Journal of Occupational Health and Safety – Australia and New Zealand*.

Bentley, T.A. and Haslam, R.A. (in press) , Epidemiological approaches to investigating causes
of slip, trip and fall injuries. In R. Haslam and D. Stubbs (eds.) *Understanding and Preventing Fall Accidents*, (Taylor and Francis, London).

Haslam, R.A. & Bentley, T.A.1999. Follow-up investigations of slip, trip and fall accidents among postal delivery workers. *Safety Science*, **32**, 33-47.

Marletta, W. 1991. Trip, slip and fall prevention. In D. Hansen (ed.) *The Work Environment, Vol. 1: Occupational Health Fundamentals*, (Michigan: Lewis Publishers), 241-261.

National Institute for Occupational Safety and Health. 2000. *Worker Deaths by Falls: A Summary of Surveillance Findings and Investigative Case Reports*. US Department of Health and Human Services, National Institute for Occupational Safety and Health, September 2000.

Occupational Safety and Health, 2001. *Workplace Accident Insurance Statistics Report 2000/01*. Department of Labour, New Zealand

MECHANICS OF MACROSLIP:
A NEW PHENOMENOLOGICAL THEORY

Scott D. Batterman, Ph.D.[1]
Steven C. Batterman, Ph.D.[1,2]
Howard P. Medoff, Ph.D.[3]

[1]*Batterman Engineering, LLC, Cherry Hill, NJ, USA*

[2]*Professor Emeritus, School of Engineering and Applied Science and School of Medicine, University of Pennsylvania, Philadelphia, PA, USA*

[3]*Associate Professor of Engineering, Pennsylvania State University, Abington College, Abington, PA, USA*

In this paper we present a new comprehensive mathematical framework for the prediction of macroslip under both wet and dry walkway conditions. The proposed slip criterion follows from a simple dimensional analysis utilizing variables known to influence the onset of slip. These variables, or parameters, are used to define a dimensionally consistent threshold force value used to predict the onset of slip. Although mathematically more complex, the same methodology is used to develop a model for the prediction of macroslip on a wet walkway surface, where a thin continuous hydrodynamic film of fluid contaminant is modeled to exist between the shoe outsole and the walkway surface. The mechanics of the shoe-fluid-floor interface is rigorously modeled utilizing prescribed constitutive behaviors in conjunction with the laws of classical mechanics.

Dry Model

The classical notion of Coulomb friction has traditionally been applied to the analysis of slip and fall accidents in an oftentimes incomplete and oversimplified manner. A viscous, or velocity dependent, shear model is proposed herein, in which the likelihood of a macroslip is a function of walking speed, stride length and other factors that affect gait and the friction (shear) forces that develop at the shoe-floor interface. For the purposes of this paper, a macroslip is defined to be one that is perceptible to the walker, with a high probability of resulting in a loss of balance. For slip on a dry walkway surface, we first consider slip at heel strike and then extend the development of the model to include slip at push-off. During a normal gait cycle, immediately following heel strike, the foot (shoe) is considered to decelerate to a zero forward velocity as it continues to roll down into complete ground contact (stance) and the push-off phase of the gait cycle. However, if sufficient friction force at the interface cannot be developed to decelerate the heel to a zero forward velocity, then a heel slip occurs. Therefore, when modeling slip at heel strike the relevant friction force is defined in terms of the coefficient of kinetic (dynamic) friction (μ_k). When modeling slip at push-off, the foot (shoe) in contact with the ground accelerates from a zero velocity. Therefore, for slip at push-off the threshold friction force, defined in terms of the coefficient of static friction (μ_s), must be overcome for slip to occur.

We note and emphasize that utility of the model is independent of testing method as long as the relevant coefficient of friction, or slip resistance measure, and anthropometric parameters are consistently defined throughout. Assuming that slip is a function of the mass of an individual, the horizontal walking speed at the center of mass and the stride length, it follows from dimensional analysis that for the onset of slip on a hard, flat, dry, level surface the proposed criterion takes the form:

$$F_\tau = \frac{C_{di}MV^2}{L},$$ (1)

where the terms in equation (1) are defined as follows: F_τ is the threshold friction force defining the onset of macroslip; C_{di} is an experimentally determined dimensionless anthropometric constant, or Anthropometric Gait Index (AGI). $i = k$ for slip at heel strike; $i = s$ for slip at push off; M is the mass of the individual; V is the forward (horizontal) walking speed at the center of mass of the individual and L is the stride length. As long as the maximum friction force generated at the shoe-floor interface, defined by the right side of equation (1), is below the threshold value (F_τ), a macroslip is not likely to occur. Slip is predicted to occur when the right side of equation (1) is greater than, or equal to, the threshold friction force, F_τ. For the purposes of this paper, F_τ is defined utilizing traditional notions of Coulomb friction, i.e., $F_\tau = \mu_i N$. We note that walking is a dynamic process with a time varying normal force. However, for the purpose of the model development, the normal force, N, is taken to be equal to the weight, W, of the individual. It is easily demonstrated by a straight substitution of F_τ into equation (1) that the condition for the onset of slip on a dry, level surface can be written as:

$$\frac{C_{di}V^2}{gL} \geq \mu_i,$$ (2)

where g is the acceleration due to gravity. Considering the case of slip at heel strike, for example, the AGI (C_{dk}) can be experimentally determined by having test subjects, in the same footwear, negotiate a walkway surface of a predetermined μ_k and measuring the walking speed and stride length at increasingly faster walking speeds until a macroslip occurs. Equation (2) can then be used to calculate the AGI for the combination of walking speed, V, and stride length, L, at which macroslip first occurs. We parenthetically note that even for the same individual, the AGI on a given surface will likely be a function of the footwear. For illustrative purposes, consider a 165 pound individual walking on a $\mu_k = 0.50$ surface who was found to commence a macroslip at a walking speed, V=4.4 ft/s, and a stride length, L=2.5 feet. Rearranging, and substituting into Equation (2), the AGI for this individual is $C_{dk} = 2.08$. Curves can then be generated for the prediction of macroslip as a function of walking speed, V, and stride length, L, for various coefficients of friction, or other slip resistance measures, and values of the AGI. This is illustrated Figures 1 and 2. As shown in the figures, the proposed model not only defines the likelihood of a slip on a given walkway, but also quantifies the likelihood of a slip when stepping between surfaces with varying coefficients of friction, or slip resistances. Curves can also be generated which demonstrate and quantify the difference in slip potential for different individuals on the same walkway surface, or for the same individual on a walkway surface in differing footwear, i.e. differing values of the AGI on the same walkway. See Figure 3.

Wet Model

Although mathematically more complex, the same methodology can be extended to model slip on a wet walkway surface. The fluid contaminant layer at the shoe-fluid-floor interface is modeled as a thin, continuous hydrodynamic film. In accord with hydrodynamic film theory, the pressure in the film does not vary with depth, and is a function of the horizontal position, x, only. The composite behavior of the interface is modeled as a Bingham Plastic-like fluid, and described by the following relationship.

$$\tau = \tau_y + \eta \dot{\gamma} \qquad (3)$$

τ is the shear stress, and τ_y is the critical shearing stress necessary to commence shearing in the fluid contaminant layer. η is the absolute viscosity of the fluid (slope of the stress-strain rate curve) once shearing commences, and $\dot{\gamma}$ is the shear strain rate ($\partial v / \partial y$) through the fluid layer. Neglecting inertial forces, the equation of motion in the contaminant (fluid) film becomes:

$$\frac{dp}{dx} = \eta \frac{\partial^2 v}{\partial y^2}, \qquad (4)$$

where p is the pressure in the film. Following the development by Fuller (1984) for a stationary tapered wedge separated from a horizontal moving plate by a film of fluid lubricant, with the appropriate change in boundary conditions, the shoe in the instant case is modeled as a moving tapered wedge separated from a stationary horizontal floor by a thin fluid film (contaminant) of varying thickness h. After integrating equation (4) twice, and applying the appropriate boundary conditions, the expression for the velocity profile through the fluid (contaminant) layer becomes:

$$v = \frac{1}{2\eta} \frac{dp}{dx} y^2 + \left[\frac{\dot{x}}{h} - \frac{h}{2\eta} \frac{dp}{dx} \right] y \qquad (5)$$

Equation (1) is modified as follows to yield the wet slip criterion:

$$\left(\tau_y + \eta \dot{\gamma} \right) A_c = \frac{C_{wi} M V_i^2}{L}, \qquad (6)$$

where, C_{wi} is the corresponding AGI to the one defined in the dry model. Here, as before, $i = k$ for slip at heel strike; $i = s$ for slip at push-off. V_i is the forward (horizontal) speed of the foot at first contact with the fluid/floor interface for $i = k$, and the forward (horizontal) speed of the center of mass of the individual for $i = s$. A_c is the area of shoe contact with the interface at the commencement of slip, and all other quantities are as previously defined. For slip at push-off, the shoe (foot) commences slip from a zero forward velocity, with the shoe and trapped fluid contaminant layer considered to initially slip together relative to the walkway surface. Therefore, there is no velocity gradient through the fluid thickness and $\dot{\gamma} = 0$, initially. The threshold friction force, F_τ, therefore, becomes equal to $\tau_y A_c$. In this case, equation (6) reduces identically to the dry slip criterion (equation (1)) and it follows that $C_{ws} = C_{ds}$.

Returning to the case of slip at heel strike and assuming that slip commences at the point of maximum pressure ($dp/dx = 0$), the velocity gradient through the fluid contaminant layer reduces to:

$$\frac{\partial v}{\partial y} = \frac{\dot{x}}{h^*} \qquad (7)$$

where h^* is the fluid (contaminant) depth at the point of maximum pressure, and $\dot{x}$ is the speed of the foot (shoe) at first contact with the fluid/walkway interface. Therefore, for slip at heel strike:

$$\frac{C_{wk} M h^* \dot{x}}{A_c L} - \frac{\tau_y h^*}{\dot{x}} \geq \eta \qquad (8)$$

By also requiring the wet criterion to identically reduce to the dry in the absence of a fluid contaminant layer, the relationship between the wet and dry AGI's becomes:

$$\frac{\dot{x}^2}{V^2} = \frac{C_{dk}}{C_{wk}} \qquad (9)$$

It is assumed that during the gait cycle the leading leg fully extends prior to heel strike, and that the forward speed of the foot/shoe at heel strike assumes the same value as the forward speed of the individual's center of mass. While this may not generally be rigorously true, it is a reasonable simplifying assumption, from which it follows that:

$$C_{wk} = C_{dk} \qquad (10)$$

It then follows from equation (8) that at the onset of slip at heel strike:

$$\frac{C_{wk} M h^* V^2}{A_c L} - \eta V - \tau_y h^* = 0 \qquad (11)$$

Although, there are numerous choices, the following form for τ_y is simple and captures the essential elements of the theory. The assumed form is:

$$\tau_y = S_{fi} \mu_i \frac{W}{A_c}, \qquad (12)$$

where S_{fi} is a measure of the walkway-shoe interaction represented as fraction of the coefficient of friction, (μ_i). All other variables are as previously defined. It is noted that τ_y can be defined in terms of either μ_s or μ_k as long as S_{fi} is consistently defined. Solving equation (11) for the walking speed, V, over a given range of stride lengths, L, and values of S_{fi}, and discarding the extraneous root, results are plotted in Figures 4 and 5, and compared to the results for the same walkway when dry. As expected, the addition of a thin fluid film lowers the slip resistance, and it is also noted that for $S_{fi} = 1.0$, the predictions of the wet model reduce to those of the dry model.

Conclusion
In this paper we have proposed and presented a rigorous model for the prediction of macroslip under both wet and dry walkway conditions as a function of anthropometry, utilizing biomechanical variables known to affect the onset of slip. Preliminary comparisons with experimental data appearing in the literature yield excellent correlation with the model predictions. However, this is a topic of ongoing research and will be presented in a future paper.

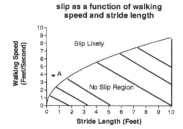

slip as a function of walking speed and stride length

Slip criterion for an exemplar individual. $\mu_k = 0.50$ and $C_{dk} = 2.08$.
Note that point A denotes a point where slip is likely to occur for this individual.

Figure 1

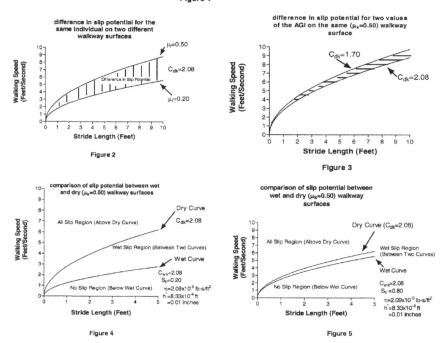

Figure 2

Figure 3

Figure 4

Figure 5

References
Fuller, Dudley D., 1984, *Theory and Practice of Lubrication for Engineers*, Second Edition, (John Wiley & Sons, Inc.)

FALL CAUSATION AMONG OLDER PEOPLE IN THE HOME: THE INTERACTING FACTORS

C L Brace and R A Haslam

Health and Safety Ergonomics Unit,
Department of Human Sciences,
Loughborough University,
Leicestershire,
LE11 3TU
UK

Falls in the home are a major problem for older people. Although personal and environmental risk factors for falling among this group are well understood, less is known about how these risks are influenced by behaviour. Focus groups and interviews were carried out with 207 older people. The findings of this investigation suggest that there are a variety of interacting factors which affect risk of falling, including intrinsic, extrinsic and peripheral influences and that behaviour is an overarching control over all of these. It can be concluded that older people who are at risk of falling need to be better educated on the individual risk factors and on the help that is available to support them in healthy aging.

Introduction

It has been well documented over the years that a third of individuals over 65, and nearly half of those over 80, fall each year. Approximately half of all recorded fall episodes that occur among independent community dwelling older people happen in their homes and immediate home environments (Lord *et al*, 1993). Fall related incidents are influencing factors in nearly half of the events leading to long-term institutional care in older people (Kennedy and Coppard, 1987). Clearly, if the incidence of falls can be reduced, people can live longer, more healthily and more independently in their own homes, with a better quality of life. Falls pose a threat to older persons due to the combination of high incidence with high susceptibility to injury. The tendency for injury because of a high prevalence of clinical diseases (e.g. osteoporosis) and age-related physiological changes (e.g. slowed protective reflexes) makes even a relatively mild fall dangerous.

Over 400 potential risk factors for falling have been identified, which are commonly split into categories of intrinsic and extrinsic risk. Intrinsic factors are age and disease related changes within the individual that increase the propensity for falls, e.g. decreased balance ability, disturbed gait, etc. Extrinsic factors are environmental hazards that present an opportunity for a fall to occur, including floor surfaces (textures and levels), loose rugs, objects on the floor (e.g. toys, pets), poor lighting etc. However, individual fall incidents are generally multifactorial.

Personal and environmental risk factors for falling among this group are well documented, although it is only recently that the influence of behaviour has been investigated in relation to these risks (Hill *et al*, 2000; Brace *et al.*, 2003). Important behavioural factors which affect the risk of older people falling in the home have been established, e.g. rushing, carrying objects. It has also been found that behaviour patterns change after a fall episode; general psychological state and experience can have an effect on the individual, affecting confidence and fear of falling, and general behaviour.

Of further interest is the extent to which the design of domestic products and areas of the home might be factors in falls. Although environment-related risk factors are reported to be causal in around one third of falls, it has only been lately that detailed work has been done to look at the design of some areas of the home environment in relation to older people, falls and independent living (Brace *et al.*, 2002).

Method

Preliminary focus groups (5) were conducted with older people (30 participants in total) to gain insight into the problem. The discussions were used to collect preparatory information on patterns of behaviour likely to affect risk of falling, informing the design of materials for the subsequent interview survey.

The main part of the study involved semi-structured interviews with 177 older people (150 households), in their own home. Quota sampling was used, based on age and gender using estimated population figures from the UK, and according to type of accommodation. Properties were selected both by age and type of housing, using national estimates of housing stock. Issues explored by the interviews included respondents' perception of factors affecting risk of falling in the home, understanding of immediate and longer term consequences of having a fall and the value and acceptability of preventative measures. The interviews involved detailed discussion of different areas of the home, and the interviewee's fall history. In addition, standard anthropometric dimensions of interviewees were recorded, along with other measurements including grip strength, ability to get off a stool without using hands, spectacle wear and measures of visual acuity and depth perception.

Interviewees were briefed both verbally and in writing about the study prior to participation. They were informed that the discussions would consider falls in the home (including the garden), examples of falls, and risk factors and safety issues that might be involved. However, they were not given any further information prior to the discussion, to avoid leading responses in any particular direction. Each interview lasted approximately two hours, with all interviews conducted by the same researcher.

Results

Intrinsic influences
Mean age of participants was 76 years (range 65-99), of whom the majority (73%) were female. Half the sample (47%) lived alone and 93% had at least one health problem related to falling, including problems with vision (35%). One or more medications were taken daily by 79% of the sample and 4 or more taken daily by one quarter (23%) of interviewees. Half the individuals (48%) had fallen at least once in the last 2 years, and

21% had experienced 2 or more falls in this period. Participants were of varying health status and inhabited a range of differing accommodation.

Although there were no significant relationships between falls and physical measures, qualitative evidence attributed intrinsic factors as the primary cause in just under one fifth (18%) of falls. Age-related factors thought to lead to increased risk of falling included the negative effects of decreased mobility, reduced balance and strength, and weakened vision. Individual capability affected by such behaviour was discussed as amplifying risk of falling. The inability to cope with chosen footwear and spectacles, the use (or non-use) of lighting and prescribed medication, and a lack of regular exercise, were examples discussed by interviewees.

Extrinsic influences
Nearly half (44%) of reported falls were attributed primarily to extrinsic factors. The design of buildings and gardens were reported to introduce risks. This was apparent when examining the areas of the home where falls were reported to occur, e.g. garden (40% of falls), stairs (23%) bathroom (8%), and kitchen (6%), due to the nature of the tasks performed in these places and subsequent behaviour (bending, reaching, etc.) and the environmental hazards present Slippery floors alone were reported to have caused nearly one fifth of falls (17%).

Choice of footwear was perceived to be a factor in fall safety, particularly the quality, thickness, grip and durability of the sole. Choice and use of footwear was reported as contributory in 10% of falls.

The design of some domestic products were reported to have directly contributed to falls (6% of cases), including oven and dishwasher doors that open downwards forming a trip hazard, or cleaning equipment that is heavy and difficult to hold.

One quarter of users of walking aids reported problems with their design and use that affected risk of falling; such devices were reported to be directly causal in 4% of incidents, and were often stated to be unsuitable for use in the home environment, due to the changes in floor surface and texture, and limited room for manoeuvre.

Combined with these extrinsic factors, behaviours involving direct use of the home environment were reported to affect fall risk. These included aspects of house maintenance, e.g. changing light bulbs, using stepladders, and 'clutter' (25%). Lack of storage space was a problem highlighted, resulting in objects being left on the floor (which was causal in 13% of falls). Often when storage was available, it was difficult to access, such as kitchen cupboards that are too high to reach without using steps etc.

Peripheral influences
Comments were made about the importance of support from family, friends and health professionals. However, this is dependent on socio-economic issues, and the proximity of family, friends and falls services. On the other hand, it was emphasized repeatedly by interviewees that older people do not want to be, or to be seen as, a 'burden on society' and wish to remain independent in their own homes for as long as possible.

It was clear that the majority of the cohort had little knowledge about the help and support that was available to them in their local area. This is something that urgently needs to be addressed, as without the advertising and subsequent awareness of fall related health and community services, many older people are missing out on useful opportunities.

Discussion

From the findings of the research, a model of the interacting influences has been proposed, Figure 1.   Although an individual has little choice over their general physical state and subsequent abilities, they do have some facility to maintain their health at its current level, e.g. by exercising, and cutting down (with help from their GP) on polypharmacy effects. However, an individual may choose to move about their home in a way that increases fall risk, due to their specific capabilities, e.g. rushing, carrying etc. Additionally, a person may choose to design and keep their home in a certain way, or impose an option on themselves, that affects their personal limitations.   The model demonstrates how falls arise from an interaction between an older person's physical capabilities (intrinsic influences), and the design, condition, suitability and use of their home environment and of aids and equipment (extrinsic influences).   The home environment and equipment are in turn influenced by the interaction with health professionals, family etc., and the older person's socio-economics (e.g. the ability to be able to afford to make changes to the home or equipment) and knowledge and understanding of fall risk. These are the peripheral influences. The latter also impact on intrinsic influences, e.g. in terms of medical support from health professionals.

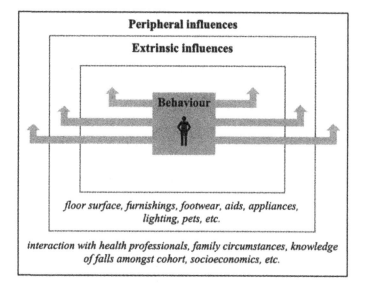

Figure 1. Influences affecting older people's risk of falling in the home

Any of these choices may be due to a lack of awareness of their personal limitations and failure to adjust their behaviour accordingly.   However, these behaviours are all dependent on physical and psychological health, fall history, socio-economic status, pressure, support from family and health professionals, and product and equipment interaction. These findings have been drawn together in the proposed model, detailing the influences in falls among older people in the home. The environment in which an older person resides and the equipment and products that are used should be designed appropriately for the individuals' capabilities in order to keep the demands of the environment and equipment as usable as possible. An individual's behaviour affects intrinsic, extrinsic and peripheral influences, each of which in turn have an impact on behaviour and each other.

Conclusions

Behaviour has been highlighted as an overarching control in fall risk and it is evident that further information is needed to direct the precise course of action for health promotion. This could involve further analysis of the specific health beliefs that older people exhibit with respect to fall risk. However, in order to combat negative health behaviours, efforts should be made to reduce the demands and challenges of products and equipment for older people to use, and the homes in which older people live in.

Most importantly, it appears that there is a need to raise awareness of the falls epidemic, amongst all stakeholders and to provide practical fall prevention advice. This approach must encourage individuals to realise that falling is not an inevitable and uncontrollable part of ageing.

Acknowledgements

Katherine Brooke-Wavell and Peter Howarth collaborated in the initial ideas for the study. The authors wish to acknowledge the support of the Department of Trade and Industry (DTI) who sponsored part of this research. The views expressed, however, are those of the authors and do not necessarily represent those of the DTI.

References

Brace, C.L., Haslam, R.A., Brooke-Wavell, K., Howarth, P. 2003, *The Contribution of Behaviour to Falls Among Older People In and Around the Home.* (Department of Trade and Industry: London)

Brace, C.L., Haslam, R.A., Brooke-Wavell, K., Howarth, P. 2002, Reducing Falls in the Home Among Older People - Behavioural and Design Factors. In: McCabe, P.T. (ed.) *Contemporary Ergonomics 2002,* (Taylor and Francis, London), pp 471-476.

Hill, L.D., Haslam, R.A., Howarth, P.A., Brooke-Wavell, K., and Sloane, J.E., 2000, *Safety of Older People on Stairs: Behavioural Factors.* (Department of Trade and Industry: London). DTI ref: 00/788.

Kennedy, T.E. and Coppard, L.C. 1987, The prevention of falls in later life. *Dan Med Bull* 1987; **34**: 1-24.

Lord, S.R., Ward, J.A., Williams, P., and Anstey, K.J., 1993, Physiological factors associated with falls in older community-dwelling women. *Australian Journal of Public Health*; **17** (3): 240-5.

A STUDY ON THE MORPHOLOGICAL FEATURES OF THE FLOOR SURFACES AND EFFECTS ON SLIP RESISTANCE PROPERTY

In-Ju Kim

School of Sport and Health Sciences
University of Exeter
Heavitree Rd, Exeter, EX1 2LU, UK

Geometric features of the floor surface could be operationally modified by surface failures and wear growths during friction measures. This would be one of the crucial factors that affect slip resistance properties. Main aims of this paper are, therefore, to analyze (1) friction and wear characteristics and (2) tribological effects on the slip resistance. A scanning electron microscope was used to obtain full morphology of wear mechanisms of the floors before and after the slip resistance tests. Micrographs clearly show that floors experienced various types of surface failure and wear. Wear behaviors were depended upon the particular combination of floors and shoes tested. Overall wear patterns also show severe material transfers and film formations on the floor surfaces during the entire rubbings. Three-dimensional approaches seem to provide unique information on wear developments and fresh ideas on the tribological characteristics of the floor surfaces for the analysis of slip resistance properties.

Introduction

Foot slippage is the most frequent unforeseen event activating falls on the same level and may also cause fall to a lower level (Andersson and Lagerlöf, 1983; Courtney et al., 2001). Prevention of such accidents requires provision of adequate friction with suitable combinations of footwear and underfoot surfaces. Although there are many factors that contribute to the falling accidents, a floor surface itself has a major important effect. The underfoot surfaces should provide safe walking conditions so that the geometry of floor surfaces should be considered from initial design stages and regularly maintained. With walking activity, however, progressive wear of the floor surfaces is unavoidable and this in turn affects the slip resistance performances (Kim and Smith, 2000). Since friction and wear features are significantly related with the surface conditions of floors, it would be reasonable to assume that slip resistance property is mainly influenced by the geometric characteristics of floor surfaces. In this study, therefore, changes of the surface topographies of various flooring specimens are qualitatively investigated by three-dimensional approaches in order to identify morphological features and obtain clear views on the nature of surface changes and wear developments of the floor surfaces during slip resistance measures.

Experimental method

The slip resistance tests were performed using a pendulum-type Dynamic Friction Tester. This tester was constructed so as to simulate the movement and loading of the foot during heel-strike and initial-slip, and quantitatively determines the slip requirement in terms of a dynamic friction coefficient (DFC). The normal force was kept around 400 N and its sliding speed was controlled at a speed of 25 cm/sec during the tests. For the floors, two totally different types of materials - metal and Perspex - were selected for their relatively predictable surface geometry and material properties such as durability and coherence. It was also assumed that these specimens could represent two extreme flooring environments such as toughness and softness. They were prepared by shot-blasting techniques to achieve graded roughness on each specimen. Four different kinds of particles - ilmenite, copper, zircon and steel - were blasted onto fresh metal plates. In the case of Perspex specimens, four different levels of blasting intensity and duration were used. The flooring specimens were labeled as *M1* to *M4* for the metal plates and *P1* to *P4* for the Perspex ones, respectively according to their roughness. A special caution was paid on each metal specimen for the prevention of oxidization. Three work boots (Polyurethane, PVC, and Nitrile Rubber) with patterned treads were used for the tests. Friction tests were conducted in turn from the smoothest surfaces (*M1* and *P1*) to the roughest ones (*M4* and *P4*) according to the roughness of the flooring specimens. The shoes were also tested in the order of hardness starting with the Polyurethane (50 Shore A°) and ending with the Nitrile Rubber shoe (70±5 Shore A°). This was intended to minimize initial damages on the heel surfaces caused by the roughest flooring specimen and to observe gradual wear effects of each flooring specimen as much as possible. Each shoe was initially rubbed 10 times on each flooring specimen and rubbed 10 times again. The surfaces of each shoe and floor specimen were thoroughly cleaned with a smooth brush to remove any loose particle after a single rubbing. To identify wear developments including particle formations and topography changes, three-dimensional images of the flooring specimens before and after the tests were obtained by a scanning electron microscope (SEM, XL 30, Philips). The SEM was operated at 10 kV setting to avoid radiation damage to the transferred polymeric particles from each shoe heel.

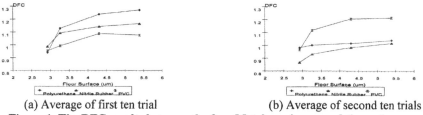

(a) Average of first ten trial   (b) Average of second ten trials

**Figure 1. The DFC results between the four Metal specimens and three shoes –
(a) mean of the first ten and (b) mean of the second ten trials**

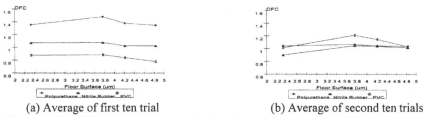

(a) Average of first ten trial   (b) Average of second ten trials

**Figure 2. The DFC results between the four Perspex specimens and three shoes –
(a) mean of the first ten and (b) mean of the second ten trials.**

Results

Overall slip resistance results
Fig. 1 shows overall DFC results between the metal plates and the three shoes. The initial slip resistance properties generally increased with the increments in the surface roughness of the metal specimens regardless of the shoe types. However, during the second ten times of rubbings, the DFC values were clearly reduced except the case of the PVC shoe. Fig. 2 shows the DFC results between the Perspex plates and the three shoes. Although the initial surface roughness of the Perspex specimens were similar to the metal ones, the DFC results show significantly different trends as compared with the results of the metal ones. For example, the *P2* plate shows the highest DFC value against all the shoes tested. Another interesting feature is that the DFC values between the three shoes and the *P4* plate were converged at approximately 1.0 after the entire tests (see Fig. 2 (b)).

SEM examinations
(1) Metal plates
    Fig. 3 shows examples of the micrographs for the worn metal plates after the entire tests. The micrographs clearly show that a diverse range and shape of wear particles from the shoe heels are found on the worn tracks of each metal surface. The greater the surface roughness the wider the ranges and shapes of wear particles. But, the particle transfers did not completely cover the peak asperities of the metal specimens so that most of the initial surface features remained without serious failures and alterations. It is also found that the transfer events occurred actively from an initial stage of the rubbings and even on the very smooth surface such as the metal specimen No. 1 (*M1*).

(2) Perspex plates
    The rubbed Perspex surfaces also show significant changes in their contact areas (see Fig. 4). Material transfers occurred in the form of small fragments and irregular shapes of polymer particles that had filled the asperity crevices in a number of locations. In the case of *P2*, the deposition of polymer fragments was much heavier than the other surfaces so that the surface seemed to provide higher DFC values against all the shoes tested because of its increased height. It is also found that the micro textures of the Perspex surfaces themselves were extensively damaged. The fresh parts of initial surfaces were heavily cracked-up after the continued rubbings so that large piece of polymer debris was able to embed into the cleavages. Thus, further rubbings under such a condition could be randomly between the deposited debris of polymer on the Perspex surface and the heel. This aspect seems to be what is basically responsible for the steady state wear.

(3) Observations of wear particles
    Figs. 5 to 6 show examples of various shapes and forms of wear particles that were formed on the peak heights and in the deep valleys of each flooring specimen after the whole tests. The main differences in the wear modes between the two groups of flooring specimens appear to be caused by the initial surface topography of each flooring specimen and mating shoe types. In the early stage of rubbings, general wear modes showed short and lumpy styles in a number of the valley areas. Whilst in the later stage of rubbings, thin and fine particles appeared to be meshed and pressed down on the surfaces. The real contact areas and adhesion between the surfaces thus seemed to take place largely between the bulk polymer and the transferred polymeric films on the floors.

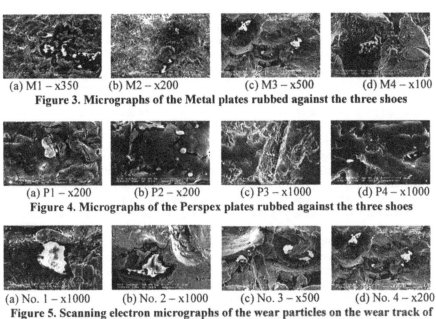

(a) M1 – x350   (b) M2 – x200   (c) M3 – x500   (d) M4 – x100

Figure 3. Micrographs of the Metal plates rubbed against the three shoes

(a) P1 – x200   (b) P2 – x200   (c) P3 – x1000   (d) P4 – x1000

Figure 4. Micrographs of the Perspex plates rubbed against the three shoes

(a) No. 1 – x1000   (b) No. 2 – x1000   (c) No. 3 – x500   (d) No. 4 – x200

Figure 5. Scanning electron micrographs of the wear particles on the wear track of wear track of the Metal specimens rubbed against the three shoes

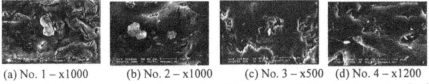

(a) No. 1 – x1000   (b) No. 2 – x1000   (c) No. 3 – x500   (d) No. 4 – x1200

Figure 6. Scanning electron micrographs of the wear particles on the wear track of the Perspex specimens rubbed against the three shoes

Discussion

(1) Formation of the surface layers

With the particle transfer, wear evolution of the floor surface is also likely to be related with indirect and/or mutual transfer of some substances such as coating materials even though its effect was not clear in the current study. That is, because the relocated wear particles from the flooring specimens into the shoe heel surfaces could cause micro-damages on the floor surfaces it is reasonable to presume that a possible source of wear transmissions is by wear particles embedded into the heel surface as well. From these possibilities, it is considered that shift direction of the wear materials between the shoe heels and floor surfaces may be classified into the following three ways:

(a) One-way transfer (shoe heel → floor surface or floor surface → shoe heel)
(b) Mutual transfer (shoe heel → floor surface and floor surface → shoe heel)
(c) Back transfer (shoe heel → floor surface → shoe heel)

Among the three prospects, one-way transfer (shoe to floor) was the dominant form of the material transferring direction in the case of the present study. But, as yet, there is no reliable theory to exactly predict the direction and relative amount of transfer for a given shoe-floor sliding system. One possibility is that material transfers should occur if the shear strength of the adhesive bond between asperities of the shoe and floor is greater than that of the transferring material. Thus, one may expect that the softer materials (shoe) mainly transfer to the harder materials (floor surface). There is clear evidence on this assumption from the current experiment but it is not always the case.

(2) Triobological characteristics of the floor surface

All the microscopic findings from this study clearly confirm the earlier conclusions drawn from the results of a comprehensive range of surface roughness measurements (Kim and Smith, 2000). This appears not only to the initial stages of sliding friction between the fresh surfaces of both the bodies, but also to the later stages of sliding friction where the surface topography was newly generated by the surface alterations and wear evolutions. No account has been taken of the fact that wear is most likely to originate from the filling of the valley areas on the floor surfaces. Slight modifications to these could affect the friction and wear mechanisms of the floors appreciably without causing many changes to the highest peak areas. Thus, whilst the flooring surfaces as a whole remain essentially random in structure, significant departures from topographic randomness may occur in those particular features responsible for wear. An analytical method to isolate these features is clearly required as a next step in relating the topographical characteristics of the surfaces to wear progression.

Conclusions

Dynamic friction tests were conducted between three shoes and two different types of flooring specimens with similar range of roughness scales under dry conditions. This test was aimed to investigate the characteristic modifications of the surface geometry of flooring specimens during repetitive sliding friction. There were clear changes in the topographic characteristics of the flooring specimens in spite of the limited amount of rubbings. All the results from this experimental investigation based on the microscopic observations clearly confirm the main hypotheses that surface topography of the floor counterface is a predominant factor to affecting the magnitude of the friction behavior and wear rate. This appears not only in the initial stages of rubbing, but also in the later stages of rubbing where the topography was generated by the sliding process itself.

References

Andersson, R. and Lagerlöf, E., 1983, Accident data in the new Swedish information system on occupational injuries, *Ergonomics*, 26, 33-42.

Courtney, T. K. and Sorock, G. S., Manning, D. P., Collins, J. W. and Holbein-Jenny, M. A., 2001, Occupational slip, trip, and fall-related injuries – can the contribution of slipperiness be isolated?, *Ergonomics*, 44, 1118-1137.

Kim, I.J. and Smith R., 2000, Observation of the floor surface topography changes in pedestrian slip resistance measurements, *International Journal of Industrial Ergonomics*, 26, No. 6, 581-601.

COEFFICIENT OF FRICTION: DOES THIS REALLY MEASURE THE SLIP SAFETY?

In-Ju Kim

School of Sport and Health Sciences
University of Exeter
Heavitree Rd, Exeter, EX1 2LU, UK

This study seeks to clarify fundamental issues on pedestrian slip safety measures. Importantly, this paper deals with the fact that a coefficient of friction (COF) index continuously varies as a result of repeated friction and wear-induced surface changes in the heel area and the floor. A new viewpoint is attempted by observing the COF index with major changes in geometrical features as a function of wear evolution between two polyurethane shoes and a ceramic tile. It is found that measured surface roughness parameters of the shoes and the floor were largely changed after the tests. Microscopic works also visibly illustrate that the heel surfaces were experienced by severe abrasion, ploughing and deep scratches whilst the floor showed massive material transfers and film formations on its surface from the early stage of rubbing. All the results clearly identify that averaged COF values are not constant with respect to repetitive friction and wear developments thus, may be insufficient to describe the intrinsic slip resistance properties.

Introduction

Slip resistance between the footwear and underfoot surface is of great importance for preventing falling accidents and has been measured as a form of coefficient of friction (COF). Hence, knowledge about the friction demand and the friction available has been recognized as the main key factor to slip safety evaluation. Since the COF measurement between the shoe and floor was adopted to determine whether a slip is to occur, however, there has been ambiguity in the interpretation of the results. It has been found that any slip resistance measurements have (1) characteristics peculiar to a specific combination of the shoe-floor-environment and (2) changed during entire service periods. Although the concept of friction is relatively simple and straightforward, its measurement, analysis and interpretation in the solution of real-world problems is quite a complex task (Irvine, 1976). One of the most important aspects, which focus on this study, is that the COF index is not a constant because friction measures are inherently noisy and change as a function of wear of the interfacing materials (Kim and Smith, 2000). In addition, friction phenomena observed between the shoe and floor are diverse and combine various sub-mechanisms (Grönqvist, 1995). Hence, there is an inherent risk in relying upon a single COF value to provide an indication of the slip safety. For more enhanced analysis of the multi-dimensional properties of the slip resistance, it is necessary to understand the tribological characteristics between the footwear and the underfoot surfaces as a function of wear evolution. This study, therefore, aims to improve our understanding on the friction and wear mechanisms and relevant tribological characteristics of the sliding interfaces between the shoe and floor. All these attempts clearly elucidate that a simple friction measure is not a proper way to represent pedestrian slip hazards any more so that friction and wear mechanisms should be thoroughly reviewed from a fundamental cause.

Experimental Method

A pendulum-type Dynamic Friction Testing Machine was used for the tests. It consists of two hydraulic systems, a force component transducer, angular displacement transducer, and a PC. The normal load of 400 N was applied to the shoe heel and its sliding speed was controlled at a relative speed of 25 cm/sec. For the test specimens, two polyurethane shoes (Nos. 1 and 2) with different sole patterns and a ceramic tile were used. Five sub-groups were designed for the tests. Each sub-group conducted 5, 10, 20, 30 and 50 rubbings, respectively with new shoes and floor samples. In order to obtain full morphological information, a range of surface roughness parameters were measured using the definitions of ANSI/ASME (ANSI.ASME, 1985). The parameters are centre line average (*Ra*), root mean square (*Rq*), maximum mean peak-to-valley height (*Rtm*), maximum mean height above the mean line (*Rpm*), maximum mean depth below the mean line (*Rvm*), average asperity slope (*Δa*), and average wavelength (*λa*). A laser scanning confocal microscope (LSCM, MRC 600, Bio-Rad) was used to measure the surface roughness. To validate roughness data and identify wear developments, initial and rubbed surfaces of the shoes and floors were thoroughly examined by a stereo scanning electron microscope (SEM, XL 30, Philips).

Results

Overall slip resistance results

Fig. 1 shows the slip resistance results plotted as a function of break-in rubbing stages. As shown, the dynamic friction coefficient (DFC) values largely decreased from 1.110 to 0.772 (over 30%) in the case of the shoe No. 1 and 1.079 to 0.646 (over 40%) in the case of the shoe No. 2, respectively after the entire tests. Although the initial DFC values were very high, their slip resistance performances drastically deteriorated and the values continuously dropped until the stage 4. Regression results also show significant relationships between the DFC index and number of rubbings ($r^2 \approx 0.80$: the shoe No. 1 and $r^2 \approx 0.86$: the shoe No. 2).

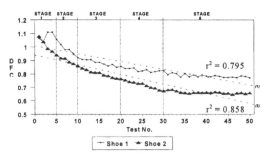

Figure 1. Slip resistance results between the two shoes and the ceramic plate

Surface topographical analysis

Tables 1 and 2 summarize the measured results of surface roughness parameters of each shoe and floor before and after the tests. All the roughness parameters were largely changed during the entire tests. In the case of the shoes, the roughness parameters were largely increased after the initial five rubbings. This increment seemed to be caused by abrasive wear so that the initial layers of the heel surfaces were roughened. With further rubbings, however, the roughness parameters were gradually decreased. This result was due to the removal of the majority of asperity crests in both the shoes. In the case of the floor, the asperity peaks and valley areas were also largely modified by massive wear and film formations. This result implied that new surfaces were evolved in the valley areas, whilst at the uppermost levels many features of the original floor surfaces also vanished.

Table 1. Summary of roughness parameters of the shoes No. 1 and No. 2

Shoe Type	Test Stage	Surface Roughness Parameters						
		Ra	*Rq*	*Rtm*	*Rpm*	*Rvm*	*Δa*	*λa*
Shoe No. 1	Initial	3.048	3.877	17.038	8.198	-8.840	0.082	233.55
	Stage 1	6.471	7.997	30.98	17.704	-12.694	0.092	455.60
	Stage 2	5.023	6.297	26.995	14.716	-12.279	0.069	457.40
	Stage 3	4.529	5.781	25.303	13.430	-11.873	0.064	444.63
	Stage 4	4.094	5.051	20.499	11.799	-8.700	0.059	435.99
	Stage 5	3.722	4.858	17.860	10.919	-6.941	0.056	417.61
Shoe No. 2	Initial	2.860	3.708	15.214	8.228	-6.896	0.069	260.43
	Stage 1	4.687	6.197	21.885	13.199	-8.686	0.089	330.89
	Stage 2	4.208	5.276	19.388	10.768	-8.620	0.065	406.76
	Stage 3	3.521	4.365	17.348	9.323	-8.025	0.056	340.36
	Stage 4	3.201	4.045	15.102	7.894	-7.208	0.052	386.78
	Stage 5	2.774	3.495	13.665	7.770	-5.895	0.045	387.32

Table 2. Summary of roughness parameters of the ceramic tile against the two shoes

Rubbed Shoe Type	Test Stage	Surface Roughness Parameters						
		Ra	*Rq*	*Rtm*	*Rpm*	*Rvm*	*Δa*	*λa*
Shoe No. 1	Initial	3.933	4.935	19.500	10.899	-8.601	0.046	531.77
	Stage 1	4.509	5.712	22.387	11.799	-10.607	0.051	572.93
	Stage 2	3.160	3.894	14.485	7.189	-7.296	0.046	573.15
	Stage 3	2.919	3.554	12.174	6.736	-6.361	0.039	525.02
	Stage 4	2.843	3.362	11.944	6.084	-5.860	0.037	575.27
	Stage 5	2.727	3.355	10.599	5.192	-5.407	0.031	576.16
Shoe No. 2	Initial	3.933	4.935	19.500	10.899	-8.601	0.046	531.77
	Stage 1	3.070	3.782	12.600	6.829	-5.771	0.039	583.13
	Stage 2	2.729	3.295	10.757	6.068	-4.689	0.038	617.19
	Stage 3	2.722	3.231	10.170	5.637	-4.533	0.031	668.55
	Stage 4	2.659	3.215	9.390	5.217	-4.173	0.030	516.98
	Stage 5	2.442	3.040	8.740	5.039	-3.701	0.029	539.81

Wear observation

Two shoes

Figs. 2 and 3 show the initial and worn surfaces of each shoe heel, respectively. Initial heel surface of the shoe No. 1 shows clean tread patterns without any protruding shape whereas the shoe No. 2 shows shiny and a number of small porous. After the entire rubbings, however, the heel surfaces were severely damaged so that their distinctive macro- and micro-tread patterns suffered massive changes. The worn heel surfaces of the shoe No. 1 show a number of paralleled tearing traces and micro-layered surface textures.

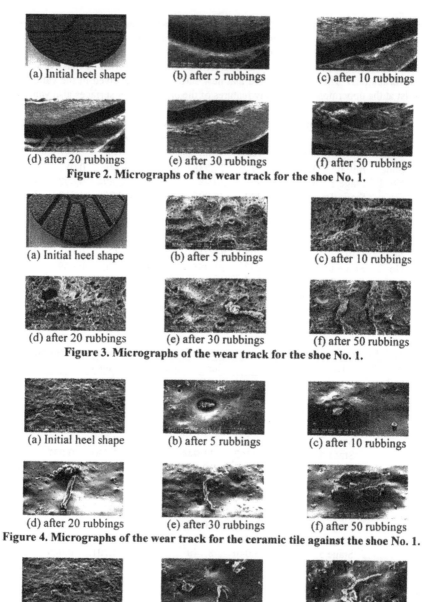

(a) Initial heel shape   (b) after 5 rubbings   (c) after 10 rubbings

(d) after 20 rubbings   (e) after 30 rubbings   (f) after 50 rubbings

Figure 2. Micrographs of the wear track for the shoe No. 1.

(a) Initial heel shape   (b) after 5 rubbings   (c) after 10 rubbings

(d) after 20 rubbings   (e) after 30 rubbings   (f) after 50 rubbings

Figure 3. Micrographs of the wear track for the shoe No. 1.

(a) Initial heel shape   (b) after 5 rubbings   (c) after 10 rubbings

(d) after 20 rubbings   (e) after 30 rubbings   (f) after 50 rubbings

Figure 4. Micrographs of the wear track for the ceramic tile against the shoe No. 1.

(a) Initial floor shape   (b) after 5 rubbings   (c) after 10 rubbings

(d) after 20 rubbings   (e) after 30 rubbings   (f) after 50 rubbings

Figure 5. Micrographs of the wear track for the ceramic tile against the shoe No. 2.

The width and depth of tearing traces gradually increased with increasing the number of rubbings. In the case of the shoe No. 2, its original micro-porosities were broken up and formed lots of cavities on its surface layer. This continuously created new surfaces so that the width and depth of tearing traces gradually increased with increasing the number of rubbing. Here it is found that wear results of the two shoes were severe than expected and occurred from the very early stage of rubbings (stage 1). The overall wear patterns also show severe abrasions, ploughings and deep scratches by fatigue wear.

Floor surface

Figs. 4 and 5 show the initial and worn surfaces of the ceramic tile rubbed against each shoe, respectively. Its basic morphology shows a number of micro-porous and flat surface textures without any specific patterns and large slopes. But, the surface was also constantly modified during the rubbings. Almost every hole was filled with various sizes of polymeric particles and some debris were adhered and freely layered on the surface. The width and length of wear particles and film formations gradually increased with the continuous rubbings. In the case of the tile rubbed against the shoe No. 2, its overall wear mode was more severe than the case of the tile rubbed against the shoe No. 1. The wear particles showed more diverse range of shapes and sizes. Different wear modes between the two cases seemed to be strongly related with the corresponding shoe type.

Conclusion

This study discussed fundamental issues on the characterization of the COF index and addressed issues in the interpretation of the COF measure, including its consistently changing aspect. Repeated sliding friction significantly changed the surface geometry of the shoe heels and the floors, with consequent effects on the slip resistance property. The measured results from the surface roughness parameters showed significant changes on the surface areas of both the bodies. Furthermore, the micrographs clearly illustrated that progressive wear was initiated in both the materials from the very early stage of rubbings. This fact evidently identifies that slip resistance property depends not just on the friction when a slip starts, but on how the friction varies as a slip progresses. Therefore, a single-threshold friction comparison may be insufficient to describe the slip resistance property between a walkway surface and a shoe sole. Factoring in wear-related material changes, as explored in this study, may provide a way of improving the reliability of walkway-safety friction determinations over the current averaged-COF-value methodology.

References

ANSI, 1985, Surface texture: Surface roughness, waviness and lay, *American Standard* ANSI B.46.1.

Grönqvist, R., 1995, Mechanisms of friction and assessment of slip resistance of new and used footwear soles on contaminated floors, *Ergonomics*, Vol. 38, No. 2, 224-241.

Irvine, C.H, 1976, A simple method for evaluation of shoe sole slipperiness, *ASTM Standardization News*, Vol. 4, April, 29-30.

Kim, I.J. and Smith, R., 2000, Observation of the floor surface topography changes in pedestrian slip resistance measurement, *International Journal of Industrial Ergonomics*, Vol. 26, No. 6, 581-601.

SURVEY ON ACCIDENTAL FALLS OF ELDERLY WORKERS WHILE COMMUTING TO AND FROM WORK

Hisao Nagata

Sunyoung Lee

National Institute of Industrial Safety
Kiyose, Tokyo 204-0024
Japan

Chinju National University
Jinju, Kyongnam 660-758,
Korea

In the building-maintenance industry in Japan where many elderly workers are employed, the ratio of accidents on the way to and from work is higher than other sectors, and accidents due to falling while commuting make up a sizeable proportion of the overall figure. Accordingly, we conducted a survey on workers in the building-maintenance industry who are 60 years old and over. We obtained 702 responses from 224 companies out of 645 totally (response rate: 34.7%). In the question "whether you have fallen and have been injured on the way to and from work," 12.1% of all respondents answered "Yes," and 87.9% of all respondents answered "No." For ratio of accidents by place, "Stairways on station premises" constituted 31.6% of all responses from the 79 effective respondents, followed by "Sidewalks" at 25.2% and "Driveways" at 11.3%.

Introduction

Projections estimate that ratio of people aged 65 years or older in Japan will soon be the world's highest, exceeding even that of Sweden. The percentage of people at least 65 years of age in Japan's population was 18% in 2001, but it is predicted to increase rapidly until 2020, after which the aging rate will continue, but at a slower pace, becoming 26% in 2015 and 35.7% in 2050, or twice the current figure.

According to the status of insurance payments for 2000 issued by the Japan Workmen's Accident Compensation Insurance Yearly Report 2000, while 603,101 new enrollees received insurance benefits, 48,537, or 8% of the total, suffered accidents while commuting. According to the overall industrial figures, the percentage of recipients among the aggregate of new enrollees who filed claims after incurring accidents, while low in mining and construction, the former being 1.3%, the latter 1.7%, public utility businesses such as electricity, gas, water supply, etc. marked the highest at 12.7%. The next highest figure was "Other businesses," such as cleaning, crematories, slaughter houses, building maintenance, and agriculture or fishery other than oceanic etc,, which accounted for 11.6%.

In general, the percentage of accidents while commuting tends to be lower in the categories of businesses, e.g. mining, construction etc. and higher in industries such as public utilities, etc. In the "Other businesses," category, building maintenance is particularly higher at 17.7%. Reviewing of the total number of occupational accidents shows that the most common form of accident, aside from traffic mishaps, is for a person to fall while commuting to and from work, with many of the victims being elderly. Consequently, this study purpose is to clarify safety issues by conducting a survey in the form of a questionnaire among workers aged 60 or more in building maintenance, who incurred the largest number of accidents while commuting, with focus on the risk of falling while commuting, using public transportation, such as trains, buses etc.

Method

Method of survey
The survey was conducted from February 14th to March 6th, 2000. From among the member list of the Japan Building Maintenance Association, 645 business establishments in Tokyo were randomly chosen and a survey in the form of a questionnaire was conducted among workers aged 60 or more. We asked each firm's personnel manager to select a maximum of 5 workers aged at least 60 years old, who had suffered injury while commuting to and from work or commuted by trains or buses etc. and to distribute a questionnaire among them. The questionnaires were returned directly to us by the pollees.

Survey queries
The pollees were asked their age, sex, occupation, years of experience, means of commuting, commuting time, evaluation of commuting stress, and perceived risky places in commuting. We asked for their descriptions of falls, the place where the falls occurred, their causes, the injuries suffered, and number of days absent from work as a result. We also sought to ascertain the bodily parts injured and the nature of the injuries, the purpose of commuting when the accidents took place, the routes used, the degree of congestion at the times of the accidents, the pace of walking at the time of the accident, and whether or not luggage, parcels, etc. were being carried when the accident happened. In conclusion, we asked the respondents for their candid opinions.

Results

Attributes of respondents
We obtained 702 questionnaires from 224 firms out of 645 selected, for a collection rate of 34.7%. Of the 702 respondents, 514 were men whose age averaged 64.6 years old, and 188 were women whose age averaged 64.4, thus men accounted for 73.2%, women 26.8%. Agewise, respondents in the 60 to 64 age group bracket represented for 58.1% and the 65 to 69 age group accounted for 33.9%, leaving 8.0% for the septuagenarians.

Means of commuting
92.6% of the 701 valid respondents answered "Train," 27.0% "Bus," 12.6% "Bicycle," 8.4% traveled "By walking only," 1.4% "Motor bike," and 1.6% "Others." (The total exceeds 100% due to multiple means of commuting).

Commuting time

13.2% of 702 valid respondents stated "Less than 30 minutes," 41.7% "30 minutes to an hour," 32.5% "One hour to 1.5 hours," 9.8% "1.5 to 2 hours," and 2.7% "Two hours or more," making 45.0%, or close to half, spending at least one hour to get to work.

Evaluation of commuting stress

To the question, "Do you feel commuting stress more than when you were in your forties?", 49.3% of the valid respondents answered "A little, 18.6% "Moderately," 6.8% "Strongly." Accordingly, 74.7% of them feel the commuting stress.

Place estimation of risk of falling while commuting

To the question, "Is there any place where you perceived the risk of falling when passing persons in the pedestrian traffic congestion?", 154 of 676 respondents, or 22.8%, answered "No." Conversely, 522 replied "Yes". 65.6% answered "On steps or stairways," 16.3% "Narrow passageways," 7.6% "Inclines," 7.5% "Escalators," and 3.0% "Others" (Multiple answers allowed).

Experience of falling, and injuries suffered

To the question, "Have you ever fallen while commuting to or from work?", 346 replied "No", while 247, or 41.7% of the valid respondents, answered "Yes." Asked "Have you ever suffered an injury in a fall while commuting to or from work?", 85, or 12.1%, answered "Yes," while 616, or 87.9%, answered "No."

Which did the accident occurred, to or from work.

According to 80 respondents who stated they had suffered accidental falls, 60.0% had the experience while commuting to work from home.

Commuting route when the accidents occurred

38.4% of 78 respondents answered "In a station or on its platform or at a bus stop," 29.5% replied "On the way from home to a train station or a bus stop," 21.8% "From the station or a bus stop to the place of work," 3.8% "While riding," and 6.4% "Others." Thus, an accident incurred somewhere between home or the jobsite and a train station or a bus stop, which accounted for 51.3% of the falls.

Place where accidental falls occurred

Based on the answers obtained from 79 valid respondents who had incurred injuries, we checked the percentage of accidents according to the specific places where they occurred. As shown in Fig. 1, 31.6% of all the answers mentioned to "Station stairways," while 25.2% replied "Sidewalks" and 11.3% "Driveways." While accidents in "Underground shopping arcades" accounted for 5.1%, which includes 3.8% occurring on "Stairways." Thus, many people hurt themselves on stairways. "Accidents suffered as a result of gaps between station platforms and trains" accounted for 5.1%. Altogether, accidents incurred in or near stations, trains, railroad crossings etc. amounted to 50.7%, while 41.6% involved falls on sidewalks, streets and arcades.

Number of days of absent from work

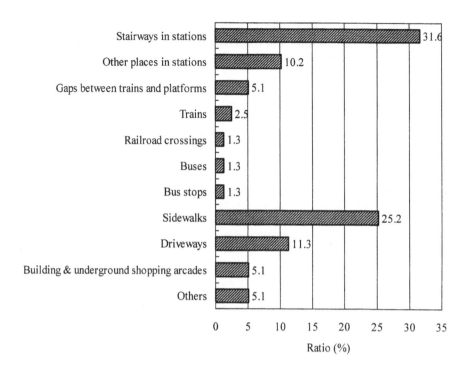

Figure 1. Place where falls occurred while commuting

According to answers from 85 valid respondents who had been injured, 40.0%, did not absent themselves from work, 20.0% were absent from "1 to 3 days," 10.0% took "4 to 7 days" off, 8.8% were absent "8 to 29 days," 13.8% were off "30 to 59 days" and 7.5% were off "60 days or more," thus 21.3% were absent from the job for a month or more.

Cause of falling
To the question, "What caused your fall?", shows, 40.8% of 76 valid respondents answered "Stumbling," 22.4% stated "Slipping, 18.4% replied "Losing one's footing," 15.8% "Swept aside or pushed, 2.6% "Unknown."

Injured part
According to answers obtained from 85 valid respondents who had been injured, "Leg or foot injury", 39.6%, "Arm or hand injury" accounted for 31.1%, "Head injury" 10.4%, "Back, waist or hip injury" 9.4% and "Others 9.5%.

Nature of injury
Answers obtained from 85 valid respondents, "Bruises" accounted for 37.6%, "Fractures" 27.7%, "Wounds, lacerations" 17.8%, "Sprains, dislocations" 15.9%, "Others" 1%.

Degree of pedestrian traffic congestion when accident occurred
Regarding ambient pedestrians traffic congestion, 55.2% of 78 valid respondents answered "Not crowded." "A little crowded" accounted for 21.8%, "Moderately" crowded" 11.5% and "Extremely crowded" 11.5%.

Pace of walking when accidents occurred
With respect to their walking conditions, 32.9% of 79 valid respondents answered "A little in a hurry", while 6.3% answered "In a considerable hurry" and 3.8% answered "In an extreme hurry." Thus, 43.0% of the respondents were in a hurry.

Carrying of hand baggage when accident occurred
To the question "Were you carrying something such as a handbag, a bag, a magazine, etc. when the accident occurred?", 26.9% of 78 valid respondents answered "No," 69.2 answered "In one hand" and 3.9% answered "In both hands." Accordingly, 73.1% were carrying something at the time they suffered an accident.

Respondents' suggestions
The item that respondents cited the most was a request for installing an escalator. In addition, they mentioned the separation of up and down sections of stairways, clarification of walking sections in the passageways, elimination of level differences in passageways, narrow passageways, etc.

Discussion

Most station platforms are nearly 1.1 meters above the track beds in Japan. Many accidents involved elderly persons, the visually impaired, etc, who were killed or injured in falls from the platforms while losing their balance. In many factories and the like, where passageways are located near devices moving at high speed, safety guardrails are undoubtedly installed. However, only a small number of station platforms have safety fences with automatic doors. While construction costs of such facilities pose a problem, a design that ensures safety is required at railway and bus related facilities, roads, etc.

Though lagging far behind Europe and the United States, Japan has been enforcing laws intended for women, the disabled and the elderly since 1975, such as the Equal Employment Opportunity Law between Men and Women (April 1999), Nurse-Care Insurance Law (April 2000), Barrier-Free Transportation Law (November 2000), including the establishment of the Quota System for Employing Disabled Persons. (June 2000), thus steadily advancing measures for such persons. As for older workers, the Law for the Employment Stability of Older Persons of October 2000 requires firms to extend employment of older workers until 65 years old. While the nation's financial burden is increasing year by year owing to the cost of nursing care for the elderly, medical care, pensions, etc., it is becoming essential to use older workers and to promote their economic independence. However, as there is no upsurge in the public interest as a whole, older workers are not used effectively though they are highly motivated. To promote the participation of the elderly in society and their independence requires establishing a viable system, as well as improving stations, streets, etc. as well as space for pedestrians from the standpoint of safety.

SERIOUS FALLS ON THE LEVEL IN OCCUPATIONAL SITUATIONS

S. Leclercq[1] and C. Tissot[2]

Institut National de Recherche et de Sécurité
[1]*Avenue de Bourgogne - Vandoeuvre - France*
[2]*Rue Olivier Noyer - Paris - France*

Falls on the level represent at least 20% of all occupational accidents leading to days lost and are as serious as other accidents. Contexts involving accidents on the level in occupational situations are highly variable and they are not well known because they are rarely analyzed in depth. This paper provides an idea of the diversity of these contexts through analysis of 459 particularly serious accident cases taken from the EPICEA database. This analysis provides information on serious accident circumstances and consequences, in particular the activity of the injured person and the accident environment. Variables such as casualty age or knowledge or ignorance of the environment are discussed. A classification ·of accident inducing situations is also presented. Research shows that a systemic approach is indeed relevant to preventing falls on the level and reveals the advantage of focusing on twin objectives in the prevention field, namely preventing loss of balance and limiting injury seriousness.

Introduction

L'Institut National de Recherche et de Sécurité (French National Research and Safety Institute) included a topic entitled "Accidents on the level: cases involving balance disturbance in occupational situations" in its 2002 research and study program. Firstly, it appeared essential to propose a definition of these accidents, for which the terminology used varies without the accidents under study being explicitly defined (Leclercq, 2002). The following definition is therefore proposed: accidents in which the casualty loses his balance during work that is not performed "at height". The casualty then recovers his balance or falls, sustaining injuries in both cases. Ground surfaces with or without changes in level, such as sidewalk curbs, steps or a slope, are considered.

The analysis presented in this paper falls within the scope of this topic. It uses occupational accident reports taken from the EPICEA database (Ho *et al.*, 1986) and offers the reader an insight into the diversity of contexts involving particularly serious falls on the level occurring in occupational situations. The results are discussed with regard to the relevant literature.

Selection of falls on the level from the EPICEA database

The EPICEA database lists cases of accidents sustained by employees covered by the French national health insurance plan, if they were fatal, serious or significant. Not all occupational accidents are recorded on the database, which is factual rather than statistical. Eighty variables describe each accident and eight of these variables will be described in the "statistical description" section through their modes, i.e. the values taken by these eight variables according to the different fall cases: for example, "use" is a mode of the variable "employee activity".

Multi-criteria selection provides a set de 459 falls on the level occurring between 1981 and 2000. These "falls on the level" were particularly serious, in fact 38% were fatal and 37% required hospitalization.

Statistical description of the 459 falls on the level

Initial analysis of all 459 falls on the level involved visualizing the distribution of each variable according to its modes, then seeking possible over-representations of other variable modes for each mode of this variable. This variable characterization was performed using SPAD© data analysis software.

Activity of employee casualty: loss of balance occurred in 14% of cases, when the casualty was walking. In every other case, the casualty was performing another activity. This result is probably specific to serious falls. Analysis of in-company falls on the level reveals a higher frequency of accidents when walking (Bentley and Haslam, 1998; Leclercq and Thouy, 2003), although another physical activity may be performed whilst walking.

Accident location: serious falls occurred:
- in one third of cases, in a workshop and most often during machining or supervision activity and in 40% of accident cases arising in a workshop, injury resulted from contact with a machine;
- in approximately 20% of cases, on work under construction and in these accidents, the casualty died in one case out of two;
- in 7% of cases, on pedestrian or vehicular traffic routes.
In 60% of cases, falls took place inside premises.

Age of employee casualty: the four age ranges between 20 and 60 years appear equally distributed amongst the 459 casualties of serious accidents. Results given in the literature concerning exposure to falls according to age are highly controversial. Literature dealing with balancing refers to the greater liability of elderly people to fall (Gabell and al., 1985 ; Pyykkö *et al.*, 1990). Bentley and Haslam (1998) did not observe any marked influence of age on fall occurrence among postal workers. On the other hand, Buck and Coleman (1985) showed that the frequency level of "slips, trips and falls on the level" occurring in occupational situations increases with employee age (between 16 and

60 years), based on U.K. Health and Safety Executive (HSE) national statistical data. Leclercq and Thouy (2003) observed that young operatives in a company are more affected by falls occurring when climbing down from a truck. Accident context, activity constraints or again personnel experience would probably explain the fact that employees in a given age range can be more affected by these accidents.

Injury origin: injury is caused:
- in 23% of cases, by crushing or jamming due to a truck, forklift or cart driven by a third party;
- in 20% of cases, by contact with a machine during machining, operation or machine adjustment;
- in 15% of cases, by contact with the ground; 28% of these injury cases occurred when the employee was walking and 19% during manual handling;
- in 6% of cases, by contact with a cutting or hot object.

These results confirm the importance of pursuing a twin objective in the area of preventing these accidents, namely preventing loss of balance and limiting injury seriousness by trying to reduce environment "aggressiveness".

Nature of injuries: sprains, pains and lumbagos represent less than 5% of injuries. More significantly, injuries are:
- in 33% of cases, bone fractures and cracks most often caused by crushing/jamming (one third of cases);
- in other cases of falls on the level, multiple injuries, bruises, wounds, amputations and burns.

Site of injuries: 30% of falls involve head and trunk injuries, particularly serious because they lead to death in 63% of cases. Furthermore:
- the hand is affected in 14% of cases concerning machining or machine adjustment activities and involving contact with the machine operating section;
- lower member (13%) and foot (4%) injuries occur when walking on a congested or slippery floor.

Employee casualty job: circumstances and seriousness of a fall on the level differ depending on the job performed by the employee casualty. Specifically:
- in the case of skilled workers, falls on the level occur mainly during structural work on site;
- in the case of machine operators and installers, falls on the level occur most often during intervention with the machine in operation. Injury, primarily to the hand, is caused by contact with the machine and requires hospitalization;
- in the case of service employees and personnel (office-based operators and employees, agency employees, cashiers, waiters, personnel employed in health care and direct services to private individuals, salesmen), falls on the level occur when moving and often involve a third party.

These results corroborate those of Kemmlert and Lundholm (1998), who showed that factors contributing to falls on the level in occupational situations differed according to activity sector.

Familiarity with activity and location: this analysis shows that in 80% of fall cases, the casualty was in a familiar location and in 70% of cases, this familiar location is associated with a familiar activity. In 60% of cases, the casualty was performing a familiar activity independently of the accident location. The 40% of cases in which the casualty was unfamiliar with the activity involve employees taken on less than one month before, who did not know the workplace and had received insufficient training.

Experience in a working environment means that an individual knows the "risk locations". This knowledge then constitutes a safety factor, which can however become a risk factor, when something changes in the environment (a step, etc.). Knowledge of the environment makes an individual less capable of detecting a change.

Typology of falls on the level in occupational situations

Typology of the 459 falls according to their circumstances was established using factorial analysis (multiple correspondence) and classification (hierarchical ascending) methods applied using SPAD© software.

The following variables were considered active: accident location, activity and employee activity purpose, phenomenon causing injury. Classification gave five different size classes of falls on the level, separated mainly according to casualty activity and fall location. These classes are described in table 1.

Table 1: fall on the level typology

Accident class (size) - name	Variable	Class characteristic modes
Class 1 (201 accidents – 44%) "machinery usage"	Activity	Drive, use, guide, supervise, clean, tidy, adjust, repair, test, measure
	Activity purpose	Installation, machine, vehicle, mobile machine, team, operation
	Location	Workshop
	Injury	Hand, amputation, hospitalization
Class 2 (127 accidents – 28%) "handling"	Activity	Handle manually, handle with mobile machine, machine, assemble, disassemble
	Activity purpose	Object, part, product, raw material, fluid network, electrical system
	Location	Storage location, freight, dispatch yard, premises interior
	Injury	Head
Class 3 (34 accidents –7%) "domestic refuse collection"	Activity	Collect, store
	Activity purpose	Refuse, raw material
	Location	Public road, building external traffic route, premises exterior
	Injury	Death
Class 4 (27 accidents – 6%) "site work"	Activity	Build, demolish
	Activity purpose	Structural, finishing work
	Location	Structure under construction
Class 5 (70 accidents – 15%) "movement"	Activity	Walking
	Activity purpose	Work location
	Location	Building internal traffic route
	Injury	Lower member

Conclusion

Firstly, it should be kept in mind that these results represent investigation of a specific set of accidents based on EPICEA data and that they cannot therefore be extrapolated to all falls on the level at work.

The typology of particularly serious falls on the level reveals different accident situations. In-company falls on the level leading to days lost are frequent and just as serious as other accidents. Moreover, they form the majority minor accidents or accidents not leading to days lost (Leclercq and Thouy, 2003). We can therefore consider that, independently of their seriousness, the diversity of contexts involving falls on the level is even greater than that observed in a set of particularly serious falls on the level.

Finally, it should be recalled that these results confirm that only a minority of serious falls on the level occur whilst moving. Yet, walking is the most common representation of activity when falling on the level. The results corroborate effectively those given in the literature by confirming that accident contexts differ according to the job performed by the casualty. It is therefore reasonable to consider that actions to be pursued in the prevention field present certainly specific characteristics associated especially with the occupational sector. The link between age and risk of losing balance is controversial in the literature. This research provided the opportunity to discuss the complexity of this factor's action on the occurrence of falls on the level.

Moreover, analysis of data provided by the EPICEA database confirms the relevance of the systemic approach adopted for analyzing these accidents, as well as the significance of pursuing twin objectives in the prevention field, namely preventing loss of balance and limiting injury seriousness.

References

Bentley, T.A. and Haslam, R.A. 1998, Slip, trip and fall accidents occurring during the delivery of mail, *Ergonomics*, **41**, n°12, 1859-1872

Buck, P.C. and Coleman, V.P. 1985, Slipping, tripping and falling accidents at work : a national picture, *Ergonomics*, **28**, n°7, 949-958

Gabell, A., Simons, M.A., and Nayak, U.S.L. 1985, Falls in the healthy elderly : predisposing causes, *Ergonomics*, **28**, n°7, 965-977

Ho, M.T.; Bastide, J.C. and Francois, C. 1986, Mise au point d'un système destiné à l'exploitation de comptes rendus d'analyse d'accidents du travail, *Le Travail Humain*, **49**, n°2, 137-146

Kemmlert, K. and Lundholm, L. 1998, Slips, trips and falls in different work groups with reference to age, *Safety Science*, **28**, 59-74

Leclercq, S. 2002, Prevention of Falls on the Level in Occupational Situations: A Major Issue, a Risk to be Managed, *International Journal of Occupational Safety and Ergonomics*, **8**, n°3, 377-385

Leclercq, S. and Thouy, S. 2003, Systemic analysis of falls on the level in a company, submitted for publication to the journal *Ergonomics*

Pyykkö, I.; Jantti, P. and Aalto, H. 1990, Postural control in elderly subjects, *Age and Ageing*, **19**, 215-221

CAUSAL THINKING IN SLIPPING AND TRIPPING ACCIDENTS

Paul Lehane

Environmental Health and Trading Standards
London Borough of Bromley BR1 3UH,

Causes of slipping and tripping accidents are often expressed in terms of the lack of frictional characterises of the flooring surface or footwear. This paper presents an initial look at the use of other mechanisms of determining the cause in a scenario based slip or trip accident by examining the stated cause against alternative causal selection strategies. The allocation of responsibility to scenario based actors is also considered with respect to general psychological models of attribution and blame.

Introduction

Tackling the unremitting burden of work related slipping and tripping accidents has become one of the key objectives for all those involved in occupational health and safety since the launch of the UK Governments Revitalising Health and Safety initiative in 2000.

The level of incidents involving slips or trips has continued to show a year on year increase since 1996/7. In 2002 there were over 29,000 accidents resulting in more than 3 days absence from work and over 10,000 accidents resulting in a specified major injury were reported to the Enforcing Authorities for the UK. Annually slips and trip account for about 41% of all reported major injury accidents and 23 % of accidents involving an absence of more than 3 days. (HSE Health and Safety Statistics 2000/01)

One of the fundamental questions that we need to ask is why we have yet to make any significant impact at a national level on the rate of slip and trip accidents.

The traditional approach to understanding slips and trips has been one of researching and tackling issues associated with the frictional aspects of the foot/floor interface. This has tended to focus attention on the measurement of slip resistance and its application to floor surfaces and to a lesser extent the effect of different footwear soling compositions.

There may be a number of reasons why this approach to slip and trip prevention has failed so far, and these include the possibility that there may be a different psychological approach to them. Lehane and Stubbs 2004 suggest that there is a life long universal exposure and experience of slips and trips whereas exposure to certain other hazards are more restrictive and usually occupationally based. e.g. machinery or chemicals.

Unexpected or negative outcomes from people's interactions with others or their environment lead to a cognitive search for a causal explanation in order to understand and if possible take control of similar situations in the future. In general psychological terms people tend to deny responsibility for "unwanted" outcomes (self serving bias) and locate the blame (causal responsibility) for the outcome (an accident) away from themselves. (Miller & Ross 1975 cited in Plous, 1993).

Lehane and Stubbs 2001 who reported different locus of causal responsibility by accident subjects and mangers for slips and trips found this self serving bias for slipping accidents but not for tripping accidents where accident subjects showed a tendency to self-attribute blame.

In order make an attribution of causal responsibility it is suggested that a person firstly has to select and choose a cause for an accident and then attribute this to the action or inaction of a person as appropriate. This could result in mutual blame by accident subjects and managers as both could view an accident as an "unwanted "outcome.

For the purposes of this paper a number of previously identified strategies for selecting the causal candidate (Hesslow, 1988) from the many which can be identified by a casual search are examined against the casual explanations given by 3 populations (accident subjects, managers and safety professionals) after reading a scenario about a slip or trip accident. The scenario related to Mary a part time supermarket checkout operator who agreed to cover for a friend on a day she did not usually work and who slipped on a spillage of milk / tripped over a box as she went to her midmorning break. The scenario was manipulated for type of accident, level of detail provided and outcome severity.

After reading the scenario respondents were asked to write a sentence giving the cause of the accident.

The list of strategies is by no means exhaustive but include the most common theories for causal selection.

Unexpected condition - Selection of something unknown.
Precipitating cause - The one condition that came into existence last.
Abnormal condition - The difference between the non-accident state and accident state.
Variability - A variable condition, an amalgam of the previous 3 causal selection criteria.
Deviation from theoretical norm - An ideal standard.
Responsibility- The action / inaction of a person.
Predictive value - Information that would have allowed us to predict an outcome.

Replaceability - Something irreplaceable is a stronger causal candidate than something replaceable.

Instrumental efficacy - Conditions that allow us to manipulate effects.

Interest - Causal selection is arbitrary based on the interest of the person giving the explanation. Different perspectives based on position or experience.

Results

332 responses where obtained to a self-completion postal questionnaire involving a slip / trip scenario. The pattern of responses included 89 Safety professionals, 124 work place managers / supervisors and 119 people who had been injured in a workplace accident (accident subjects) covering 167 slips and 165 trips.

This paper presents the results for two aspects of the stated cause. 1. The classification of causal statements against the causal selection strategies identified above and 2. The identification of the scenario based actor that the causal statements referred to (causal actor).

Classification of Causal Statements
The causal statements were coded against the 10 selection strategies and are shown in table 1.

The distribution of the results between slips and trips revealed no significant differences (Binomial test). To test if the choice of the causal selection theory was random a Chi Square test was performed on the 5 most frequently used categories (*unexpected condition, precipitating cause, abnormal condition, deviation from theoretical ideal and responsibility). Pearson Chi-Square value 39.958, 8df (degrees of freedom), was significant at p=.0005 indicating that the use of these causal theories was not random but the strength and direction of any relationship is not identified by this test.

Identification of causal actor
In light of the significant use of "responsibility" as a causal selection strategy by managers and accident subjects the 3 scenario actors most commonly targeted for such an attribution were identified. (Table 2)

Discussion.

Whilst these results represent an initial and superficial analysis of the stated causes of slip and trip accidents against previously identified causal selection strategies, they do indicate some trends which it is suggested would prove interesting areas for further analysis and research.

There appears to be a difference in the use of causal selection strategies adopted by the three populations. The most commonly used strategy by both managers and accident subjects is that of "responsibility" across both slipping and tripping accidents whereas

Table 1 Classification of causal statements overall and by slip and trip separately

Theory	Safety Professionals		Managers		Accident subjects		Total	
*Unexpected condition	1 (1%)		4 (3%)		17 (14%)		22	
	Slip	Trip	Slip	Trip	Slip	Trip	Slip	Trip
	0	1	0	4	10	7	10	12
*Precipitating cause	12 (14%)		16 (13%)		17 (14%)		45	
	Slip	Trip	Slip	Trip	Slip	Trip	Slip	Trip
	6	6	10	6	11	6	27	18
*Abnormal condition	46 (52%)		29 (24%)		17 (14%)		92	
	Slip	Trip	Slip	Trip	Slip	Trip	Slip	Trip
	20	26	17	12	10	7	47	45
*Theoretical ideal	10 (11%)		29 (24%)		11 (10%)		50	
	Slip	Trip	Slip	Trip	Slip	Trip	Slip	Trip
	6	4	14	15	8	3	28	22
*Responsibility	19 (21%)		42 (34%)		55 (46%)		116	
	Slip	Trip	Slip	Trip	Slip	Trip	Slip	Trip
	8	11	18	24	26	29	52	64
Predictive value	0		1 (1%)		2 (2%)		3	
	Slip	Trip	Slip	Trip	Slip	Trip	Slip	Trip
	0	0	1	0	0	2	1	2
Variability/ Replaceability / Instrumental efficacy /Interest	0		0		0		0	
	Slip	Trip	Slip	Trip	Slip	Trip	Slip	Trip
	0	0	0	0	0	0	0	0
Unknown	1 (1%)		3 (2%)		0		4	
	Slip	Trip	Slip	Trip	Slip	Trip	Slip	Trip
	1	0	1	2	0	0	2	2
Column total and %	89 (100%)		124 (100%)		119 (100%)		332 (100%)	
	Slip	Trip·	Slip	Trip	Slip	Trip	Slip	Trip
	41	48	61	63	65	54	167	165

Table 2 The three most commonly targeted actors for "Responsibility"

	1st		2nd		3rd		Total in top 3	
Manager	Supervisor 10		Mangers 9		Accident Subjects 6		25	
	Slips	Trips	Slips	Trips	Slips	Trips	Slips	Trips
	3	7	3	6	4	2	10	15
Accident Subject	Accident Subject 14		Supervisor 14		Other worker 11		39	
	Slips	Trips	Slips	Trips	Slips	Trips	Slips	Trips
	4	10	11	3	4	7	19	20
Safety Professional	Employer 7		Supervisor 4		Other worker 3		14	
	Slips	Trips	Slips	Trips	Slips	Trips	Slips	Trips
	3	4	3	1	0	3	6	8

safety professionals appear to use the "Abnormal Condition" as a basis for determining causal selection.

The potential for managers and accident subjects to attribute causal responsibly to each other is a possibility given the results in Table 1 thus perpetuating a culture of mutual blame. This was examined in Table 2 which indicated that tripping accidents are more likely to be subject to a "responsibility" causal selection than slipping accidents (trips 35: slips 29) and that accident subjects are more likely to use "responsibility" as a causal selection strategy than managers (accident subjects 39: managers 25). The existence of mutual blame is a possibility but is not likely to be substantial given the numbers involved as a proportion of the populations who responded.

The tendency reported by Lehane and Stubbs 2001 for accident subjects to judge themselves as being causally responsible for their own tripping accidents is again found in that 10 of the 14 accidents subjects who allocated responsibility to themselves did so for tripping accidents, however the difference was not significant at p= . 05 (chi Sq 2.571, 1df, p= .109)

The fact that managers also seem to allocate responsibility to themselves almost as much as to supervisors is interesting but may be a function of the level of seniority of the respondents in the population called managers. Whilst generally referred to as managers it was open to any supervisor or manger to complete the questionnaire. At this level of analysis it is not possible to distinguish between them, and given the effect of the self serving bias the mutual desire to distance oneself from a situation where some criticism or blame may arise and thus supervisors may seek to blame managers and visa versa in much the same way as it appears that accident subjects seek to lay responsibility away from themselves generally (Self 14: other 25 (14 supervisors + 11 other workers).

The greater our understanding of how people arrive at and use causal attribution in the case of slipping and tripping accidents the better we will be able to design and implement improved investigation techniques and preventative strategies. Further research into the psychology of slipping and tripping accidents is therefore highly relevant.

References

Department of the Environment, transport and Regions, London, 2000 *Revitalising Health and Safety*, June 2000

Health and Safety Statistics 2000/01, HSE Books, HMSO Norwich

Hesslow G. 1988 The problem of causal selection. In D.Hilton (ed) Contemporary science and natural explanation –commonsense conceptions of causality, (Harvester Press, Brighton) 11-32

Lehane P. and Stubbs D. 2001, The perceptions of managers and accident subjects in the service industries towards slip and trip accidents, *Applied Ergonomics* **32**, 119-126

Lehane P. and Stubbs D. 2004, The investigation of individual incidents. In Haslam R. and Stubbs D. (eds) *Understanding and preventing fall accidents*. (Taylor and Francis, London) in press.

Plous S. 1993, The psychology of judgement and decision making, McGraw Hill inc New York.

THE INFLUENCE OF OUTSOLE OIL RESISTANCE ON SLIP RESISTANCE CHARACTERISTICS IN WINTER CONDITIONS

Carita Aschan, Erkki Rajamäki, Mikko Hirvonen, Tarmo Mannelin

Finnish Institute of Occupational Health, Department of Physics
Topeliuksenkatu 41 a A
FIN-00250 Helsinki
Finland

Oil resistance of outsoles is defined as an obligatory requirement for professional footwear in European standards. Based on this requirement, non-oil resistant outsole materials that have been considered to be more slip resistant in winter conditions than oil resistant materials can't be used as outsoles in professional footwear. The aim of this study was to find out whether there is a difference in slip resistance characteristics between oil resistant and non-oil resistant outsoles, and therefore a need to change the CEN standards in case of oil resistance. The Portable Slip Simulator was used to measure slip resistance of different outsole materials in simulated winter conditions. According to the results, oil resistant outsoles showed poorer slip resistance characteristics, and therefore the current European standards for professional footwear should be altered in order to make oil resistance as an additional requirement.

Introduction

Safety footwear fitted with steel toecaps (European standard EN 345:1992) are commonly used in many workplaces, both indoors and outdoors. In European standards EN 345:1992 and EN 346:1992 oil resistance of outsoles is defined as an obligatory requirement for professional footwear. Based on this requirement, non-oil resistant outsole materials, such as TR (thermoplastic rubber) or natural rubber, that have been considered (Grönqvist and Hirvonen, 1995; Gao and Abeysekera, 2002) to be more slip resistant in winter conditions than oil resistant materials, can not be used as outsoles in professional footwear. However, in many cases the risk of slipping is much higher than that caused by fuel oil.

The change of CEN standards in case of requirement for oil resistance has been proposed by Sweden and the UK (CEN/TC161/WG1+2/WG2 N510; CEN/TC161/WG1+2/N024). However, research on the subject is needed before it is possible to consider / carry on the changes in CEN working groups CEN/TC 161/WG 1&2. Therefore, the aim of this study was to find out whether there is a major difference in slip-resistance characteristics between oil resistant professional footwear and non-oil resistant footwear, such as common winter shoes and rubber boots, when measured in simulated winter conditions.

Materials & Methods

Materials

The project was started by selecting the footwear to be used in slip resistance measurements. When selecting the footwear e.g. outsole material and cleating had to be taken into consideration. Following commonly used pairs of footwear and outsole materials (23 in total) were selected: oil resistant safety footwear with PU, TPU and NBR outsoles, separate outsole samples made of NR, TR, SBR, NBR, PU and EVA, ordinary winter shoes with TR outsoles and rubber boots as well as winter and leisure time shoes, the outsoles of which were made of different rubber compounds. The specifications of the outsole materials are presented in Table 1. Oil resistance of all the samples were measured according to European standard EN 344:1992.

Table 1. The specifications of the outsole materials

Sample no.	Material (according to the manufacturer)	Oil resistant
1	SBR	no
2	NR	no
3	NR	no
4	SBR	no
5	NR	no
6	NBR	yes
7	PU	yes
8	NBR	yes
9	Rubber compound? [a]	no
10	Rubber compound? [a]	no
11	Rubber compound? [a]	no
12	Rubber compound? [a]	no
13	TR	no
14	TPU	yes
15	1-layer PU	yes
16	EVA	no
17	Rubber compound	no
18	1-layer PU (ESD)	yes
19	1-layer PU	yes
20	2-layer PU	yes
21	TR	no
22	NBR	yes
23	1-layer PU	no

[a]detailed information not available; pair of footwear provided from local department store

DCOF measurements

Slip resistance, i.e. dynamic coefficient of friction (DCOF), of the selected outsoles was measured using the Portable Slip Simulator of FIOH (Aschan *et al*, 2003). The measurements were performed in climatic chamber using the following measurement parameters:

- normal force 500 N,
- normal force build-up rate 4 500 Ns^{-1} leading to normal force build-up time of approximately 110 ms when F = 500 N,
- horizontal sliding velocity 0.2 ms^{-1},
- contact angle 5°, and
- evaluation of DCOF between 100 and 300 ms after achieving the full normal force.

The result of the measurement was the mean value during the measurement period of 200 ms.

Smooth ice on different temperatures (0°C, -5°C and -20°C) was used as a testing surface. Prior the measurements, the samples were stored in the measurement temperature in the climatic chamber for 8 hours. For comparison, steel surface with glycerol as a lubricant, as defined in European draft standard for the determination of slip resistance of professional footwear (prEN 13287: October 2003), was also used in the DCOF measurements performed at room temperature (20°C).

Statistical analysis

One-Way Analysis of Variance (ANOVA) was used to compare differences between the DCOF values measured with oil resistant and non-oil resistant footwear. DCOF values measured with different surfaces were compared by using Pearson's product moment correlation coefficient (r).

Results

Oil resistance of the selected footwear are presented in Table 1. From 23 pairs of footwear in total 9 pairs were found to be oil resistant. Their outsoles were made of NBR, PU and TPU. The non-oil resistant outsoles were mainly different rubber compounds, such as NR, TR or SBR, and one sample of EVA and PU of each.

The results of DCOF measurements performed on different icy surfaces and steel surface lubricated with glycerol are presented in Figures 1 and 2. In all conditions, non-oil resistant footwear was found more slip resistant than oil resistant ones, in generally. Statistically significant differences between these two groups were found when measured on ice surface of -20°C and on steel surface with glycerol as a lubricant.

The ranking of the measured footwear varied in different conditions. No significant correlation was found between the results measured on steel surface and different icy conditions (0°C, -5°C and -20°C). Results obtained with the icy surfaces of 0°C and -5°C as well as 0°C and -20°C correlated statistically significantly (r = 0.645, p<0.001 and r = 0.586, p<0.01, respectively). Similarly, a significant correlation was found between the results measured on icy surfaces of -5°C and -20°C (r = 0.573, p<0.01).

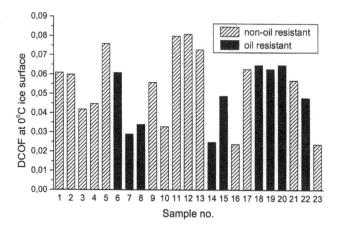

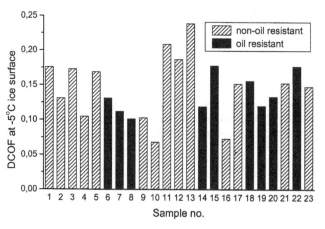

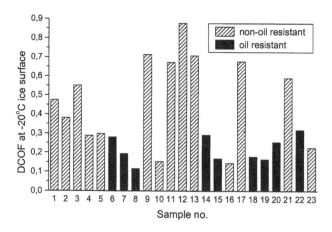

Figure 1. DCOF values of the footwear samples on icy surfaces

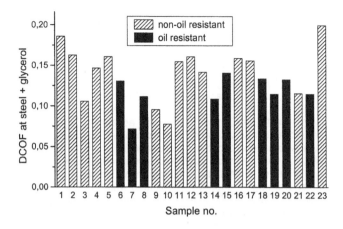

Figure 2. DCOF values of the footwear samples on steel surface

Conclusions

According to the results of this study, the current European standard with its obligatory requirement of oil resistant outsoles seems to restrict the use of more slip resistant outsole materials. Therefore, it is concluded that the CEN standard for professional footwear should be altered in order to make oil resistance as an additional requirement to allow slip resistant materials to be used when the slipping risk is higher than the risk caused by fuel oil.

References

Aschan, C., Hirvonen, M., Mannelin T. and Rajamäki, E. 2003, Prevention of slipping accidents - development of the portable slip meter II. In: *Proceedings of the XVth Triennial Congress of the International Ergonomics Association*, August 24 - 29, 2003, Seoul, Korea

CEN/TC161/WG1+2/WG2. 2002, Document N510

CEN/TC161/WG1+2. 2003, Document N024

European draft standard prEN 13287: October 2003, Safety, protective and occupational footwear for professional use - Test method for the determination of slip resistance

European standard EN 344:1992, Requirements and test methods for safety, protective and occupational footwear for professional use

European standard EN 345:1992, Specification for safety footwear for professional use

European standard EN 346:1992, Specification for protective footwear for professional use

Gao, C. and Abeysekera, J. 2002, The assessment of the integration of slip resistance, thermal insulation and wearability of footwear on icy surfaces, *Safety Science*, **40**, 613-624

Grönqvist, R. and Hirvonen, M. 1995, Slipperiness of footwear and mechanisms of walking friction on icy surfaces, *International Journal of Industrial Ergonomics*, **16**, 191-200

The Effects of Age and Stress Associated with a Fear of Falling on Gait Adjustments

Thurmon E Lockhart
Virginia Polytechnic and State University
Blacksburg, VA 24061

Previous research has shown that with advancing age, there are increasing incidences of slips and falls. Additionally, studies suggest that older adults who have experienced a previous fall are more likely to suffer future falls. A study was conducted to investigate if stress associated with a fear of falling contributes to the increased incidents of falls among older adults. Both age groups were evaluated while walking over dry and slippery floor surfaces. First, a baseline measure was collected for normal gait prior to any exposure to slipping. A second measure was collected following a slip from a contaminated floor surface. The investigation compared physiological stress with biomechanical parameters of walking for twenty-eight participants in two age groups: (18-35) and (65 or older). Biomechanical parameters included: step length, required coefficient of friction (RCOF), slip distance, and heel contact velocity. Physiological stress was assessed utilizing Salivary Amylase level. Overall, the results indicated that there were differences between older and younger adult's biomechanical parameters of walking, and their physiological stress associated with an inadvertent slip. Findings suggest that older individuals required an additional step to properly adjust gait for a contaminated walking surface. Furthermore, it is concluded that some stress may be beneficial and may lead to an increased awareness of their surroundings, and appeared to help facilitate appropriate gait adaptation when hazardous conditions were encountered.

Introduction

Falls are among the most common and serious problems facing older adults. Falling is associated with considerable mortality, morbidity, reduced functioning, and premature nursing home admissions (Robbins, Rubenstein, & Josephson, 1989). Both the incidence of falls and the severity of fall-related complications rise steadily after about age 60. In the age 65-and-over population, approximately 35% to 40% of community dwelling, generally healthy elderly persons fall annually. After age 75, the rates are higher (Rubenstein & Josephson, 2002). Fall-related injuries account for 6% of all medical expenditures for persons age 65 and older in the U.S. The total cost of all fall injuries for people age 65 or older in 1994 was $20.2 billion (Englander, Hodson, & Terregrossa, 1996). By 2020, the cost of fall injuries is expected to reach $32.4 billion (Englander et al., 1996).

Research has shown that the populations of older adults who have experienced a previous fall are 60-70% more likely to suffer future falls versus those who have not fallen previously (Carpenter, Frank, Silcher, and Peysar, 2001). This observation highlights the relationship between fear of falling and gait control. Fear of falling may be a result of deteriorated balance capabilities and decreased balance confidence. Alternatively, fear of falling may influence changes in strategy or execution of postural control, which could be indirectly related to decreased postural performance and gait characteristics that can negatively impact slip induced falls (Carpenter, Frank, Silcher, and Peysar, 2001).

Gait changes associated with aging may affect the outcome of slip and fall accidents. A review of the biomechanical literature indicates that there are several differences in the gait characteristics of older and younger people. Older adults tend to walk slower, have a shorter step length, and a broader walking base. This results in a gait cycle with a longer stance or double support time. On slippery floor surfaces, people of all ages tend to shorten their step length to reduce horizontal foot forces to reduce the likelihood of slipping. Most slips that lead to falls occur when the frictional force ($F\mu$) opposing the movement of the foot is less than the shear force (F_h) of the foot immediately after the heel contacts the floor (Perkins and Wilson, 1983). Particularly, at the heel contact phase of the gait cycle, there is a forward thrust component of force on the swing foot against the floor. This results in a forward horizontal shear force (F_h) of the ground against the heel. Moreover, a vertical force (F_v) results as the body weight and the downward momentum of the swing foot (and leg) make contact against the ground (Lockhart et al., 2003). The ratio (F_h/F_v) has been used to identify where in the gait cycle a slip is most likely to occur (slip initiation). Analyzing this ratio, Perkins found that dangerous forward slips were most likely to occur shortly after (< 50 –100 ms) the heel contact phase of the gait cycle (peak 3). Currently this ratio (F_h/F_v at peak 3) is termed "Required Coefficient of Friction (RCOF)" (Redfern and Andres, 1984).

Research indicates that older adults who fell previously modify their gait patterns (shorter step lengths, slower heel contact velocity, and reduced RCOF) in an attempt not to fall (Lockhart, 1997). This change in gait characteristics suggests that older adults who have previously fallen should be less likely to fall. However, epidemiology findings clearly indicate that this is not the case. Therefore, the relationship between older adults who have fallen versus older adults who have not fallen needs to be examined. Additionally, the effects of stress associated with a fear of falling needs to be investigated to identify if stress is a factor that contributes to older adults who have fallen being more likely to fall again. Therefore, in order to effectively reduce slip and fall accidents in society and industry, there is a need to examine the behavior and psychological characteristics of humans walking on a slippery floor surface to ascertain if there is a relationship between stress, biomechanics of the gait, and the occurrence of slips and falls. It was hypothesized that (1) older adults would have a higher level of stress after slipping and/or falling than their younger counterparts, and (2) Older adults would modify their gait characteristics in a way that would adversely affect slip-induced falls, specifically higher heel contact velocity and higher RCOF.

Method
Fourteen younger (18-35 years) adults (7 male and 7 female) – [mean height 172.81 cm] and [mean weight 72.52 Kg] and 14 older (65 and over) adults (7 male and 7 female) – [mean height 168.49 cm] and [mean weight 72.59 Kg] participated in this experiment. Prior to participating in the experiment, participants were required to have had a medical exam to ensure that they had no physical problems, which may lead to further injury from participation in the study. During the experiment, the participants walked across the baseline floor surface (vinyl tile) for 10 –15 minutes. Within a subsequent 10-minute session, slippery conditions were randomly introduced by changing the floor surface utilizing a manual floor changer, and measurement of the participant's posture and ground reaction forces were recorded. First trial data was used for the data analysis.

These parameters were measured utilizing a 6-camera 3-D motion analysis system, 2-force plates, analogue-to-digital converter and microcomputer, and a fall arresting rig. Physiological stress level was measured through Salivary Amylase (SA). Amylase is an enzyme that hydrolyzes starch to oligosaccharides and then slowly to maltose and glucose. Salivary amylase concentrations are predictive of plasma catecholamine levels and can be used as a measure of stress (Chatterton, Ellman, Hudgens, Lu, & Vogelsong, 1996). Four saliva samples were collected from each participant [baseline, pre-slip, slip, and post-slip] as follows: baseline was collected prior to the participants walking on the test track; pre-slip was collected prior to introducing the slippery floor surface; slip was collected immediately after the participants slipped; and post-slip was collected as the participants stood on the test track looking at the floor surface they previously slipped on prior to starting the adjusted gait phase of the study. Additionally, for the baseline and pre-slip samples, the participants were not aware of the alternate slippery floor surface. Standard shoes with rubber soles were supplied to each participant to control for coefficient of friction (COF). Normal heel contact velocity and adjusted heel contact were calculated as follows: The horizontal velocity (x) of the heel before the heel contacts the floor was measured. The position data was used to calculate the heel velocity. V_x of a sequence of data was obtained using the finite difference method. The linear finite difference equation is using the difference of the foot displacements of last 1/120 second (Δt) before and after the heel contact divided by the elapsed time ($2\Delta t$). Peak RCOF was calculated from the ratio between the horizontal and vertical ground reaction force (F_h/F_v) on the non-oily vinyl floor surface with the Lab View data collection software system. This ratio indicates where in the walking step a slip is most likely to occur (slip initiation). Adjusted RCOF was calculated from the graphed kinematic data (Figure 1). The (F_x) for force plate two was used to determine the point of heel contact, and based on this point of contact, the relationship between the horizontal and vertical ground reaction forces were calculated by the ratio (F_x/F_z) for force plate two. Slip Distance (SD): The horizontal distance traveled by the foot after contact with the floor was measured utilizing the Lab View data collection software system. Saliva samples were sent to the Department of Ob/GYN at Northwestern University to be analyzed. Measurement of amylase concentration in saliva includes the observation of chemical color changes according to standard photometric procedures developed by Northwestern (Chatterton et al., 1996). The concentration of amylase is then determined from a table of values relating time and temperature to amylase activity. The SA levels were analyzed using a 2x2 two-way (age x gender) multivariate analysis of variance (MANOVA). Additionally, separate ANOVA was performed on all significant variables. Dependent measures (e.g., heel contact velocity, RCOF, slip distance, step length) were analyzed by 2x2 two-way (age x gender) repeated measures analyses of variance (ANOVA).

Results
In the interest of space, only age related effects will be presented. Consistent with previous finding (Lockhart, 1997, Lockhart et al., 2003), results show that older adults reduced their step length. In general, the result of the two-way ANOVA indicated a statistically significant ($p \leq 0.05$) SL difference between the age groups: normal gait ($F_{3, 24} = 3.1702$, $p = 0.0139$) and adjusted gait ($F_{3, 24} = 4.9265$, $p = 0.0362$). Additionally, there was a significant difference between the two groups normal gait RCOF ($F_{3, 24} = 11.0666$, $p = 0.0028$) however, the adjusted gait RCOF was not significantly different between the two age groups ($F_{3, 24} = 0.0379$, $p = 0.8473$). Although older adults heel contact velocity was

slower than their younger counterpart, the two-way ANOVA indicated no statistically significant difference between the age groups: normal gait ($F_{3, 24} = 1.2450$, $p = 0.2756$) and adjusted gait ($F_{3, 24} = 1.8841$, $p = 0.1831$).

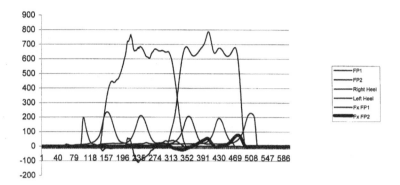

Figure 1. Adjusted gait kinematic profile data.

The results of the SA test indicated that older adults had a statistically significantly higher stress level after a slip than their younger counterpart ($F_{3, 24} = 5.7894$, $p = 0.0242$). Additionally, older adults had a statistically significant higher stress level than their younger counterpart prior to walking over the known slippery floor surface [adjusted gait] ($F_{3, 24} = 6.4170$, $p = 0.0183$). As expected, there was no statistically significant difference between the two groups pre-slip stress level ($F_{3, 24} = 0.5309$, $p = 0.4733$) or baseline stress level ($F_{3, 24} = 2.1954$, $p = 0.1514$).

Discussion
The purpose of this study was to investigate mechanisms associated with slips and falls among older adults. Specifically, this study analyzed factors influencing the initiation phase of inadvertent slips and/or falls utilizing friction demand characteristics of old and young adults. Additionally, the effects of age on stress level after incurring a slip and/or fall, and the effects of stress on gait characteristics were evaluated.

The findings from this study indicated that both younger and older adults are susceptible to inadvertent slips when a slippery floor surface is randomly introduced. However, older adults had a significantly higher-level of stress as measured by salivary amylase. Salivary amylase level was significantly related to slip severity, and should increase as slip distance increase. However, this was not the case, older adults slipped less than younger adults, but their salivary amylase level was higher. This may be due to other factors; one factor may be a fear of falling.

Results from the present study support previous research by showing that gait parameters are adjusted for contaminated walking conditions for all participants.

Persons who have a prior knowledge of a contaminated walkway adjust gait parameters by reducing RCOF, heel velocity and step length (Cham and Redfern, 2002).

A very interesting finding for the adjustment condition lies not on the contaminated force plate as hypothesized, but on the dry force plate preceding it. For the preliminary adjustment step, heel contact velocity and friction utilization were considerably lower for older participants than for younger participants. The adjustment from the dry force plate to the contaminated force plate was similar for both age groups.

These findings add to previous research by showing both groups reduced friction utilization and heel contact velocity from normal gait to adjusted gait, but that older individuals reduced both of these parameters well before the younger individuals during this adjustment condition. An age group difference in adjustment strategy is evident in the preliminary adjustment prior to stepping on the contaminated floor surface for the older participant group. It was evident that older participants required an extra step in order to adjust gait parameters for the slippery floor condition. This information could be very important to take into consideration when designing environments specifically focused on an older adult population. Design consideration could be taken specifically in regards to risk communication and hazard recognition. Furthermore, conventional design considerations, such as flooring materials, environment, and lighting, may be of importance as well.

ACKNOWLEDGEMENTS

This publication was supported by Cooperative Agreement Number UR6/CCU617968 from Centers for Disease Control and Prevention (CDC/NIOSH, K01 - OH07450), and Jeffress Foundation (J604). Its contents are solely the responsibility of the authors and do not necessarily represent the official views of CDC/NIOSH and Jeffress Foundation.

Carpenter, M.G., Frank, J. S., Silcher, C. P., & Peyser, G. W. (2001). The influence of postural threat on the control of upright stance. Experimental Brain Research, 38, 210-218.

Cham, R., & Redfern, M. (2002). Changes in gait when anticipating slippery floors. Gait and Posture. 15, 159-171

Chatterton, R. T, Vogelsong, K. M., Lu, Y., Ellman, A. B. & Hudgens, G. A. (1996). Salivary amylase as a measure of endogenous adrenergic activity. Clinical Psychology, 16.

Englander, F., Hodson T., and Terregrossa, R., 1996, Economic dimensions of slip and fall injuries. Journal of Forensic Science, 41, 733–46.

Perkins, P.J., & Wilson, M.P. (1983). Slip resistance testing of shoes - New developments. Ergonomics, 26(1), 73-82.

Lockhart, T.E. (1997). The ability of elderly people to traverse slippery walking surfaces. Proceedings of the Human Factors and Ergonomics Society 41st Annual Meeting.

Lockhart, T.E., Woldstad, J.C., and Smith, J.L., (2003), Effects of age-related gait changes on biomechanics of slips and falls. Ergonomics, 46:12 1136-1160.

Redfern, M.S. and Andres, R.O. 1984, The analysis of dynamic pushing and pulling; required coefficients of friction, in Proceedings of the 1984 International Conference on Occupational Ergonomics, Toronto, 569-572.

Robbins, A., Rubenstein, L., and Josephson, K., 1989, Predictors of falls among elderly people. Results of two population-based studies. Arch Intern Med, 149, 1628-1633.

Rubenstein, L., and Josephson, K., 2002, The epidimiology of falls and syncope. Geriatr Med Clin, In press.

INCLUSIVE DESIGN

REAL WEBSITE DESIGN FOR REAL PEOPLE

Sue King and Helen Graupp

iSys, RNIB
105 Judd Street
London, WC1H 9NE

In the world of computer access for blind and partially sighted people there are three main user groups: blind and partially sighted people together with those with some degree of impaired vision. They all experience problems with websites — most of which can be easily fixed. Guidelines exist that tackle such problems, but very often they are aimed at professionals, and are frequently, ironically, difficult to use. Sue King at iSys, RNIB has developed a website to increase awareness and encourage the non-professional developer, to design sites with accessibility in mind. The site, Access Wise, is effective, easy to use and appealing. It focuses on real people, using photographs and sound files to enable real understanding of how blind and partially sighted people experience websites and how small adjustments can improve their experience.

Introduction

There has been a push in recent years to make websites accessible. This has come about predominantly because of pressure from the DDA (Disability Discrimination Act). There are legal reasons why many websites should be accessible, but, more importantly, there are ethical reasons why all sites should be accessible as an inaccessible site will exclude many members of society. In addition it's worth noting that websites that are accessible tend to be easier to use and are consequently more appealing to all users, not just those who are disabled. If a site is accessible and usable by visually impaired people, the chances are it is also easier for a sighted user who may use a palmtop or something similar, or users with older modems who may surf with the graphics switched off. The website, Access Wise, which has been developed by the iSys department at the Royal National Institute of the Blind (RNIB), aims to "encourage amateur web developers to ensure that there are no barriers on their site which hinder access by visually impaired people".

Background
The site was developed because although there are a great many Internet sites dealing with accessibility there seems to have been a mystique built up around it suggesting

that it may be difficult to achieve. It was felt that this is likely to be off-putting to amateur developers who may think that they don't have the requisite skills and who may be confused by the language and technical detail on some of these sites. Therefore the Access Wise site is directed at non-technical, nonprofessionals who may be designing, for example, a personal web page or a website for a club or school. Naturally, although the site does not attempt to support professionals developing a complex interactive site, for instance, it is hoped that some of the simple demonstrations on Access Wise may help them empathise with the experience of people using screenreaders, magnification or simply non-standard computer settings.

The website content

The website provides information about visual impairment in a way that's relevant to the web developer. It describes how revolutionary the Internet is for disabled people, as it can "provide access to resources that were previously not easily available, if at all. But only so long as the website is accessible." For instance, thanks to the Internet, a blind colleague who works in the music industry now has access to the Groves Dictionary of Music. A braille version of this publication would run to more than a thousand volumes so searching for information would be very time-consuming, storage would be a problem and it would be cost prohibitive. The site puts the emphasis on the web designer, pointing out that design decisions made will determine whether disabled people can or will use it.

Aiding understanding of the user
To assist understanding of the problems experienced by visually impaired people, explanations and examples in the form of sound files, description and photographs of people and screens are used. In addition to the general introduction, there is a brief explanation, with examples, of the difference between accessibility and usability. For instance, a sound file demonstrates how unusable a page may be if a screenreader is going to relentlessly and tediously read all the opening and closing tags of a particular code. The section is still technically accessible but very evidently unusable. More detailed pages on Access Wise are devoted to particular elements of a page, i.e. colour, images and text. Photographs show what can happen if background and text colours are changed and sound files demonstrate the different results if images have non- or inappropriate- Alt text compared with appropriate Alt text.

Helping the developer prioritise
Consideration has also been given to usability of the Access Wise site and effort has been made to enable the user find what they need quickly and easily. As many users won't be prepared to spend much time on the site a Quick Info link is provided near the top of the navigation links list. This goes to a single page that lists the most important issues to check and links to different pages and detailed issues so people can prioritise their needs at a glance. Thus providing an alternative way of navigating the site to the main links list. Another page is devoted to telling developers how they may check their sites for accessibility themselves, as mechanical accessibility checkers alone cannot be relied on.

Evaluation

An MSc student at UCL, London evaluated the Access Wise site as part of a final year thesis. The aim of the evaluation was to see whether amateur developers were able to easily improve the accessibility of their sites using the guidance of Access Wise. Three amateur web developers agreed to participate in an evaluation of the Access Wise site by having their existing personal websites evaluated by experts in accessibility and usability, before they had looked at the Access Wise site. The developers were then given password access to the Access Wise site and had a month in which to review their sites and make adjustments to the design using the site for guidance. These sites were all different in content and subject matter: one was for a college, another a fan site and the third a personal home page about the area where the person lived. The participants were interviewed afterwards as to how usable and helpful they had found the site.

Results

The evaluation showed that overall participants found the site useful and easy to use although a few difficulties were highlighted and as a result the site structure was altered and changes made. The main difficulty was that all participants found the links to external sites confusing and caused them to become lost. The solution used was to make all external sites open in a new window and also remove some of the less important links. Overall the evaluation produced a positive response from those involved. All participants liked the photographs of visually impaired people using the Internet but thought that these would be better integrated into the site rather than kept in a separate section. This suggestion has now been implemented. Participants also wanted more graphics and examples and this, too, is being implemented.

Improvements that can benefit both the developer and the end user
The accessibility of all participants' sites were shown to have improved when they were assessed by accessibility experts after the developers had changed their sites using Access Wise for guidance. All of the amateur developers involved found the experience of using Access Wise to improve their sites, an educational and extremely beneficial exercise. For example, one of the participants runs a fan site and prior to using Access Wise had been disappointed that it never appeared in a Google search. However, since changing the site from a frames to a non-frames based structure it appears first in a Google search. So not only was it made much more accessible there was also benefit to the developer. Additionally, since using Access Wise to improve the site, it has been recommended by the Telegraph newspaper and BBC website as the number one site for its field. To see the site go to http://wwwDavidWalliams.com

Conclusions

It is in the nature of a website that it is never 'finished' but will evolve as need arises. Access Wise is no exception and it is hoped that people will use the Feedback page on the site to assist this process. It is important for all of us not to think of accessibility as "someone else's need". Most people can expect their vision to start deteriorating after the age of 50 so the issue of accessibility isn't some new and fleeting politically correct fashion. An inaccessible site excludes many ordinary people so it is in every developer's interest to help continually help to reinforce good practice as the Internet evolves, for their own sakes as well as for others.

Design for real people
Think about all the websites that interest you now. If your eyesight fails, your interests won't suddenly change with it. Visually impaired people have the same needs and interests as everyone else. Our blind associates will have Internet needs as diverse as anyone else and will want to surf the Internet on subjects as varied as Terry Pratchett, poetry or to buy CDs. The fact that they won't be accessing the Internet by the standard techniques is no reason to exclude them from taking part in the fun and the services on offer. It is up to web developers to meet the challenge of the different needs people have and to ensure that their sites are technically accessible and, above all, usable. This is not difficult. It just requires awareness and a bit of effort and the kind of guidance that gives the developer some straightforward help.

References

Access Wise website address: http://AccessWise.RNIB.org.uk

Harwell, C. 2003, *An Evaluation of the RNIB 'Access for all' website.* University College London Interaction Centre (UCLIC) thesis.

HEARING LOSS:
INCLUSION THROUGH DESIGN

Mary Sheard, Neil Thomas and Jo Kewley

RNID
19-23 Featherstone Street,
London, EC1Y 8SL.

There are around 9 million deaf and hard of hearing people in the UK. This paper outlines some of the causes of hearing loss, and the way in which these can affect what residual hearing remains. The types of challenges faced by deaf and hard of hearing people, and the range of technology available to overcome these problems, are discussed. The aim of this paper is to provide an overview of the causes and effects of deafness. It will also provide some insight into the factors that designers and ergonomists need to consider when working with this population, whether the aim is an inclusive product or an assistive device.

Causes and consequences of hearing loss

One in seven people in the UK have some level of hearing loss; that is currently about 9 million people. Of these, around 698,000 are severely or profoundly deaf, and about two thirds of these are unable to hear well enough to use a voice telephone even with amplification. Most people with a hearing loss are older adults with 55% of over 60's being deaf or hard of hearing, although the majority of this group has only a mild to moderate loss. The effect of a hearing loss varies greatly from person to person, with the level of hearing loss and age of onset being key factors.

There are two main types of deafness; conductive and sensorineural. Conductive losses are due to a problem in the outer or middle ear. Possible causes are wax in the ear, "glue ear" (as a result of middle ear infection) and otosclerosis (a condition affecting about 2% of the population, where there is extra bone growth in the middle ear). Conductive losses affect the loudness of sound, and in many cases can be treated. Sensorineural losses are caused by problems in the inner ear or the nerve that leads from the ear to the brain. People with this type of loss may have been born deaf, have age-related hearing loss, or have lost their hearing through exposure to noise or infection. Sensorineural losses are far more prevalent than conductive losses, and there is currently no cure. This type of hearing loss affects both the loudness and the quality of the sound, making it difficult to distinguish between sounds. This makes it particularly hard to understand speech. People with a sensorineural loss will not achieve perfect hearing with a hearing aid, but can still benefit a great deal from the technology. The effectiveness of hearing aids is also dependent upon factors such as the degree and type of deafness, expectations and the appropriateness of the aid.

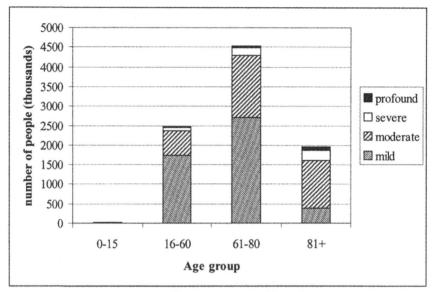

Figure 1. Levels of hearing loss in different age groups

Around 1 in 1000 babies are born deaf. This may be for a variety of reasons; genetic causes, that the mother had rubella during pregnancy, premature birth or lack of oxygen during birth. Early identification of hearing loss is essential so that the child and their family can learn communication and language skills, and interact appropriately. In addition, preparations for the provision of suitable educational opportunities must be made. While some deaf children learn sign language and are educated in sign language, others are brought up orally, learning to lipread and to speak. Increasingly, however, children learn and use both spoken and sign languages.

Gradual loss often occurs with ageing, or after repeated exposure to loud noises. Sudden loss of hearing in adulthood is rare, and can be caused by infection, head trauma, tumour or as a side effect of some powerful drugs. When hearing is lost post-lingually (after the age of three, when language has been acquired) the individual does not face the same challenges as a pre-lingually deaf person, however many new communication tactics must be learnt to make the most of different listening situations. In addition, a number of adjustments often have to be made to various aspects of everyday life. Some of these are discussed below.

According to the medical model of disability, the inability of people with disabilities to join in society is due to their impairment. The social model would argue that the way that society is organised is the factor that prevents their inclusion. Based on the former model, the best way to resolve the problems experienced by deaf people is to improve their hearing. This might be through medical or surgical interventions, or by fitting a hearing aid or cochlear implant. The social model would suggest restructuring society to remove or reduce barriers faced by people with disabilities. This might involve overcoming prejudice and stereotypes, changing procedures, installing specific equipment and making information, buildings and transport accessible. In reality, these two models are extremes on a continuum, and most practical situations are a blend of the two.

For many people, a hearing aid or cochlear implant to amplify the sound that they hear, or to improve the quality, can be very beneficial. Some people, with certain types of hearing loss, may benefit from surgery to relieve the problem and thus improve their hearing. Hearing aids and cochlear implants cannot provide normal hearing though, so even with this help, deaf and hard of hearing people, and other members of society, will succeed better if they find ways of overcoming barriers rather than expect a cure.

Everyday problems and solutions

There are many everyday challenges that are faced by deaf and hard of hearing people, from how to wake up in the morning without being able to hear an alarm clock to being able to socialise with a group of people in a noisy environment. The solution to each problem will be different for each individual, and may be dependent upon technology, or upon the understanding and consideration of other people.

To take the example of being unable to hear the alarm clock, products are available. These offer loud audible alarms, flashing lights and vibrating pads that can be placed under the pillow. Available as both home and travel versions, and in a variety of styles, this type of alarm clock is affordable and widely available. However, some deaf people will find that a human alternative – asking their partner to wake them – is just as effective! The technical solution is not always the first choice for everyone.

Being aware of visitors at the door also poses great problems and security risks for a large number of deaf and hard of hearing people. Again, devices are available; doorbells linked to a strobe light in the living room; systems that allow the user to place the loudspeaker in their main room rather than on the back of the door; vibrating pagers worn on the belt. Again, these products will not be taken up by everyone. In particular, people who live in flats with shared door entry systems may not be able to install their own doorbell. In this case they may have to resort to giving visitors the number of their mobile phone (with vibrate alert) in advance, so that they can send a text message when they arrive. This solution is clearly far from ideal.

To help with television viewing there are a number of options. Subtitles are now available on the majority of programs on terrestrial television, and will become more widely available on all channels over the next few years. These enable the viewer to read what is said in the program, and can also give additional descriptions, such as indicating that a gunshot was heard. Headphones, either corded, infrared or radio, can allow the hard of hearing person to increase the volume at which they listen without imposing it on other family members. Portable listening aids may also be used to listen to the television. Including a built in microphone as well as a lead to connect it to the television, this device is also suitable to help with hearing conversations.

Smoke alarms, at home, in public places and in hotels, are a real concern for people with a hearing loss. While not hearing the doorbell may be inconvenient, not being awoken in case of a fire can have much more serious consequences. Again, the technical solution is a specialist product. Smoke alarms that include a strobe light and vibrating pad that may be placed under the pillow are available, though the cost is much higher than that of a simple audible smoke alarm. When staying in a hotel or with a friend, a special alarm will not usually be available. In this case the deaf or hard of hearing person normally has to ask someone to wake them if the alarm goes off. This reliance on human memory for something so important frequently results in deaf people being forgotten as others evacuate the building. This area is one that causes very high levels of concern among deaf and hard of hearing people at present, but which may be impacted on by the Disability Discrimination Act (DDA).

Problems related to hearing loss may be reduced by removing barriers to communication in employment and services. Legislation, such as the DDA in the UK is working towards this. Increased deaf awareness in the general population, coupled with the provision of suitable products, services and support, will help deaf and hard of hearing people to participate more fully in society, and alleviate many of the problems and stresses linked to their hearing loss.

Communication

Communication can be particularly difficult for people with a hearing loss, although with understanding and the provision of suitable support it need not be a great barrier. While, in general, people who are pre-lingually deaf often learn sign language and use this as their first or preferred language, people who lose their hearing in adulthood already have spoken language and will tend to continue to use this as their preferred language. British Sign Language (BSL) is not a direct signed replacement for spoken English; it is a completely separate language with its own grammar. BSL developed as a natural language in Britain; other countries have their own sign languages that developed independently. Around 50,000-70,000 people in the UK rely on BSL as their first or preferred language.

BSL users face quite specific problems. Background noise, that would disrupt conversation for a hard of hearing person, such as music in pubs or engine noise on a plane will not hinder communication among a group of BSL users. The greatest challenges are faced when BSL users wish to communicate with hearing people. Buying a train ticket or conducting a transaction in a shop can be very difficult, with both parties using a combination of tactics to communicate; speech, lipreading, written notes, gesture and mime. Longer, more complex interactions such as job interviews often require the services of a human aid to communication, such as a BSL-English interpreter. With English as a second language, text communication and access to resources such as the Internet are restricted. In particular, web sites and printed publications that do not use simple, clear English, or which cover very complicated material, are extremely hard for many deaf BSL users to tackle.

For people who are hard of hearing or deaf and use English, speech, lipreading and writing are usually the preferred means of communication. Face to face, most use lipreading to support their residual hearing. This can be a very effective solution, particularly in one to one situations. However, when conversing in a group it can be difficult to keep up with who is talking, and impossible to keep track when several people speak at once. This problem is further exacerbated in informal social venues such as pubs. Here there is often background noise, which masks people's speech, and low light levels that make lipreading more difficult. However, simple deaf awareness can help to reduce many of the problems.

Telephones pose a major problem for people with a hearing loss. In fact, it could be argued that the invention of voice telephones created one of the single greatest barriers to deaf and hard of hearing people. A number of solutions are available. Some telephones have built in amplification, allowing the user to adjust the volume to a suitable level. Alternatively, amplifiers that can be attached to a normal telephone provide a more portable solution. Many landline telephones also have an induction coil. This means that hearing aid users can switch to the T setting, and receive sounds directly from the telephone, eliminating background noise. Mobile phones cause great problems for hearing aid users, as the signal from the phone causes interference problems. However, even with this range of options, some 450,000 people in the UK are unable to manage with a voice telephone. These people have the option of using a text telephone, also referred to as a textphone or a Minicom. Textphones have a keyboard and small screen that presents scrolling text. They operate via a normal phone line, enabling the user to communicate with other textphone users. To communicate with those who have only a voice telephone, the textphone user must go via a relay service such as RNID Typetalk in the UK. Here, an operator acts as intermediary, typing the words spoken by the voice phone user, and speaking the information that is entered on the textphone.

More recent technological developments are beneficial to deaf and hard of hearing people. The Internet and mobile phones facilitate text communication. Both technologies have been widely taken up by both hearing and deaf people, providing equal access to text communication for everyone. Unlike textphones, which are a specialist technology, the Internet and mobile phones with SMS facility are widely owned and used by all groups of society. This integration has the potential to make communication far easier for deaf and hard of hearing people, assuming correct considerations are made during service development.

Inclusive design

An inclusive design approach can make many products more accessible to deaf and hard of hearing people. It is rarely (possibly never) the case that the needs of every user can be incorporated into every product, and as such there will always be a market for specialist products. However, economies of scale mean that mainstream products that meet the needs of deaf and hard of hearing people are always preferable over specialist design solutions. Such specialist products are not only expensive when compared to more mainstream products, but often fall behind while technology and product development in other areas forges ahead. When a specialist product is chosen as the way forward it should only be as a last resort.

In addition to the benefits that will be incurred by deaf, hard of hearing, and other disabled people, an inclusive design approach can have a number of advantages for manufacturers. Many people, particularly older adults, have a disability of some sort, and hearing loss is very common. To release a product to the market that is not accessible and usable by such a large group of people may result in lost revenue and a reduction in market share. In addition, features that make a product accessible to people with disabilities will often also make the product more attractive to users in general.

When involved in designing products or services that should be accessible to deaf and hard of hearing people, there are a number of factors to be taken into account. The information given in this paper provides an insight into the types of deafness that there are, and some of the ways that hearing loss can impact on a persons life. Rather than making assumptions based on this or other information, it is vital to involve members of different groups of potential users early in the design process, so that their requirements are considered and included wherever possible.

Resources

RNID web site: www.rnid.org.uk
    This site includes information about technology development work at RNID, fact sheets
    about deafness and an equipment database, as well as links to many other relevant sites.
Solutions Catalogue: printed brochure of equipment for deaf and hard of hearing people.
    Email solutions@rnid.org.uk to request a copy.

The role of ergonomics in the development of assistive technology

D H O'Neill and A R Frost

Silsoe Research Institute
Silsoe, BEDFORD MK45 4HS

The purpose and scope of assistive technology (AT) are outlined and these help identify where ergonomics can both contribute to the development of equipment and reduce the risk of negative experiences by users of the technology. By considering the various elements of AT systems, the technical requirements and the user requirements can be defined for various types of disability and, if necessary, differing degrees of disability. The requirements for user-friendliness, the use of interactive control circuits to achieve this and the need for fail-safe mechanical reliability are summarised. Participatory ergonomics, understanding and including the needs of all stakeholders, is essential to the successful development and commercialisation of AT products.

Introduction

The term Assistive Technology (AT) refers to any device or equipment, irrespective of its complexity, which is used to increase, improve or maintain the functional capabilities of individuals with disabilities. The wide nature of disabilities, coupled with the many options for dealing with them, implies that AT can be very diverse, as shown in Table 1. Different types of disability warrant different kinds of AT but the AT will be effective only if it can be used to enhance the person's capabilities. For example, someone who has lost both legs might benefit more from prosthetics than from a wheel-chair, whereas someone who has lost the use of their legs (paraplegic) may be better suited to a wheel-chair[1]. Not all types of AT facilitate mechanical actions – a hearing-aid, for example – but mechanical assistance to compensate for loss of limb action (e.g. strength, range of movement, motor control) is the main application of AT in common parlance, especially with reference to enabling people to maintain an independent lifestyle in their own homes.

---

[1] FES (Functional Electrical Stimulation) is not yet "on the shelf" as an AT.
(see http://fesnet.eng.gla.ac/CRE)

Table 1. Examples of assistive technology

AT need (mode of activity)	AT equipment	Intelligence needed
Locomotion (use of lower limbs)	stick	no
	grab rails	no
	frame	no
	wheel-chair	optional
	stair-lift	optional
	hoist	optional
Manipulation / dexterity (use of upper limbs)	tool giving mechanical advantage	no
	tool giving extended reach	no
	controllers (switches, levers, buttons)	optional
Domestic chores (activities of daily living – ADL)	curtain openers	no
	window openers	no
	light switches	no
	door controllers	optional
	kitchen / bathroom accessories	optional
	kitchen / bathroom adaptations	no
	robot cleaners	optional/yes
Eating / drinking	special cutlery	no
	robot	optional/yes

This extension of independence aspect of AT is generating considerable interest at the moment. Various analyses have demonstrated convincingly that demographic changes are leading to an older population in Europe (similarly in USA and Japan). See for example the Age Shift (via www.foresight.gov.uk) which was discussed by Professor Tom Kirkwood at his Donald Broadbent Memorial Address to the Ergonomics Society in 2002. This age shift means that, as families disperse and older people become progressively less independent, more assistance will have to be made available in society generally, but, furthermore, there will fewer younger people to provide this care and assistance. In 1990 the ratio in the UK population of people aged between 16 and 65 years to those over 65 was about 4.5; it is predicted in the year 2030 to be just over 3 (from Bradley, 2003).

AT is seen as one means, but a very significant means, of enabling older people to retain their independence longer and impose less demand on the rest of society, particularly those in the overstretched caring professions (health, social services etc). In their 2002 report "The Age Shift – Priorities for action", the UK Government's Foresight Panel on Ageing Population[2] stated that it is important that policies ensure that all older people (including the frail old) are provided with appropriate technological support systems. Their recommendations included, *inter alia*, research into assistive technologies and inclusive design and to continue with the EQUAL Programme (Extending QUAlity Life). This Programme is sponsored by the EPSRC (Engineering and Physical Sciences Research Council) and aims to improve the lives and living environments and hence the quality of life of older people and disabled people by promoting interdisciplinary physical science and engineering research and its application.

---

[2]   www.foresight.gov.uk/servlet/Controller/ver=1547/userid=2/PrioritiesForAction.pdf

Elements of AT systems

Over the last ten years or so, the commercial emergence of electronic-based technologies (robotics, cybernetics) has created the possibility of developing many AT devices, some of which can be identified in Table 1. In "black box" terms, any AT system has no more than three component parts, with the simplest systems having only one of the three components. A walking stick, for example, is a simple device providing a mechanical output – a reactive force from the ground delivered to the hand. More complicated systems may have sensors, or actuators, which prompt delivery of a mechanical response (or a series of mechanical responses), such as a button or lever to set a wheel-chair in motion or a controller to open curtains. The most complicated systems have analysers or integrators to process the information from the sensor(s) to make a decision on, and probably also control, the mechanical output. These machines which appear to have intelligence are often described using the prefix "smart-".

The three black boxes, therefore, are a i) mechanical sub-system, ii) a sensing sub-system and iii) an information processing sub-system. All three sub-systems may interact in some way or other with the person who is being assisted. The more remote the mechanical action (e.g. operating curtains) or the less (physically) invasive the sensing system (e.g. camera in the ceiling), the less critical is the design of the interface in ergonomics terms. There may, however, be ethical considerations in use of sensing systems but these are beyond the scope of this paper. If the information processing sub-system incorporates options for user adjustment, the user interface must be carefully designed to avoid misunderstandings and minimise errors.

Of the equipment listed in Table 1, the most exacting design is required of the feeding robot although this may not necessarily be the most mechanically or technologically complex. Clearly, the task that it has to do is the most sensitive and the actions delivered can permit the least margin of error. On this basis, it could be argued that a simple robotic arm activated by a single switch would be the preferred design as there would be fewer things to go wrong. However, users of such devices can become frustrated that the machine is unable to be more helpful or that they do not operate correctly if part of the mechanism has become misaligned. Incorporating some intelligence is, therefore, an attractive option.

Technical requirements of AT systems

The technical requirements increase as the systems become more complex but the key characteristic must always be that the system responds when (and only when) it is instructed to do so and the response is reliably consistent and correct. The production specification should also incorporate failsafe features and an appropriate factor of safety. Safety issues for the more complicated types of system have been well summarised by Fei et al (2001). In their paper devoted to safety issues of medical robots, they identified seven core principles for systematically analysing and controlling safety issues. They divided errors into four types – pure hardware faults, pure software faults, hardware faults triggered by software and software faults triggered by hardware. Human or operator errors were additional to these and should be dealt with differently. Tzafestas and Tzafestas (2001) insisted that the importance of safety can not be overstressed and they defined two safety goals, mainly from the user's perspective: 1) a human-friendly robot should be able to operate without posing a threat when humans are inside the robot's workspace and 2) in an unstructured environment which may involve humans, any action autonomously taken by the robot must be safe even when the robot's sensor information about the environment is uncertain or false.

User requirements of AT systems

In addition to the requirements of the preceding paragraph, which a user could quite justifiably expect as a matter of course, there are more specific needs, or expectations, of the interface – i.e. the input and output functions. As has been mentioned earlier, the types of disability vary widely and so, too, does the degree of disability. These will inevitably influence user requirements and must be borne in mind when considering specific cases. Nevertheless, it is still possible to make some useful generalisations. On the input (activation) side, the switch (be it a button-type, lever-type etc) may be binary (on/off) or proportional; it may have to be larger than for conventional users and it may have to be able to discriminate between intended or accidental use. These latter points are particularly relevant for the more severe forms of disability such as cerebral palsy or motor neurone disease. In a paper on smart homes for the future, Machate (1999) suggested that the best input devices for the elderly and people with special needs would be: (portable) touch screens, speech recognition or even gesture recognition. A touch screen may get complicated with too many menus and may not be suitable for people with movement difficulties. Any speech recognition system would have to learn how the user speaks.

Requirements of output devices depend, to a large extent, on the degree of intimacy with the user. The more intimate the task(s), the more intense is the user's concern on how they are performed. If, for example, a robotic arm is giving assistance, the speed at which the arm moves (including acc- and deceleration phases) or how close it comes (or a combination of both) will affect attitudes to the use of the system. A robot with some programming facility could be set up to operate in a way that meets the user's preferences, provided that these had already been established through careful experimentation and could be properly coded. To be more versatile, the robot's memory could be expanded to accommodate various users with different preferences. An even more intelligent robot might even be able to sense the user's responses to its actions and adjust the actions accordingly. This option, which modern technology should be able to provide would obviate the need for different users' preferences to be held in memory. Nevertheless, very thorough user trials would still be needed to validate the technology. Even if a challenge such as this takes a while to be met, the incorporation of some sensing facility to locate the robotic arm with respect to the user rather than to some arbitrary point in the system's space, should elicit a positive response from the user

The overall experiences of using the whole AT system also warrant consideration. As implied by the comments above, systems must be user-friendly but, additionally, it would be highly beneficial if they could be made pleasurable to use. The topic of emotive or pleasurable reactions to manufactured goods is relatively new but the hedonic values associated with a person-product relationship should not be underestimated and are worthy of research for certain types of product design (Simon and Benedyk, 2000). Making AT pleasurable to use would, in the authors' opinion, be highly desirable. The key emotions associated with the pleasurable use of a product are security, comfort, confidence, pride and satisfaction (Jordan and Cervaes, 1995).

Ergonomics in AT systems

For a system to be ergonomically sound, all the requirements stated, or implied, above must be satisfied. In developing any technology, there is a considerable "top-down" push because the technologists know what can be done. In the AT sector there is probably a bigger knowledge gap between the technologists and ultimate users so it is even more important to embrace the "bottom-up" approach and undertake thorough, participatory user trials at all stages of equipment development and take account of the views and attitudes of all the other major stakeholders, particularly carers. Failure to do this will certainly result in non-adoption (e.g. see Hawley, 2003 and Orpwood, 2003). But ergonomics has more to offer than properly conducted user development and acceptance trials – we need to compile data on the boundaries of robotic performance that are found acceptable by disabled people. Unless it can be proven that such boundaries are not related to the nature and degree of the disability, data will have to be collected across the whole spectrum. The ergonomist is in an ideal position to understand and integrate the needs of all the stakeholders – the disabled users, their carers, the medical and social services professionals (including their roles as the likely purchasers), the therapists and the technologists.

References

Bradley, D. 2003, Mechatronics, telehealth and healthcare. In *Mechatronics in Medicine, Healthcare and Rehabilitation – Applications and Solutions*, Proceedings of IMechE seminar, Loughborough University, 21 November 2003, (IMechE, London)

Fei, B., Ng, W.S., Chauhan, S. and Kwoh, C.K., 2001, The safety issues of medical robots. *Reliability Engineering and System Safety* **73**, 183-192

Hawley, M.S., 2003, Integrating advanced assistive and rehabilitation technologies into health and social care. In *Mechatronics in Medicine, Healthcare and Rehabilitation – Applications and Solutions*, Proceedings of IMechE seminar, Loughborough University, 21 November 2003, (IMechE, London)

Jordan, P.W. and Servaes, M., 1995, Pleasure in product use: beyond usability. In S.A. Robertson (ed) "*Contemporary Ergonomics 1995*", 341-346. Taylor & Francis, London. ISBN 0 7484 0328 0

Machate, J, 1999, Being natural: on the use of multimodal interaction concepts in smart homes. In H.J. Bullinger and J. Zeigler (eds) *Human-Computer Interaction: Communication, Cooperation and Application Design*, Lawrence Erlbaum Assoc, NJ", 937-941

Orpwood, R., 2003, User involvement – the key to success. In *Mechatronics in Medicine, Healthcare and Rehabilitation – Applications and Solutions*, Proceedings of IMechE seminar, Loughborough University, 21 November 2003, (IMechE, London)

Simon, J. and Benedyk, R., 2000, Addressing pleasure in consumer products through ergonomics. In P.T. McCabe, M. Hanson and S.A. Robertson (eds) "*Contemporary Ergonomics 2000*", 390-394. Taylor & Francis, London. ISBN 0 748 40958 0

Tzafestas, S G and Tzafestas, E S, 2001. Human-machine interaction in intelligent robotic systems: a unifying consideration with implementation examples. *Journal of Intelligent and Robotic Systems* **32**, 119-141

SENIORS COPING WITH WINTER

Tay Wilson

Psychology Department
Laurentian University, Ramsey Lake Road
Sudbury, Ontario, Canada   P3E 2C6
tel (705) 675-1151   fax (705) 675-4889

Problems encountered by Sudbury seniors when trying to cope with winter were studied by means of the Social Information Generation technique (Wilson, 1981) administered to ten groups including extendicare residents, and seniors of Finnish, Croatian, Ukranian, Francophone and mixed backgrounds. Findings were classified under several topics including snow clearance, attitude towards winter, winter activity, indoor activities, problems with waiting for and getting on buses and automobile issues. Specific recommendations for improving Sudbury as a winter city for seniors were made including the possibility of selective snow clearing at some bus stops near seniors residences.

Introduction

The aim was to examine the life of Sudbury seniors, particularly with regard to their coping with winter with the goal of providing a set of empirically derived descriptive ideas upon which better to assess social and technical interventions and to develop a suitable and again empirically grounded topic upon which to encourage joint community action to improve quality of life of seniors in the Sudbury Region. It is first contended that the needs and values of people are not immediately self-evident, particularly to planners and that it is necessary to generate social information that will help to determine these needs. Second, about 150-300 well-chosen social variables once validated could and should be routinely and effectively implemented in a specific technological decision making context since consideration of only a few variables can lead to great distortion while the adoption of a thousand or more variables in a given instance seems likely to overpower present research and social intervention capabilities.

Method

In order to respond to the above requirements, the method used in this study was the social information generation (SIG) technique (Wilson & Neff, 1981, p155ff) in order to obtain more precise descriptions of lifestyle (social factors) of community residents and relate these descriptions to particular planning interventions to improve quality of life. The procedure, having identified a particular problem area, is to recruit individuals of the affected community in small discussion groups (8-15 people) which are asked to talk about the implications of the problem on their lives. Rather than focus down upon a specific dimension, individuals are encouraged to think divergently in ever-expanding waves of implications which touch more and more aspects of their lives. Thus, the technique is, in many ways, the obverse or even opposite of "focus groups" which, as the name indicates, are devoted to narrowing the topic of examination.

In Social Information Generation, the discussion leader briefly broaches the topic and then, insofar as possible, allows the individuals to introduce and expand upon topics as they will. For each group, the material is tape recorded and typed verbatim. Each utterance is then coded into one or several one-word categories by a judge. The utterances are then sorted according to category. The judge then re-reads all the utterances under a given category and develops an overall rubric describing the content. This output is the set of social ideas representing the content of the discussion. At this point, the ideas and their supporting quotes from each of the ten groups are combined and semantically sorted into a single master set of ideas. The final output of the process is the set of ideas and the implicated utterances. This result is taken as the clearest and richest available representation of individuals' understanding of the topic area which, in this case, is life for seniors particularly in the winter. The material from each group is returned to that group for any additions or comments; appropriate changes are made on the basis of this feedback to the final output. This set of ideas with supporting quotes form the basis for policy and program design.

Ten groups were chosen from a list of seniors' clubs and organizations spread over the whole Sudbury regional basin. Because of Sudbury's great ethnic diversity, considerable effort to involve diverse ethnic groups was made. Two groups were from outside the city proper: Levack (Onaping Falls) in the west and Falconbridge in the east. One group was chosen from an Extendicare residence for seniors requiring considerable care. One group was chosen from a low-rise suburban Finnish seniors residence. One group was from a centrally located high-rise seniors residence. Two groups were francophone. There was one group each from the Croatian, Ukrainian, and Italian communities. All the groups were run after winter had set into the community during the end of November and the beginning of December.

Results

The 190 derived ideas or social factors from the SIG study are grouped into ten clumps, as follows. The first set of ideas centre upon enjoyment of winter and problems with snow, clothes, mobility, and health in the winter. The second clump of ideas talks about the comparison of activity for seniors in summer and winter. Walking is a major sub-component of these activities. Within the activity clump, there is a grouping of activities for mobile people who live in seniors' residences and activities for seniors who are very much involved in clubs and live so far outside of senior residences. The third clump of ideas talks about the field of communication with television being a major discussion item. Radio, telephone, newspaper, and use of library are discussed in lesser volume.

The fourth clump discusses people taking trips primarily on their own and often in the winter time. Transportation mode is the next clump including driving one's car alone and cooperative transport arrangements. A considerable amount of material on the use of the bus is clumped together combined with some discussion of railway and taxis. The fifth clump is shopping with some discussion of delivery of goods and use of cafeterias in various locales. The sixth major clump is an attempt to express the mental world of seniors as empirically derived from the discussions. In this section is discussed or described seniors' attitudes towards specific aspects of the social world in which they live. As would be expected, one subclump of major attitudes concern the government. The seventh clump of ideas deals with seniors' homes: those who live in private homes and those who may be deciding or who have just decided to move into apartments. The eighth major clump concerns the design desiderata for seniors' residences. The ninth clump of items concerns the general subject of crime as it affects seniors and the subtopic of insurance. The tenth clump reveals some ideas on the topic of health. The tenth and last clump of ideas presents a description of an "extendicare" residence by the residents themselves.

Aspects of the above idea clumps bearing particularly upon coping with winter are here summarized in more detail. Consider winter snow, clothes, and health in winter. Most seniors have lived in the North for the vast parts of their life. They therefore take winter as a natural season and try to adapt to and enjoy it. Most seniors do enjoy winter although it exacerbates many problems that they already have as older people. Dry cold as opposed to damp cold is preferred. Adaptations are necessary including a change in clothing. Overshoes or boots are a particular problem both with regard to design in terms of getting them on and off, forgetting accompanying shoes when one goes out and the grip that they provide in ice and snow.

Although most seniors show strong mental health by putting a good face on the problems of winter, there is no doubt that special adaptations and stresses do occur at this time. Snow and ice can be real problems. If one owns one's own home, shovelling the driveway is a problem. Ploughs go by and fill in the end of the driveway where it is shovelled and it is difficult to get vehicles out. Seniors like to go out in the morning and they need to have their driveways clear then. Improperly cleared sidewalks with pieces of ice are major dangers for seniors for whom a fall can be a crippling or life-threatening event. The worry of falling is compounding by the desire by the vast majority of seniors to go outside daily for a walk which they consider rightly as a major ingredient of their continuing health and well-being. This desire is probably particularly strong among those seniors who live in residences in which the provided rooms are not as expansive as those of people who might still live in their own homes. For those who try to get their driveways cleared, it appears to be difficult to obtain reasonably priced, reliable assistance.

Consider winter activities. Many seniors are active both winter and summer. Activities are different in the winter from those in the summer. There is an indication that social get-togethers are more common in the winter. Walking is a critically important activity. Most seniors like to walk in the morning, probably on about the same route everyday and not at night. As mentioned earlier, fear of slipping on ice is justified and strong and perception of the quality of sidewalk cleaning in their walking areas is not good. In summer, some worry about skateboards and bicycles on sidewalks. Crossing roads can be a major problem and it has been observed that cars go through pedestrian crossings set up specifically for seniors when the light is red.

Beyond walking, many seniors are still engaged in cross-country skiing which seems to be an activity that one can carry on very late in life. Christmas time is a very busy time and a major visiting time. Visits from relatives are otherwise less frequent in the winter than in the summer. Real summer is accurately perceived to be rather short in Sudbury, perhaps six to eight weeks. However, many go to their camps, tent, fish, swim, and golf. Those who live in homes work outside in their yard; some seniors make use of the parks in the summer but the major city park, Bell Park, is perceived to be primarily for young people.

For those who live in seniors residence, there are many organized social activities, winter and summer. This round of activities is perceived to be a major reason why, for many, life in the residences is much better than the lives that they had in their homes: they were very lonely. Most seniors appear to stay home at night and engage in indoor activities. There is a major round of activities during Christmas season where visits are looked forward to from relatives and friends and various schools and other types of groups.

For those who live outside senior residences, the senior club can be a major centre of activity. Activities are well advertised in the newspaper and there are many local seniors clubs. Those who belong to the seniors clubs often seem to centre much of their social lives around them. Again, activity is much greater in the winter than in the summer at these small seniors clubs. Activities include bowling, group suppers, dancing, all sorts of hobbies, group aerobics, swimming, and other sports and social get-togethers. Many go down to the club once or twice a week just to socialize and perhaps play cards. An important, generally summer activity for seniors involved in these organizations is a round of trips by bus. The trips can be to local areas where guides take them to good locales and describe activities for them or they can be quite extended trips across the country or down into the United States. For a considerable number of seniors, the church and related social activities are central parts of social life.

Consider travelling. Many seniors go south to the United States in the winter as snowbirds. There is some disagreement as to which month is best to go. A major worry about going to the United States is health care costs. There is a concern about what happens to their home while they are gone. Going to Toronto or Montreal is not a major or common destination trips of seniors. It appears that from a city point of view, they get pretty well everything they want from Sudbury. Many drive winter and summer. Driving is less pleasurable and more dangerous in the winter. There are complaints about the speed that people drive in the city and the highway 69 to Toronto is perceived as dangerous and many choose not to drive on it. Seniors choose their time in which to drive downtown, avoiding the rush hours. They are concerned about parking, the expense of running cars, and in general like large cars. In at least one regional town, a senior run cooperative volunteer driving program to assist seniors has been established which is very much appreciated and seems to work rather well.

The bus service is perceived to be generally good and the drivers treat seniors well. There are complaints about insufficient service on Sunday, particularly centred around going to church. The greatest complaint concerns the cleaning of snow around the bus stops. Seniors find it often very dangerous to get on and off the bus because of snow banks and ice. A bus that went around joining hospitals together was liked before it was changed. It is believed that some seniors are isolated in their homes or apartments because of the lack of special buses to shuttle them back and forth to activities. Friendliness in regional towns is appreciated. Delivery of shopping can be an issue particularly in carrying things rather long distances and up and down stairs in small apartments. Many people go downtown everyday for the social activity, to meet friends in coffee shops and talk. Smoking in these areas is generally disliked and perceived to be a major health problem for those, for most seniors and particularly for those who have breathing-implicated problems.

Consider seniors' homes. Many seniors believe that living in the community where they have lived for many years and in which they know many people is critically important and overweighs any advantages that are perceived in moving to another location for climate or cost reasons. Perhaps most of the seniors like the community in which they have lived for years and years. Generally, as long as health permits, many seniors do not want to move from their home but recognize that there are circumstances under which it would be better and are accepting of the eventuality. They further recognize that one should make this decision in a timely manner and that some people wait too long but further that it is they themselves that should make this decision, even though it is difficult and complicated. Few make or perceive that it is necessary to make any adaptations to their homes to make them more livable as seniors. The one particular exception to this rule is, for some, the introduction of hand rails in bath tubs and in some stairs. One should not be too hasty in selling one's house and moving into an apartment. Location, timing, and cost are important in making the decision and it should be recognized that the time of selling and moving is a stressful time. Seniors do not want to share their homes with other people.

The vast majority of residents in seniors residences appear to be happy that they made the decision to leave their homes and move to such a residence. Many feel that they should have moved earlier than they did. Reasons for moving from their former homes are loneliness, insecurity, and the inability to keep their place going and better location for services and visits. It is the loneliness and insecurity that is dealt with so well in the seniors' residence if they are able to participate in activities and make friends.

Concerning design of seniors' housing, it is suggested that the housing not be high rise in many situations; that self-care and partial care and more intensive care facilities be located together so that seniors can flow easily from one to the other and not be worried about what happens to them as they develop from one type of individual to another. It is suggested that a senior day centre or club should be designed as an integral part where possible of senior residences so that people outside of the community interact easily with people inside the residences. Residences should be designed so that people can, if they so choose, live as one big happy family. Location is important. Access to a lake where people can walk to water is desirable in many cases in the Sudbury area. The location on a major transportation route, near shopping, drug store, and hospitals is desirable. Some want residences set up in present areas of high concentration of seniors (e.g., Donovan) so seniors can move into a social location where they have developed friends and skills through many years. Some like a suburban location with a country feel providing that there is good bus service to the centre shopping area.

Consider crime. Many do not walk outside at night alone. Many worry about being assaulted or having purses snatched even though few in the discussions had any experience of themselves or friends having had such an experience. On the other hand, many have personally experienced break-ins in their homes, particularly after they have retired. These experiences have been unsettling and in some cases have led to people moving into senior residences.

Discussion

It is to be emphasized once again that the reader is directed to the 190 specific ideas and their supporting quotes (Wilson, 1990) to obtain a better understanding or working model of seniors in Sudbury and how they cope with winter.

References

Wilson, T. and Neff, C. 1983, *The Social Dimension in Transportation Assessment.* (Gower, Aldershot, England).

Wilson, T. 1990, *Sudbury Seniors: Coping with Winter.* Unpublished report. Psychology Department, Laurentian University, Sudbury, Ontario, Canada.

DESIGNING FOR PEOPLE WITH LOW VISION: LEARNABILITY, USABILITY AND PLEASURABILITY

Chandra M Harrison

Human Interface Technology Laboratory
University of Canterbury
Christchurch
New Zealand

A new low vision reading aid is under development that uses advanced technology and aims to improve usability and pleasurability while maintaining the learnability of existing reading aids. Existing video magnifier users, other low vision people and normally sighted matched-age controls completed a reading test using existing machines and a mock-up control panel of the redesigned unit to determine if the aims had been reached. Improved usability was achieved in the redesigned product and once users mastered the more complex interface most found the redesigned product more pleasurable. However, learning time was greater than with the existing technology. Results also indicate that while people with low vision experience a similar level of technology-related anxiety to the control group, they are more likely to engage with technology that enhances their quality of life.

Introduction

Low vision reading aids such as video magnifiers have remained essentially unchanged for 20 years despite improved computer technology. The traditional and predominant product is a Closed Circuit Television (CCTV) system that magnifies printed text placed under the camera and displays the magnified text using a computer monitor or television screen. Users navigate the text by means of an x-y table under the camera. These reading aids provide independence for people with low vision who would otherwise have to rely on other people to read them any printed material.

CCTVs have a simple and learnable interface; they also have several limitations. Discomfort can result from using the product for extended periods and viewing can be physically, visually and mentally exhausting. Physical requirements for dexterity and coordination to use the x-y table can be further complicated among the elderly, the predominant user group, who may also have other health issues such as arthritis. Due to the magnification required reducing resolution and viewing on a screen, the visual load can be high (Harpster *et al*, 1989). Users also need to remember the beginning of the previous line for long periods, which can cause a heavy mental workload.

Technological innovations currently available could eliminate these limitations. Optical Character Recognition (OCR) software has now progressed to a stage where it could enhance text to reduce the visual fatigue. Processed text could be presented automatically, in either a single line or as a single column of text, eliminating the need for the x-y table and assisting with the flow of text. Synthesised speech could also be utilised and would prolong the life of the machine by accommodating worsening eye conditions. However, increasing functionality should not be at the cost of usability, pleasurability or learnability. The goal then is to develop a new system that takes advantage of new technology, improves usability in terms of reading speed and comprehension, reduces physical demands by eliminating the x-y table, and maintains the learnability of the existing technology.

Recent product development literature has focused on enhancing the pleasure of products by assessing users' needs (Jordan, 2002). To have pleasure, however, products must not only appeal to the user but there must also be an absence of negative emotion. It is therefore important to not only determine what aspects make a product pleasurable for a specific user group, but also to determine what aspects of a product elicit negative emotion. Technology-related anxiety or frustration with consumer electronic products is common for many people, especially the elderly (Rosen & Weil, 1995). Identifying what causes the frustration or anxiety can help reduce it, making a more pleasurable product. In addition, identifying why some people avoid technology may assist with designs that encourage use.

The main user group for low vision products is elderly (75% of CCTV users are elderly, 15% are in employment and 10% in education). There is an increase in the elder population and the elderly are more likely to experience technology-related anxiety. Therefore, it is important to make the experience of using accessibility products such as low vision reading aids as pleasurable as possible. One example of how to increase pleasurability is to determine if participants could use the redesigned machine for longer than the CCTV. People with low vision do not usually read text for prolonged periods. Instead, they use talking books and use the reading aids intermittently for smaller tasks. However, surveys show that they would like to read books but cannot because of the limitations of the CCTVs or the severity of their eye conditions. In addition, identifying specific actions that induce frustration or anxiety and altering the design to reduce negative emotions will enhance the pleasure.

Learnability, usability and pleasurability are reliant on each other. Pleasurability is not possible if frustration results from using a product. If a product takes an excessive amount of time to learn users may not preserver to determine if the product is usable. Therefore, the amount of time taken to achieve competency, the level of task achievement, frustrating actions and the degree of positive emotion that accompanies the tasks needs to be assessed to determine if a product is worthwhile. While other aspects of the control panel were assessed (e.g. optimum control layout, biomechanical issues) during a study for the company developing this product, research discussed in this paper looks specifically at whether learnability was maintained and whether usability and pleasurability were increased in the redesigned low vision reading aid.

Method

Participants included 16 expert users of the existing CCTV technology (experts), 14 people with low vision who were not current users or who had only recently begun using the machines (novices) and a matched aged control group of 15 people whose sight was sufficient to qualify to drive a car in New Zealand (control). Participants ranged in age from 9 to 90 years old with the average age being 71 years, with 16 males and 29 females. Three participants were unable to participate in the testing due to the severity of their eye conditions (1 expert and 2 novices), three others were unable to complete both tests (1 novice, 1 expert and 1 control). Their partial results were excluded from the reading speed analysis as it was possible to obtain only one reading rate. Their responses to the questionnaires are included.

Once recruited, participants completed a telephone questionnaire to gather demographic information, current usage of CCTV machines and other low vision aids, exposure to technology and current levels of negative emotion towards technology. The questionnaire was developed during a pilot study and combined aspects from Rosen and Weil (1995) along with questions to gauge technology exposure such as "how many hours do you spend using a computer per week." It also included questions to assess negative emotion such as "have you ever felt frustrated when using a new electronic product."

Each participant was visited twice in their own home where a reading test was administered, once using a CCTV machine, either their own if they were experts or with a *SmartView 8000*, and once using a mock-up of the control panel of the redesigned product now known as *myReader*, running off simulation software on a Compaq Evo laptop. *MyReader* incorporates processed text, automatic scrolling and uses a Liquid Crystal Display. The x-y table is eliminated and navigation is via trackball or scroll wheel. The order of product exposure was counterbalanced.

Learnability was assessed by timing how long participants took to achieve a predetermined level of competency. Novice and control participants were instructed in the function of each control of both machines, how to navigate text and asked to read two paragraphs of text. They were then tested on their knowledge of the controls and if 100% accuracy was not obtained they were asked to read a further piece of text and tested until they were competent. Expert CCTV users were asked to read two paragraphs of text and name the various controls on their CCTV machines and provided with the same instruction as others on the *myReader* machine.

Initial assessment of usability was on a performance only basis,[1] using comparisons of reading speed and comprehension with an existing CCTV product and the *myReader* mock-up using a similar paradigm to Harland *et al* (1998). Once competency was attained, participants were asked to read one of two chapters of a children's book, *Swallows and Amazons*, by Arthur Ransom (counterbalanced exposure). They were asked to read at their normal speed and told that they would be given several multi-choice comprehension questions at the end of each 10-minute period. Reading speed was determined as the average number of standard-length words per minute (wpm). Standard-length words were determined by the total number of characters including spaces and punctuation, divided by six (Carver, 1990). Comprehension was the percentage of correct answers on multi-choice questions.

Pleasurability was based on the subjective assessment of participants, problems encountered with actions and their willingness to continue reading (endurance). In a post-test interview, participants were asked to discuss the experience of using both machines and at the completion of both tests were asked for their preference and reasons for it.

---

[1] Further assessment of biomechanical elements will be conducted at a later stage, using behavioural analysis of video footage of the interactions between participants and the two machines. Data logging from *myReader* will also be used to assess biomechanical differences and confirm optimum button placement.

Repeated errors and negative comments regarding frustrating functions were analysed to determine aspects of the interface that could be altered to enhance pleasurability. Endurance was determined by the difference in time participants were willing to read. Given that average use of existing CCTVs does not exceed 10 minutes at one time, it was envisaged that people may have been able to read for longer with the redesigned machine and this measure of endurance would also gauge pleasurability. Their level of technology-related anxiety gathered from the questions was also used to analyse their preference

Findings

Learnability
The time taken to achieve competency was greater for the CCTV machines (approximately 5 minutes) and the *myReader* (approximately 10 minutes). A difference was to be expected given the increased functionality and complexity of the interface and that experts would have very low learning times for the CCTV. Due to problems experienced with the simulation software and mock-up, further analysis of the video footage is required to eliminate time not directly related to achieving competency. While this may reduce the difference between CCTV and *myReader* times, it is expected *myReader* will still have a greater learning time. The design has also been modified to eliminate the number of steps required to initiate reading which could also reduce the learning time.

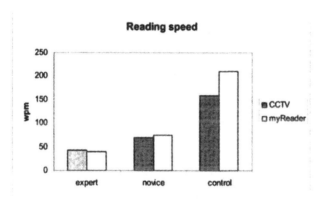

Figure 1. Reading speed as words per minute

Usability
Three participants were excluded from the reading speed analysis as it was only possible to obtain only one reading rate. There was an overall increase in reading speed of 18% for the *myReader* unit (see figure 1). The speeds achieved are comparative to those found in previous research (Harland *et al,* 1998). Experts' speed decreased 3%, novices increased 8% and controls increased speed with *myReader* by 25%. Those unfamiliar with CCTVs show an improvement in reading speed after a relatively short exposure. It is likely that with further exposure experts may also show an increase in reading speed.

Comprehension for CCTV was 62% and 63% with *myReader* (see figure 2). While an increase in comprehension was not experienced, there was also no degradation with the automatic scrolling, which in itself supports the increased functionality. The overall level of comprehension achieved was not as good as reported in previous studies (Harland *et al*, 1998). As there were no significant differences between low vision and control participants, we could assume this may be due to the questions.

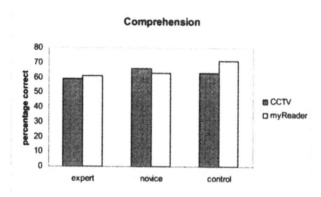

Figure 2. Percentage of correct comprehension questions

Pleasurability
In the post-test interviews 31 of the 42 (74%) participants tested preferred the redesigned product (figure 3). The group with the greatest preference for the simpler interface of the CCTV was the control group, who, it could be argued, have the least to gain from the more complex technology. Two of those stating they preferred the CCTV technology had reported discomfort during its use, highlighting the limitations of subjective assessment.

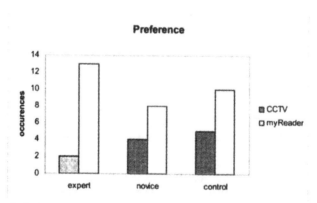

Figure 3. Subjective preference for CCTV or myReader reading aid

Analysis of the questionnaires revealed that people with low vision experience a similar level of technology-related anxiety as the control group. Many reported that they preferred to do without technology if they did not require it and had limited exposure. However, 15 participants had a cell phone, mostly for emergencies. Participants with low vision were more likely to engage with technology that would enhance their quality of life. Those with a support person or not interested in reading were less likely to engage.

Only two of the 42 people tested were willing to continue after 10 minutes with either product indicating endurance has not increased. However, many reported being less fatigued by the *myReader*. It is suggested that being observed may have reduced participants' willingness to continue and longer-term usage of the machine may result in increased endurance. Beta testing of a working *myReader* prototype is planned. This testing will involve obtaining a base line performance, leaving the equipment with participants for four weeks and then retesting to determine if gaining greater proficiency may also increase the length of time users are comfortable reading at one sitting.

User comments revealed that frustration was caused by a lack of understanding of terminology used in labelling of controls and simulation software notices. While the computer processes the text the message "loading" appeared. This was changed to "please wait" which was better received. A control labelled "next" was misunderstood causing some frustration. Increased memory load caused by an extra step to access processed text also caused frustration. The company has since eliminated this step. Other issues that effected the pleasurability of the control panel included a high profile control which was often bumped. The company has since lowered the profile of the dial.

This research forms the basis of a larger study to determine optimal design characteristics of the *myReader*. Because *myReader* was under initial development at the time of the study only a part of the functionality was assessed. The reading task was chosen due to the ease of simulation and the desire expressed by low vision people to be able to read faster and for longer.

Conclusion

The redesigned low vision reading aid offers greater functionality than the existing CCTV machines. While it does take longer to master the control panel of the machine, users achieve faster reading, their comprehension is unaffected and they report greater pleasure with using the *myReader*. This sample did appear to avoid technology for a variety of reasons. However, technology such as low vision reading aids can assist these people in maintaining independence and those in need were willing to accept the new technology. It is important to further investigate why this demographic (the elderly) in particular avoids the technology to assess if accessibility products can be designed to avoid these issues. The findings will assist the company in the further development of the product.

References

Carver, R.P. (1990). *Reading rate: A review of research and theory.* (Academic Press, San Diego).

Harland, S., Legge, G.E. and Luebeker, A. 1998, Psychophysics of reading: XVII. Low vision performance with four types of electronically magnified text. *Optometry and Vision Science,* **75** (3), 183-190.

Harpster, J., Frievalds, A., Shuman, G. and Leibowitz, H. 1989, Visual performance on CRT screens and hard-copy displays. *Human Factors,* **31**, 247-257.

Jordan, P. 2002, *Designing pleasurable products.* (Taylor & Francis, London).

Rosen, L.D. & Weil, M.M. (1995). Adult and teenage use of consumer, business, and entertainment technology: Potholes on the information superhighway? *Journal of Consumer Affairs,* **29**(1): 55-84.

MODELLING
&
USABILITY

PHOTOGRAMMETRY FOR ERGONOMICS: CAPTURING 3D POSTURE, BODY FORM AND WORKSTATION DESIGN FROM PHOTOGRAPHS

Malcolm T. Cope

Health & Safety Laboratory, Broad Lane, Sheffield, UK. S3 7HQ
www.hsl.gov.uk

Photogrammetry is a method of calculating the 3-dimensional (3D) position of a point in space from a set of photographs or video. This paper examines the potential usefulness of photogrammetry to the ergonomist, using only relatively low-cost equipment and software, for capturing the 3D posture of a subject and his/her work area, and for capturing shaped surface topography. The method of photogrammetry used here is most useful where a relatively quick-to-set-up, inexpensive and non-invasive method of 3D measurement is required.

Ergonomics, measurements and photogrammetry

Photogrammetry is a method for deriving distance, angle and shape measurements from photographs by using the physical parameters of the camera and a set of mathematical algorithms. The output from photogrammetry is commonly a 3D model which can be measured or used in a computer aided design (CAD) system. This can be useful for ergonomics work which involves measuring human postures and work areas, or where models need to be constructed in human modelling systems, 3D modelling software and simulators (Meister, 1995; Porter et al, 1995; Wilson, 1997). Photogrammetry can be used to support or sometimes replace traditional instruments such as tape measures, inclinometers, goniometers, laser scanners and 3D marker systems. The mathematics used for photogrammetry today are very complex, but developments in photogrammetry software over the last 10 years make photogrammetry a practical measurement method in many industries (Fraser 1997).

The potential benefits of using photogrammetry are:

- Faster data collection in the field for complex measurements
- No need to physically touch/disturb the items to be measured
- If further measurements are needed, a reduced risk of needing to return to site
- The photographic data often has other uses during a project

Recognising the potential benefits of photogrammetry, the Health & Safety Laboratory (HSL) decided to investigate its use for its own work, which includes workplace and work task assessments, ergonomics research projects, incident investigation, computer human modelling and virtual reality (VR) simulation. Before employing photogrammetry, though, it was necessary to understand the levels of accuracy and reliability of the method, and how to get the best results for the type of work that HSL carries out.

Scenarios to test photogrammetry for ergonomics work

Body posture capture

Traditional instruments for collecting postures are visual estimation, 3D marker systems and goniometers. For many ergomonics assessments a visual assessment of posture is sufficiently accurate and reliable, but accuracy is affected by lens distortion and parallax errors. However, more accurate measurements of posture might be required for biomechanics calculations or for re-creation of the posture within a human modelling package, for example. 3D marker systems and goniometers provide good accuracy but tend to require more time and intervention than simply taking photographs, and this can sometimes be a problem in field situations. If simple posture measurements could be captured from photographs, this would eliminate some problems associated with the other methods for some situations. If the measurement results are in a format which can be imported into a human modelling package then this would also be an advantage.

Workstation dimensions and equipment layout

Workstation dimensions are normally collected using tape measures and protractors. This necessitates drawing a diagram or making notes for identifying the dimensions later on. This task can become very tedious and error prone on very complex workstations, especially where it is difficult to find or physically access reference points for measurements. This might happen on complex machinery or workstations with curved shapes, for example. Some workstations are also difficult to measure without disturbing the work progress or stopping machinery, which can be awkward in field situations. Capturing a workstation's dimensions from photographs is therefore an attractive idea, especially if the resulting measurements can be readily imported into 3D modelling software.

Capture of a 3D shaped surface

Capturing the 3D surface profile of body topography or objects such as seats and handles into a computer model can be sometimes useful. Creating a computer reconstruction of a surface is normally carried out using 3D laser scanners or surface digitising devices. These methods are still relatively expensive, and tend to cover a limited range of size of object. An alternative to these methods is therefore of interest. There are photogrammetry systems currently available which will automatically reproduce a surface as a point cloud of 3D data points, but these systems are still rather expensive. The cheapest method of capturing a surface using photogrammetry involves only a digital camera and relatively cheap software, and that is the method which is investigated here.

Test 1: Capture of body posture

To test the capture of a posture, a participant was photographed from different angles using a 1.3 megapixel digital camera. Figure 1(a) shows one of the photographs. At the same time, an optical motion capture system was used to record the posture, and the markers can be seen as white dots in Figure 1(a). The motion capture system was capable of very high 3D positional accuracy, down to 1mm or better, and so provided a measurement against which the photogrammetry could be checked. The photographs were then processed using the photogrammetry software in two ways. Firstly, the 3D position of the motion capture markers were calculated. Secondly, the position of the limbs, body and head of the participant were estimated without using the motion capture markers. The aim of the second step was to establish how accurately the posture could be calculated without using body markers on the subject.

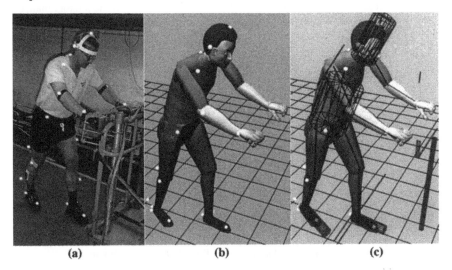

Figure 1. (a) Photograph of the test posture with the motion capture markers; (b) the motion capture markers imported into a human modelling system; (c) photogrammetry data imported into the human modelling system

The 3D coordinates for each body marker as calculated by both the photogrammetry and the motion capture system were then imported into a spreadsheet for comparison. Just over 74% of the photogrammetry measurements were within 1cm of the reference measurements from the motion capture system, with 99% falling within 2cm of the reference marks.

The motion capture data was imported into a human modelling system and the virtual body markers matched up to the corresponding body points on the virtual human (Figure 1b). The estimates of body posture which were derived from photogrammetry were also imported into the human modelling system (Figure 1c), and a visual comparison of the postures showed a correlation which was close enough for most purposes, and better than simple estimation from photographs.

Test 2: Workstation dimensions and equipment layout

For testing the collection of workstation measurements, 3 photographs were taken of a computer workstation, using a 1.3 megapixel digital camera. To improve the accuracy of the photogrammetric calculations, some round 'donut' markers were placed at points in the scene where no strong visual features existed. One dimension on the workstation was measured with a tape measure in order to provide a reference scale for the photogrammetry software. Figure 2a shows the workstation and the 'donut' markers.

Figure 2b shows the resulting 3D points and lines after processing the photographs within the photogrammetry software. Points *b*, *c*, *d* and *e* in Figure 2b are the 'donut' reference markers which helped to process the photos. The software that the author used in this case allowed the measurement of distances within the software. Various measurements were extracted from the photogrammetry model and compared with measurements using a tape measure, the result of which can be seen in Table 1. The accuracy of the photogrammetry varied from 99.07% 100% of the tape measure values, which represented an actual error of between 0 and 6mm in this case.

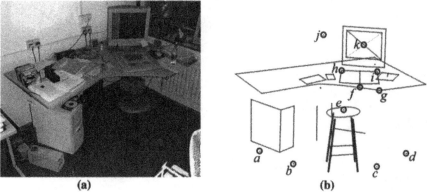

(a)   (b)

Figure 2. (a) The workstation, (b) The photogrammetry model

Table 1. Comparison of photogrammetry and tape measurements

Dimension*	Measurement (mm)		Accuracy
	Tape	Photogrammetry	(%)
Desk height (g)	857	859	99.80
Stool height (e)	647	641	99.07
Screen to desk edge (k-f)	680	681	99.85
Tablet width (h-I)	409	408	99.76
Screen width	411	411	100.00

Letters in brackets refer to marked points in Figure 2a.

Test 3: Capture of a shaped surface

A plastic model of a head was chosen as a shape to capture by photogrammetry (Figure 3a). Shaped surfaces have few or no visible marks to act as measurement points and so it was therefore necessary to create a set of points by projecting a pattern of white dots onto the surface. A set of six reference points were also created around the head using printed 'donut' paper shapes. The distance between two of the reference markers was measured as a reference scale. Three digital photographs were taken of the head and then processed within the photogrammetry software. This resulted in a 'point cloud' of 3D data points defining the surface of the head (Figure 3b).

A continuous surface mesh was then produced from the 'point cloud' which gave it a solid appearance. This process was possible within the photogrammetry software itself or by exporting the point cloud to a capable 3D modelling program. A 3D laser scan of the head model was then imported into a computer 3D modelling system along with the photogrammetry mesh so that the two could be directly compared. Figure 3c shows the 'solid' photogrammetry surface with the data points from the laser scan shown in black dots. The comparison was carried out by comparing the two data sets slice-by-slice, working down from the top of the head, with a 1mm background grid. Figure 4 shows one of these slices. The photogrammetry surface was generally within 1mm of the laser scan.

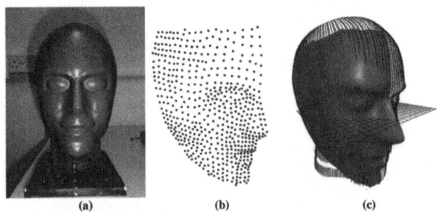

(a)   (b)   (c)

**Figure 3. (a) Plastic model head, (b) Photogrammetry 'point cloud',
(c) Photogrammetry 3D surface, with laser scan data overlaid**

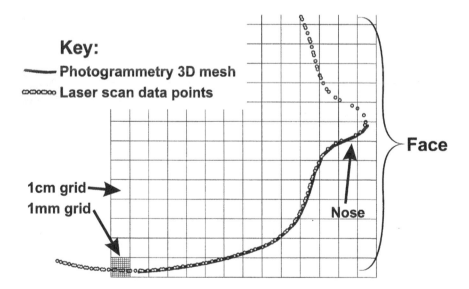

**Figure 4. Comparison of a 'slice' through the photogrammetry head surface data
and the laser scan data (the grid in Figure 3c shows the position of the slice).**

Summary discussion

The results of the posture test showed lower accuracies than expected, but good enough for many purposes and certainly an improvement over guessing the posture from the photographs. Accuracy can be significantly increased by a) using a few 'donut' reference marks in the scene, b) minimising movement of the participant between taking the photos, c) practice in taking the best photographs and using the photogrammetry software, and d) use of a higher resolution digital camera. One drawback of using a single camera to take the pictures is that the subject must hold the posture while the photographs are taken, or the photographer must wait until the posture is repeated while it is photographed from another angle.

The results of the workstation measurements showed very good accuracies from the photogrammetry. This was very encouraging considering that the photographs were taken within only a few minutes. This shows that photogrammetry will be an advantage where a short data collection time in the field is required with minimal disturbance of the subject. This can be especially important at incident investigation sites. Only a few measurements were actually extracted from the photogrammetry in this case, but many more could be taken if required. For reconstruction of a workstation in a human modelling system, photogrammetry is ideal because the processing of the photographs results in a partial 3D model which can be used as the basis for constructing a full model. An important advantage which has been found is that photogrammetry results in a model of the true 'as-built' shape of equipment, which can often differ from CAD drawings supplied by manufacturers. This has been found to be especially true of older industrial equipment.

The surface capture of the head was found to be very accurate, with most measurements within 1mm of the scanned surface data. This could be improved by using a higher resolution camera. One drawback of the method is that it cannot cope well with highly convoluted surfaces, but this is no more of a drawback than with laser scanning systems. The method also has applications in measuring the damage or deformation to object surfaces, which can be useful in forensic human factors work.

The costs and benefits of photogrammetry to the ergonomist can be summarised as:

- Shortened time on-site
- Relatively low intervention
- High quantity of data available
- Little physical access required
- Scalability: from heads to buildings

- Longer data processing time off-site
- Only visible objects measureable
- Knowledge of 3D modelling preferable
- Variety of camera positions required
- Obtaining the best results takes practice

References

Fraser C.S. 1997. Some thoughts on the emergence of digital close-range photogrammetry. President's medal address to the Photogrammetric Society, London. http://www.sli.unimelb.edu.au/vms/vms_papers/pdffiles/london97_paper.pdf

Meister D. 1995. Simulation and modelling. In J.R. Wilson & E.N. Corlett (eds.) *Evaluation of human work: Second edition*, (Taylor and Francis, London), 202–228

Porter J.M, Freer M, Case K. and Bonney M.C. 1995. Computer aided ergonomics and workspace design. In J.R. Wilson & E.N. Corlett (eds.) *Evaluation of human work: Second edition*, (Taylor and Francis, London), 574–620

Wilson J R. 1997. Virtual environments and ergonomics: needs and opportunities. *Ergonomics,* **vol 40, no. 10,** 1057–1077

ADVANCED HUMAN BODY MODELLING FOR HUMAN CENTRED DESIGN: TRENDS AND SOLUTIONS

Niels CCM Moes
Imre Horváth

Delft University of Technology, Dept OCP/DE
Section Integrated Concept Advancement
Landbergstraat 15, 2628 CE Delft, the Netherlands
C.C.M.Moes@IO.TUDelft.nl, I.Horvath@IO.TUDelft.nl

The research for human body modelling for application in the shape design of artefacts for physical interaction is calling for technologies of increasing power. Three main aspects for research were recognized: geometric modelling, the modelling of physical-physiological behaviour, and the implementation in the design of the shape of the contact area. First research in this field is mentioned.

Introduction

The massive amount of publications on modelling human bodies or body parts shows a strong need for human body models. Most of the reported models are conventional ones supporting for instance ergonomics, industrial design, medicine and animation (Dirken, 1997). Each field puts its specific requirements for the models. In the field of physical ergonomics efforts were made to develop human body models for capturing complex relationships between external loads and physiological effects. It has given impetus to the research in advanced human body modelling, which tries to capture complex body characteristics, such as assemblies of tissues, uncertainty of a shapes (Carter and Heath, 1990), effects of dynamic loads on stress and deformation (Levine et al., 1990; Goossens, 1994), the relationships with the human body type (Kernozek et al., 2000), structural changes in person-artefact interaction, biophysical changes of body parts and tissues (Dahlin et al., 1986), and complex constitutive behaviour (Malinauskas et al., 1989; Zhang et al., 1997). Although several authors expressed the wish to have all-embracing body models covering all these kinds of characteristics, due to the complexity of problems, lack of data, and processing power they had to apply and accept certain simplifications (Hubbard et al., 1993).

What has been achieved related to these characteristics? Only a few characteristics have been seriously considered for investigation and incorporation in conventional body models, for instance geometric aspects (Zajac et al., 1986; Zhu et al., 1998) and the constitutive behaviour (Vannah and Childress, 1996) of the anatomical tissues. Currently the nominal shape of the tissues can be reproduced by geometric models (Ramirez, 1992; Todd and Wang, 1996), but generic rules are still missing. Models to simulate the constitutive behaviour are still based on purely elastic behaviour that have been developed for rubber materials (Moes, 2002), although there are publications on multi-phase modelling (Oomens et al., 1987). It is also a problem that conventional body models show a lack of compatibility due to differences in dimensionality, geometric simplifications, and ability to handle quantities (e.g., linear measures of distances circumferences), physiological quantities (e.g., the amount of subcutaneous fat), and the overall characteristics (somatotype, gender) of the individual (Moes et al., 2001). Two additional investigations are needed for the generic model: the validation of the geometric model and the expansion for a wider stratification and application.

Behaviour

Modelling the behaviour of the tissues and the body under load needs a geometric model, the knowledge of the loads, the physiological tissue properties that can be influenced by mechanical loads, and the mechanical characteristics of the tissues. The synthesis of geometry, behaviour and external loads enables the computation of the effects on physiological processes such as the flow of fluids (blood, interstitial fluid). Figure 2 gives a schematic view of this process.

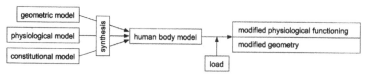

Figure 2.   The synthesis of the constituents of the human body model and the modification by loads.

Constitutive modelling
Since the deformation of tissues during artefact usage can be extremely high the constitutive behaviour of tissues must be adequately modelled by highly non-linear formulas. Current elastic models describe the material behaviour unsufficiently. Even the rheologic models, implemented in state of the art FE software, were not developed to simulate the total complexity of the mechanical behaviour of organic tissues. To advance for this level the following investigations seem required: (i) getting insight in the complex material behaviour of tissues under various types of mechanical loads, (ii) the development of adequate constitutive models to represent this complex behaviour, (iii) modelling the kinematic and kinetic behaviour (biomechanics) of the meso and the holonic assemblies to enable the integral mechanical behaviour of the body, (iv) the constructive validation of the models for the tissues or the body in the natural, functional context, (Moes, 2002).

Effects of internal loads on physiological functioning
The assessment of the relationships between internal loads and the physiological effects of these loads requires the investigations of (i) the effects of the hydraulic pressure on physiological processes such as the flow of the interstitial fluid, (ii) the effect of forces and pressures on the transportation systems, such as blood vessels, which includes the forces exerted by the smooth muscles of the vessel walls and the pressure exerted by the surrounding tissues, (iii) the effect of pressure on the blood supply of tissues, and (iv) possible tissue damage from mechanical load.

Implementation in the design process

Assuming that a human body model is build and validated, then it must be applied in the shape design process. Figure 3 shows the implementation modelled as a optimization process. Based on posture and personal characteristics a body model is generated (Moes and Horváth, 2002b). The result of the external load is an internal stress and strain distribution. The effects on the physiological functioning of the tissues is evaluated by the testing for large deformations. The majority of these models were based on the finite elements technology, see, for instance, (Todd et al., 1990; Chen and Zeltzer, 1992; Bidar et al., 2000; Lemos et al., 2001; Moes and Horváth, 2002).

Trends

Due to the ever increasing achievements of hardware and software a trend is recognized to develop support for human centred design with knowledge intensive computational human body models. If a sufficient amount of knowledge seems to exist regarding physiological processes, functional anatomy, biomechanics, biophysics, continuum mechanics and finite elements methods, why does the development of advanced human body models then lag behind? As far as we can see, the main barriers for the knowledge synthesis are (i) the incompatibility of available knowledge and model requirements, (ii) the inconsistency of the developed constitutive models, (iii) the lack of generic methods to handle the variability of humans and circumstances of using artefacts. Based on our literature survey we came to the conclusion that the development of an advanced human body model needs the following fields of investigation: (i) modelling the geometry and the assembly, (ii) modelling the mechanical and physiological behaviour of tissues under external load, and (iii) implementation of the assembled model in the shape design process.

Geometry

The geometry of a human body model can be developed in different levels of simplification, figure 1. The ideal solution is an assembly of (i) the micro structures (for instance contractive muscle fibres as active force elements, or blood vessels inside the tissues) into tissues (meso structures), (ii) the tissues and their geometrical relationships into body parts (macro structure), and (iii) the body parts into a complete body (holonic level). In the holonic level different postures can be implemented. Because of the uncertainty and the incompleteness of measured shape data, and the natural anatomical variability of the shape, the tissues must be represented by vague expressions (Moes et al., 2001). The mathematical fundamentals for the assembly and the contact conditions of vaguely defined tissues have been elaborated in (Rusák, 2003).

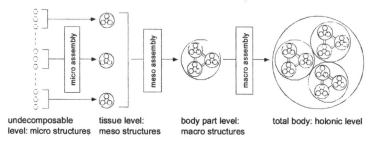

undecomposable   tissue level: body part level: total body: holonic level
level: micro structures   meso structures   macro structures

Figure 1.   Micro, meso and macro assembly.

To enable the adequate description of the characteristics of the tissues for application in advanced human body models such elaboration requires a specific organization of applied anthropometric, anatomical and physiological research, and the support by descriptive statistics and the vague discrete interval modelling (Rusák, 2003). This needs the knowledge of the unloaded shape and the spatial orientation, and the influence of gravity, muscle activation, passive elongation, and the contact properties of tissues. To enable the generic character of such vague modelling the measured shape must be related to anthropometric criteria, for which purpose an ergonomics optimization functional must be defined (Moes and Horváth, 2002a). The algorithm for the shape optimization is based on an iterative improvement of the fulfilment to the criteria, see figure 3.

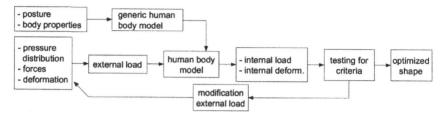

Figure 3.   The implementation of the body model in the shape generation process.

Relationships between external measures and internal loads
An important research aspect is to find the relationships of external load and the internal load. This knowledge is of primary importance in e.g., the field of decubitus research and the research of designing prostheses, where the external load can be measured without too much difficulty, but where it has also generally been accepted that the possible tissue damage occurs as a result of processes inside the body (flow of fluids) and not at the body surface (pressure distribution in the contact area). Within the field of sitting research the effects of the pressure exerted by a particular shape of a seat has been expressed in terms of the distribution of internal tissue deformations, compressive and shear stresses, but a systematic, quantified correlation has not yet been found (Moes, 2003).

Conclusions

A need exists for advanced human body modelling. Although much knowledge is available, it is not yet sufficient and consistent for application in such modelling. Research in the fields of geometry, physiology, mechanical modelling and optimization is needed for the further development and the implementation. Currently, basic research has been set up in our department to investigate the fundamentals for possible solutions. The feasibility of certain aspects of the conceptual solution has been demonstrated in several publications, see above. The foreseen perspectives are the development of human body modelling of increasing complexity, validity and applicability, including the physical and eventually the psycho-physical influences on physiological functioning.

References

Bidar M, Ragan R, Kernozek T, and Matheson JW (2000). Finite element calculation of seat-interface pressures for various wheelchair cushion thicknesses.

Carter JEL and Heath BH (1990). *Somatotyping – development and applications*. Cambridge University Press.

Chen DT and Zeltzer D (1992). Pump It Up: Computer Animation of a Biomechanically Based Model of Muscle Using the Finite Element Method. *Computer Graphics*, 26(2):89–98.

Dahlin LB, Danielson N, Ehira T, Lundborg G, and Rydevik B (1986). Mechanical effects of compression of peripheral nerves. *J Biomech Engin*, 108:120–122. Transactions of the ASME.

Dirken H (1997). *Product-ergonomie*. Delft University Press, first edition. in dutch.

Goossens RHM (1994). *Biomechanics of Body Support*. PhD thesis, Erasmus Universteit, Rotterdam, the Netherlands.

Hubbard RP, Haas WA, Boughner RL, Canole RA, and Bush NJ (1993). New Biomechanical Models for Automobile Design. *SAE Transactions*, 930110:164–171.

Kernozek TW, Amundson A, Hummer J, and Wilder P (2000). Effects of Body Mass Index on Seat Interface Pressures of Elderly that were Institutionalized.

Lemos R, Epstein M, Herzog W, and Wyvill B (2001). Realistic Skeletal Muscle Deformation using Finite Element Analysis. *Proceedings of the 14th Brazilian Symposium on Computer Graphics and Image Processing*, pages 192–199.

Levine SP, Kett RL, and Ferguson-Pell M (1990). Tissue Shape and Deformation Versus Pressure as a Characterization of the Seating Interface. *Assistive Technology*, 2(3):93–99.

Malinauskas M, Krouskop TA, and Berry PA (1989). Noninvasive measurement of the stiffness of tissue in the above-knee amputation limb. *Journal of Rehabilitation Research and Development*, 26(3):45–52.

Moes CCM (2002). Estimation of the Nonlinear Material Properties for a Finite Elements Model of the Human Body. In Horváth Imre, Peigen L, and Vergeest Joris S.M., editors, *Proceedings of the TMCE2002*, pages 451–467, Huazhong Univ. of Science and Techn, Wuhan, Hubei, P.R. China. HUST Press.

Moes CCM (2003). Sitting stresses inside the body. In McCabe Paul T., editor, *Contemporary Ergonomics 2003*, pages 549–554. The Ergonomics society, Taylor & Francis.

Moes CCM and Horváth I (2002). Estimation of the non-linear material properties for a finite elements model of the human body parts involved in sitting. In Lee DE, editor, *ASME/DETC/CIE 2002 proceedings*, pages (CDROM:DETC2001/CIE–34484), Montreal, Canada. ASME 2002.

Moes CCM and Horváth I (2002a). Optimizing the Product Shape for Ergonomics Goodness Index. Part I: Conceptual Solution. In McCabe Paul T., editor, *Contemporary Ergonomics 2002*, pages 314–318. The Ergonomics society, Taylor & Francis.

Moes CCM and Horváth I (2002b). Optimizing the Product Shape for Ergonomics Goodness Index. Part II: Elaboration for material properties. In McCabe Paul T., editor, *Contemporary Ergonomics 2002*, pages 319–322. The Ergonomics society, Taylor & Francis.

Moes CCM, Rusák Z, and Horváth I (2001). Application of vague geometric representation for shape instance generation of the human body. In Mook DT and Balachandran B, editors, *Proceedings of DETC'01, Computers and Information in Engineering Conference*, pages (CDROM:DETC2001/CIE–21298), Pittsburgh, Pennsylvania. ASME 2001.

Oomens CWJ, van Campen DH, and Grootenboer HJ (1987). In Vitro Compression of a Soft Tissue Layer on a Rigid Foundation. *Journal of Biomechanics*, 20(10):923–935.

Ramirez ME (1992). Measurement of Subcutaneous Adipose Tissue Using Ultrasound Images. *American Journal of Physical Anthropology*, 89:347–357.

Rusák Z (2003). *Vague Discrete Interval Modelling for Product Conceptualization in Collaborative Virtual Design Environments*. PhD thesis, Delft University of Technology, Fac. Industrial Design Engineering.

Todd BA, Thacker JG, and Chung KC (1990). Finite Element Model of the Human Buttock. In *Proceedings of the Resna 13th annual conference*, pages 417–418, Washington DC.

Todd BA and Wang H (1996). A visual basic program to pre-process mri data for finite elements modelling. *Comput. Biol. Med.*, 26(6):489–495.

Vannah WM and Childress DS (1996). Indentor tests and finite element modeling of bulk muscular tissue in vivo. *J. of Rehab. Res. and Devel.*, 33(3):239–252.

Zajac FE, Topp EL, and Stevenson PJ (1986). A dimensionless musculotendon model. In *IEEE Eighth Annual Conference of the Engineering in Medicine and Biology Society*, pages 601–604.

Zhang M, Zheng YP, and Mak ATF (1997). Estimating the effective young's modulus of soft tissues from indentation tests – nonlinear finite element analysis of effects of friction and large deformation. *Med. Engin. Phys.*, 19(6):512–517.

Zhu Q, Chen Y, and Kaufman A (1998). Real-time Biomechanically-based Muscle Volume Deformation using FEM. *Eurographics*, 17(3):275–284.

ANTHROPOMETRIC PROCEDURES FOR DESIGN DECISIONS: FROM FLAT MAP TO 3D SCANNING

Hongwei Hsiao, Ph.D.

*National Institute for Occupational Safety and Health,
1095 Willowdale Road, Morgantown, West Virginia, 26505, USA*

Anthropometric principles have been applied to many applications in various industries, for reasons of product value, efficacy and safety. This paper presents four anthropometric approaches for product design decisions. A univariate method has been widely used in designing for extremes, such as in determining door heights. A bivariate method has been useful for products that primarily involve dual essential dimensions, such as in shoe-sizing applications. For many other applications, a multivariate accommodation approach is necessary to account for both the body size variance and proportional variability. A 3D shape-quantification approach is advisable for determining sizing schemes and size ranges of personal protective equipment, such as fall protection harnesses. A 3D digital feature-envelope approach is instrumental for placement of control components in workspaces, such as in a farm-tractor-cab accommodation.

Introduction

Anthropometry is the study of human body size and proportions. It contributes to product value, efficacy, and safety in numerous applications. Anthropometric information is traditionally reported as means and standard deviations for various body segments (Roebuck, 1993). While the traditional anthropometric approach is useful for some simple applications, recent literature has shown that the point-to-point anthropometric information currently being used seems to be insufficient for many design applications (Hsiao *et al*, 2003). For instance, dimensions measured and tabulated by traditional methods are not linked to one another; knowing shoulder width would not enable a designer to create an accurate representation of shoulder location related to the cab space for a vehicle-design application. The newly available 3-D scanning and shape-quantification technologies make the assessment of 3D anthropometric information for product-design decision a much more feasible undertaking. This paper presents traditional univariate and bivariate methods and two emerging 3D anthropometric procedures for design decisions. Examples are provided for better understanding of general anthropometric principles and the specific approaches.

Procedures for design decisions

In product design practices, a six-step anthropometric procedure is typically used for design decisions (Hertzberg, 1972). These steps include (1) determining the body dimension that is of essential importance (e.g., stature, hand width, etc.), (2) determining the population to be considered (e.g., sex, age, occupation etc.), (3) selecting the percentage of the population to be accommodated (e.g., for safety, cost-benefit, or other concerns), (4) obtaining the necessary reference materials to determine the appropriate statistics (users may need to collect their own data.), (5) computing the specific dimensions, and (6) adjusting as necessary for clothing and other equipment.

Design for extremes with univariate analysis
In some design applications, the extreme value of a single anthropometric measurement plays the critical role in design decisions. An example is to specify the height for a doorway to avoid unintentional head injuries in manufacturing plants. The solution can be drawn by following the above-mentioned six steps. The essential dimension is stature. The population to be considered is U.S. civilians. The percentage of the population to be accommodated is 99.9 percent of the male population, reflecting the paramount importance of safety concerns; this will cover almost all females as well. The reference information X (mean stature) is 175.3 cm and S (standard deviation) is 7.1 cm (Webb Associates, 1978). The Z value (a coefficient whose value varies with the percentage of population to be covered) for $Z_{99.9}$ is 3.09. Therefore, $Stature_{99.9}$ is 175.3 + 3.09 * 7.1 = 197.2 cm. By adjusting the height for shoes (2.5 cm) and headgear (5.0 cm) (Hertzberg, 1972; Das and Grady, 1983), we find that the desired opening should be 197.2 + 2.5 + 5.0 = 204.7 cm.

Design with bivariate analysis
In other product design applications, the design decision may involve some dimensions that can be simplified to two critical measurements. Shoe design is in this category. Most people are familiar with typical shoe sizing schemes, which consider foot length and foot width. Some people can find good-fitting shoes easily; others have difficulty in finding even a barely fitting one. Why? Assume that the U.S. population mean for foot length is 27 cm with a standard deviation of 1.6 cm and the mean foot width is 13 cm with a standard deviation of 1.2 cm. Based on the assumption that the data are normally distributed, the ellipse in Figure 1 would cover 95% of the dataset. Assuming that a space of 0 to 1 cm between shoe length and foot length is the comfort range and that a space of 0 to ¼ cm between shoe width and foot width is tolerable, we will have a number of two-dimensional configurations: a 28L by 13.25W, a 29L by 13.25W, and many other shoe sizes (Figure 1). However, we will notice that, based on the dataset, the central size fit only 7% of the persons. Some sizes only fit 2%, or even 0.2%. Comfort tolerance and cost probably dictate the available sizes in the market.

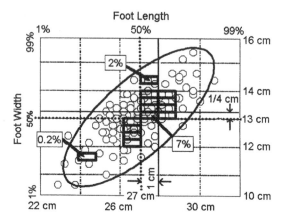

Figure 1 Use of foot length and width in determining shoe size

Design for effective sizing scheme incorporating multivariate analysis
In some design applications, the proper accommodation of a product to the user population is affected by multiple anthropometric variables. Fall protective harness design is an example. If we choose three key anthropometric measurements and make 3 levels for each measurement, we will have 27 sizes. As additional measurements are added to the design, the complexity of the sizing scheme very quickly becomes unworkable. Fortunately, mathematical or statistical techniques, such as principal component analysis methods, can be used to reduce the complexity involved with all relevant measurements into a manageable number of new "principle component" variables. With the six-step anthropometric procedure in mind, harness-design practices face three major challenges: (1) determining essential body dimensions (step 1), (2) obtaining the necessary data (step 4), and (3) computing the specific dimensions (Step 5). In a harness-sizing evaluation study, body weight and 7 other parameters were found to affect harness-size selection (NIOSH, 2003). A 3D extension of Elliptic Fourier analysis (Kuhl and Giardina, 1982) was used to quantify construction worker body shapes through multivariate statistics (Figure2). It was concluded that stature, hip circumference, and weight would be useful anthropometric dimensions for the correct assignment of a harness size. Based on the results obtained from the shape comparisons of the current sizes, it was concluded that three sizes were sufficient to accommodate most of the variation. An inverse EFA was then performed to predict the best-fit harness size, and to draft a sizing chart as follow: Small (weight < 58 kg, stature < 166 cm, hip circumference < 88 cm), Standard (Weight 58-105 kg, stature 166 – 180 cm, hip circumference 88-108 cm), and Large (weight > 105 kg, stature > 180 cm, hip circumference > 108 cm) (NIOSH, 2003).

Design for complex workspace incorporating multivariate analysis
In some complex system applications, such as in the design of farm tractor cabs, having multiple cab sizes to accommodate user population is not cost effective. Optimizing the control-component arrangement in the workspace would be a reasonable practice. Using the same six-step anthropometric procedure as for product design, we will notice the same three

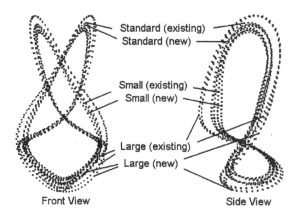

Front View   Side View

Figure 2 Mean outlines for the three existing and redesigned harness sizes, based on multivariate predictions of Elliptic Fourier analysis coefficients

major challenges as in the harness design case: determining critical body dimensions, obtaining the necessary data, and computing the specific dimensions. In a study on anthropometric criteria for the design of tractor cabs, knee height (sitting) and 8 other parameters were found to affect the accommodation rating (Hsiao *et al*, 2004). A principle component analysis approach was utilized to foam a set of 15 anthropometric model variables used in the cab design process, through significant composite variables, derived in a mathematical reduction process of key anthropometric dimensions. These 15 representative models are useful in the design process, especially in using digital human modeling techniques, to ensure that the finished product will accommodate the desired population.

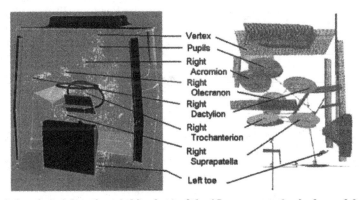

Figure 3 Landmark locations (white dots) of the 15 representative body models related to the tractor workspace (left picture) and the 95% ellipsoid representations of the feature envelopes for some landmarks (right picture; side view of the workspace).

While the identified 15 representative models are useful for examining tractor cab accommodation, the design process is incomplete without considering the use of feature envelopes of body landmarks in optimizing the layout of cab components. For instance, designers need to know where the knees are with respect to the seat, to position the steering wheel. Visually, a feature envelope can be thought of as an ellipsoid enclosing a cloud of three-dimensional data points representing the variability in a landmark location. Figure 3 shows some of the 14 critical landmarks for the 15 representative models for tractor design. Alternatively, designers can digitize the critical landmarks of all subjects in a national 3D survey database. A set of centroid coordinates of the key body landmarks and the 95% confidence semi-axis-length for each landmark location can be developed through a principle component analysis approach to guide tractor designers in their placement of tractor control components to best accommodate the user population.

Summary

In product design applications, a six-step anthropometric procedure is useful for design decisions. This paper presented traditional univariate and bivariate approaches and two emerging 3D anthropometric methods, based on the six-step anthropometric procedure, for various product design decisions. Examples of door, shoe, harness, and tractor designs were provided for better understanding of the general principles and the specific approaches.

References

Das B. and Grady R.M. 1983, Industrial workplace layout and engineering anthropometry, In: Kvalseth TO, ed. *Ergonomics of Workstation Design.* (Butterworths, London)

Hertzberg H.T.E. 1972, Engineering anthropometry. In: Van Cott HP, Kincade RG, eds. *Human Engineering Guide to Equipment Design*, rev. ed. (U.S. Government Printing Office, Washington, DC), 468-584

Hsiao H., Bradtmiller B., Whitestone J. 2003, Sizing and fit of fall-protection harnesses, *Ergonomics*, 46, No. 12, 1233-1258

Hsiao H., Whitestone J., Bradtmiller B., Zwiener J., Kau T., Whisler R., Gross M., Lafferty C. 2004, *Determining Anthropometric Criteria for the Design of Tractor Cabs and Protection Frames*, (NIOSH technical report, Morgantown, West Virginia)

Kuhl F.P. and Giardina C.R. 1982, Elliptic Fourier features of a closed contour, *Computer Graphics and Image Processing*, 18, 236-258

NIOSH, 2003, *Development of a Method for the Design and Sizing of Whole Body Fall Protection Harnesses*, (NIOSH technical report, Morgantown, West Virginia)

Roebuck J.A. 1993, *Anthropometric Methods: Designing to Fit the Human Body*, (Human Factors and Ergonomics Society, Santa Monica, CA), 53-56

Webb Associates, 1978, *Anthropometry source book*, Vol. I (National Aeronautics and Space Administration, Washington DC)

AN EYE ON USABILITY STUDIES

Alastair Gale, Ruth Filik, Kevin Purdy & David Wooding

Applied Vision Research Unit, Institute of Behavioural Sciences, Kingsway House, Kingsway, Derby, Derby DE22 3HL

Usability investigations typically utilise relatively small numbers of participants which are enough to yield the key information relevant to the vast majority of potential user issues. Adding appropriate eye movement recording of these individuals whilst conducting such usability investigations can increase the information yielded from such studies and so can lead to better design and acceptance of products and materials by end users within a shorter development time. The advantages, difficulties and potential pitfalls of using such an approach are covered. Examples are presented from a number of studies demonstrating how eye movement recording contributes to the evaluation of the design of: web sites; printed advertising, marketing and instructional materials.

Introduction

Usability can be described as the extent to which a product can be used by specified users to achieve specified goals with effectiveness, efficiency and satisfaction in a specified context of use (ISO, 1998). Usability testing is now endemic in a number of different fields (including websites, software, new products) and at different stages of product/software development. A typical approach is to utilise a small group of participants, representative of the target user population of interest, which is then studied as they individually undertake some prescribed representational tasks. Various types of data can be recorded concerning the individuals' behaviour which can involve video taping of the interaction, questionnaires, cognitive walkthroughs etc. Although eye movement recording of how a potential user interacts with the software/product can be used (e.g. Goldberg & Wichansky, 2003) it is not commonly employed.

Where part of the usability interest is in whether the individual recalls particular information or not then post-trial memory recall tests are useful in examining what information individuals can remember from that which has been presented to them. For instance in designing advertising billboards or traffic information signs an easy approach is for an individual to drive past the sign and then to be asked what information they recall from it. In order for someone to remember informational content it is necessary for him or her to have first visually attended appropriately to that part of the display/material.

This is where visual search recording can help. As the eye generally moves without much overt or conscious control it is difficult, if not impossible, for an individual to state where they have actually looked with any real accuracy. Thus, for instance, in a control room situation one would know how readily the user had operated various controls but would have no real insight into how much difficulty they had had in first locating or selecting that control from others prior to operation without utilising eye movement recording.

Types of eye movements

In humans the retina has variable acuity across its surface with the highest acuity at the fovea, which is only some 2mm in size. Consequently in order to examine something in detail (i.e. to visually attend to something) then the eye has to move so that the image of the object of interest falls upon the fovea. In order to maintain vision the retina has to be constantly stimulated and so the eye is never still and makes a range of movements, some of which are voluntary and others involuntary, as follows:

Saccade - The most conspicuous eye movement. These movements are very rapid and conjugate (both eyes move together in the same direction). They are ballistic in nature and under voluntary control. Most saccades are less than 15^0 in amplitude. Saccades made to a particular target typically overshoot or undershoot the target location thereby giving rise to further corrective small saccadic movements.

Pursuit – these are conjugate movements which follow slow moving objects in the visual field. A moving object is usually necessary to initiate these movements which are not typically under conscious control.

Smooth - are similar to pursuit and can be made without a moving stimulus.

Compensatory - are smooth movements which compensate for head or body movements so as to partially stabilise an object on the retina during such movement.

Vergence – the two eyes move in opposite directions so as to permit the acquisition of near or far objects through the binocular fusion of the two images of that object.

Nystagmus - a rhythmic movement of the eye comprising an alternating slow and a fast phase. There are various forms of nystagmus such as; optokinetic which can be produced by moving repetitive visual patterns, vestibular is produced by stimulating the semicircular canals as the head is rotated.

Miniature – a range of movements, typically less than 1^0 visual angle in amplitude and include: flicks, drift, high frequency tremor, and irregular movements. The effect of such small movements is to constantly shift the image of a fixated object across the retina thereby stimulating the retinal cells.

Internationally, current eye movement research is discussed through the eye movement email list at http://www.jiscmail.ac.uk/lists/eye-movement.html .

Recording eye movements

There exist a range of eye movement recording techniques, which have remained largely unchanged for many years (c.f. Young & Sheena, 1975). However, more recent developments have seen radical approaches that utilise the detection of head pose information together with 'eye-in-head' orientation which offer new opportunities for future research possibilities. Increasingly there are good and robust commercial systems available and an independent database (EMED - Eye Movement Equipment Database) of equipment currently available from some 44 manufacturers is available on our website (http://ibs.derby.ac.uk/emed/) to aid in equipment selection.

Each technique has advantages and disadvantages and all of them require some degree of initial calibration requiring participant co-operation - which typically involves the participant in fixating known locations in the environment. For instance, the most accurate method uses the dual Purkinje image technique which has very high spatial and temporal resolution but can require accurate head stabilisation by using a headrest combined with a bite bar where the participant bites into dental wax so providing skull rigidity. Whilst the accuracy of such an approach is high it has to be considered that (1) the participant is well aware of the nature of the investigation, (2) one does not normally look at the world in such a restricted manner and (3) also the accuracy of the data can be very much higher than the investigation actually requires. Thereby almost the first thing that the investigator may do is to degrade the data to match the investigation at hand.

Lightweight head-mounted recording systems allow the participant to walk around the environment or travel in a vehicle but s/he is well aware of the recording process which then may/may not affect how they consciously or subconsciously look at the environment during the study. It is logically hard to place a device on someone's head, adjust it, calibrate the participant and then fully expect them to go about a task as they would do 'normally'.

To overcome such participant awareness there are available some inconspicuous remote oculometer systems which allow monitoring of visual search behaviour with the equipment placed some distance from the participant and therefore requiring no participant attachments. Thus, virtually 'secretive' eye movement recording could take place although, as with all eye movement recording, ethical requirements necessitate that the participant first be informed of the use of such equipment. Typically with these systems there is a trade off between the degree of allowable participant head/body movement and the actual experimental measurement accuracy.

A common approach in usability studies is to utilise a corneal reflection based technique where some recording is made of incident infrared illumination on the eye and such systems can be problematic in bright or variable (as in driving) sunshine where the level of infra red illumination is changing.

Eye movement recording systems typically provide some form of x,y co-ordinate output of measured gaze location which is provided as a computer data file and sometimes as a display overlaid on a video record of the user trial. In cases where the participant moves about the environment then the analysis of where the individual has looked, which is complicated as their head position constantly changes, can involve time consuming and inaccurate manual frame-by-frame video analysis although techniques are available to speed up this process considerably. Where the participant is fairly stationary with respect to a display then fast on-line data analyses of eye position are possible.

Usability and eye movements

In usability investigations it is typically saccadic movements which are of chief interest due to them being assumed to be the outcome of both cognitive and environmental factors (Gale, 1997). Depending upon the nature of the investigation then recording systems with an accuracy of about 0.5^0 visual angle are usually suitable. This then yields a wide choice of possible techniques to employ. However, simply recording eye behaviour appropriately does not mean that the resultant data will be useful or at all meaningful. For instance, typically, such accuracy would inform the investigator that a small area of text on a web page (for instance) had been visually attended to but not necessarily which specific words within that text block. Additionally, the published accuracy of the device by the manufacturer may be obtained under 'best' conditions and may not be what is actually achieved by an investigator in a particular empirical setting. There are several eye movement parameters of interest and these include: the number of eye fixations; fixation durations; scan path length and duration; fixation density in areas of interest, and the transition probability between regions of interest of a layout.

Issues concerning eye movements in usability studies

Measuring that a participant's eye is directed at a particular target location should mean that the individual is visually attending to that target. Usually this will be a correct assumption but the eye is a directly coupled system and so it can be directed towards a particular environmental location with the individual actually attending elsewhere. This gives rise to a class of errors variously called 'look but failed to see' errors and is an important topic, particularly in domains such as image interpretation (Gale, 1997) or driving. Therefore in interpreting eye movement data it is important that the investigator has sufficient experience.

Eye movement systems should essentially give a two dimensional co-ordinate output of where a participant is 'looking', where this is typically taken as the point of regard of the participant and is assumed to be that portion of the visual world that falls onto the fovea.

Eye movement techniques yield an estimate of this point of regard albeit with some degree of error partly depending upon the particular technique. This error is due to the nature of the recording method employed (e.g. corneal reflection systems have a limiting factor due to the physiological structure of the eye) together with any inherent limitations in the manufacturer's implementation of the recording device and software.

The fovea is approximately 1.5^{0} visual angle in diameter (a ready estimate is the size of one's thumbnail held at arm's length). However, it is not always necessary for an object to fall precisely upon the fovea in order for an individual to visually attend to it. Consequently, an area somewhat larger than the fovea can be used in measuring visual attention to an object/location.

The retina can be subdivided into a range of different areas such as the fovea centralis, fovea, parafovea, perifovea. In various areas of eye movement research researchers often utilise the useful field of view (UFOV). This is taken to be a 'visual attentional' area, typically assumed to be circular in shape, but not necessarily so, which can vary in size (i.e. constrict or expand) depending upon a number of factors, both environmental and subjective. To research a particular domain the UFOV can be measured empirically – there is a large body of research on this within the domains of driving and image perception investigations. Slightly differently, in research on reading, researchers typically distinguish between foveal and parafoveal vision with the parafovea referring to an area extending some 5^{0} from the centre of the fovea (e.g. Rayner et al., 2003). Often in reading research the parafoveal area is considered to be a horizontal lozenge shape with the main interest being in that part of the parafoveal area to the right of the currently fixated word/letter (assuming left to right reading behaviour).

Examples of the use of visual search

There is a wealth of evidence concerning the individual nature of eye movements and much current research is aimed at identifying more generic group data analytic approaches from individual eye movement records (e.g. Wooding, 2002). Typically with eye movement recording, users are given a specific task to do, as opposed to simply 'browse' through a website or other material. We have successfully utilised eye movement recording as an adjunct to usability investigations in a number of areas:

Web sites: eye movement recording can ascertain whether users actually attend to specific graphic devices which designers include as navigational aids. Where there are multiple ways of navigating through a site to find a particular item then recording the mouse and keystrokes informs which particular link is actually used but adding visual search recording can elucidate whether the users have easily determined to use that particular navigation aid or not

Informational materials and product packaging: the content of textual materials such as patient information leaflets is typically arrived at through having a group of users read through the leaflet and then ask them a series of questions where they have to correctly recall a certain percentage of that material. Studying eye movements and can inform us on how users process instructional material and product information on packaging and labelling. Our current research interests include the influence of textual characteristics (Filik et al., 2003) on eye movements and visual search during product identification, specifically the influence of these factors on the likelihood of error in the identification and selection of drug products.

References

Filik, R., Purdy, K. J., Gale, A. G., & Gerrett, D., 2003. Using eye-movements to investigate medication errors. Poster presented at *ECEM 12*, Dundee, August 20-24th.

Gale A.G.: Human response to visual stimuli. 1997. In W. Hendee & P. Wells (Eds.) *Perception of Visual Information - second edition*, (New York, Springer Verlag).

Goldberg J.H. & Wichansky A.M. 2003. Eye tracking in usability evaluation: a practitioner's guide. In A. Hyona, R. Radach & H. Deubel (Eds.) *The mind's eye: cognitive and applied aspects of eye movement research*. (Elsevier, Amsterdam)

ISO 9241-11 (1998): Guidance on Usability

Rayner, K., White, S.J., Kambe, G. Miller B. & Liversedge S.P. 2003. On the processing of meaning from parafoveal vision during eye fixations in reading. In A. Hyona, R. Radach & H. Deubel (Eds.) *The mind's eye: cognitive and applied aspects of eye movement research*. (Elsevier, Amsterdam).

Young L.R. & Sheena D. 1975. Survey of eye movement recording methods. *Behavior Research methods & Instrumentation*, 7: 397-429.

Wooding, D.S., 2002. Eye movements of large populations: II. Deriving regions of interest, coverage, and similarity using fixation maps. *Behavior Research Methods, Instruments & Computers*. 34(4), 518-528.

APPLYING ERGONOMICS TO SUBSTANTIATE THE USABILITY OF HUMAN-MACHINE INTERFACES

Ed Marshall and Andrew Shepherd

Synergy Consultants Ltd
Yewbarrow, Hampsfell Road,
Grange-over-Sands,
Cumbria, LA11 6BE

The development of the human-machine interface (HMI) for system control has been driven by three key factors: Technological change, the requirement to reduce staffing and the realisation that the HMI can be a principal ingredient in the causation, and also the prevention, of accidents. Ergonomics has gained much prominence through the investigation of incidents involving human error. However, this means that management may have come to consider that ergonomics is restricted to addressing safety concerns and justifying Safety Cases. This paper emphasizes the importance in terms of economy and system effectiveness of minimizing operational problems by considering and evaluating HMI usability at the design stage, as well as attending to accident scenarios after the design is complete. The paper considers the problems of implementing a process for assuring good ergonomics in the HMI. It outlines techniques for encouraging design engineers to assess the HMI. and to undertake task analysis.

Introduction

The development of the human-machine interface (HMI) for system control has been driven by three key factors: Technological change, the requirement to reduce staffing and the realisation that the HMI can be a principal ingredient in the causation, and also the prevention, of accidents. Indeed, it can be argued that ergonomics has gained much prominence through the investigation of incidents involving human error.

However, this means that consideration of ergonomics is often restricted to the justification of Safety Cases, which consider rare and sometimes implausible scenarios and fails to consider everyday usage of the HMI. We would argue that, although ergonomics has contributed to the prevention of disastrous accidents, the reluctance to study the usability of the HMI leads to sub-optimal operation, frustration on the part of the operators and thence to more human errors and subsequent loss of production. It would be economical and effective to minimize operational problems by considering and evaluating HMI ergonomics at the design stage, as well as attending to the big accident scenarios after the design is complete.

Too often, engineers and managers, concerned with shaping working environments and organizing human resources, are reluctant to consider ergonomics as an input to their design decisions, regarding it as a 'frill', or a distraction and, at worst, 'just common sense'. Application of ergonomics is often piecemeal, for example, utilizing guidelines which can justify design preferences, but overlook important issues because there is a lack of understanding about how the system will be operated. Ergonomics assessment can enable designers to appreciate how aspects of tasks and personnel interact, the information necessary to meet operational goals, where safety takes priority over production, and the system consequences should anything go wrong. These concerns are fundamental to the design and management of complex systems.

This paper considers the problems of implementing a process for assuring good ergonomics in the HMI. It will report techniques recently devised for permitting design engineers, with no formal knowledge of ergonomics, to assess the HMI. and will address those design issues that should involve professional ergonomists. Finally, it will consider how ergonomics should be delivered in order that it can assume a more productive role in the design and management of HMI systems.

The requirements for ergonomics inputs

The application of sound ergonomics principles, when redesigning HMIs, is beneficial in terms of safety, efficiency and operating reliability. It is important because the reliability of people employed to work in a system directly affects system reliability and performance. Concerns with the health, safety and wellbeing of workers relates to the manager's duty of care.

Human beings are important components of complex systems because, within limits, they are physically and mentally versatile, so engineering does not need to solve all functional problems. People are skilled and can do difficult physical tasks where automation has not been devised or is uneconomical. They can also carry out tasks entailing judgment and decision-making, often with greater flexibility than machines, although this may be accompanied by greater unreliability. One of the most potent benefits to be gained from employing human beings is their versatility. This often enables them to overcome flaws in system design and deal with unforeseen circumstances; these skills evolve with experience of the system rather than depend on systems analysts anticipating every operational nuance. There is a danger that managers take this versatility for grated and place too much reliance on the human operator overcoming system deficiencies.

Thus the implementation of ergonomics within HMI design and system development has profound benefits. These benefits need to be well understood by designers and may be summarized as follows.

- More fluent operation.
- Reduction in the likelihood of human error.
- Reduction in frustration
- Improvements in user attitudes
- Safer and more efficient process operation

Moreover, inappropriate design of work – including inappropriate design of the workplace, operating tasks, work-load, information displays and controls, training and support – can contribute to system failure. Managers and designers must balance their aspirations for system performance with reality concerning human limitations. They must also take the necessary steps if they wish to optimize human performance.

The reasons for reluctance

Although managers and engineers usually appear willing to acknowledge the use of human factors where it is obviously crucial to safety, there is often reluctance to appreciate it as a factor that *contributes* to making design decisions regarding general usability. This response is understandable when taken in the wider context of managerial responsibilities, where remaining within budget and meeting development and operational targets may be paramount.

In process control tasks, software engineers often observe human computer interface design guidelines to create screens that conform to standards of character height, colour consistency, glare and screen flicker rate. Yet they often fail to take proper account of the tasks that operators actually have to carry out. Failure properly to understand tasks can mean that the information required in order to support operating decisions is not sufficient or provided in a satisfactory form. Thus, the HMI does not adequately support the tasks. Neglect of these issues during design is understandable if their managers do not regard them as important when setting targets and appraising performance

It is indisputable, however, that resolving a design problem that has been identified after completion of any phase of system development is often very costly. It may require substantial rework and investment in additional equipment. It will incur delays in meeting project targets and will cause the system to be unavailable when it should be making a profit. Human factors issues raised *during* system development or the normal phases of operation are rarely greeted with enthusiasm by managers who have other, more obvious, demands for their attention that their own managers require them to solve. In any case, concerns about human factors do not mean that a system failure will necessarily occur. It may even be assumed that conscientious and experienced staff will avoid severe problems or take them in their stride and resolve them before serious consequences arise. Managers and engineers may prefer to take the risk and rely on human resources and management to deal with remaining problems through personnel selection, training and conscientious supervision and operation.

Ways to involve ergonomics in the design process

Encourage management
It is this apparent managerial risk-taking, which can frustrate the human factors specialist trying to engage constructively in system design and development. If wider organizational pressures mean that managers and engineers are reluctant to acknowledge this contribution, it becomes an issue for human factors to consider how best to support managers and engineers in taking human factors into account at a time in system development where problems can be identified and solved with minimum expense and inconvenience. This principle of *timeliness* in human factors needs to be appreciated to the extent that project managers take this into account when planning project stages.

Make engineers aware
Ergonomists must strive to make engineers aware of the advantages of implementing human factors and to demonstrate the various tolls and techniques. A major step in this direction has been the recent publication of a human factors textbook for engineers by the Institute for Electrical Engineers (Harvey and Sandom, 2003). In particular interested engineers should be encouraged to learn the application of key ergonomics techniques. For the control and operation

of process tasks, these include task analysis methods and techniques for the design and assessment of HMIs.

Participation in task analysis

Task analysis is a powerful tool for assessing the operability of a system,. Hierarchical task analysis (HTA) – see Shepherd, 2001, for a comprehensive description – is a technique which has been used extensively in process operation and is accessible to non-ergonomics specialists. The authors have found it valuable to institute task analytic exercises at a number of installations. Shepherd and Marshall (2003) describe exercises where engineers learnt and applied HTA with the help and supervision of professional ergonomists. An approach, when modeling a multi-facetted operation task, is to convene a joint activity to enable interested parties to collaborate. This is an unusual way to conduct task analysis, but can be very effective. The analyst uses a white board, flip-chart, or a laptop computer with projection facilities so that everyone can share the information and the emerging task description. The analyst, in this case a professional ergonomist, operates as a *facilitator*, often focusing attention on one workshop member at a time, whilst taking account of comments made by others. In this way contributing parties can represent their discipline and appreciate why operating decisions are taken. It is noticeable in such workshops how different people, with their different perspectives gain from this contact as they often start to understand aspects of the system they were unfamiliar with, and appreciate where compromise is necessary.

In an unpublished study, similar techniques were exploited to assist engineers at a nuclear installation to assess operational tasks in order to determine manning requirements particularly in terms of the availability and location of process plant interfaces. In this exercise, operational engineers were given a short workshop course on HTA. Each engineer then accepted ownership of a plant segment and carried out an HTA, with appropriate support from the ergonomists.

Participation in HMI Assessment

A research project has recently been undertaken as part of the UK Industry Management Committee's Generic Nuclear Research Programme with the objective of developing a toolset for providing ergonomics guidance to assist Control and Instrumentation engineers in designing new HMIs or modifying existing HMIs in nuclear plant control rooms (Gregson *et al*, 2003).

As a first step, a survey was carried out to identify the extent to which ergonomics issues had been incorporated into control room HMI refurbishment schemes at 16 UK sites both within the Nuclear Industry and in related industrial applications. It was clearly established that, whilst the benefits of considering ergonomics was generally recognized by design engineers, there was little evidence of any formal ergonomics input into any of the system refurbishments that were surveyed.

To encourage the incorporation of ergonomics into HMI designs, it was apparent that guidance should be provided to designers in a clearer and more practical way. For a discussion of the usability of ergonomics guidance see Gregson and Gait (2004) in this volume. A *toolset* was developed which uses linked flowcharts to determine systematically the key ergonomics factors in the proposed HMI and identifies any requirement for specialist ergonomics intervention. The *toolset* also provides a spreadsheet-based assessment of the proposed design, which is linked interactively to a large body of ergonomics guidance gathered from a wide range of respected sources. The key objective was to provide an appropriate level of guidance to support and encourage designers or engineers, with no professional ergonomics expertise, so they can incorporate quality ergonomics into the design process.

Conclusions

Where engineers focus on the functional requirements of design and fail to attend to operating requirements, problems relating to human performance are likely to emerge later on as a result of a safety review, an operating problem or an accident. Rectifying such problems incurs greater cost the longer they are left. Even where the engineer is aware of the potential problems for human operation, unfamiliarity with human factors knowledge or methods can compromise design. Many human factors problems encountered can be attributed to design decisions being taken without full consideration of the implications for human performance.

Human factors specialists gain from adopting an operational perspective since it takes better account of the context in which the task is carried out, including, the environment in which people work and the time pressures they experience. It takes account of wider system goals and activities and the other duties that a worker must fulfil. It takes account of the system values, including the costs and consequences of error. By adopting an operational perspective, human factors specialists and their clients can collaborate on a common agenda.

It is important for management to understand and take account of human factors throughout the design process and approaches should be adopted that secure the involvement of engineers with ergonomics. Such ergonomics input will aid substantially in the identification of problems that could be resolved early on in order to reduce the likelihood of serious difficulties emerging that will be costly to change later. The timeliness of human factors should thus be an essential consideration in project management.

Ideally, organizations will employ and respect their own in-house ergonomics expertise. However, in large engineering development projects, it may be appropriate to buy in expertise for the duration of the project. Whichever way ergonomics is utilized, all members of a design team should appreciate the role that human factors plays in contributing to system development so that their own specific design decisions are appropriately informed and that designers properly understand the compromises they must make both to accommodate ergonomics principles and to compromise their aspects of design with those of colleagues. In this way, human factors can justify its input by complying with system goals of safety and operability. Engineers and managers, responsible for system performance, can then be satisfied that their real needs are being addressed.

References

Gregson, D., Marshall E.C., Gait A., and Hickling E.M. 2003, *Providing Ergonomics Guidance to Engineers when Designing Human-machine Interfaces for Nuclear Plant Installations* Paper Presented at OECD Workshop on Modifications at Nuclear Power Plants –Role of Human Factors in Paris 2003.

Gregson, D. and Gait, A. 2004, *Ergonomics Guidelines – A Help or a Hindrance?* In this Volume.

Harvey, R.S. and Sandom C.W. (Eds.) 2003, *Human Factors for Engineers Textbook.* (Institute for Electrical Engineers, London).

Shepherd, A. and Marshall, E.C. 2003, *Task Specification in Designing for Human Factors in Railway Operations* Paper presented at 1st European Conference on Rail Human Factors, York, October 2003.

Shepherd, A. 2001. *Hierarchical Task Analysis*. (Taylor and Francis, London).

IMPROVING THE USABILITY OF MOBILE PHONE SERVICES USING SPATIAL INTERFACE METAPHORS

Mark Howell[1], Steve Love[1], Mark Turner[2] and Darren Van Laar[2]

[1]*Department of Information Systems and Computing, Brunel University, Uxbridge, Middlesex, UB8 3PH*
[2]*Department of Psychology, University of Portsmouth, King Henry Building, King Henry I Street, Portsmouth, PO1 2DY*

This study compared the usability of 3 different metaphor-based versions of a speech activated mobile city guide service. The effects of individual differences on attitude towards and performance with the mobile service were also explored. A Wizard of Oz methodology was used to provide the service functionality. All participants completed tasks with a standard control service, and then with one of the 3 metaphor-based services (shopping, office filing system, and transport system). Subjective and objective measures were recorded and analysed to compare services. The office filing system service was rated as being the most usable, and the transport system service least usable. Correlation data suggested that verbal ability and previous telephone experience may be important factors to consider when developing metaphor-based services.

Introduction

Mobile phones allow users convenient access to information using automated mobile phone services, which can be speech activated to allow the user hands free interaction. However, problems arise with these services because speech is produced in a slow linear nature, and the menu options are arbitrarily assigned to options. This results in slow interaction and a burden on short-term memory, which can cause a reduction in usability. It is possible that the use of a spatial interface metaphor may alleviate some of these usability problems by allowing users to visualize the service structure, and to navigate using their spatial properties.

In this study, a hierarchically structured, speech-based mobile city guide service was designed, which allowed users to access city information by listening to auditory service messages, and then to navigate to the required information by verbally selecting menu options. A standard version, using numbered menu prompts, and 3 different metaphor-based versions of the city guide service were evaluated. The 3 metaphors were 'shopping', 'office filing system', and 'transport system', and were derived from a previous study conducted by Howell *et al* (2003), which investigated the types of metaphors that most closely matched users' conceptions of the structure of speech-based automated phone services.

The current study formed part of an iterative process of designing the dialogue and structure for the 3 metaphor-based services, which would later be tested in a mobile context.

The 3 main objectives of the study were (1) To investigate whether the use of different interface metaphors led to an improvement in usability. (2) To investigate whether there were any differences in usability between the 3 metaphor-based services. (3) To investigate whether verbal ability, spatial ability, and previous telephone and computing experience affected participants' performance with and attitudes towards the metaphor-based services.

Interface metaphors

The case made for the use of metaphors is that they reduce the time and effort necessary for new users to learn to use a system (Carroll and Mack, 1985). An effective interface metaphor is one that is appropriate, explicit, and quickly understood, and will lead the user to develop a mental model of the system that is closely related to the system image. Humans have well-developed spatial abilities as a result of navigating real world spatial environments, and a number of studies have reported that users also like to organise computer-based information spatially (e.g. Jones and Dumais, 1986). The only previous work conducted on the use of speech-based spatial metaphors for hierarchically structured automated telephone services was by Dutton *et al* (1999). They found improvements in usability for a department store metaphor-based service suggesting that interface metaphors may indeed provide a suitable conceptual model for the design of automated mobile phone services.

Individual differences

The range of human performance on computing tasks is much greater than on most other work tasks, and can play a major role in determining whether humans can use an interactive system to perform a task effectively (Egan, 1988). When using automated phone services verbal ability may affect how well a user will comprehend the information provided by the service at each stage of the interaction. Love *et al* (1997) found that the performance of low verbal ability users was significantly poorer with an automated music.catalogue telephone service than high ability users.

High spatial ability is known to correlate with improved navigation of hierarchically structured graphical user interfaces (Vincente *et al*, 1987). However, spatial ability has been found to have a limited effect on performance with automated telephone services, largely because they present information serially and do not contain spatial cues. The spatial interface metaphors used in the present study require users to form an internal visualisation or 'cognitive map' of the service structure to aid the completion of tasks. The use of such mental devices is thought to be closely related to spatial ability (e.g. Thorndyke and Hayes-Roth, 1982), suggesting spatial ability might also be correlated with measures of usability.

Prior domain knowledge has emerged as an important predictor of performance with a system (Egan, 1988). Maglio and Matlock (1998) investigated people's metaphorical conceptions of the World Wide Web and discovered that novice web users tended to use mixed metaphors more often than experienced users. Previous domain experience may therefore impact metaphor preference and usage, and led to the investigation of previous mobile phone, fixed line telephone, and computing experience as part of the current study.

Methodology

Design

The design was a 2 (trial) x 3 (metaphor) mixed factorial design. The between subjects factor was the version of the service, and the within subjects factor was the trial. There were 4 different versions of the service, 3 of which were metaphor-based (shopping, office filing system, and transport system), and one of which was designed without reference to any metaphor. The non-metaphor service was designed by simply pairing numbers with menu options, and will be referred to as the standard service.

Participants

Twenty participants took part in the study, consisting of 1 male and 19 females with ages ranging from 20 to 26. All participants had prior experience of using automated telephone services. The participants were divided into the 3 experimental groups with 6 participants assigned to each of the shopping and office filing system services, and 8 participants assigned to the travel system service. For the control trial (trial 1), all 3 groups completed tasks using the standard service, and for trial 2 each group used one of the metaphor-based services.

Apparatus

A Wizard of Oz (WOZ) methodology was used for the experiment. The technique involved the experimenter simulating the functionality of a fully implemented system to create the illusion that the user was interacting with a real telephone service. Each of the services was designed as an HTML page, and the experimenter clicked on hyperlinks to play the appropriate sound files to the participants at the other end of the phone line. The participants both listened to the service prompts and gave responses using a telephone headset.

Data collection

Subjective attitudes towards each implementation of the service were recorded using a Likert style usability questionnaire balanced for both positively and negatively worded statements. Two objective measures of task performance were collected during the participants' interaction with the services: successful task completion, and time to complete the tasks as a percentage of the optimum path prompt time. The second of these measures was the actual time taken as a percentage of the total prompt time if the optimum path (lowest number of nodes) was used, but will simply be referred to as 'total time' for the purposes of this paper.

Procedure

Each participant was tested individually within a 1-hour session. Firstly, participants completed the AH4 test (Heim, 1970) to measure verbal and spatial ability, and a technographic questionnaire to gather data about age, gender, previous mobile phone, fixed line telephone, and computing experience. Secondly, participants were called on the landline telephone in their experimental cubicle, and had to complete a practice task followed by 3 tasks using the standard service. Each task required the participants to find a specific piece of information, for example, 'Find the names of 2 wine bars that close at 11pm and then exit the service'. Participants were then asked to complete a usability questionnaire. Task times were recorded using a stopwatch. Finally, participants were called a second time to complete 3 tasks using one of the metaphor-based services. Again, task times were recorded, and on completion of the tasks participants were asked to complete a second usability questionnaire.

Results

Subjective measures

Paired samples t-tests were performed to investigate differences in subjective measures of attitude between the standard service and the 3 metaphor-based services. The only significant difference was between the transport system service and the standard service (Table 1), suggesting that participants perceived the usability of the transport system service to be significantly lower. To investigate differences in usability between the 3 metaphor-based services, a one-way ANOVA was performed. Although no significant differences were found, the office filing system service scored the highest usability mean, which was only 2% lower than the mean score for the standard service, compared to a 4 % lower score for the shopping service group, and a 20% lower score for the transport system service group.

Table 1. T-test results for the 3 experimental groups

Experimental group	Trial number	Mean	SD	t	df	Sig. (2-tailed)
Shopping service	1	3.57	0.84	0.42	5	n.s.
	2	3.42	0.82			
Office filing system	1	4.42	0.77	0.19	5	n.s.
	2	4.33	0.92			
Transport system	1	4.30	0.83	2.38	7	<0.05
	2	3.44	0.74			

Objective measures
Differences in 'total time' between the standard service and the 3 metaphor-based services were explored using paired samples t-tests. Although no significant differences were found, the time taken to complete tasks was lower for all of the metaphor-based services than it was when using the standard service. In order to investigate whether there was a significant difference in 'total time' between the 3 metaphor-based services, a one-way ANOVA was performed. No significant differences were apparent, but the greatest improvement in performance was shown by the office filing system service, with participants taking 22% less time to complete tasks than when they used the standard service, compared to 17% less time for the shopping service, and 6% less time for the transport system service. Successful task completion rates were high across all 3 groups because participants tended to persevere until they found the relevant information. However, participants using the office filing system service were the only group to achieve 100% successful task completion across all 3 tasks.

Individual differences
The technographic data was calculated as 3 separate variables: mean mobile phone, fixed line telephone, and computer scores. Verbal and spatial ability, and the 3 technographic variables were then correlated against the mean attitude scores, and the 'total time' mean scores. No significant correlations existed for the shopping service. For the office filing system service, there was a significant negative correlation between verbal ability and attitude ($r=-0.848$; $n=6$; $p=0.03$), indicating that as verbal ability increased so attitudes towards the usability of the service became more negative. For the transport system service, there was a significant positive correlation between fixed line telephone and 'total time' ($r =0.788$; $n=8$; $p=0.02$), indicating that as fixed line telephone experience increased the time taken to complete tasks also increased. No other significant correlations were found.

Conclusions

Of the metaphors examined, the filing cabinet service emerged as being the service that was perceived most positively, and generated the best performance levels, with the transport system service demonstrating the lowest levels of usability. However, none of the metaphor services were rated as being significantly better than a standard phone service. Since contemporary (numbered menu) interface designs for automated phone services are so well established, it will be necessary for users to invest time and effort to learn and accept new metaphor-based services. It may therefore be considered encouraging that the usability of the office filing system service was rated so highly given the novel nature of this system and the limited exposure afforded to participants in the present study. It can be speculated that with more trials, and more time to explore and learn the metaphor, the office filing system service may have emerged as being more usable than the standard service. Future studies will therefore evaluate the usability of the office filing system service over longer periods of time, and in different mobile contexts.

The correlation between verbal ability and attitude for the office filing system service may suggest that high verbal ability participants became engaged by the service, but were frustrated by the limited input options, possibly preferring a more conversational style of interface. The poorer performance with the transport system service by participants with high fixed line telephone experience may be explained with reference to mental models. Such participants may have strongly developed mental models of automated telephone services, and were not capable of fully accommodating the transport system service design, which represented the most detailed metaphor system and the most radical departure from standard telephone systems. This may have led to poorer performance times and attitudes. These associations highlight the need to investigate the effects of individual differences on the usability of metaphor-based systems.

References

Carroll, J. and Mack R. 1985, Metaphor, computing systems and active learning, *International Journal of Man-Machine Studies*, **22**(1), 39-57.

Howell, M.D., Love. S., Turner, M. and Van Laar, D.L. 2003, Interface metaphors for automated mobile phone services. *Proceedings of HCI International '03*, 10th International Conference on Human-Computer Interaction, 128-132.

Jones, W.P. and Dumais, S.T. 1986, The Spatial Metaphor for User Interfaces: Experimental Tests of Reference by Location Versus Name. *ACM Transactions in Office Information Systems*, **4**(1), 42-63.

Dutton, R., Foster, J. and Jack, M. 1999, Please mind the doors–do interface metaphors improve the usability of voice response services? *BT Technology Journal,* **17**, 172-177.

Egan, D. 1988, Individual differences in Human-Computer Interaction. In Helander, M (ed.) *Handbook of Human-Computer Interaction*, Amsterdam: Elsevier Publishers, 543-568.

Love, S., Foster, J. and Jack, M. 1997, Assaying and isolating individual differences in automated telephone services, *Proceedings of the 16th International Symposium on Human factors in Telecommunications (HFT'97)*, 323-330.

Vincente, K., Hayes, B. and Williges, R. 1987, Assaying and isolating individual differences in searching a hierarchical file system, *Human Factors*, **29**(3), 349-359.

Thorndyke, P. W. and Hayes-Roth, B. 1982, Differences in spatial knowledge acquired from maps and navigation. *Cognitive Psychology*, **14**, 560-589.

Maglio, P. P. and Matlock, T. 1998, Metaphors we surf the web by. In *Workshop on Personal and Social Navigation in Information Space*, Stockholm, Sweden, 138-149.

Heim, A. W. 1970, *AH4 Group Test of General Intelligence*, UK: NFER.

INTERFACE DESIGN

FREE-SPEECH IN A VIRTUAL WORLD: A USER-CENTRED APPROACH FOR DEVELOPING SPEECH COMMANDS

Alex W. Stedmon, Sarah C. Nichols, Harshada Patel, & John R. Wilson

Virtual Reality Applications Research Team (VIRART)
School of 4M
University of Nottingham
Nottingham NG7 2RD

In many speech recognition applications, users are required to follow strict rules and vocabularies that can sometimes affect system performance. In principle, if a speech metaphor is well designed then it can be intuitive, taking the burden of interaction away from physical input devices and increasing opportunities for multi-modal interaction. In order to investigate the concept of developing intuitive speech commands, a trial was conducted where participants were allowed to speak freely, using any words they wished, for navigation and object manipulation within a virtual reality training application. The underlying principle was that free-speech is more intuitive as it does not have to be pre-learnt. By analysing the commands people naturally use, it might then be possible to identify the most and least common used commands as well as develop generic command vocabularies for future use.

VIEW of the future

Virtual and Interactive Environments for Workplaces (VIEW) of the Future is a project that seeks to develop best practice initiatives for the appropriate implementation and use of virtual environments (VEs) in industrial settings for the purposes of product development, testing and training. As part of the VIEW project a special interest group has been established representing key areas ranging from the technical integration of speech recognition hardware into VEs, through to the potential human factors issues associated with using speech as an input device in virtual reality (VR) applications.

Speech as an input device

Input devices can be defined as the medium through which users interact with a computer interface and, more specifically in the context of the present research, a VE (Stedmon *et al*, 2003). Currently, there is an increasing variety of input devices on the market that have been designed for VR use, which range from traditional mouse and joystick devices, to wands, data-gloves, and speech.

Speech offers the systems designer a ready input modality for Human-Machine Interaction (HMI), however, much of the impetus for the development of speech recognition technology has come from the belief that speech exists as an untapped resource. This is a contentious issue, for speech is already an active or semi-active mechanism in most working environments (Linde & Shively, 1988). Another important issue is that using a speech recognition system is demanding in itself and as such it may be necessary to "review the notion of 'eyes free/hands free' operation as [sole] justification" for such systems (Baber *et al*, 1996). It is only when the underlying human factors issues begin to be understood, that the usability of speech as an input device can be enhanced.

Free-speech in a virtual world

Of particular importance in the VIEW project is the need to examine which input devices are best suited to conducting specific tasks within VR applications. With such a variety, users may select an inappropriate input device which could compromise the overall effectiveness of a VR application and the user's satisfaction in using the application.

In future VR applications, speech might be used as a replacement for input devices such as mouse, joystick or keyboard. As such, speech could allow for hands-free interaction and might be useful for activities such as short cuts and menu commands; and, furthermore, speaker independent recognition systems could support multi-user applications in collaborative VEs allowing many users to interact with each other and a VE at the same time (Stedmon, 2003).

In principle, if a speech metaphor is well designed then it can be intuitive, taking the burden of interaction away from physical input devices and increasing opportunities for multi-modal interaction (Stedmon *et al*, 2003). In many speech recognition applications, users are required to follow strict rules and vocabularies that can sometimes affect system performance. Indeed, the match between the language people employ when using a computer system and the language that the system can accept, or the 'habitability' of a system, is a key issue in system usability (Hone & Baber, 2001). Furthermore, in situations where acute time pressure is a potential stressor, there is a danger that users may revert to pre-learnt vocabularies with the result that a speech recognition system might not recognise, or may even misinterpret, illegal syntax or unfamiliar vocabulary (Fitts & Seeger, 1953).

In order to investigate the concept of developing intuitive speech commands, a trial was conducted where participants were allowed to speak freely, using any words they wished, for navigation and object manipulation within a VR training application. The underlying principle was that free-speech is more intuitive as it does not have to be pre-learnt. By analysing the commands people naturally use, it might then be possible to identify the most and least common commands used, as well as develop generic command vocabularies for future use.

Participants
6 male and 6 female participants took part in the trial. Participants ranged from 20 - 52 years with a mean age of 31.6 years. All participants had English as their first language and normal or corrected to normal vision. Most participants were naive VR and speech recognition system users.

Apparatus
The VR system comprised of an 800 MHz laptop PC, running Superscape software, with a data-projector and a front projection screen. Participant input was via a hand held microphone, whilst experimenter interaction with the VE was via a joystick. The VE used for the trials was 'Netcard', which has been developed by the University of

Nottingham to assess the effectiveness of VR as a training tool for computer network card replacement.

Design & Procedure

Each participant conducted the trial once. A number of task-related measures were collected including: time taken to complete the task; total number of commands used; number of different commands used; and mean number of words in each command.

Within the 'Netcard' VR application, participants were required to change a PC netcard using speech to control their actions. In order to complete the task, participants had to navigate to a desk with a computer on it; remove the computer cover; take out the old network card and replace it with a new card; and replace the cover. Participants were given instructions for completing the task on an external overlay directly outside the VE and were also assisted by the use of picture icons (also on the external overlay) in order to see what they were holding. Participants believed they were speaking to a computer via a speech recognition system, however, an experimenter was concealed behind a screen who carried out the spoken commands to interact with the VE. Participants were only informed that they were speaking to the experimenter at the end of the experiment.

Results

In total, 348 different commands were used by the participants to complete the task. Descriptive statistics for the task related variables are shown in Table 1 below.

Table 1: Descriptive statistics for task related variables

Task Variable	Units	Mean Score	St Dev	Range
Time to complete task	mins:secs	35:18	15:17	18:42 – 61:19
Total no of commands	commands	237.5	87.16	128 - 415
No of diff commands	commands	55.5	13.56	37 - 72
No of words per command	Words	2.16	0.55	1.74 – 3.62

Discussion

From the total number of commands that the participants used to complete the task, it is apparent that when users are allowed to use free-speech the range of commands is very wide. Looking in more detail at the task related data, the time taken to complete the task seemed to relate to navigation errors and orientation problems, rather than object manipulation. Furthermore, the total number of commands used was an indication of how accurately participants completed the trial, as some of the tasks were only possible with a particular viewpoint in the VE. The number of different commands indicated that participants repeated commands they had used previously in the trial and so built up a command structure that they could re-use. The standard deviations for these task related variables was high and this indicates the subjective nature by which participants developed vocabularies in order to complete the trial. It was only the mean number of words per command that showed a low standard deviation, which indicated that participants used a similar style of speech (short and direct commands) when they believed they were speaking to a computer.

By analysing the most common and least common commands it is possible to begin to develop ideas regarding generic command structures. Movement in the VE was continuous, and so the most common command used was 'stop'. If movement in the VE was discrete then this command would hardly have been used although actual navigation commands would have increased as there is a trade-off in using speech for continuous or discrete movement in a VE. The most common navigation commands were 'move', 'look', 'turn', 'walk', 'go' combined with supplementary commands such as 'forwards', 'backwards', 'right', or 'left'. The least common navigation commands were 'side step', 'zoom in', 'shuffle', and 'straife'. Whilst these commands are still quite general, they were not used very often and were therefore not as intuitive as the more common commands.

In relation to object manipulation, the most common commands were 'pick up', 'put down', 'open', combined with supplementary commands such as 'screwdriver', 'screws', 'door'. The least common commands for object manipulation were 'take out', 'hold', 'use'. Such commands use relative terms to other items in the VE and so, in general, participants may have refrained from using such terms because they believed they were talking to an inanimate object such as a computer.

Overall, it would be difficult to develop a specific command vocabulary due to the degree of variance in the commands that participants used. This trial would seem to support the idea that a general command structure could be developed based on the most common commands that users might therefore find most intuitive. Speech is very much context specific and so task vocabulary, by its nature, will be application specific.

Kalawsky (1996) argues that the design of an input device should match the perceptual needs of the user. As such, the integration of input devices should follow a user needs analysis to map their expectations onto the attributes of the overall VR system. Jacob et al, (1993) recommends that the device should be natural and convenient for the user to transmit information to the computer and further, Jacob et al, (1994) suggest that input devices should be designed from an understanding of the task to be performed and the interrelationship between the task and the device from the perspective of the user.

Building on a sound understanding of user needs, it is important, therefore, to analyse the task in the correct level of detail so that the VR system and the VE that is developed supports user interaction and overall application effectiveness. As such, Barfield et al, (1998) argue that a VE input device should account for the type of manipulations a user has to perform and that the device should be designed so that it adheres to natural mappings in the way in which the device is manipulated, as well as permit movements that coincide with a user's mental model of the type of movement required in a VE.

The results from this trial indicate that speech was considered to be beneficial for discrete tasks, object manipulation and 'short cuts' whilst anecdotal evidence suggests that speech may not be best suited for specific actions such as navigation (Stedmon et al, 2003). As such, a combination of speech with other input devices might offer users with a more flexible and integrated set of interaction tools for VR applications of the future.

Conclusion

Within the VIEW project a programme of research includes the exploratory testing of existing applications and interaction devices as part of an iterative design process in order to provide guidelines for the development of new interfaces; as well as more traditional experiments that will ultimately feed into the creation of an interactive design support tool for VE developers. This will integrate the knowledge gained from the project (e.g. guidelines, code of practice, best use guidance, examples) for the specification, development and use of VEs in general, particularly workplace oriented VE applications. This research will not only be an exploration of the key issues of using speech recognition technology in a VE but will hopefully lend itself to the development of protocols for VE applications, through the lifespan of the VIEW Project, and beyond.

Acknowledgements

The work presented in this project is supported by IST grant 2000-26089: VIEW of the Future.

References

Baber, C., Mellor, B., Graham, R., Noyes, J.M., & Tunley, C., (1996). Workload and the use of Automatic Speech Recognition: The effects of time and resource demands. *Speech Communication*, 20, 37-53.

Barfield, W., Baird, K., & Bjorneseth, O. (1998). Presence in virtual environments as a function of type of input device and display update rate. *Displays*, 19, 91-98.

Fitts, P.M., & Seeger, C.M. (1953). S-R compatibility: spatial characteristics of stimulus and response codes. *Journal of Experimental Psychology*, 46, 199-210.

Hone, K.S., & Baber, C. (2001). Designing habitable dialogues for speech-based interaction with computers. *International Journal of Human-Computer Studies*, 54(4), 637-662.

Jacob, R., Leggett, J., Myers, B., & Pausch, R. (1993). An agenda for human-computer interaction research: interaction styles and input/output devices. *Behaviour and Information Technology*, 12(2), 69-79.

Jacob. R., Sibert, L., McFarlane, D. & Mullen Jr., P. (1994). Integrality and separability of input devices. *ACM Transactions on Computer-Human Interaction*, 1(1), 3-26.

Kalawsky. R. (1996). Exploiting virtual reality techniques in education and training: Technological issues. *Prepared for AGOCG*. Advanced VR Research Centre, Loughborough University of Technology.

Linde, C., & Shively, R. (1988). Field study of communication and workload in police helicopters: implications for cockpit design. In, *Proceedings of the Human Factors Society 32nd Annual Meeting*. Human Factors Society, Santa Monica, CA. 237-241.

Stedmon, A.W., Patel, H., Nichols, S.C., & Wilson, J.R. (2003). A view of the future? The potential use of speech recognition for virtual reality applications. In, P. McCabe (ed). *Contemporary Ergonomics 2003*. Taylor & Francis Ltd. London.

Stedmon. A.W. (2003). Developing Virtual Environments Using Speech as an Input Device. . In, C. Stephanidis (ed). *HCI International '03. Proceedings of the 10th International Conference on Human-Computer Interaction*. Lawrence Erlbaum Associates.

THE FLIGHTDECK OF THE FUTURE: FIELD STUDIES IN DATALINK AND FREEFLIGHT

Gemma Cox[1], Sarah C Nichols[1], Alex W Stedmon[1], John R Wilson[1], & Helen Cole[2]

[1]*Virtual Reality Applications Research Team (VIRART), University of Nottingham Nottingham NG7 2RD, UK*

[2]*Human Factors Rolls Royce Naval Marine RAY-3W-203 PO Box 2000 Derby, DE21 7XX*

New technology, such as communication via datalink, or pilot mediated air traffic control (ATC), will have an impact on ATC-Pilot collaboration. The Flightdeck and Air Traffic Control Collaboration Evaluation (FACE) project, funded by the Engineering and Physical Sciences Research Council (EPSRC), takes a systems perspective to investigate the human factors requirements for the flightdeck of the future in relation to the two specific elements of flightdeck IT: datalink and freeflight. The project employs a joint cognitive systems (JCS) approach to the collection of data. This paper summarises a number of approaches to data collection and reviews their suitability for use within the FACE project. In addition to this, transcript analyses are reviewed in relation to key themes identified by preliminary fieldwork and literature reviews.

Introduction

Within the flightdeck-air traffic control (FD-ATC) system a distributed, collaborative, decision-making network exists whereby the goals of safety and efficiency are mutual but the preferred tactics/procedures used by each part of the team may be different (Stedmon *et al*, 2003). From this perspective, a joint cognitive system emerges incorporating a number of operators and a number of systems (Hollnagel, 2001).

Using a joint cognitive system (JCS) approach the representation of knowledge is examined as well as the propagation of knowledge through the system. As such, the external representations of knowledge can be used to assess what the internal representations (or the cognition of the individual actors) may be. Communication between air traffic controllers (ATCOs) and between ATCOs and the flightdeck can be viewed as a system of representations and these communications can only be understood as a function of the system that they exist within (Fields *et al*, 1998).

Methods

In order to identify how such changes will impact the human factors issues identified in preliminary work, it is important to apply appropriate research methods. It follows that most JCS analyses depend on detailed observation of the functional system under investigation, along with interviews with the actors within the JCS. The following section summarises a number of approaches to data collection and analysis.

Direct observation - there are many different approaches to observation; these include the degree of structure to the information being gathered. The way the information is gathered can be either informal, which takes the form of note taking, diaries, etc; or formal, where a large amount of structure is imposed on the information that is gathered. High reliability and validity are easier to achieve with formal information gathering methods, but the informal methods produce a more complex and complete account of the group being observed. The foremost advantage of observation is its directness; the evaluator does not ask about people's views, feelings, or attitudes, instead he/she watches the actions of those people. Despite being a very useful tool, a major problem with direct observation is the extent to which the presence of the evaluator affects the situation under observation.

Video analysis - these have a number of advantages over real time observation, the video may be analysed repeatedly in order to extract the relevant information, and the data derived from analysis is precise and verifiable (Laws & Barber, 1989). Another advantage of video analyses is their objectivity; as the same situation can be viewed by a number of evaluators to ensure that objectivity has been achieved. A potential problem is that any single analysis is dependent on the objectivity of the observer.

Video analysis is the procedure by which particular significant events may be extracted and complex behavioural actions may be translated into manageable data sets for statistical analysis (Laws & Barber, 1989). Analysing, coding and transcribing video recording is extremely time-consuming, although there are computer supported tools for video analysis to facilitate the job of the analyst and reduce the high ratio of analysis time to actual observation time.

Interviews - there are many different types of interview including structured, semi-structured, unstructured, and contextual. A structured interview is a formal, planned version of an interview, which is highly directed by the interviewer. A semi-structured interview is still directed by the interviewer's desire to cover specific topics, but the exact wording and sequencing of questions is given more freedom. There are different styles of unstructured interview, during non-directive interviews the informant controls/has total control over the areas covered. The focused interview gives the interviewer some level of control, and is usually conducted after some kind of situational analysis (Robson, 1993). The structured transcripts produced may be easier to analyse than those resulting from unstructured interviews but preparing for the interview requires much more work from the interviewer (Shadbolt, 1995). A general interview guide is then developed focusing on the major research areas. Unstructured interviews require much more skill on the part of the interviewer as they need to be able to probe deeper into issues of particular interest and be flexible with the progression of investigation. A contextual interview is a specific interview technique where the system investigated is used as a bridge to augment the interview (Holtzblatt & Jones, 1995). The contextual interview approach entails interviewing the interviewee in context, with them actually carrying out their daily work activities. The benefits include giving the user a concrete frame of reference, so they do not have to generalise from their experiences and speak in abstract terms.

Questionnaires and surveys - questionnaires are written interviews that are completed by respondents. This saves a considerable amount of researcher time and enables the questionnaire to be sent out to hundreds or even thousands of people in a short amount of time. Problems with questionnaires include their superficiality as there is little or no checking that the responses are serious (Robson, 1993). They need to be very carefully constructed to result in meaningful data so that questions and scales are easily interpreted and reduce ambiguity.

Table 1 reviews the suitability of the above methods in the context of research into ATC-flightdeck collaboration. The data within table 1 was elicited during an expert review session where those present had experience of applying such techniques, and illustrates methodological characteristics that may affect the feasibility of application of such approaches in field data collection. This table was in turn used as a basis for selection of methods within the FACE project.

Table 1: Matrix of method utility for FACE

	Direct observation	Video recordings	Structured Interviews	Semi-structured Interviews	Informal Interviews	Contextual Interview	Questionnaire survey
2+ people	Yes if more than one evaluator	Yes	No	No	No	Yes	Yes
2+ domains	Very difficult	Very difficult	No	No	No	Very difficult	Yes
2+tasks	Yes	Yes	Yes	Yes	Yes	Yes	Yes but must design questionnaire well
Communication	Yes	Yes	No	No	No	Yes	Difficult
Dynamic information	Possible if more than one evaluator	Yes	No	No	No	Yes	No
Time	Real time, depends on analysis	Real time, need to allow x10 of real time for analysis	Need to allow x3 of real time for analysis	Need to allow x10 of real time for analysis	Need to allow x10 of real time for analysis	Need to allow x10 for analysis	Quick if organised, analysis quick
Access required	Yes	Yes, but can be done by agent	Yes for people but don't need access to site	Yes for people but don't need access to site	Yes for people but don't need access to site	Yes	No but need some way of distributing questionnaires

Transcript analyses

The above techniques are all useful in collecting data, however the mapping of data to theory is more complicated related to communication or work within the joint cognitive systems. Hutchins & Klausen (1996) demonstrate the use of ethnographic techniques for evaluating the cognitive and communicative processes that take place within the aircraft system. The distributed nature of these processes is examined and the propagation and re-representation of information around the system is illustrated.

Within the FACE project transcripts of FD-ATC communications have been evaluated in order to examine the propagation and breakdown of knowledge through the JCS. The transcripts provide both the FD-ATC communications as well as the cockpit voice recording (CVR) giving the intra-crew communications on the flightdeck. The shift from verbal communication via the radio telephone (R/T) channel to a more visually based datalink is one of the key issues that the FACE project is investigating. The analyses of transcripts and reviews of the literature have indicated that informal work practices such as the flightdeck crew's overhearing of FD-ATC communications are instrumental to the co-ordination of work within the JCS (Roger & Ellis, 1994; Hutchins & Klausen, 1996).

Verbal communication via the R/T channel supports the distributed access of information within the JCS. Sharing access of information from ATC allows the co-ordination of expectations on the flightdeck, which in turn contributes to the formation of shared mental models. Without access to this information it would be virtually impossible for distributed actions to be co-ordinated and this must be incorporated into datalink technology in order to preserve this sharing access of information (Rogers & Ellis, 1994).

Analyses of the transcripts have also supported the present literature confirming that misunderstandings in R/T communications are very common (Cushing, 1994). The following example from an EgyptAir transcript (National Transportation Safety Board, 2000) illustrates a misunderstanding occurring:

ATC:	*EgyptAir nine ninety heavy turn left then proceed direct to SHIPP*
FO:	*direct SHIPP, nine nine zero heavy.*
FO:	*turn right direct SHIPP, execute sir?*
Captain:	*did he tell you turn left?*
FO :	*yes, establish.*

This example also highlights the need for shared access to information on the flightdeck. Here the first officer (FO) has misheard the ATC communication but due to the shared access of this verbal information on the flightdeck, the Captain is aware of the anomaly and is able to realign their expectations and the shared knowledge of the members of the flightdeck.

Analyses of transcripts allow the researchers to identify problems with the existing technology (misunderstandings/mishearing of R/T communications) as well as predict potential problems of future technologies (intra-crew communication) with the overall aim of supporting the collaboration and co-ordination of work activities within the JCS.

Future work

Examination of the transcripts has identified critical issues such as intra-crew communication and R/T misunderstandings that require further structured examination. Whilst being useful, the transcripts only describe a small part of the overall JCS structure. Simulator trials will be conducted within the project in order to provide a richer data source and will examine the communicative pathways within the JCS. Simulator trials will also allow knowledge propagation within the functional system to be investigated including non-verbal communications and transformation of information between different modes (such as verbal communications converted to key presses), resulting in a mapping of the shared and individual knowledge and the means by which it is represented, accessed, communicated, adapted and used within the JCS (Rogers & Ellis, 1994). The combination of simulator studies, real-world observation techniques and experimental trials provides a strong basis for investigating the impact of new technological initiatives on the present JCS.

References

Cushing, S. (1994). *Fatal Words: Communication Clashes and Aircraft Crashes*. The University of Chicago Press, Chicago

Fields, R.E., Wright, P., Marti, P., & Palmonari, M. (1998). Air Traffic Control as a Distributed Cognitive System: A Study of External Representations. *Proceedings of the Ninth European Conference on Cognitive Ergonomics*, Edited by T.R.G. Green, L. Bannon, C.P. Warren and J. Buckley. European Association of Cognitive Ergonomics (EACE), Le Chesnay, France.

Hollnagel, E. (2001). Cognition as Control: A Pragmatic Approach to the Modeling of Joint Cognitive Systems. *Special issue of IEEE Transactions on Systems, Man and Cybernetics A: Systems and Humans.*

Holtzblatt, K., & Jones, S. (1995). Conducting and analyzing contextual interview. In R. M. Becker & J. Grudin & W. A. S. Buxton & S. Greenberg (Eds.), *Human Computer Interaction: Toward the year 2000* (2nd ed., pp. 241-253). San Francisco, California: Morgan Kaufmann Publishers, Inc.

Hutchins, E., & Klausen, T. (1996). Distributed cognition in an airline cockpit. In Y. Engestrom & D. Middleton (Eds.), *Cognition and Communication at Work* (pp. 15-34). Cambridge: Cambridge University Press.

Laws, J. V., & Barber, P. J. (1989). Video analysis in cognitive ergonomics: a methodological perspective. Ergonomics, 32 (11), 1303-1318.

National Transportation Safety Board (NTSB).Vehicle Recorders Division. Specialist's Factual Report of Investigation (2000), *DCA00MA006* by Albert G. Reitan.

Robson, C. (1993). *Real World Research*. Oxford: Blackwell Publishers Inc.

Rogers, Y., & Ellis, J. (1994). Distributed Cognition: An Alternative Framework for Analysing and Explaining Collaborative Working. *Journal of Information Technology* 9 (2), 119-128.

Shadbolt, N., & Burton, M. (1995). Knowledge Elicitation: A Systematic Approach. *Evaluation of Human Work*, Edited by J.R. Wilson and E.N. Corlett. Taylor & Francis, London, 2nd Edition.

Stedmon, A.W., Nichols, S.C., Cox. G., Neale, H., Jackson, S., Wilson, J.R. & Milne, T.J. (2003). Framing the Flightdeck of the Future: Human Factors Issues in Free flight and Datalink. *HCI International '03. Proceedings of the 10th International Conference on Human-Computer Interaction*. Lawrence Erlbaum Associates.

Dyslexia and Voice Recognition Software - Really the Perfect Match?

Anne Nelson and Caroline Parker

Intuitive Internet, 13 Gillespie Drive, Helensburgh, G84 9BL
anne@intuitiveinternet.com

Centre for Research into Systems and People, Computing Division, Glasgow Caledonian
University, Glasgow G4 0BA

Voice Recognition (VR) Software is commonly recommended as an assistive software tool for people with dyslexia. This paper investigates the current use of these software tools, both in the home and in education. A study in 1999 found that there were severe usability problems with the technology (O'Hare and McTear, 1999). Since then however there have been considerable improvements. This study replicates the earlier one to see whether it has improved and what the main usability issues are. The results suggest that despite voice recognition software being a more attractive and viable option than before, it still has many serious usability issues to overcome before it can be recommended.

Introduction

Voice Recognition Software (VRS) would appear to be an obvious choice for supporting users who have difficulty with spelling and writing, or using a standard keyboard. Using a microphone (usually mounted on a headset) it will translate spoken words into written text on the screen, and can read it back to you, as well as performing menu commands and various other instructions. It is often cited as a potentially powerful tool for children with Dyslexia (Miles, 1998; iANSYST Ltd)

VRS has been around for over 30 years, but has until recently been held back by its need for greater than average processing power. However the tremendous reduction in price of the technology has now made its introduction into schools feasible. How useful is it in practice? The most recently documented studies testing the suitability and accuracy of VRS were conducted in 1997 and 1999, respectively, using IBM's **VoiceType** V3.0 dictation package: (O'Hare and McTear, 1999), and **InCube** software: (Zhang, *et al*, Oct 1999). The study by Zang et al. however is not relevant to schoolchildren as the age range of the subjects was 20-30 years and the speed of development of the software now means that the O'Hare results are out of date.

This study examines the usability of the more recent VR technologies by replicating the O'Hare study with the most recent ViaVoice software. It also examines in more detail problems encountered by children with dyslexia when using computers, for everyday tasks, at school and at home.

Method

Replication of O'Hare study

Eleven pupils of comparable chronological (12-14yrs) and reading ages (9-10) with those who had undertaken the original study were selected. Testing was carried out over a period of three days, over two consecutive 55 minute sessions, with additional time for set-up and explanation of the tests and software. A short pilot study was undertaken.

The original format of the O'Hare and McTear study conducted in 1997, i.e. samples of text and pictures, was followed as closely as possible. All tasks were monitored for time taken and printed for analysis, Errors were counted and all the data was compared with the original study. **The two main tasks were:**

Structured Text Input (broken down as follows) -
- Hand-written dictation of a piece of text
- Typed dictation of the same piece of text into the computer using a standard keyboard
- Voice-dictation of the same piece of text into the computer using the VRS (following enrolment into the system and training each student to create their own personal voice model)

Spontaneous Text Input
- The children are shown a title for a story and a collection of pictures
- They are then asked to invent their own version of the story and spontaneously dictate it using the VRS

Web-based survey

A Web-based survey was also conducted, the purpose of which was to obtain information about VRS from dyslexic people of all ages and walks of life. The web address of the survey was posted on a large number of online forums and message boards.

Results

Replication study

Results from the practical testing showed that the latest version of the software was faster for voice input than the version used in the previous study. It also showed that voice input was, on average, up to eight times faster than typing, and five times faster than handwriting. However, the original study showed higher accuracy levels for voice input. This was mainly due to time factors relating to the level of training for each student's voice model within the software.

Closer examination of the voice dictation times shows just how much faster dictation times are now with the new version of the software. Average times recorded across all students show that dictation with the latest version was almost twice as fast as the older version (79 rather than 47 words per minute). Though slightly faster overall at all three tasks than their predecessors, the new students still only produced an average of 15 words per minute in the handwritten task, whilst achieving a slow 9 words per minute in the typing task. As far as accuracy is concerned, though, the new software performed worse than the original, averaging 55% accuracy, against 79% in the original study. Handwritten text was the most accurate, averaging over 85% accuracy in both studies.

One of the hardest things for the children to do in relation to the software was train it, training requires reading aloud from text on the screen. Given that many of these children struggle with reading (Orton, 1937/1989) and that reading from the screen tends to be harder than reading from the page (Elkind, *et al.* 1996) this is not a surprising result.

Web Survey Results

The survey generated 220 responses, 59% female, and 41% male, made up of a variety of age ranges. The majority (46%) were from the 22-40 range, while only 12% were under 16. 68% of those who responded had been formally diagnosed with dyslexia, and the other 32% seemed in no doubt that they were. Several comments from this group highlighted the fact that they were only made aware of their own symptoms after a child or sibling had been formally diagnosed.

Computer use
88% of those surveyed use computers every day. When asked what type of user they considered themselves to be, 75% were either experienced or advanced users, with only 7% regarding themselves as beginners. 73% also use computers at home *and* in the workplace.

It therefore seems that even though *all* of the respondents were either formally diagnosed (68%) or at least known to have symptoms of dyslexia, this has not led to an avoidance of computer use, rather, judging by the additional comments received; in most cases computers have made work (and life) much easier for them.

More than half of those surveyed (56%) had never tried any specialist tools, such as text editors, readers or voice recognition software. When asked *why* no specialist tools had ever been used, 58% said it was because they were unaware of their existence. This could also partly be caused by 52% - including those formally diagnosed – never receiving any kind of formal support for their dyslexia. Of the 44% who had tried specialist tools to assist them, the majority seem to have to have found them fairly easy to use *and* helpful. This would suggest that, were these tools more readily available, they would provide significant help. This is further highlighted by answers given to the question, *'What would encourage you to use/try any software support tools in the future?'* 40% indicated that more publicity and greater availability would encourage them, while 33% would require support and training on how to install and use the software.

Cost was also a priority, with 26% indicating they would try it if it were cheaper, or greater availability of trial versions.

Voice recognition use
Opinions of Voice Recognition software were less positive, however. Of the 20% of respondents who *had* tried VRS, 62% found them to be *very unhelpful* and *very difficult* to use, in comparison to specialist tools in general which 64% of users found either *very* or *quite* helpful or easy to use.

General discussion

The results are mixed. While the new software appears to have improved in terms of the speed with which it can recognise text the accuracy still leaves a lot to be desired. Both the increase in speed and degradation of accuracy appears to be related to the new mechanism that IBM has adopted for speech input. In the old system 'discrete' speech had to be used, slowing the reader down, whereas the new system allows for 'natural' or continuous speech. Practicing isolated speech gave the original students a chance to focus on the pronunciation of each word, encouraging them to speak more clearly. This would undoubtedly produce more accurate results. The new students, being asked to speak in their usual voice produced less accurate results as the software substitutes words it does not recognise with others, which in most cases have completely different meanings. This did however lead to much hilarity and was a great source of entertainment for most of the students.

Background noise is still difficult to filter, and has an effect on accuracy. If the noise is there consistently, the software *will* adapt to it, but in a classroom situation, where noise levels are changing constantly, this could be a problem.

Although voice dictation appears to be considerably less accurate from this short study, it does however become much more accurate over time. Taking that into consideration along with the extremely slow input times for typed and handwritten text, and even allowing additional time for correction of voice dictated text, it would still be possible for some dyslexic users to produce an accurate, word processed document in less than half the time it takes to type The productivity levels achieved by each student for both typed and voice dictated tasks shows just how significantly faster voice dictation is. However, by analysing each student's results individually and observation during the tests, it was clear that voice dictation would be extremely productive for some students; but not necessarily for others.

Summary

To summarise, the voice recognition software testing shows that:

- The method of 'discreet speech' used in the older version of the software produced much greater accuracy than the newest version, using 'natural speech'

- The interface design for the training procedure in the newest version is still too inflexible to be used by the majority of those with dyslexia – if they cannot read the training dialogue from the screen, they cannot record their voice model, therefore they cannot use the program
- On the positive side, the newer version was much faster than typing or handwriting for all of the students tested
- The children who produced the slowest times and least accurate documents in the handwritten and typed tests, produced the greatest accuracy and fastest times in the voice recognition tests
- Given more time to train the voice model accurately, and more practice reading aloud, the new version would produce much higher accuracy than in the initial short tests
- The children *enjoyed* using the software more than typing or handwriting

The survey results show that:
- Over half of those surveyed were unaware of the existence of any specialist software available to them
- Those who had tried some form of specialist software, generally found it helpful
- Voice recognition software was, on the other hand, found to be unhelpful and difficult to use

In conclusion then these findings show that VRS could be more viable now as an aid to children with dyslexia than it was before, however, time, patience and persistence of both student and teacher are needed. A greater level of publicity and support for these tools is also needed if those who could benefit from it are to be given the chance to do so. Apart accuracy the main issues which affect VRS usability for dyslexic school age children are:
- The inability to change the background colour on the screen
- The inability to enlarge or change the font used on the screen
- No option to have the training dialogue read aloud – a feature available within the system to read portions of dictated text back to you – but only *after* you have trained it!
- A sensitivity to background noise

Doctors, scientists, and even the Government are now being seen to be actively supporting children and adults with dyslexia in their quest for normality. It seems it may be up to developers to turn these findings into realistic goals.

References

BBC News Online, July '02, Figures from Gartner Dataquest

Elkind, J. Black, M.S. Murray, C. Nov 1996 'Computer-based Compensation of Adult Reading Disabilities' *Annals of Dyslexia*, (The Lexia Institute) **46**

Johnston, D.J. & Myklebust, H.R. 1967, *'Learning Disabilities: Educational Principles and Practices'* (Grune and Stratton, New York)

iANSYST Ltd. http://www.iansyst.co.uk/ *Computers and technology for people with disabilities*, Cambridge

IBM Research Projects/Publications 'Think Research' *www.research.ibm.com* [online]

Miles, M. Martin, D. Owen, J. Feb 1998, 'A Pilot Study into the Effects of Using Voice Dictation Software with Secondary Dyslexic Pupils' (Devon LEA)

Orton S.T. 1937/*1989 'Reading Writing and Speech Problems in Children and selected papers'* (Austin, TX: Pro-Ed)

O'Hare, A. McTear M.F. Aug 1999 'Speech Recognition in the Secondary School Classroom: an exploratory study' *Computers and Education, 33*, 1,

Stephens, C.A. (1997) 'The Application of Voice Recognition Software by Tertiary Students Who Have Specific Learning Difficulties' *Speech and Image Technologies of Computing and Telecommunications*, IEEE Tencon Conference '97

Seymour, Prof. P.H.K. 1986, *'A Cognitive Analysis of Dyslexia'* (Routledge & Kegan Paul)

Zhang, W. Duffy, V.G. Linn, R. Luximon, A. Oct 1999, 'Voice Recognition based on Human-Computer Interface Design' *Computers and Industrial Engineering, 37*, 1-2

BIOMECHANICS

PUSH AND PULL STRENGTH: A COMPARISON OF ISOKINETIC AND ISOMETRIC DYNAMOMETRY RESPONSES

Jonathan P. James and Andrew I. Todd

Department of Human Kinetics and Ergonomics
Rhodes University,
Grahamstown, 6140, SOUTH AFRICA

The present study used two methods to assess strength capabilities. Firstly, speed-related isokinetic measurements were recorded for eight male participants (Mean Age: 20.25yr ±1.16). In phase two isometric work-simulating responses were collected for 16 male participants (Mean Age: 20.63yr ±2.31). Furthermore benchmark and effort level consistency data for work outputs were established for an adapted Work-Simulation isokinetic strength test where there appears to be limited publication. Isokinetic testing was conducted on a CYBEX 6000® dynamometer and involved two work-simulating push-pull tests at the following velocities: $30°.s^{-1}$ and $60°.s^{-1}$. These tests were designed to simulate push-pull actions which are frequently evidenced by Ergonomists working in industry, for example, in the pushing and pulling of carts or manual handling devices (MHD). Isometric testing was completed using the Chatillon™ Hand-Held Dynamometer (CSD 500 Model) for whole-body and arms only pushing and pulling with participants standing and seated respectively.

Introduction

The present study aims to contribute in an area that has not been widely researched, namely push-pull strength evaluations using both dynamic and static work-simulating dynamometry. According to Resnick and Chaffin (1995) changes in work processes and task design have resulted in a greater prevalence of pushing and pulling tasks in industry. Baril-Gingras and Lortie (1995) argued that as much as 50% of industrial manual materials handling (MMH) is completed through pushing and pulling actions. Operators completing push-pull tasks are predisposed to a strong likelihood of injury as a result of poor working postures, repetitive movements and excessive force outputs. These factors also contribute to an increased incidence of slip, trip and fall accidents.

Pushing and pulling of carts or manual handling devices (MHDs) is predominantly a dynamic activity. Resnick and Chaffin (1995) stated that operators cannot maintain optimal working postures for the duration of a given dynamic push-pull task, and inertial components affect the forces which are generated. Workers have also been shown to adopt different working postures in dynamic push-pull exertions from those adopted in static testing (Lee *et al*, 1991). An evaluation of dynamic strength output would therefore appear to be more accurate than a static test when assessing push-pull risk factors which are more common in the work environment. Investigating dynamic and static push-pull exertions demonstrated differences in working posture for the two, and Mital *et al* (1995) contended that isokinetic measurements are more appropriate measures of strength than static maximal exertions. Isokinetic testing devices allow for dynamic strength evaluation and the adapted CYBEX 6000® Work-Simulation System (WSS) package is appropriate for the measurement of human muscular capabilities.

However, despite the accuracy of isokinetic testing devices, the practical relevance to industry is limited by lack of portability of the testing device. Portable isometric dynamometers, such as the Chatillon™ Hand-Held Dynamometer (CSD-500 Model), have much greater potential for use *in situ* and static strength measurements are still more commonly used to evaluate worker capabilities. The aim of this paper is to allow for some baseline comparisons between dynamic and static strength methodologies with a focus on some of the key areas for consideration by the ergonomist using isometric (hand-held) dynamometry *in situ*.

Methodology

This experiment was conducted in two phases.

Phase 1: Isokinetic Strength Evaluation
The first phase consisted of the collection of isokinetic strength responses for pushing and pulling using the WSS package for the CYBEX 6000® Isokinetic dynamometer. Speed-related isokinetic measurements were recorded using an adapted push-pull test and a modified UBXT seat (see Figure 1). Isokinetic testing involved two four-repetition maximal work-simulating push-pull tests at test velocities of $30°.s^{-1}$ and $60°.s^{-1}$. These tests were designed to simulate push-pull actions which are frequently evidenced by ergonomists working in industry, for example, in the pushing and pulling of carts or MHDs. Effort level consistency data for total work outputs were established for the adapted isokinetic test following the work of Charteris and James (2000).

Figure 1. Isokinetic strength testing setup for pushing and pulling

Phase 2: Isometric Strength Evaluation
In phase two isometric push-pull responses were collected with participants seated on the CYBEX 6000® UBXT while completing the maximal isometric efforts using the Chatillon™ Hand-Held Dynamometer (Model CSD 500) as the testing apparatus. Participants were required to complete the isometric testing protocol using the same seated posture as was used in the isokinetic study to facilitate comparisons between sets of static and dynamic strength data. The testing protocol required three maximal isometric efforts for pushing and pulling

from the participants. Figure 2 shows the setup used for the seated isometric push-pull testing where participants were required to complete the pushing and pulling tests at a height equal to 60% of stature.

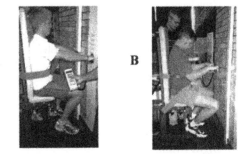

Figure 2. Isometric strength testing setup for pushing (A) and pulling (B)

Informed Consent

Informed consent was administered to all participants (who were free from recent or ongoing injury) who volunteered to partake in this research.

Demographic and Anthropometric Data

Table 1. Participant demographic and anthropometric data for both phases.

		Age (yr)	*Mass (kg)*	*Stature (mm)*	*BMI*
Phase I participants	Mean	20.25	80.95	1828	24.24
(n=8)	SD	1.16	7.20	59.95	2.65
	CV	5.75	8.90	3.28	10.93
Phase 2 participants	Mean	20.63	75.43	1800	23.25
(n=16)	SD	2.31	9.97	51.08	2.81
	CV	11.18	13.22	2.84	12.11

Results and Discussion

Phase 1: Isokinetic Strength Evaluation

Isokinetic test responses showed significant differences ($p < 0.05$) between pushing and pulling for peak torque and total work outputs. The highest values (see Table 2) were recorded for the arms-only pulling peak torque evaluations (6.09 Nm.kg^{-1} at 30°.s^{-1} and 5.24 J.kg^{-1} at 60°.s^{-1}).

Table 2. Peak torque (Nm.kg^{-1}) and total work (J.kg^{-1}) isokinetic responses.

	30°.s^{-1}				60°.s^{-1}			
	Push		Pull		Push		Pull	
Measure	Torque	Work	Torque	Work	Torque	Work	Torque	Work
Mean	4.89	3.09	6.09	4.14	4.26	3.10	5.24	3.83
SD	0.64	0.52	1.18	0.83	0.58	0.46	0.90	0.72
CV (%)	13.03	16.83	18.35	20.10	13.55	14.71	17.26	18.87
ELC (%)	94.26		95.03		90.86		94.65	

ELC = Effort Level Consistency (following Charteris and James, 2000).
(Mean Work/Maximum Work) x 100

Participants were considerably stronger when completing the pull as opposed to the pushing action. Mital *et al* (1995) argued that dynamic tests of push-pull strength require the participant to maintain stable posture during the maximal exertion, thereby resulting in a failure to take advantage of the leverage provided by the lower extremity. The stable base of support provided by the adapted CYBEX 6000® UBXT seat (push-pull at 90°) aided the participants in the production of higher than expected peak torque and total work outputs. The stability provided by the seat also resulted in the consistency of effort-level for work outputs being recorded at an average of 94% across the selected push-pull tests for the sample.

These findings indicate that a worker provided with a stable base of support would be in the best position to exert maximal force to move the required object in industry. The feasibility of providing a stable support base is however questionable due to space constraints and dynamic movement requirements in most push-pull task areas.

Phase 2: Isometric Strength Evaluation

The strength values recorded for the seated push-pull isometric tests (at 90°) were significantly higher than maximal values reported in other research for two handed standing or seated static efforts (Chaffin *et al* 1983, Mital *et al* 1995 and MacKinnon 1998). Table 3 shows the peak and average forces collected which provide the ergonomist with benchmark values from an above average cohort of participants as a basis for comparison when assessing individual workers *in situ*.

Table 3. Pushing and pulling responses (N) collected for isometric testing.

	Push		Pull	
Measure	Peak	Average	Peak	Average
Mean	739.70	681.49	861.66	800.53
SD	197.59	187.25	163.20	147.58
CV (%)	26.71	27.48	8.94	18.43

Although hand-held dynamometry does not produce the same level of accuracy as isokinetic strength testing it does provide the ergonomist with a number of advantages when used in industry. The portability, ease of use and simple operation of the apparatus makes isometric evaluation particularly appealing for use in the field.

Isokinetic versus Isometric Responses: Some Considerations

A comparison of isokinetic and isometric responses should provide a useful guideline for workplace evaluations. The present study demonstrated that in a controlled laboratory test isometric values were significantly greater than isokinetic responses. These findings are similar to those of Mital *et al* (1995) where isometric and isokinetic methods were used to assess vertical plane strength.

The application of static strength values in dynamic situations remains problematic. Operators who demonstrate the minimum strength required to complete a push-pull task in an isometric or static test, do not necessarily have the appropriate level of strength to complete a dynamic push-pull activity. The postures adopted during the completion of dynamic tasks are frequently sub-optimal, thereby predisposing the worker to injury. The evaluation of push versus pull ratios is one possible method that can allow for the comparison of static and dynamic strength responses. Isokinetic push versus pull ratios were 0.80 for torque and 0.75 for work at $30°.s^{-1}$, and the faster isokinetic speed ($60°.s^{-1}$) showed ratios of 0.81 for torque and 0.81 for work. The isometric push versus pull ratios were 0.86 (peak) and 0.85 (average) respectively. All ratios thus highlighted the dominance of pulling over pushing when considering strength expression. These findings would suggest that push-pull tasks should therefore be designed with pulling as the major component of the task when high force outputs are required. However, it must be noted that these push-pull actions were completed using a symmetrical action and were not one-handed pulls as are frequently evidenced in industry (James and Todd, 2003).

Conclusions and Recommendations

These findings have relevance to the Ergonomist assessing push-pull tasks and assisting in the re-design of working areas. Ergonomists need to ensure that work-related musculoskeletal disorders do not result from poorly designed workplaces where high force outputs are required while completing single handed pushing or pulling actions. Push-pull tasks need to be designed in such a manner as to minimise the force required and to allow the operator to adopt the most suitable working posture using symmetrical pushing or pulling styles to complete the task.

MacKinnon (1999) has cautioned that a small sample does not necessarily allow for generalized statements regarding pushing or pulling guidelines applicable to industry. Further research needs to be conducted to assess the differences in dynamic and static strength evaluations on a much larger sample. Static testing methods, for example, hand held dynamometry tests are more frequently used *in situ* and could well lead to workers appearing to have the required levels of strength to complete the majority of push-pull tasks. Based on the findings of this research, it is clear that maximal static efforts need to be used cautiously when considering dynamic occupations. Operators may well exhibit high levels of static or isometric strength, but this does not mean that they will be as proficient in dynamic work, particularly if these tasks require more commonly evidenced single handed pulling or pushing.

Reference List

Baril-Gingras, G. and Lortie, M. 1995, The handling of objects other than boxes: univariate   analysis of handling techniques in a large transport company, *Ergonomics,* **38**, 905-925.

Chaffin, D.B., Andres, R.O. and Garg, A. 1983, Volitional postures during maximal push/pull exertions in the sagittal plane, *Human Factors*, **25**, 541-550.

Charteris, J. and James, J.P. 2000, Replication of maximal work output levels in able-bodied workers and candidates for disability assessments: benchmark data and guidelines, *ergonomics SA,* **12**, 13-17.

James, J.P. and Todd, A.I. 2003, Push-pull force evaluations in an IDC Automotive Industry, *CD-Rom Proceedings of the XVth Triennial Congress of the International Ergonomics Association and The 7th Joint Conference of the Ergonomics Society of Korea/Japan Ergonomics Society.*

Lee, K.S., Chaffin, D.B., Herrin, G.D. and Waikar, A.M. 1991, Effect of handle height on lower-back loading in cart pushing and pulling, *Applied Ergonomics,* **22**, 117-123.

MacKinnon S.N. 1999, A psychophysical approach to determining self-selected movement frequencies for an upper-body pulling activity, *ergonomics SA,* 11, 6-16.

MacKinnon, S.N. 1998, Isometric pull forces in the sagittal plane, *Ergonomics*, **29**, 319-324.

Resnick, M.L. and Chaffin, D.B. (1995), An ergonomic evaluation of handle height and load in maximal and submaximal cart pushing, *Applied Ergonomics,* **26**, 173-178.

CHANGES IN THE SPINE KINEMATICS DURING A SIMULATED POSTAL WORKER'S TASK

Neil E. Fowler [1] and **Andre L.F. Rodacki** [1, 2]

[1] *The Manchester Metropolitan University,*
Department of Exercise and Sport Sciences,
Stoke-on-Trent, ST7 2HL.
[2] *Universidade Federal do Paraná,*
Departamento de Educação Física.,
Curitiba, Paraná, Brazil.

The aim of this study was to quantify the kinematics of the spine and stature loss induced by asymmetric load carriage. Six males acted as participants. Participants walked on a treadmill for 8500 m with and without a standard Royal Mail bag containing a load which was reduced gradually during the task. Motion of the spine was recorded using two optoelectric cameras. Spinal measurements were performed throughout the task. The loaded condition produced a spinal shrinkage two-fold greater than that observed in the unloaded condition. Increased forward leaning (up to 6°) and lateral bending of the spine (up to 12°) was observed when the heaviest loads were carried. The data provided evidence against mailbags designs in which the workers can not alternate the side of carriage.

Introduction

The efficiency of the postal service is closely related with the physical work performed by the postal carriers, who are responsible for the transport and delivery of the postal items. In general, this physical task is manually performed by walking door—by-door and the postal load is conventionally transported in a mailbag, which is often carried asymmetrically by means of a strap placed over one of the shoulders.

Lateral bending of the trunk to counteract the asymmetric placement of the load on the spine has been suggested as an important risk factor for a number of low-back disorders (Marras & Granata, 1997; Murray & Miller, 1984). Lin et al. (1996) and Ayoub & Smith (1999) are the only studies found that have investigated the effects of different mailbag designs over the letter carrier's spine. The assessment performed in these studies focused on a few kinematic variables used to describe shoulder and hip deviation (with respect to the horizontal) and the torsion of the whole spine, but did not provide a more detailed analysis of the changes in the spinal profiles of the thoracic and lumbar regions.

Several studies have reported that lateral bending of the trunk is associated with increased spinal stress and risk of low-back pain development (Anderson et al., 1980; DeVita et al., 1991; Marras et al., 1993; Murray & Miller, 1984). Wells et al. (1984), showed higher rates of back injury in postal workers in comparison to other workers who perform similar tasks, but do not carry any load (e.g. gas and electricity meter readers). Murray & Miller (1984), reported a 70% rate of low-back discomfort among American postal workers suggesting that this represents a significant occupational health problem.

The aim of the present study was to quantify the kinematics of the spine and stature loss induced by a dynamic task that simulates the asymmetric load carriage performed by a postal worker.

Methods

Six healthy and physically active male participants (mean ± SD; age = 26.3 ± 4.6 years; height = 182.3 ± 8.2 cm; body mass = 88.3 ± 12.2 kg) with no history of low-back disorders gave their written informed consent to participate in this study. Ethical approval was granted by the Ethics Committee of the Manchester Metropolitan University.

Walking was performed using a modified treadmill (Woodway, model ELG-2, USA) that allowed the walking speed to be adjusted to the preferred pace of each participant throughout the activity. Participants were requested to walk for 8500 m with and without a standard Royal Mail bag (model MB36) containing a dummy load. The initial load of the bag corresponded to 17.5% of body mass. The load of the bag was reduced gradually during the task (10% reduction of the initial load in each stage of the task).

In order to perform the kinematic analysis of the spine, data were collected during the last 50 m of every 750 m during the first 4500 m (i.e., 750, 1500, 2250, 3000, 3750 and 4500 m). The kinematic analysis during the last 4500 m of the task was performed within the last 50 m of every 1000 m (i.e., 5500, 6500, 7500 and 8500 m). Two optoelectric cameras (ELITE, BTS, Milan, Italy) sampling at 100 Hz were placed behind the participant to allow for 3D reconstruction of the spine. Thirteen markers were placed on the participant's back: eight over the spinal processes of the vertebrae C7, T4, T7, T10, T12, L2, L4 and S2; two between L4 and S2, and one marker over each shoe (at the level of the heel).

Spinal shrinkage measurements were performed at the end of each stage of the task (except the first stage, i.e., 50 m), immediately after the kinematic analysis. Spinal length was compared to the measurements performed prior to the beginning of each experimental session (baseline) to determine shrinkage.

Results

An assessment of the kinematic characteristics of the spine during each stage of the task in the unloaded condition (50, 750, 1500, 2250, 3000, 3750, 4500, 5500, 6500, 7500 and 8500 m) revealed that no significant changes (p > 0.05) occurred in the selected set of angles used to quantify changes in posture (LB_{TRU}, LB_{LUM}, LB_{THO}, SL_{TRU}, SL_{LUM}, ST_{THO}) during the unloaded condition in any phase of the gait cycle (0, 25, 50 and 75%). This assessment was performed using a number of two-way ANOVAS for repeated measures on each dependent variable. The coefficient of repeatability (Chronbach's alpha) of these trials were as great as 0.98. Therefore, mean values of all stages were calculated (ensemble averages) and selected to represent the unloaded condition.

A number of two-way ANOVAs were applied for each variable to detect whether significant differences occurred across the stages of the task. Six dependent variables (LB_{TRU}, LB_{LUM}, LB_{THO}, SL_{TRU}, SL_{LUM}, ST_{THO}) were compared across the phases of the task in four periods of the gait cycle (0, 25, 50 and 75%). A two-way ANOVA for repeated measures (2 load condition x 3 task phases) was also used to determine changes in spinal length and the velocities of the task. Significant differences were detected by applying a Neuman-Keuls test. A Kolgomorov-Smirnov test was applied and confirmed data normality. All statistic analyses were performed in the Statistica ® package software, version 5.5 (StatSoft Inc.®, Tulsa, USA) and the significance level was set at $p < 0.05$.

Results

The velocity in the unloaded and loaded conditions was similar in all three phases of the task ($p > 0.05$). No significant interactions between the two conditions were found ($p > 0.05$). An increased forward leaning of the trunk (SL_{TRU}) was observed at the beginning of the task during the loaded condition in comparison to the unloaded condition. This leaning ($5 \pm 3.2°$ in relation to the unloaded condition) was gradually decreased as the task progressed and the load was reduced in such a way that, in the late phase of the task, no significant differences between the unloaded and loaded conditions were found The anterior leaning of the trunk at the thoracic level (SL_{THO}) also followed a similar trend with a significant increase in SL_{THO} (up to 5°). No significant differences ($p > 0.05$) were detected in the anterior leaning of the trunk at the lumbar level (SL_{LUM}) between the unloaded and loaded conditions. The mean difference in SL_{LUM} between the unloaded and loaded conditions was of approximately $3.2 \pm 2.1°$.

Lateral bending of the trunk (LB_{TRU}) increased in comparison to the unloaded condition in a direction opposite to that in which the bag was held (i.e., to the left side). This increase (up to 12° during the first stages phase of the task) was observed from 50 to 4500 m, but was not found after 5500 m. The lateral deviation of the trunk was found to occur in the lumbar region only.

The spinal shrinkage responses were not linear and resembled a negative exponential curve. The greatest rate of shrinkage occurred during the early phase of the task in both conditions. This is more evident in the loaded condition. The loaded condition produced a spinal shrinkage two-fold greater than that observed in the unloaded condition at the end of the task.

Discussion

The increased forward leaning of the spine at the beginning of the task, when the heaviest loads were carried, can be interpreted as a postural adjustment to counteract the effect of the load, which tends to displace the subject's centre of mass away from the mid-line of the body. It has been suggested that leaning the trunk in one plane coupled with movements in other planes increases the risk of low-back disorders (Kelsey et al., 1984; Murray & Miller, 1984; Noone et al., 1993). The analyses also revealed that most adjustments in the configuration of the spine during the gait cycle in the thoracic area occurred in the sagittal plane, while changes in the lumbar area occurred in the frontal plane. This emphasises the importance of assessing changes in the thoracic and lumbar spinal profiles separately rather than treating the spine as a whole unit. Because changes in the thoracic area were observed during an asymmetric task, it can be assumed that asymmetric load carriage causes increased forward bending of the spine in the thoracic area, irrespective of the side of the load. Therefore, the forward bending of the thoracic area can be pointed out as a predisposing factor that may contribute to the development of idiopathic postural problems (e.g. kyphosis). It may also constitute an aggravating factor in subjects in which an augmented anterior-posterior thoracic curvature is already present. It is well established in the literature that long-lasting loads may produce adaptive responses of musculoskeletal components and predispose to idiopathic postural problems, pain and may lead to a number of disabilities (Bobet & Norman, 1984; Chafin, 1973).

The kinematic analyses performed in this study have important implications over the design of the mailbag. The mailbags in which the load is fixed in only one side of the body are likely to increase the long-term effects over postural deviations. Therefore, mailbag designs that incorporate a belt to distribute the load around the waist (and do not allow the subjects to alternate the side of the load) may have the drawback of increasing the unilateral stress of the spine.

In the loaded condition, the first part of the task (when subjects carried the heaviest loads) was characterised by a marked stature loss rate. The rate of height loss in the unloaded condition was almost constant through the task. In the early stages of the task, when the intervertebral discs are fully filled with fluid and the leaning of the trunk in both planes of motion is the largest, the risk of injury is great due to the increased intradiscal volume and pressure (Adams et al., 1987). As the task progressed, further stature loss was observed until 3500 m when no further large decreases in stature occurred although it was not possible to evidence that equilibrium deformation was achieved to suggest that some critical limit of shrinkage was reached during the final stages of the task. Probably, the weight of the bag, which was gradually reduced as the experiment progressed, was not large enough to cause further decreases in stature.

Conclusion

The changes observed in the kinematic profiles of the spine in the sagittal and frontal planes suggested that an increased forward leaning of the trunk associated with an increased lateral bending towards the unloaded side occurred during the task. When a more detailed analysis of the increased leaning of the spine was taken into account, it was noticed that the lateral adjustments occurred predominantly at the lumbar area. The adjustments in the sagittal plane occurred predominantly at thoracic level and may constitute a risk factor for idiopathic postural problems such as kyphosis. This study did not provide evidence that lateral deviation of the spine (scoliosis) is a major concern among letter carriers because they are free to alternate the load carriage side. The data of the present study provided further evidence against mailbags designs in which the workers can not alternate the side of the mailbag. These models are likely to produce postural problems (e.g., scoliosis) when repeated frequently, in a daily basis. Intervertebral disc problems are also likely to occur among postal workers due to the stress that occurs due to the load of the bag. Epidemiological studies are required to substantiate the arguments proposed in this study.

References

Adams, M. A. Dolan, P. and Hutton, W. C. 1987, Diurnal variations in the stress on the lumbar spine, *Spine* **12**, 130-137.

Anderson, G. B. J. Ortengren, R. and Schultz, A. 1980, Analysis and measurement of the loads on the lumbar spine during work at a table, *Journal of Biomechanics*, **13**, 513-520.

Ayoub, M. M. Smith, J. L. 1999, Evaluation of satchels for postal letter carriers. *International Journal of Industrial Ergonomics*, **23**, 269-279.

Bobet, J. and Norman, R. W. 1984, Effects of load placement on back muscle activity in load carriage, *European Journal of Applied Physiology*, **53**, 71-75.

Chafin, D. B. 1973, Localised muscle fatigue: definition and measurement, *Journal of Occupational Medicine*, **15**, 346-354.

De Vita, P. Hong, D. and Hamill, J. 1991, Effects of asymmetric load carrying on the biomechanics of walking, *Journal of Biomechanics*, **24 (12)**, 1119-1129.

Kelsey, J. L. Githens, P. B. White, A. A. Holford, T. R. Walter, S. D. O'Connor, T. Ostfeld, A. M. Weil, U. Southwick, W. O. and Calogero, J. A. 1984, An epidemiological study of lifting and twisting on the job and risk for acute prolapsed lumbar intervertebral disc, *Journal of Orthopaedic Research*, **2**, 61-66.

Lin, C. J. Dempsey, P. G. Smith, J. L. Ayuob, M. M. and Bernard, T. M. 1996, Ergonomic investigation of letter-carrier satchels: Part II. Biomechanical laboratory study, *Applied Ergonomics*, **27 (5)**, 315-320.

Marras, W. S. Lavender, S. A. Leurgans, S. E. Rajulu, S. L. Alread, W. G. Fathallah, F. A. and Ferguson, S. A. 1993, The role of dynamic three-dimensional motion in occupationally-related low back disorders: the effects of workplace factors, trunk position and trunk motion characteristics on risk of injury, *Spine*, **18**, 617-628.

Marras, W. S. Granata, K. P. 1997, Spinal loading during trunk lateral bending motion. *Journal of Biomechanics*, **30**, 697-703.

Murray, G. W. Miller, D. C. 1984, The postal posture problem. In Proceedings of the *International Conference on Occupational Ergonomics*, 559-563.

Noone, G. Mazumdar, J. Ghista, D. N. and Tansley, G. D. 1993, Asymmetrical loads and lateral bending of the human spine, *Medicine and Biology in Engineering and Computing*, **31**, Supplement, S131-136.

Wells, J. A. Zipp, J. F. Schuette, P. T. and McEleney, J. 1984, Musculoskeletal disorders among letter carriers, *Journal of Occupational Medicine*, **25 (11)**, 814-820.

MUSCULOSKELETAL
DISORDERS

THE EFFECTS OF MUSCULOSKELETAL PAIN ON WORK PERFORMANCE: WHICH PAIN MEASUREMENT TOOL SHOULD BE USED

Ozhan Oztug, Peter Buckle

Robens Centre for Health Ergonomics
European Institute of Health and Medical Sciences
University of Surrey
Guildford, GU2 7TE

Work-Related Musculoskeletal Disorders are a significant burden on sufferers, employers and moreover national economies. These disorders usually follow a pathological process that may lead to impairment or disability and cause individuals to be absent from their work. Many researchers have studied absenteeism as a productivity indicator. However the effects of musculoskeletal symptoms (pain) on workers performance, when they are present, has not been studied extensively. This paper describes the selection of the most suitable pain measurement tool to be used in assessing the levels of musculoskeletal pain in a sample of assembly line workers. The selected criteria were validity, reliability, ease of administration, ease of scoring and ratio data as outcome. As a result the Borg CR10 scale has been found to best satisfy all the criteria.

Introduction

Work-Related Musculoskeletal Disorders (WMSDs) are described as inflammatory and degenerative diseases and disorders that result in pain and functional impairment, and may effect the neck, shoulders, elbows, forearms, wrists and hands (Buckle and Devereux, 2002). Musculoskeletal disorders (MSDs) are the most common type of work-related ill-health problem in Great Britain. Apart from their impact on health, the symptoms of MSDs may effect the productivity of those sufferers. This issue has been addressed mostly by considering the sickness-absence records as outcomes. However the effects of the symptoms when the workers are present at work has received little attention (Hagberg *et al*, 2002; Yu and Ting, 1993). Therefore in this study the effects on work performance amongst workers who have differing levels of musculoskeletal pain will be addressed. The selection of a suitable pain measurement tool is described in this paper.

Pain measurement

The International Association for the Study of Pain (IASP) defines pain as 'an unpleasant sensory and emotional experience associated with actual or potential tissue damage, or described in terms of such damage' (Merskey *et al*, 1979). As it is stated by Caraceni *et al* (2002), pain is a subjective sensation that can be described according to several relevant futures such as quality, location, intensity, aversiveness, emotional impact and frequency. Melzack and Casey (1968) proposed a multidimensional model for pain that groups the features mentioned by Caraceni *et al* (2002) under three distinct dimensions (Price, 1988). 1) The sensory discriminative dimension composes experiences such as location, quality, and intensity of the painful sensation, and other spatial and temporal characteristics. 2) The cognitive-evaluative dimension is related to the interpretation of perceived pain, as what is taking place and what might take place in relation to this sensation. 3) Finally the affective-motivational dimension is the felt sense of perceived and interpreted pain in relation to one's desire to avoid harm and/or one's expectation of avoiding harm.

As Borg (1998) emphasised, there are many objective physiological correlates of pain (e.g. skin conductance, heart rate, temperature, etc.), but most importantly the subjective perception of the individual must take the priority as a starting point in studies and then interpretation.

Self-report pain assessment tools may be divided into two broad categories, unidimensional and multidimensional (Caraceni, 2002). The unidimensional pain assessment tools measure only the intensity attribute of pain whereas the multidimensional tools assess also other factors that affects its perception (e.g. quality and temporal sequence of pain, the affective contributions, patient belief system) (Ho *et al*, 1996). Some of the most widely used tools in this category are the McGill Pain Questionnaire (MPQ) (Melzack, 1975; Melzack, 1983), Brief Pain Inventory (BPI) (Cleeland and Ryan, 1994), and the Memorial Pain Assessment Card (MPAC) (Fishman *et al*, 1987). All these were suggested by Caraceni *et al* (2002) as, well-validated tools. However these tools are complex and difficult to use and interpret. Their length makes them impractical to use in settings where rapid assessment is necessary (Ho *et al*, 1996). For this reason the multidimensional tools could not be used in work settings where time is a critical factor and the workers can not be distracted from their work for more than a few minutes (e.g. machine paced jobs, assembly lines, etc.). In addition to this, the selected population is also an important factor in deciding whether to use a unidimensional or multidimensional tool. In the proposed investigation a unidimensional tool was required, as the study population (acute) does not need more complex evaluation tools (Ho *et al*, 1996).

Visual Analog Scale (VAS)

The VAS is a most commonly used pain measurement tool in both research and clinics. It consists of a 10-cm line bounded with two descriptors as "no pain" and "worst pain possible" (Ho *et al*, 1996; Lee, 2001; Caraceni *et al*, 2002). The assessment is done by placing a mark on the line, which is then quantified by measuring its distance from the "no pain" end.

The VAS has been shown to be a valid and reliable method to assess the intensity of pain (Huskisson, 1983). As Huskisson (1983) states, the tool has some advantages such as its sensitivity, simplicity, reproducibility, and universality (e.g. independent of language). Briggs and Closs (1999) emphasised that the VAS may be sensitive in cases of interventions, such as when pain is measured before and after change for the same persons but that it may not produce reliable results when used across different patient groups. This is because each patient may interpret the scale differently. Another disadvantage of the method is the conceptual complexity where the ability to translate sensory experience into a linear format is required (Briggs and Closs, 1999). As reported by Lee (2001) the noncompliance rates for the method ranges from 7-25%.

Verbal Rating Scale (VRS)

The VRS is one of the unidimensional methods that consist of a list of anchors for different levels of pain experience. The tool is administered by asking the respondents to choose a word among a set of descriptors such as 'none', 'mild', 'moderate', or 'severe'. The VRS is an easy and rapid to use but as with other tools, there are some problems with its use. One of these is that the intervals between the descriptors are usually not similar. Hence a change from 'none' to 'mild' may not represent the same change from 'moderate' to 'severe'. As Briggs and Closs (1999) stress, this is sometimes considered as a weakness of the method due to the fact that it limits the statistical analysis to nonparametric methods. In addition to this the tool may not be responsive to significant changes in pain intensity since it consists of a few categories. On the other hand increasing the number of categories may not increase the responsiveness. This is because the responders may not be able to differentiate between 'unbearable' and 'excruciating' (Lee, 2001). Another weakness of the method is its demand for ability to read and interpret the words outlined. But the method is shown, as having higher compliance rates than VAS since it is easier to use (Briggs and Closs, 1999).

Numerical Rating Scale (NRS)

The NRS is another simple and easy to use tool where the respondents are asked to rate their intensity of pain on a scale of 0-10 or 0-100 with 0 representing 'no pain' and the 100 representing 'worst possible pain' (Jensen *et al*, 1986). As emphasised by Jensen *et al* (1986) the method has some advantages over the other methods. It is very simple to administer and score, and provides the opportunity to be administered either in written or verbal form. The compliance rates are also shown to be higher than the VAS (Lee, 2001). Lee (2001) stressed that whilst the method suggested might be treated as ratio data, this hadn't been established yet.

Borg CR10 Scale

The Borg CR10 scale is a psychophysiological method that has been developed to measuring various sensory perceptions such as perceived exertion and pain. Borg (1998) stresses that the tool is commonly used in assessing and quantifying the intensity of pain such as angina and musculsokeletal pain. The tool has been found to be highly reliable and valid (e.g. high correlation with VAS that is already accepted by the IASP as a valid tool for pain measurement). In addition, one of the characteristics that discriminates it from the other scales is that the respondents have been provided with the opportunity to report any pain that exceeds the previous maximum experiences. However the VAS, VRS and NRS do not have this characteristic. Also the suitability for verbal administration is another advantage of the method.

Criteria and assessment

There are several issues that should be considered before selecting a pain measurement tool. As stressed by Caraceni *et al* (2002) those issues could be classified under two headings. These are the type of study, and the intended study population. In selecting a pain measurement tool the most important qualities to be satisfied are the validity and the reliability of a specific method. However issues such as the characteristics of the pain (e.g. acute, chronic), and suitability of the tool in settings where self-administration is not possible are other issues that should be considered. In addition, the conceptual simplicity of the tool together with the amount of effort demanded from the responders in completion are factors that affect the compliance rates (Jensen et al, 1986). Also in cases where time is critical, the scoring time would also be selected as an important criterion. Hence the following criteria has been specified and assessed in this study.

A-Validity, B-Reliability, C-Ease of administration (physical restrictions, communication difficulties, conceptual simplicity), D-Ease of scoring, E-Ratio data as outcome

Table1: A decision matrix for the selection of a pain measurement tool.

Scale/Criteria	A	B	C	D	E
VAS	✓	✓	✗	✗	✓
VRS	✓	✓	✓	✓	✗
NRS	✓	✓	✓	✓	?
CR10	✓	✓	✓	✓	✓

✓ : Criterion satisfied; ✗ : Criterion not satisfied; ?: Decision not possible

As a result of the assessment the Borg CR10 Scale found to satisfy all the selected criteria and would be used in assessing the musculoskeletal pain in work settings.

Discussion

The selection of a pain measurement tool usually depends on the population characteristics, settings and also the purpose, such as clinics or research. Every method has advantages and disadvantages depending on the settings of usage. In this study the most important issues that were considered as primary requirements were the validity and the reliability of the methods. These were satisfactory for all the methods. In addition to this the ease of administration which consist of physical restrictions (e.g. psychomotor requirements), communication difficulties in settings where physical response is restricted (e.g. work settings such as assembly line) and conceptual simplicity were considered as the secondary requirement in the assessment. In settings such as assembly lines, where the work is machine-paced or physical response (e.g. placing a mark on VAS) is restricted the tools that allow verbal communication are preferable. The VRS, NRS and CR10 scales satisfy this criterion. In addition to this it is also suggested that the VAS is a conceptually more complex tool compared to the others. Hence more guidance would be required when administering it. The third criterion was the ease of scoring. For VAS, metric measurement is required to quantify the reported pain experience. For multiple limb assessments a separate VAS is essential for every painful body-parts to be reported. For this reason the tool is impractical. On the other hand the VRS, NRS, and CR10 scales are very practical and would give direct scores. The ratio data outcome was the last criterion to be satisfied. The scales with this characteristic allow the usage of the parametric methods, which is considered an advantage since the non-parametric methods may increase the risk of type-II errors (Ho et al, 1996). This criterion is satisfied by the VAS and CR10 scales whereas this hasn't been established for the NRS.

References

Borg, G. (1998) *Borg's perceived exertion and pain scales*, Leeds: Human Kinetics.

Briggs M. Closs S. J. (1999) A descriptive study of the use of visual analogue scales and verbal rating scales for the assessment of pain in orthopaedic patients. *Journal of Pain and Symptom Management*, **18(6)**, 438-446.

Buckle, P.W. and Devereux, J.J. (2002) The Nature of Work-Related Neck and Upper Limb Musculoskeletal Disorders. *Applied Ergonomics* **33**, 207-217.

Caraceni, A., Cherny, N., Fainsinger, R., Kaasa, S., Poulain, P., Radbruch, L., De Conno, F. and Steering Commitee of the EAPC Research Network (2002) Pain measurement tools and methods in clinical research in palliative care: Recommendations of an Expert Working Group of the European Association of Palliative Care. *Journal of Pain and Symptom Management* **23**, 239-255.

Cleeland, C.S., Ryan, K.M. (1994) Pain assessment: The global use of the Brief Pain Inventory. *Annals Academy of Medicine Singapore* **23(2)**, 129-138.

Hagberg, M., Tornqvist, E.W. and Toomingas, A. (2002) Self-reported reduced productivity due to musculoskeletal symptoms: Associations with workplace and individual factors among white-collar computer users. *Journal of Occupational Rehabilitation* **12**, 151-162.

Ho, K., Spence, J. and Murphy, M.F. (1996) Review of pain measurement tools. *Annals of Emergency Medicine* **27**, 427-432.

HSE (2001) Work-related upper limb disorders statistics information sheet. 02,

Huskisson, E.C. (1983) Visual Analogue Scales. In: Melzack, R., (Ed.) *Pain measurement and assessment*, pp. 33-37. (New York: Raven Press)

Jensen, M., Karoly, P. and Braver, S. (1986) The measurement of clinical pain intensity: A comparison of six methods. *Pain* **27**, 117-126.

Lee, J.S. (2001) Pain measurement: understanding existing tools and their application in the emergency department. *Emergency Medicine* **13**, 279-287.

Melzack, R. and Casey, K.L. (1968) Sensory, motivational and central control determinants of pain: A new conceptual model. In: Kenshalo, D., (Ed.) *The skin senses*, pp. 423-443. (Illinois: Thomas)

Melzack, R. (1975) The McGill Pain Questionnaire: Major properties and scoring methods. *Pain* **1**, 277-299.

Melzack, R. (1983) The McGill Pain Questionnaire. In: Melzack, R., (Ed.) *Pain Measurement and Assessment*, pp. 41-47. (New York: Raven Press)

Merskey, H., Bonica, J.J., Carmon, A., Dubner, R., Kerr, F.W.L., Lindblom, U., Mumford, J.M., Nathan, P.W., Noordenbos, W., Pagni, C.A., Renaer, M.J., Sternbach, R.A. and Sunderland, S.S. (1979) Pain terms: A list with definitions and notes on usage. *Pain* **6**, 249-252.

Price, D.D. (1988) *Psychological and neural mechanisms of pain*, New York: Raven Press.

Yu, I.T.S. and Ting, H.S.C. (1993) Musculoskeletal discomfortand job performance of keyboard operators. In: Salvend, G., (Ed.) *Human-Computer Interaction: Software and Hardware Interfaces*, pp. 1058-1063. (Amsterdam: Elsevier)

Development of a database for the analysis of and research into occupational strains on the spinal column

Dirk M. Ditchen, Rolf P. Ellegast

Berufsgenossenschaftliches Institut für Arbeitsschutz (BIA)
Alte Heerstr. 111
53754 Sankt Augustin
Germany

The described database, which is used within the German Berufsgenossenschaften (the institutions for statutory accident insurance and prevention, or BGs), allows different objectives to be pursued. On the one hand, it provides an instrument for the uniform handling of occupational disease; on the other, it aids in collecting stress and strain data from occupational practice for evaluating workplaces and developing suitable preventive measures.

Introduction

Strains placed on the human musculoskeletal system, especially on the spinal column, have become more and more important to occupational life. With the incorporation of three new occupational diseases related to damage caused to the spinal column in the German Ordinance on Occupational Diseases in 1993, the BGs were obliged to provide compensation for damage to the spinal column caused by professional activities and to step up preventive measures taken in this area.

In response to this, the BG Institute for Occupational Safety and Health (BIA), the central institute for research and testing for the German BGs, in co-operation with several BGs, decided to develop a database for musculoskeletal strains based on the already existing OMEGA database system.

There are two main intentions in developing this kind of database. On the one hand, the enlarged OMEGA database is planned to provide a standardised instrument for the 35 German industrial BGs in the adjudication process of occupational disc-related diseases of the lumbar spine. The most significant in this regard is occupational disease No. 2108 ("Intervertebral disc-related diseases of the lumbar spine caused by the long-term lifting or carrying of heavy objects or caused by long-term activities in extreme trunk-flexed postures", or BK 2108).

On the other hand, the enlarged OMEGA database is also meant to be used for prevention purposes. To do so, strain data from a number of occupational situations are to be archived in the database and then prepared in a manner to enable industry-specific prevention concepts to be developed.

The quality of these data can range from simple questionnaire data to sophisticated measuring data, eg from the CUELA system (Computer-assisted registration and long-term analysis of musculoskeletal load) - a person-centred measuring system developed at BIA to establish the extent of strain placed on the spinal column at the workplace (Ellegast and Kupfer, 2000).

This huge data pool will help to develop new registers and methods for assessing occupational strain in the future.

Methods

For conceiving the database, the so-called OMEGA system was used. It is based on the Delphi programming language and has been in use within the BGs for several years now. This OMEGA system has already proven its suitability in the risk categories "hazardous substances" and "noise", and it already contains some 1.4 million data sets in these fields. Decentralised data input and data query are possible for different users with this system, while central management of the software is also ensured by the developers.

A team of experts from several statutory accident insurance institutions was assembled to conceive the contents of the spinal column databank.

Software "Anamnesis of occupational disease No. 2108"
In order to achieve uniform processing of the anamnesis in the procedure to BK 2108, a program was developed for calculating the "total occupational dosage". This program is oriented to the standardised exposure assessment procedure based on the so-called Mainz-Dortmunder Dose Model (MDD) as published by Jäger, et al., (1999) and Hartung, et al., (1999).

The program simplifies and improves occupational anamnesis in many various ways. Aside from automated dosage calculation, the software offers research into different library data compiled by the BGs over the years. These library data depict typical work situations, each including the load weights handled, the frequency, and the respective periods of time per work shift for various tasks. This method allows jobs performed many years ago to be reconstructed and included in the calculation of the total occupational dosage.

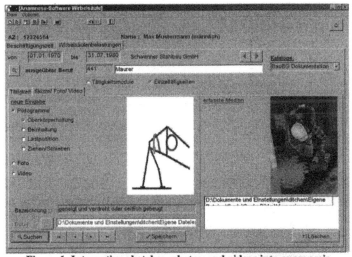

Figure 1. Integrating sketches, photos, and videos into anamnesis

In this respect the function for researching within different load weights is a helpful option, too. Those load weights are stored as measured values in the database to provide valid information on the masses dealt with.

Sketches, photos, or videos can be added to depict specific load handling or bodily postures (see Figure 1), and these illustrations can partly be printed in the anamnesis report also.

Finally, the software provides the option of archiving the data electronically and exchanging the data with other researchers.

Prevention database

The much greater portion of the database is dedicated for use as a database on measurement and prevention data. Here, the database is intended to serve in creating a large data pool where strain data from various occupations and different musculoskeletal risk factors can be brought together. Aside from the data required in determining the occupational disease on the manipulation of heavy objects (lifting and carrying) or working in extreme trunk-flexed postures, other factors are also to be considered here. This means, for instance, that load handling, such as by way of pulling and pushing, are described and evaluated. The different bodily postures assumed for a specific task can also be depicted and evaluated with fine differentiation, just as is the case for working in static postures or under highly dynamic conditions.

Two different levels are considered when depicting these strains. On the one hand, individual tasks can be documented in the database (eg: load handling); on the other hand, entire work shifts should be depicted where possible in order to evaluate the total strain on a particular worker. This implicates not only the actual physical strain factors but also aspects of the workplace environment and work organisation (eg: break rules).

In order to meet these different requirements, different computer screen forms are used to enter data into a database assigned to a specific evaluation procedure. Table 1 shows the different evaluation procedures currently integrated into the program.

In addition to these procedures which each produce special data sets that often diverge from one another, it is also possible to produce a pure description of tasks and work shifts.

Table 1. Integrated evaluation procedures

Procedure	Description	Authors
LMM	Leit-Merkmal Methode	Steinberg and Windberg (1997)
MDD	Mainz-Dortmunder Dose Model	Jäger, et al. (1999), Hartung, et al. (1999)
NIOSH	National Institute for Occupational Safety and Health	Waters, et al. (1993)
OWAS	Ovako Working Posture Analysing System	Karhu, et al. (1977)

CUELA measurement system

The data sets discussed so far - as has already been mentioned - each require manual entry using computer screen forms. As a rule the data here are either from questionnaires or from workplace observations. Yet the OMEGA database "Spinal column strain" also provides the option of importing measurement data automatically through an appropriate interface. For this reason, a special measuring system - the CUELA system (see Figure 2) - was distributed among users along with the creation of the database.

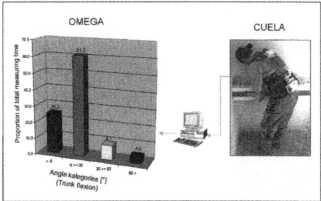

Figure 2: Data transfer CUELA - OMEGA

The CUELA measuring system was developed by BIA to enable measurements of the strains on the musculoskeletal system that are so common to numerous occupational tasks. CUELA is a person-centred measurement system composed of modern sensor technology that can be worn directly on the body. Therfore it can be applied at the workplace under real working conditions. The corresponding WIDAAN software allows for an automated evaluation of the measurement data according to occupational science and biomechanical evaluation criteria along with the automated data transfer to the OMEGA database. On this basis, it is possible to make statements on the measures needed for avoiding occupation-related health risks.

First Results

The anamnesis software is now used by some 80 users at 22 different BGs, and the system has proved itself.

The CUELA measurement system has been continuously refined by BIA since 1995, and it is now used in the different branches. It is presently being distributed within the BGs, so that the first measurement data sets are expected here at the end of 2004.

Future Implications

The uniform application of the OMEGA database system in collecting data from different risk areas (hazardous substances, noise, vibration, spinal strain) will make it possible to assure the provision of task- and occupation-related compilations of stress and strain data in terms of holistic strain profiles. The long-term aim of the project is a holistic risk assessment of workplaces.

References

Ellegast, R. and Kupfer, J. 2000, Portable posture and motion measuring system for use in ergomomic field analysis. In: Landau Ergonomic Software Tools in Product and Workplace Design (Ergon, Stuttgart)

Jäger, M.; Luttmann, A.; Bolm-Audorff, U.; Schäfer, K.; Hartung, E.; Kuhn, S. Paul, R. and Francks, H.-P. 1999, Mainz-Dortmunder Dosismodell (MDD) zur Beurteilung der Belastung der Lendenwirbelsäule durch Heben oder Tragen schwerer Lasten oder durch Tätigkeiten in extremer Rumpfbeugehaltung bei Verdacht auf Berufskrankheit Nr. 2108. Teil 1: Retrospektive Belastungsermittlung für risikobehaftete Tätigkeitsfelder. In Arbeitsmedizin, Sozialmedizin, Umweltmedizin 34 (1999) 101-111

Hartung, E.; Schäfer, K.; Jäger, M.; Luttmann, A.; Bolm-Audorff, U.; Kuhn, S.; Paul, R. and Francks, H.-P. 1999, Teil 2: Vorschlag zur Beurteilung der arbeitstechnischen Voraussetzungen. in Berufskrankheiten-Feststellungsverfahren. In Arbeitsmedizin, Sozialmedizin, Umweltmedizin 34 (1999) 112-122

Karhu, O., Kansi, P. and Kuoriuka, I. 1977, Correcting working postures in industry: A practical method for analysis, Applied Ergonomics, 8, 199-201

Steinberg, U. and Windberg, H.-J. 1997, Leitfaden Sicherheit und Gesundheitsschutz bei der manuellen Handhabung von Lasten. In Schriftenreihe der Bundesanstalt für Arbeitsschutz und Arbeitsmedizin, Sonderschrift, 43

Waters, T. et al 1993, Revised NIOSH equation for the design and evaluation of manual lifting tasks. In: Ergonomics, 36, 749-776

CONTEMPORARY ERGONOMICS CONSULTANCY PRACTICES: REDUCING THE RISK OF MSDs

Z J Whysall[1], R A Haslam[1], and C Haslam[2]

[1] *Health and Safety Ergonomics Unit, Department of Human Sciences*
Loughborough University, Leicestershire, LE11 3TU

[2] *Institute of Work, Health and Organisations, University of Nottingham*
Nottingham Science & Technology Park, Nottingham, NG7 2RQ

Despite the importance of reducing work-related musculoskeletal disorders (WRMSDs), there appears to have been little evaluation of consultancy interventions aimed at reducing risks leading to these conditions. Behaviour change theory suggests that if change is to take place, recipients need to hold positive attitudes and beliefs relating to the desired behaviour. To investigate the extent to which this is accommodated by current ergonomics consultancy practices, fourteen ergonomics consultants were interviewed to explore the consultancy process, the factors that are assessed, perceived barriers to change, and the evaluation of outcomes. Consultants' recommendations generally focused on physical aspects of the work environment and did not take explicit account of employees' knowledge or attitudes. Evaluation of outcomes was rare. Implications of these findings for improving the efficacy of interventions are discussed.

Introduction

Musculoskeletal disorders (MSDs) remain the most common form of work-related ill health in Great Britain (HSE, 2002), highlighting a pressing need to examine why their prevalence is so high and furthermore, how attempts to alleviate risks can be made more effective. Undoubtedly, the complex causation of MSDs poses a significant obstacle to their control. As a result, approaches to tackling MSDs often need to be directed at various different levels. As a general framework for the management of upper limb disorders (ULDs) in the workplace, HSE (2002) proposed a seven staged management cycle:
- Understand the issues and commit to action
- Create the right organisational environment
- Assess the risk of ULDs in the workplace
- Reduce the risks of ULDs
- Educate and inform the workforce
- Manage any episodes of ULDs
- Carry out regular checks on programme effectiveness

The extent to which ergonomics consultants incorporate the aspects of such frameworks into their work is unknown, and there have been few attempts to evaluate the effectiveness of such interventions (Griffiths, 1999).

When advising on MSDs, it is frequently the case that consultants will indicate that changes are required, such as modifications to equipment, workplace layout and/or working practices, yet acceptance of such advice depends on attitudes and beliefs regarding MSDs held by the client and their employees. As argued by Haslam (2002), application of the Stage of Change model (Prochaska and DiClemente, 1982) to ergonomic interventions may increase their efficacy by identifying the attitudes, beliefs and knowledge, which can act as barriers to change. At different stages individuals are receptive to different types of information or advice.

Given the complexity of change processes, this study sought to investigate the current practices of ergonomics consultants advising on MSDs. Of particular interest were the consultants' judgments of the effectiveness of their recommendations for change, the barriers they consider to inhibit the achievement of successful implementation of their advice, and the extent to which the outcome of their work is evaluated.

Method

Fourteen in-depth interviews were conducted with ergonomics consultants, exploring the consultancy process. Interviews typically lasted 45-60 minutes. The discussions were tape recorded and transcribed. Interviewees were ergonomics consultants from 13 different consultancy organisations in the UK, each with varying degrees of experience. The only specification was that the consultant deals with projects addressing MSDs.

The data were analysed by sorting the material into emergent themes as described by Dey (1993). The analysis was led by the original guiding questions, considering the consultancy process, the type of information gathered in the course of their investigations, whether consultants seek feedback or conduct any type of evaluation, and the extent to which consultants experience resistance from within the client organisation.

Results

Descriptive statistics
The average length of experience of consultants was 10 years, ranging from 3 to 30 years. Five were independent consultants, 8 were employed within consultancy firms, and 1 operated as an associate to another organisation. All but 2 interviewees practised in a context of professional ergonomics accreditation either through registration held by themselves or through the registration of their employer.

The Consultancy Process
Compliance with HSE inspections, or having identified a specific problem themselves were identified as the most common motivators for clients to request help. Only three interviewees reported receiving requests from organisations that had not yet experienced problems, in the words of one senior consultant representing a larger consultancy firm, companies that are *'wanting to look after people'*. A crucial part of the initial contact phase, identified by all interviewees, is the negotiation of a contract (formal or informal), above all to establish the likely cost and duration of the project. Two consultants highlighted the importance of clarifying expectations at this early stage, particularly to break down preconceived ideas of what an ergonomics consultant does. A consistent message regarding general consultancy procedures was that there is no set process followed when undertaking an assignment; *'no fixed linear response'*. Several reasons were suggested for this lack of standardisation, including the diversity of the work; *'What is required depends very much on the type of project...'* Consultants

tended to rely on an *'informal checklist'* developed as a result of experience. The importance attributed to building up a rapport with the client, highlighted by four interviewees, might also contribute to the lack of standardisation, it being perceived as obstructing the development of a good rapport. Perhaps also related to the development of a good rapport, the importance of meeting clients' expectations was highlighted by several interviewees as imposing a strong influence on the consultancy process and resulting recommendations. In the words of one, *'...it's all very well being right, but if you're not getting the work...'*

Work Assessment

The factors that are assessed, and information gathered common to all consultants included an overview of the work and key problems with the contact within the client organisation (typically health & safety manager), a tour of the workplace, detailed analyses of specific tasks, using a tool such as RULA (McAtamney and Corlett, 1993), photographic and video evidence of operators performing tasks, and talks with individual operators. Photographs and tools that enabled the quantification of information (such as RULA or REBA) were highlighted as being particularly *'good for explaining problems to management'*, and enabling the prioritisation of action.

One interviewee stated that recommendations are *'tailored to the level of existing knowledge in an organisation'*, on the basis that it is *'no good telling them something they already know'*. Generally though, the consultants did not appear to make any systematic attempt to explore stakeholders' underlying understanding, other than asking what had already been tried. An associate consultant was of the opinion that *'...their occupational health dept should have covered this in the initial stages.'* Another consultant did not believe it was necessary to make people understand the issues, but felt that highlighting the potential costs of claims is sufficient to motivate companies to take action. As one co-director of a consultancy firm explained: *'We're not psychologists, so it's very much to do with physical aspects.'* Several interviewees did emphasise the importance of talking to operators to clarify factors that may not be directly obvious, such as the degree of repetition in the job, the frequency of breaks, the level of discomfort, 'aches and pains', and how employees feel their job could be improved. As described by one female consultant:

> *'We would not assess attitudes as such, but do give them* [operators] *the opportunity to say what they think should be changed...but there are also certain things that they do not think of...'*

Regardless of the amount of information drawn from a workplace assessment however, four interviewees indicated that it is often immediately apparent what the problem is, and what a particular workplace is like. The process was referred to as relatively implicit, *'a feel thing'*, developed as a result of experience.

Recommendations

The tendency to offer clients a range of potential solutions was common. This appeared to be due to two main factors: the complexity of ergonomics problems, and consultants' recognition that clients are unlikely to implement all of the recommendations that are made, due to limitations in factors such as time, cost, expertise, and motivation:

> *'Although most ergonomics problems are very complex...essentially companies want definitive answers, they want to know what they can do that is achievable.'*

Overcoming barriers

One specific barrier to promoting change highlighted by several interviewees, was that the consultant's contact in an organisation is rarely the person responsible for the 'purse strings'. Only one of the fourteen consultants identified contact with senior management as a common occurrence. In this case, it was because the consultant, co-director of a firm with 10 years experience, made this a requirement in their contract with the client.

Clients' motivation to request help was also highlighted as significant in determining whether or not recommendations are implemented, as it was thought that *'...they often have the intention (or not) right from the beginning.'* For example, if a client was motivated by an adverse HSE inspection, they may be more likely to want to do as little as necessary to be seen to have done something. Finally, despite the lack of assessment of employees' or managers' knowledge, consultants suggested a lack of understanding as a further reason why recommendations are rarely fully implemented. As remarked by one sole-practitioner:

> *'Generally people have no idea what an ergonomics assessment is... This lack of understanding can make it less likely that recommendations are implemented properly... the consultancy report goes straight over most people's heads.'*

The importance of clients' understanding when recommending ergonomics changes was highlighted by the view that clients will usually do something to tackle the problem *'if the solutions are easy and inexpensive.'* Clearly, it would appear that these clients do not understand the complexity of ergonomics, and consequent benefit of large scale changes.

Evaluation
When asked whether they evaluated their interventions, and if not, how they know whether they have been successful or not, the common response was that 'we don't often get the chance to follow up'. General agreement existed regarding the lack of evidence for the benefit of consultancy approaches, many consultants doubtful that their recommendations are carried out in the manner intended. One interviewee suggested that this is either because consultants are not putting the message across clearly, or not highlighting cost benefit enough. A reason for the lack of evaluation given by a number of consultants was uncertainty of its practical benefits. One senior consultant explained that:

> *'We rarely evaluate, but then again there are pros and cons for doing this. On the one hand clients may see it as a thorough approach if...you say that you will come back and assess changes, others may just decide to go with someone who knows what they do will work.'*

For the most part though, lack of evaluation was attributed to factors outside of consultants' control, specifically that *'... companies seem very disinterested in evaluation. Once they have made their changes, they're on to the next thing.'* Where evaluation is conducted, it tended to be more for sales purposes, to initiate further work. As a result, when asked "how do you know that your proposed solutions are effective?", the consensus was that *'we don't really!'* This response was frequently accompanied by one of two qualifying statements, either the notion that it is *'better to get them* [clients] *to change some aspects than nothing'*, or *'the fact that we get asked back into certain companies is a good indicator'*.

Discussion and Conclusions

It was evident that ergonomics practice with respect to MSDs represented in the sample interviewed focused heavily on the physical aspects of work. This is at odds with the notion of ergonomics as a 'holistic approach to understanding complex and interacting systems' (Wilson, 2000), and the widely recognised importance of the contribution of psychosocial factors to MSDs. In emphasising physical issues, current ergonomics interventions may overlook critical facets of the work system. At least part of the reason for this may be the lack of techniques in the repertoire of ergonomics methodology available to consultants for assessing psychological and systems environments (Haslam, 2002).

The apparent lack of explicit attention to their clients' attitudes and beliefs has implications for the change process. As outlined in the introduction to this paper, it is important that recipients of advice and recommendations hold appropriate attitudes and knowledge if efforts to achieve change are to be effective. Of course it is also likely that knowledge, attitudes and beliefs will vary between different groups within an organisation, across employees, supervisors, managers, and directors (Haslam, 2002). At a managerial level, attitudes, beliefs, and understanding are highly significant to consultancy, given that individuals in these roles exert a strong influence on whether recommendations will actually be implemented or not. A failure to appreciate the complexity of ergonomics, and potential benefits that can be gained from effective implementation of ergonomic solutions, might underpin the tendency for organisations to implement changes themselves and to select from the recommendations the least expensive or easiest changes. Thus, assessing these factors is not only likely to increase the likelihood that changes are implemented, but also that they are implemented effectively. Further research is needed to evaluate and collate evidence on the benefits that might accrue from ensuring clients' knowledge, attitudes, and beliefs support the behaviours consultants are trying to elicit.

Another important finding from this study is the lack of evaluation performed, leading to uncertainty over the extent to which interventions are effective, also impeding consultants' ability to refine and improve their practice. The importance of evaluation and feedback for learning has been emphasised in both the organisational learning and systems thinking literature (e.g. Greve, 2003; Senge, 1990). Moreover, in today's culture of clinical governance and evidence based practice, there is a growing need for practitioners to prove the efficacy of their methods.

In conclusion, with MSDs continuing to be the most common form of work-related ill health in the workplace, this study has revealed important directions that should be examined in the attempt to make interventions to reduce the risk of MSDs more effective.

Acknowledgements

This work forms part of larger study supported by a grant from the Health and Safety Executive. We thank all of the consultants who allowed themselves to be interviewed.

References

Dey, I. 1993, *Qualitative Data Analysis: A user friendly guide for social scientists*. Routledge, London.

Greve, H.R. 2003, *Organizational Learning from Performance Feedback: A Behavioral Perspective on Innovation and Change*. Cambridge University Press, Cambridge, UK.

Griffiths, A. 1999, Organisational interventions: Facing the limits of the natural science paradigm. *Scandinavian Journal of Work, Environment & Health*, **25**, 589-596.

Haslam R A. 2002, Targeting ergonomics interventions - learning from health promotion. *Applied Ergonomics*, **33**, 241-249.

Health and Safety Executive. 2002, *Upper Limb Disorders in the Workplace*. HSE Books, Sudbury, UK.

McAtamney, L. and Corlett, E.N. 1993, RULA: A survey method for investigation of work-related upper limb disorders. *Applied Ergonomics*, **24 (2)**, 91-99.

Prochaska, J.O., and DiClemente, C.C. 1982, Transtheoretical therapy: Toward a more integrative model of change. *Psychotherapy: Theory Research and Practice*, **19**, 276-288.

Senge, P.M. 1990, *The Fifth Discipline: The art and practice of the learning organisation*. Random House Business Books, London.

Wilson, J.R. 2000, Fundamentals of ergonomics in theory and practice. *Applied Ergonomics*, **31**, 557-567.

Addressing Musculoskeletal Disorders in the New Zealand Log Sawmilling Industry

David Tappin[1]
Marion Edwin[2]
Tim Bentley[3]
Liz Ashby[4]

[1]Centre for Human Factors and Ergonomics, Forest Research, Auckland, New Zealand
[2]Optimise Ltd, Rotorua, New Zealand
[3]Department of Management and International Business, Massey University at Albany, Auckland, New Zealand
[4]Centre for Human Factors and Ergonomics, Forest Research, Rotorua, New Zealand

The paper presents findings from a government funded, industry-based study of timber handling tasks in New Zealand log sawmills, and discusses some of the issues encountered in conducting a long term study of this type. Twelve months' of accident register records from a sample of log sawmills were first analysed to determine the tasks most commonly linked with musculoskeletal disorders (MSD). Work system assessments were then conducted on two high risk tasks in a total of four log sawmills. Possible interventions were identified and priorities for implementation were developed following discussion with each mill. Informal implementation assistance was provided to two mills over the next 18 months, after which re-assessment of the two work systems was conducted in two sawmills with the main findings outlined in this paper.

Introduction

New Zealand produces approximately 20 million m^3 of plantation timber per year, of which 20% is processed by log sawmills into sawn lumber (NZ Forest Industry, 2003). Log sawmills employ around 50% of the workforce involved in this first stage timber processing (MAF, 2001).

The log sawmilling industry has been aware of the prevalence of MSD through their own injury statistics, however little has been done to assess the extent of the problem or identify causes and possible solutions. Anecdotally, efforts from within the industry have focused on increased mechanisation, not only as a way to improve system productivity but as a means of reducing worker exposure to perceived high-risk manual handling tasks.

In NZ, the Accident Compensation Corporation (ACC) scheme provides 24-hour no-fault personal accident insurance cover for both work and non-work injuries. ACC injury data from the four year period 1994/1995 to 1998/1999 (Laurs, 2000) placed log sawmilling as the wood processing sector with the highest level of new claims (42%). Of these claims, most were soft tissue injuries: strain/sprain (51%), along with back injuries (17%), and laceration (14%). More recent data from 1999/01 (ACC, personal communication) showed log sawmilling new claims to be 52% of the total for the wood processing sector, with similar proportions for injury types found as for previous years claims. Similar data from the British

Columbia Workers' Compensation Board (1999) lists overexertion claims in sawmills as causing 27% of all time-loss injuries and repetitive motion at 5%. Mill labourers and labourer material handlers comprised 51% of the workforce affected.

Recent initiatives from the Occupational Safety and Health Service of the Department of Labour (OSH) and ACC Injury Prevention to reduce industry injuries in high-risk sectors has focused on log sawmilling as one of their target industries. Regional inspection audits and the establishment of industry-based health and safety groups has led to an increasing awareness of injury prevention strategies among log sawmillers. This has provided a valuable forum for discussing, developing and implementing outputs from the study itself.

This government funded study commenced in 2001, with the aims of: determining the prevalence of musculoskeletal problems among log sawmilling workers; identifying high risk log sawmilling tasks; and designing and evaluating measures to prevent or alleviate musculoskeletal problems in these tasks.

Accident Register Survey

In New Zealand it is a requirement of the law for every workplace to hold an accident register. This is a potential source of industry information at a level beyond which ACC data can usually provide, such as identifying injury incidence for specific tasks or work areas involved in the injury. Twelve months' of accident register records from 37 log sawmills (representing approximately 26% of the NZ log sawmill workforce and 45% of the annual production volume), were collected and analysed to identify: (a) which tasks were linked to reported injuries; (b) the nature of these injuries; (c) the opinions of the health and safety staff in the mills involved.

Millhands (30%), Tablehands (26%) and Sawyers (23%) had the highest percentage of reported MSD injuries, with timber handling specifically mentioned in the task descriptions of 60% of all injuries. Back (37%) and upper limb (35%) were the body areas most commonly involved in all injuries. The opinions of the health and safety staff surveyed largely supported the accident data, with timber handling on the sorting tables being seen as creating the most significant injury risk (involving tablehands and many millhands). Filleting timber was also seen as a high-risk task.

Despite obvious weaknesses with this data, such as not being able to determine frequency or severity rates, there were tangible benefits derived from this phase of the study independent of the data. These included establishing an ongoing contact with a large number of log sawmills and other industry groups, and increasing understanding of the industry processes, technology and politics through this contact. Tappin, et al (2003) provides more information on this survey.

Analysis Phase

Method
Timber handling tasks at four sorting tables (two green timber tables, two dry timber tables) were assessed. These mills had all shown interest in the study during the accident register survey, and were therefore thought more likely to put in place interventions arising from the study. Yard filleting tasks were also assessed at two of these mills at a later date.

Data was collected over two main site visits totaling several hours at each mill, enabling different shifts, personnel, production volumes and outputs to be assessed. Assessment methods included: company archival data collection, semi-structured interviews with management and workers involved in the tasks, task verification/participation, physical workload analysis (REBA, RPE, Standardised Nordic Questionnaire, NZ Manual Handling Code assessment tools), and the collection of physical measurements (static anthropometric data, workplace dimensions, force measurements for handling timber).

Findings

In the initial work systems assessment, all physical workload analysis tools indicated a high number of MSD injury risk factors present in both the table and filleting work systems. Many aspects of the work design contributed to this, including piece rate payment, work compression, minimal task rotation, and limited task training. Analysis of anthropometric data with workplace dimensions indicated a number of significant mismatches in the workspace geometry of both tasks. Edwin, Tappin, & Bentley (2002) reports on preliminary work conducted with sorting table tasks.

Perhaps the most significant finding was that in all four mills, while there was a willingness to identify and address any risks present, there was also quite a low level of understanding about what the potential MSD risks were. On subsequent interaction with the industry, this would appear to remain a valid observation. Unsurprisingly therefore, prioritised interventions and further involvement were well received by all four mills, despite being critical of many of the systems in place at the mills.

Intervention Phase

Method

From assessment findings, interventions were developed and presented to each mill. These were prioritised according to their likely contribution to the occurrence of MSD and could be grouped into organisational design, physical design, and training design categories. Emphasis was placed on the increased likelihood of success through implementing a range of interventions from across these categories rather than focusing on just a few, or those that seemed easiest to implement (Karsh, 2001). Each mill then worked their selection of interventions into their own management plan, amending them to fit in with logistic, engineering, and business priorities. In some cases the implementation of interventions involving capital outlay or downtime occurred a considerable time after they were first presented, to fit in with planned plant overhaul or maintenance schedules. The number of mills involved in the ongoing study was reduced from four to two. After the initial presentation of interventions, 2-3 monthly contact was maintained through site visits, telephone contact and industry group meetings.

To enable as many interventions to be evaluated as possible, reassessment of the filleting and table work systems occurred at the two mills 9 and 18 months respectively after the initial findings were presented to the mills. The methodology was repeated as closely as possible to the original format.

Findings

Interventions, while specific to each mill, were largely based around workspace geometry improvements, reduction in forces required, workflow and workload management, and task technique training.

At the time that the mills were embarking on implementing their interventions, a health and safety group for log sawmillers and other timber processors was being established in the central North Island (the largest timber harvesting and processing area in New Zealand) by ACC and OSH. This group provided impetus to the study in two main ways – by encouraging changes within mills to reduce MSD and other health and safety risks, and by becoming a forum for the refinement of industry resources developed out of the work conducted. A similar group has consequently been initiated in the South Island.

Interventions that exhibited some tangible measure of success when evaluated included:

(i) Reduction in the force required to begin moving boards off tables through better surface maintenance and reduced area in contact with the boards. In one mill this force reduced on average by almost 50% for three common board dimensions.

(ii) Improvement in workspace geometry in one mill so that the need for reaching forward and double handling were eliminated, and transfer distances between source and destination were brought within a more comfortable range for a larger number of workers.

(iii) Workflow improvements in both mills through raised awareness, better communication between mill departments, and availability of trained additional staff if required. In one mill, planned mechanisation had been implemented but with consideration also given to reducing MSD risks. As a result, exposure to repetitive, rapid heavy handling had been eliminated, task rotation steps had increased, and work space increased in manual handling work areas.

(iv) Preventive maintenance programmes helped ensure ongoing risk reduction, particularly for high use areas such as table and chain maintenance.

(v) Staff perceptions in both mills were very positive about changes made, mainly due to feelings of ownership from involvement in the change process.

Conclusions

There are a number of limitations in the approach followed. Significantly, only four mills were involved, reducing to two later in the study. These mills were also approached on the basis of their willingness to be involved, with industry representativeness being a secondary consideration. This has been partly offset by the more recent involvement of a larger number of mills through the industry health and safety group. Between-mill and before-after comparisons were also limited by differences in work systems and individual sawmill priorities for implementation. Further limitations are listed in Edwin et al, 2002.

Conducting this study in industry has helped to raise the otherwise low profile of MSD among log sawmills as well as providing a willing test bed in which to trial and refine potential interventions, and a forum for refining and disseminating information of use to industry. Conducting industry-based studies does however mean that the study process is difficult to control and monitor. Issues such as business priorities, the pace and nature of implementation, and industry pressures can all have a significant impact.

Difficulties in conducting industry-based studies include:

• Even the most useful suggested interventions will need to be weighed against other business considerations. In this study these included; mill capacity; production targets; staffing availability, quality control standards; engineering resources and available capital.

• The timing of implementation is beyond the researchers control, despite the advised priority of the intervention. Two mills had to wait several months for a maintenance shutdown before making some physical changes. Such delays can create further problems.

- Depending on the degree of participation and involvement, the potential exists to get things wrong. Although improvements were made in one mill to workspace geometry, a significant component was excluded – with a likely reduction in overall effectiveness.

- Changes in company structure, personnel, and processes are likely to impact on the study in some way. For example, reducing rapport or credibility within the organisation, or making significant physical changes to the mill plant.

- Factors external to the industry may also have an effect. In NZ unfavourable exchange rates have made it harder to sell sawn lumber. There is less money to spend and priorities inevitably shift away from health and safety.

- A 'Hawthorne Effect' is very likely to remain in place as contact with mills continues over time.

The benefits of conducting industry-based studies include a growing level of rapport with the companies which may result in a greater level of possible participation – further involvement in implementation, and gathering more useful information from a wider number of staff, for example. Industry credibility may increase over time as they see the study as being 'real life' and offering practical, usable advice. In this study, the development of industry reports has been well received and has helped smooth the way for further initiatives in the industry developed from the original study. This will include the establishment of a log sawmill injury surveillance scheme.

Overall, the benefits of conducting this type of study within an industry setting outweigh the difficulties, as ergonomics principles are only relevant if ultimately applied to the end user. Demonstrating effectiveness of interventions might be better achieved through discrete on-off studies, but applicability to 'real life' situations relies on having the flexibility to use less tangible but realistic means and measures such as those employed at times throughout this study.

References

Edwin, M., Tappin, D., Bentley, T. 2002, Musculoskeletal disorders in the New Zealand Log Sawmilling Industry, *Proceedings of the 11th Conference of the New Zealand Ergonomics Society*, (NZES, Palmerston North), 112-117.

Karsh, B.T, Moro, F.B.P., Smith, M.J. 2001, The efficacy of workplace ergonomics interventions to control musculoskeletal disorders: a critical analysis of the peer reviewed literature, *Theoretical Issues in Ergonomics Science* 2(1), 23-96.

Laurs, M. 2000, *Analysis of ACC claims data for the period 1994/95 to 1998/99 for: Forestry and Logging, & Log Sawmilling and Wood Product Manufacturing*, ACC Report, Rotorua, New Zealand.

MAF, 2001, *New Zealand Forestry Statistics 2000*, Ministry of Agriculture and Forestry, Wellington, NZ.

New Zealand Forest Industry, 2003, *Facts and Figures 2003/04*, NZ Forest Owners Association, Wellington.

Tappin, D., Edwin, M., Moore, D., 2003, Sawmill Accident Register Records – Main Findings of a Survey from 37 Mills, *COHFE Report 4(5), 2003*.

Workers Compensation Board of British Columbia (1999). *Forest Products Manufacturing: focus report on preventing injuries to workers*. WCB, British Columbia, Canada.

EVALUATION OF SOFTWARE MONITORING FOR THE PREVENTION OF WORK-RELATED UPPER LIMB DISORDERS

Douglas R.S. Pringle, Elizabeth A. Kemp, Chris H.E. Philips, Duncan Hedderly, Brett Dickson, Michael L.K. Chan, Kim Findlay

Massey University,
Private Bag 11 222,
Palmerston North,
New Zealand

This project aimed at identifying and evaluating software monitoring tools for the prevention of work-related upper limb disorder or Occupational Overuse Syndrome (OOS). The project was undertaken in two broad phases: an initial study to identify and evaluate available software products which prompt rest breaks followed by a more detailed evaluation of selected products. The initial study involved a feature analysis and limited evaluation of ten products, which produced a short list of three. The detailed evaluation involved 24 subjects who each used the three pieces of OOS prevention software for a week. Data was collected through questionnaires and interviews. Two products scored equally well suggesting that some degree of choice is necessary if software is to meet the needs of the user.

Introduction

The objective of the research was to identify and evaluate break reminder software. Break reminder software usually monitors keyboard use in order to assist in the prevention of Occupational Overuse Syndrome (OOS). OOS is an umbrella term used in New Zealand to cover a range of conditions, which in the occupational setting are caused by prolonged muscle tension and repeated actions usually of the upper limbs. Similar umbrella terms are work-related upper limb disorder (WRULD) or repetitive strain injury (RSI).

A popular model for OOS is based on ischaemic muscle contraction and anaerobic metabolism (Large *et al*, 1990). Muscles are nourished by blood, which travels through capillaries inside the muscle. A muscle held tense in static contraction squeezes the blood out of the capillaries, making them smaller and slowing the flow of blood. The slower blood supply reduces nutrition and removal of the lactic acid waste products. The lactic acid stimulates pain receptors. This muscle pain may activate a self sustaining pain cycle as neighbouring muscles tense in sympathy to provide relief to the muscles initially held in static contraction.

A key component in preventing this pain is to allow replenishment of blood supply to muscles by using frequent short rest breaks or micropauses as recommended by Konz (1995). A micropause can be as short as a few seconds. Micropauses have been demonstrated as beneficial in office settings (Rohmert, 1973; Henning *et al*, 1996). Henning *et al*, (1996) found that self-management of rest break schedules by increasing user autonomy reduces the health risk associated with computer use. Considerable personal discipline is required of users if they are to take breaks of their own volition. An alternative is to encourage breaks through the use of break reminder software. A reminder to take a break is seen as being preferable to enforcing a break for example by disabling the keyboard. Enforcing breaks is likely to provoke stress and discomfort arising from the failure of users to achieve their task related goals (Darby, 1998). Darby agues that few people without an OOS problem would want to use break reminder software.

The intention of the research was to evaluate software-monitoring tools that prompt users to take breaks. In part a commercial motivation existed to ensure appropriate "value for money", but more importantly there was a need to determine user preference to ensure appropriate software monitoring tools were made available to University staff. Potentially all University staff using computers were the target audience. The research involved an initial study to identify and evaluate software products, which prompt rest breaks or micropauses, followed by a more detailed evaluation of selected software products, including user perception of using software products in every day work. Both academic and administrative personnel were involved.

Methodology

Screening and Feature Evaluation of Products

Software monitoring tools that prompt rest breaks can be evaluated on the basis of features such as the method of break pattern detection (the cues used by the software), the sophistication of the algorithms used, the nature of the prompts provided, the quality of the user interface, the assistance provided for the user, the quality of reporting, and other technical aspects of the software, for example performance and security.

An initial search of the web and other literature identified almost thirty products, which prompted rest breaks as a means of preventing and managing OOS. This list was reduced to ten in a 'pre-selection' phase, based on information provided by the software vendors.

A further screening process was undertaken to reduce the number of products to be evaluated in depth to a short list of three. The features of each software monitoring tool were assessed using the product as supplied by each vendor. One person completed feature evaluation with assigned weights to the relevant criteria (Jordan and Machevsky, 1990), so that all software-monitoring tools could be listed in order.

The features were grouped into two broad categories: user-centred issues, and technical issues. The features were weighted such that an overall product score out of 100 would be produced on completion of the feature analysis, with each of the two broad feature categories contributing 50%. Some of the features are inter-related. The weightings were based partly on experience gained in the pre-selection phase.

User-centred issues are concerned with the quality of the product as perceived by the user. The major component in this category is the nature of break enforcement, which includes a measure of the intrusiveness and quality of the prompting plus the quality of associated system output. This reflects the primary requirements of the product are to prompt the user to take breaks and to use these breaks effectively. The level of customisation and control available to the user, and the quality of other system outputs are also part of this category.

Technical issues include the approach adopted by the product in break pattern detection, factors associated with the setting up and on-going support for the software, and issues relating to performance, robustness and security. The major component in this category is the method of break pattern detection to reflect the primary purpose of the product.

As a result of the feature analysis, the top three products (referred to as A,B, and C) were selected for detailed evaluation.

Methodology for Detailed Evaluation

The detailed evaluation involved two groups of 12 subjects who each used the three pieces of OOS prevention software for one week. The first group consisted of 12 people randomly selected from a list of staff with reported OOS symptoms and known in the study as the *Symptom* group. These were matched as far a possible by a person from the same section (chosen by the manager) as they did a similar job but with no reported OOS symptoms (*Control* group). Participation was voluntary. Subjects who performed prolonged repetitive computer work were selected, although it was not the aim of this research to focus on "high need" or data intensive environment such as occurs in typing pools. Each group was composed of nine general staff and three university lecturers. Gender in each group was predominately female, with 3 males in the *Symptom* group and 4 in the *Control* group.

The study design used both quantitative and qualitative data to enhance credibility and the integrity of the findings (Patton, 1990). The group size was selected to ensure sufficient subjects to give statistical reliability. The detailed evaluation methodology was evaluated using a pilot study involving 3 people who were not part of the main study.

As subjects worked with each software product they kept a daily log in which they recorded the software that they used and any problems that arose. At the end of each week, they filled in a questionnaire consisting of nine positive and nine negative statements relating to OOS prevention software. "Halo" effects were minimised by using some negatively worded questions (Sekaran, 1984). Statements incorporated in the questionnaire were based on Software Usability Measurement Inventory (SUMI). SUMI measures the users' perception of the usability of software (Porteous *et al*, 1993). A within-subjects analysis of variance was carried out on questionnaire results, to determine which construct features distinguished a person's most preferred package from the other packages.

The order of software was randomized with every combination used to remove REcency effects. A debriefing interview was used to rank the software products in order of preference with the reasons for their choice. The interviews were semi-structured since the same issues were explored with all subjects but the order in which they were dealt with varied (Scott *et al*, 1991). All the interviews were taped and notes were taken. The conceptual framework of factors of interest (reasons for preference, wanting OOS software on the machine, view of quotes etc.) was established using FACET (Kemp and Gray, 1998). Then the text of interviews, answers to open-ended questions and the log was searched using the package NUDIST (Quality Solutions and Research Pty, 1997) for specified words or patterns.

Results and Discussion

The questionnaire responses suggest that attributes of a package fall into three categories. Those features that do not show consistent differences between packages were; help information, menu organisation, disrupting other software, needing support from a technical person, inexplicable behaviour, and the visual presentation. The attributes which are associated with the preferred piece of software included such things as; ease of use, helpful instructions and reminders about micropauses, as well as statements that are related to affect such as 'would recommend to others', 'felt tense', 'wanted to stop using', etc. The third category are those features which do seem to be linked to specific packages, regardless of preference – such as whether training is needed, user needs are taken into consideration, how easy it is to work with other software while the package is running.

Products B and Product C which both incorporated micropauses, had a large number of supporters (particularly in the *Control* group.) The responses to the questionnaires and the interview data help explain why Product A received fifteen third places overall. Product A had significantly lower scores than Products B and C with respect to efficiency and affect. Affect relates to people's emotional response to a package and is crucial when determining preferences. Product A received marginally worse ratings from all subjects for the questionnaire item "Would you like to use this product every day? The reasons why Products B and C were generally chosen first and second emerged during the de-briefing interviews as well as in the logs that were kept. Product B was seen by its supporters as providing low-key reminders to take breaks/micropauses. They also enjoyed its quotations. Product C was viewed as a very professional product that easily allowed users to tailor it to their requirements. The exercise range was seen as excellent.

It is not the reported preferences that are important in this study, but the overall findings. As products of this type are often upgraded no final recommendation may be possible. Most of the participants in the study were willing to consider running OOS prevention software often with the proviso that the application be the one preferred by them. The implication of this is that acceptance of the introduction of OOS prevention software by an organisation may depend for its success on giving users a choice. It might be that users as mentioned by Darby (1998) feel very frustrated if they feel that a particular application is imposed on them. More than one application, therefore, should be made available. Perhaps the overall conclusion that can be drawn from this study is that it can be counter productive to force a worker to use an application for OOS prevention that they dislike.

The other interesting findings in this study were that some of the *Symptom* group felt the need to skip exercises under the pressure of work. This was surprising as given their history one would have expected that they would welcome the opportunity to do something to assist with their condition. Perhaps, though, it is their very unwillingness to take a break that might have led to them developing OOS in the first instance. The support of the *Control* group for running OOS prevention software and doing exercises was unexpected.

Conclusion

The study described originated in the requirements of the University to select one piece of OOS prevention software but widened into a broader study of issues associated with software. The research was multidisciplinary and involved expertise in areas such Software Engineering, Human-Computer Interaction, Statistics, Health and Ergonomics.

An extensive research process was followed which had several stages: pre-selection of OOS prevention software, detailed feature evaluation of ten pieces of software, and an in depth study of three pieces of software. The in situ evaluation by the 24 subjects allowed for the collection of both qualitative and quantitative data. The multi-stage, multi-method research process followed was necessary for a realistic assessment of software products that run in the background. The analysis of the questionnaire and interview data provided a rich picture of the subjects' perspective whilst organisational and ergonomic issues were taken into account in the earlier stages.

The 24 subjects used the OOS prevention software for three weeks. On the whole this proved to be an educative experience for the subjects who were exposed to different approaches to taking a break. Whilst there was no overall preference, both Product B and Product C, which allowed micropauses as well as breaks to be taken scored ten first places. The main finding of the research, however, is that, where possible, people should be allowed to choose the software which best meets their needs. Ninety two percent of each group were willing to run their preferred OOS prevention application. In these circumstances they seem to achieve their goals and do not feel a sense of frustration.

An organisation will do better to offer a range of software rather than prescribing one particular product. This implies the need to allow employees to try out a range of products to allow the employee to select one that is seen as most appropriate.

References

Darby, F. 1998. Keyboard interrupts. *Safeguard*, (Colour Workshop Ltd, Auckland, New Zealand) (47), 35-36

Henning, R.A. Callaghan, E.A. Ortega, A.M. Kissel, G.V. Guttman, J.I. and Braun, H.A. 1996. Continuous feedback to promote self-management of rest breaks during computer use. *International Journal of Industrial Ergonomics* (18), 71-82

Jordan E.W. and Machevsky, J.J., 1990. *Systems Development: Requirements, Evaluation, Design, and Implementation*, (PWS-KENT Publishing Company, Boston)

Kemp, E.A. and Gray D.I. 1998. The pros and cons of using FACET for analysing qualitative data in Calder, P. and Thomas, B. (Editors) *Proceedings of OzCHI'98, IEEE Computer Society* (Los Alamitos, California), 208 – 214

Konz, S.A. 1995. *Work design: Industrial ergonomics*. 4th Edition, (Publishing Horizons, Inc. Arizona). 221-224

Large, R Butler, M. James, F. and Peters, J. 1990. A systems model of chronic musculo-skeletal pain. *Australia and New Zealand Journal of Psychiatry* (24), 529-536.

Patton, M. 1990. *Qualitative analysis methods,* (Sage Publications, Beverley Hills)

Porteous, M. Kirakowski, J. and Corbett, M. 1993. *SUMI User Handbook, Human factors Research group*, (University College, Cork)

Rohmert, W., 1973. Problems of determination of rest allowances. Part 2: Determining rest allowances in different human tasks. *Applied Ergonomics* 4.3: 158-163

Scott, C.A. Clayton, J.E. and Gibson E.L. 1991. *A Practical Guide to Knowledge Acquisition*, (Addison Wesley, Reading, Mas)

Sekaran, U. 1984. *Research methods for Managers*, (John Wiley & Sons, New York)

OCCUPATIONAL HEALTH AND SAFETY

HEALTH MANAGEMENT IN THE CONSTRUCTION INDUSTRY

C L Brace[1,2] and A G Gibb[2]

APaCHe - A Partnership for Construction Health
[1]Department of Human Sciences , [2]Department of Civil and Building Engineering
Loughborough University, LE11 3TU, UK

The construction industry is a dangerous business. Many safety initiatives have been implemented within the vocation, but for many years health has been the 'poor relative' of safety in 'health and safety' considerations. In order to understand how to tackle the growing epidemic of ill health, health and safety directors were interviewed about their health management techniques and their acceptance of current practice in related industries. The methods of best practice which were acceptable, usable systems were assimilated and a health management toolkit was developed.. The next stage will involve piloting the model and improving upon the design in an iterative process.

Introduction

The construction industry is a perilous business and it is inevitable that failing safety issues that cause workplace deaths should reach the headlines. However, far less appreciated are the serious and widespread difficulties associated with work-related ill health. Ill health is a major problems for construction workers. It can affect ability to work and have an impact on an individual's life. Very often there is a delay between exposure to hazardous materials and activities, and the onset of health problems.

Ill-health kills and maims large numbers of construction operatives. Every year many thousands of construction workers suffer from work-related ill health. This is due to exposure to hazardous substances used such as asbestos, silica and cement, as well as exposure to manual handling activities, and noise and vibration in the working environment. Recent data illustrate these hazards; the UK's self-reported work-related illness survey found an estimated 134,000 construction-related workers report a health problem caused by their work, resulting in an estimated 1.2 million days lost in a workforce of 1.5 million (Gibb, 2002). In particular there were 96,000 cases of musculoskeletal disorders; 15,000 cases of respiratory disease; 6,000 cases of skin disease and 5,000 cases of noise induced hearing loss. Hand arm vibration syndrome (HAVS) has also been identified as a health hazard as shown by recent research (Gibb, 2002).

Aim of the study

The researchers were tasked to develop a toolkit for managing health within one area of construction – civil engineering. It was paramount that the toolbox consist of *practical* guidance on *simple* management strategies for reducing the incidence of ill health amongst employees and sub-contractors. The toolkit was to be suitable for use in any contracting company, and to be made freely available as a resource to all those who could benefit in UK construction. The key components would be identification of main ill health effects of civil engineering activities; practical (but specific) actions that could be taken to reduce the incidence of selected ill health effects, including heath screening, case management, health surveillance etc.; formulation of realistic key performance indicators (KPIs) and monitoring arrangements (including standardised reporting arrangements, such as spreadsheets, proformas etc.), and; development of training materials as appropriate. At the same time, it was felt that a realistic baseline for current performance needed to be established and the liaison with other federations to seek common solutions was paramount. After these developments the material will be piloted with a range of civil engineering companies in order to trial and improve on the initial design and to create a full working model.

Potential benefits of the study

The principal benefit of the toolkit would be the establishment of meaningful occupational ill-health management arrangements for construction contractors. This would directly address one of the most difficult implementation issues facing the construction industry's Revitalising Health and Safety in Construction programme. This ultimately would help drive the industry's targets on reducing ill health amongst its workforce. The active management of health issues features heavily in the agendas of all the construction umbrella organisations, and in key initiatives, such as Accelerating Change and Rethinking Construction. Improving health and safety conditions in construction are also an essential component in helping to solve the industry's retention and recruitment problems. Another potential benefit would be the reduction of Employers Liability Insurance premiums for contractors adopting the developed approach.

Method

In-depth interviews (n = 12) were conducted with civil engineering health and safety management from major organisations, at locations across the UK. Each interview lasted approximately 2 hours and were conducted to identify main ill health effects, record baseline data, and understand the use of health related KPIs and monitoring arrangements. A selection of management from other major organisations in the wider construction sectors were also consulted (n= 15), in order to establish best practice from related disciplines. Analysis and development of the key themes was used to provide a framework for the toolkit. At each development, the interviewees were contacted and their opinions sought on the viability and relevance of the proposed design.

A Model for Toolkit Development

Dialogue with health and safety management yielded useful feedback. After management interviews with contractors both large and small, hand arm vibration syndrome (HAVS), manual handling, dermatitis (and hand injuries), and noise induced deafness were identified as the main ill health effects (key health issues) within the civil engineering sector. There were a range of pro-health activities happening, with differing uptake and awareness amongst the sector. Having considered the themes that emerged from the management consultation stage, and after reflection on information from other federations, it is proposed that the components below should be contained in the health management toolbox, further explained in Figure 1.

Component

1. Site Health Log Book

Similar to a Site Accident Book, this proforma will be completed after a reported episode of ill health, whether the problem results in periods of time off work or a case of mild discomfort.

2. Start up questionnaire

Questions will be asked about the general health of the employee on start up, for the records of the organisation. This information should be kept within the organisation for a 50 year period. Information will be provided for management/supervisors on the best ways of doing this.

3. Information on how to register with a GP

This information is available as an information pack for supervisors/management. After studying this, they will be able to give a toolbox talk (TBT) (education session) to site operatives, about what a GP is there for, the importance of registering with a GP and how to go about registering. The TBT will also cover other arrangements that are available in the local area, e.g. NHS direct, and what to do if an employee has a work related health problem.

4. Information letter to local GPs

The aim of this letter is to inform/educate local GPs on the health risks that site operatives may be exposed to. Therefore, if a site operative attends a health clinic with a work related problem, the GP concerned may be more likely to pick up on issues. This letter should be personalised by an organisation as required.

5. Information on key health issues

Educational documents for managers/supervisors will be produced to highlight the key issues, the latest facts and figures and how to avoid health problems, e.g. personal protective equipment, job design, specialised tools/equipment.. Having studied these, managers/supervisors will be able to perform health TBTs for site operatives. Proformas for records of attendance at TBTs will also be delivered.

6. Self-health checks for operatives for key health issues

Self administered health checklists for uncovering symptoms of ill health will be produced. These will be completed after operatives have been educated by the TBTs.

7. Information on health screening

This information will be available for managers/supervisors who may be thinking about health screening for their workforce. It will highlight the types of medical screening that are available from health care providers and the approximate costs involved. Contact details of health care providers will also be supplied.

8. Periodic updates

This information will be available to companies as and when there are updates in the area of occupational health. This will include information about legislation, research findings and new equipment that has an impact on health. The information will be distributed in a user friendly format that can be applied easily to the workplace.

9. Trade Union assistance

It is envisaged that information will be available for employees and employers with input from the Trade Unions, about what usually happens if a worker has a health problem. The information will be targeted at the range of individuals in the company. It is the intention that this material will be able to answer queries and concerns of the workforce regarding job security and health.

10. Key performance indicators

Member organisations will be given information on the use of KPIs, and how to collect the information. They will also be issued with proformas to keep track of the following indicators: workforce GP registered; workforce sickness absence; workforce welfare; workforce health; civil claims strike rate;

Implementation visits

Before implementation of the toolbox, the research team will be available as necessary to visit the organisation concerned to highlight the importance of the key issues, run through the toolkit and to answer any questions about it's use. It is predicted that it would be useful for representatives from several areas of the business to attend (as applicable) including: senior management/directors, personnel/human resources management, site supervisors, health professionals, health and safety representatives, and union representatives. This is in order that the message about managing health is joined up throughout the organisation and that everyone understands their role in how to improve health.

Follow-up contact

It is anticipated that a member of the research team will contact each participating organisation during the initial year of toolbox implementation. This will initially be over the telephone or by email with visits to an organisation as required. This contact is to 'troubleshoot' any problems, to understand how the toolbox is working in practice, to continuously and iteratively improve it, and to ensure consistency and maintain engagement by participating organisations.

This contact should also assist with (anonymous) performance monitoring, the compilation of periodic progress reports, the feedback of learning between CECA members, the assessment of supply chain response, and the maintenance of KPI data.

Conclusions

A model for health management has been developed based on interviews with health and safety management and the guidelines in the literature. The next phase will involve piloting the model and improving upon the design in an iterative process.

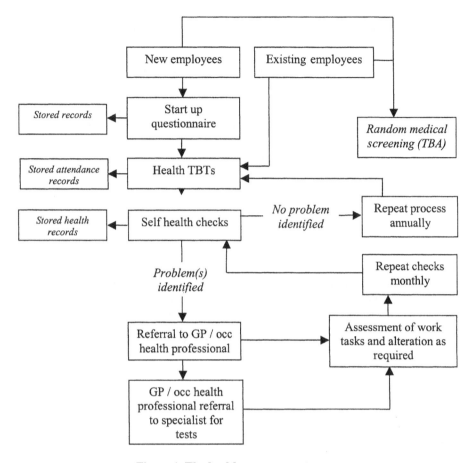

Figure 1. The health management process

Acknowledgements

The authors wish to acknowledge the support of the Civil Engineering Contractors Association (CECA) who sponsored this research. The views expressed, however, are those of the authors and do not necessarily represent those of CECA.

References

Gibb, A G F (2002), Health, Safety's Poor Cousin – Keynote presentation, *CIB W99 Triennial International Conference.* Hong Kong, May 2002, awaiting publication by Spon, Rowlinson, S (ed).

Ergonomic Intervention at Sewing Industry Workplaces

Rolf P. Ellegast, Christian A. Herda

Berufsgenossenschaftliches Institut für Arbeitsschutz (BIA)
Alte Heerstr. 111
53754 Sankt Augustin
Germany

The German statutory accident insurers initiated an ergonomic intervention study to examine the potential factors behind the elevated rates of sickness-related absenteeism in the German sewing industry attributed to diseases of the musculo-skeletal system. In the first part of the study, the actual state of the musculo-skeletal load situation at different sewing workplaces was analyzed using a newly developed measuring system. On the basis of the resulting measurements an ergonomic sewing workplace was designed and then field-tested in different areas of the sewing industry. The load reduction at the new ergonomic workplace was indicated by new measurements.

Introduction

The German sewing industry has experienced elevated sickness levels with the consecutive absenteeism for years. The main type of diseases found in industrial sewing are musculo-skeletal diseases, and particularly those effecting the spinal column and the shoulder-arm system. The risk factors in industrial sewing jobs for the musculo-skeletal system have been explored in several scientific studies (e.g. Blader, et al. 1991; Westgaard and Jansen 1992). The highly repetitive and often one-sided strains on the hand, arm, and shoulder musculature as well as the typical, intensely forward-leaning posture while seated were described. The key question for this study was: Does an ergonomic design of sewing workplaces lead to a reduction in the known risk factors and thus to a reduction in the typical strains arising from sewing? A field study was initiated using a newly designed mobile motion analysis system to explore the real conditions at such workplaces.

Methods

A newly developed measuring system, named CUELA (Computer-assisted registration and long-term analysis of musculo-skeletal load) (Ellegast and Kupfer 2000) devised by BIA was used to determine the bodily posture and the related musculo-skeletal loads. In the CUELA device, electronic angle and angular velocity sensors mounted at the joints and on other parts of the body provide the necessary information on position and angle in order to enable kinematic reconstruction of the test subject. Table 1 shows the detected body angles of the joints and bodily regions recorded using this measurement technology.

Table 1. Measured body angles and measuring sensor using the CUELA system

Joint/bodily region	Degree of freedom	Measuring sensor
Head	Inclination, flexion/extension	Inclinometer
Cervical vertebrae	Flexion/extension	Calculated
Dorsal vertebrae/ Lumbar vertebrae (seperately)	Inclination, flexion/extension, side inclination	Inclinometer, gyroscope
Pelvis	Inclination (sagittal)	Inclinometer, gyroscope
Hip joint	Flexion/extension	Potentiometer
Knee joint	Flexion/extension	Potentiometer
Scapula	Depression/elevation, anterior/posterior	Potentiometer
Shoulder joint	Flexion/extension, ad-/abduction, inner/outer rotation	Potentiometer
Elbow joint	Flexion/extension	Potentiometer
Forearm	Pro-/supination	Potentiometer
Hand joint	Flexion/extension, radial/ulnar duction	Potentiometer

The system includes a data storage unit with a flash card and its own battery supply worn on the body of the test subject. It is able to record movement data over several hours with a sampling rate of 50 Hz. After the measurement is concluded, the flash card can be read directly from a computer for further analysis. The measurements are accompanied by video documentation. Synchronising the video film with the measurement data later allows for a simple association of specific load points to related work situations.

A specially developed CUELA software makes it possible to display the posture at any random time during the measurement on a three-dimensional computerised figure, the measured bodily angles as time angle graphs and the corresponding video images automatically (see Figure 1). In addition to that the software allows for an export to statistical software.

In the following, the assessed risk factors for the musculo-skeletal system and the related evaluation schemes are specified: Extreme joint angle positions (evaluation as in ISO 11226, prEN 1005-4, Drury 1987, McAtamney, et al. 1993), static postures (evaluation as in EN 1005-1), repetitive movements (evaluation as in Silverstein, et al. 1986, Kilbom 1994), force exertion (conducted using EMG by the Munich university of applied sciences).

The intervention study was conducted at a total of four businesses in the sewing industry in the fields of footwear, technical textiles, clothing, and stuffed animal production. Two workplaces were selected per business for the study, for a total of eight seamstresses participating in the study as voluntary test subjects.

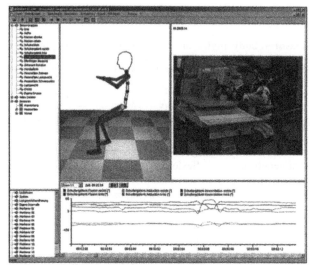

**Figure 1. Display of the measured data and the work
situation – CUELA-software**

The procedure of the study was organised as follows: First the above described load factors were measured over a period of three hours during normal sewing work tasks before intervention at the various workplaces. After identifying the main points of musculo-skeletal load the workplaces were ergonomically redesigned. The ergonomic work equipment was installed in the companies, and after a period of adjustment the equipment was studied once more during a normal work shift. An especially designed questionnaire recorded the subjective attitudes of the seamstresses to their new workplaces.

Results

Figure 2 shows the results of the status analysis for the average trunk flexion angle as an example for the four sewing areas footwear, technical textiles, clothing, and stuffed animal production.

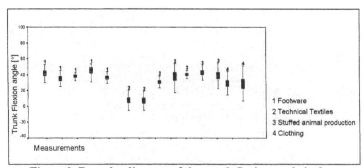

**Figure 2. Box-plot diagram of the trunk flexion angle in four
different sewing areas**

Figure 2 is a box-plot diagram, in which the value groups are depicted in an inner box whose boundaries are defined by the 25th and 75th percentile of the measured body angle distribution (here: trunk flexion angle). The median (50th percentile value) of the distribution is given as the central value in the box. Figure 2 shows clearly that static back postures lie mainly within a range of trunk flexion angles between 30° and 40°.

Similar load characteristics also resulted from the analysis of the neck flexion angles. In the shoulder-arm region, a higher proportion of repetitive activities was found in which the arms are often raised to the front and to the sides and the lower arms were turned inwards (pronation).

These results led to the following requirements for the new ergonomic sewing work place: Individual adjustability of the table and chair to the different anthropometrics of the individual worker, tilt adjustment of the work surface, the option of working in a sitting or standing position (dynamic work), more extensive leg room, adjustability of the table surface to the weight and size of the piece(s) being sewn, the creation of variably adjustable armrests. The developed ergonomic sewing workplace meets all above listed requirements. It was installed in different sewing companies and musculo-skeletal load factors were once more measured using the CUELA system after the workers had a chance to get used to it.

Figure 3 illustrates the average flexion angle of the back in a box-plot diagram before and after the ergonomic redesign at sewing facilities for stuffed animal production. It is clear that the spinal column postures improved considerably during sewing work at the ergonomically redesigned workplace than was the case at the conventional workplace. Similar positive results were obtained for the postures of the cervical spine and of the shoulder-arm system. As for the risk factor "repetitiveness", which is largely caused by the sewing tasks themselves, no significant improvements could be identified. The seamstresses were very receptive of the ergonomic redesign of their workplaces.

The results of the study show that an ergonomic workplace design can lead to a considerable reduction in extreme postures in the area of the spinal column and of the shoulder-arm system. The ergonomic sewing workplaces have been installed in several sewing workshops and have been well received by the seamstresses in the meantime. The findings of the study will in future be used for developing ergonomic operating instructions and work rules for sewing production companies.

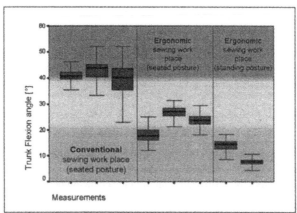

**Figure 3: Box-plot diagram of the average flexion angle before and
after redesign at sewing facilities for stuffed animal production**

Acknowledgement

The authors wish to thank the Institute of Ergonomics of Munich university of applied sciences (Prof. Lesser), the engineering studio Schwan Frankfurt (Mr. W. Schwan) and the institution for statutory accident insurance and prevention in the textile clothing and leather industry for their cooperation and support in this project.

References

Blader S.; Barck-Holst U.; Danielsson S.; Fehrm E.; Kalpamaa M; Leijon M.; Lindh M.; Markhede G. 1991, Neck and shoulder complaints among sewing-machine operators, *Applied Ergonomics*, **22**, 251-257

DIN EN1005-1 2002, Menschliche körperliche Leistung Teil 1: Begriffe. (Beuth, Berlin)

Drury, C. G. 1987, A Biomechanical Evaluation of the Repetitive Motion Injury Potential of Industrial Jobs. Seminars in *Occupational Medicine*, **2**; 41-49

Ellegast, R. and Kupfer, J. 2000, Portable posture and motion measuring system for use in ergomomic field analysis. In: Landau *Ergonomic Software Tools in Product and Workplace Design* (Ergon, Stuttgart)

ISO/CD 11226   1995, Ergonomics - Evaluation of working posture, International Organization of Standardization.

Kilbom, Å. 1994, Repetitive work of the upper extremity: Part I – Guidelines for the practinoner. *International Journal of Industrial Ergonomics*, **14**, 51-57

McAtamney L. and Corlett, E. N. 1995, RULA: a survey method for the investigation of work-related upper limb disorders. *Applied Ergonomics*, **24**, 91-99

prEN 1005-4 2002, Menschliche körperliche Leistung Teil 4: Bewertung von Körperhaltungen und Bewegungen bei der Arbeit an Maschinen. (Beuth, Berlin)

prEN 1005-5 2003, Sicherheit an Maschinen - Menschliche körperliche Leistung - Teil 5: Risikobeurteilung für repetitive Tätigkeiten bei hohen Handhabungsfrequenzen. (Beuth, Berlin)

Silverstein, B. A.; Fine, L. J.; Armstrong, T. J. 1986, Hand wrist cumulative trauma disorders in industry. *British Journal of Industrial Medicine*, **43**, 779-784

Westgaard R.H. and Jansen T. 1992, Individual and work related factors associated with symptoms of musculoskeletal complaints. Different risk factors among sewing machine operators. In: British Journal of indutrial medicine, **49**, 154- 162

Manual Handling in the Construction Industry: Finding a Format for Change

Phil Bust, Alistair Gibb, Roger Haslam

Loughborough University,
Loughborough,
Leicestershire, LE11 3TU,
United Kingdom.

As part of Loughborough University's APaCHe (A Partnership for Construction Health and Safety) program a 12 month project, sponsored by the Construction Health and Safety Group, was undertaken to investigate the manual handling of highway kerbs. As well as examining the physical issues of the task and risk transfer of any new initiatives a detailed assessment of the operation within the construction project system was undertaken. This covered design and training issues as well as the effect of product manufacture and the role of the specifying bodies. It is planned to carry out further research in this area to develop a strategy that can be used by the construction industry to reduce manual handling risks in other areas.

Introduction

The construction industry in the United Kingdom, with the demise of the coal and heavy industrial manufacturing industries is one of the dwindling bastions where large strong men are valued. As such it continues to carry out work practices that have long since disappeared, removed from other areas of work by regulations designed to protect workers from musculo-skeletal injuries and work related upper limb disorders.

Figure One: Manual Handling of Kerbs

To address this problem the Health and Safety Executive (HSE) have targeted areas of manual handling within construction and worked with the industry to put together controls. This has resulted in manufacturers reducing the weight of cement bags from 50kg to 25kg and designers no longer specifying large concrete block units in walls. The manual handling of concrete highway kerbs is now an HSE enforcement priority.

A meeting organised by the HSE was held at the Contractors Confederation offices in London in December 2003 for the key stakeholders for kerb installation. It was agreed at the meeting that the industry would aim to have controls in place to replace manual handling of kerbs in large works by June 2004 and smaller works by January 2005. To assist with this the HSE plan to introduce guidelines early in 2004.

In the face of enforcement and without guidelines contractors represented by the Construction Health and Safety Group have sponsored this work. It is hoped that this will provide the necessary information for the industry to move from the traditional manual handling methods to either mechanically assisted methods of lifting the heavy kerbs (usually 70kg) or adopting alternative solutions by changing the specification, materials or design.

Method

Loughborough University have for some years through their construction health and safety unit been using the experience of the construction and human science departments to carry out research into health issues in construction. It was decided to look at all of the manual handling operations in the lifecycle of the kerbs – from the manufacture through delivery, storage, installation and removal – and investigate the current organisation of the work practices to see if redesign or reorganisation of these could provide any benefits.

A desk study was carried out which involved a search of library documents, industry publications, data bases and a general internet search for related research information and details of alternative methods. Eventually a number of site visits were undertaken to observe kerb installation work which was recorded on video and with digital photographs. Those responsible for the organisation of the work and the workers carrying out the installation were interviewed.

Steering group meetings were developed with the addition of different experts into focus group meetings to discuss specific areas of the research such as manufacture, design and training. Steering group meetings were recorded on audio tape which was then transcribed so that themes and salient points could be identified.

The video and digital photo recorded information was used to further observe the work. Then to construct hierarchical task analyses (HTA) to identify small elements of the work which could then be tabulated with risk of injury noted and appropriate control measures listed. Postural loads were calculated for the separate sub tasks using REBA (Hignett and McAtamney 2000) postural analysis tool.

Findings

The search for related information revealed that no research work had been carried out on this specific task. However, there was research work on manual handling in construction (Niskanen 1993) and also looking at ergonomics in highways work. A tool has been developed (Buchholz,1996) in response to ergonomic hazards only being characterised crudely. PATH (Posture, Activities, Tools and Handling). There is however no mention of the handling of highway kerbs.

A number of types of lifting equipment were found designed specifically for highway kerbs and there were alternative types of kerbs. The kerb alternatives were not necessarily designed to remove manual handling.

Early visits to observe kerb installation involved purely manual operations with no use of mechanical lifting. It was noted that the installation process included other operations that represent a hazard to the workers – using large rubber headed hammers to tap the kerbs into place and shovelling of concrete. Initial reports that mechanical lifting clamps were not being used proved to be wrong as equipment was found to be used so much that it had worn out and been repaired. Vacuum lifting equipment had begun to be put on trial to assess it's suitability to replace the manual operations.

With the introduction of vacuum lifting equipment, the type which had been used in factories for decades, came problems as not all the parties involved in the process were working with their use in mind. Manufacturers were not despatching kerb units in a way that was assisting with the use of the equipment but instead to prevent damage in transit. Designers were not all stating in contract documents that mechanical lifting of kerbs was required so contractors that allowed for it's use were at a disadvantage when pricing for work.

Although manufacturers were not forthcoming with any plans to change the specification of the concrete kerb units it was discovered that trials had been carried out on thinner (100mm instead of 125 mm) kerbs and also that shorter kerbs (either 500mm or 600mm) had been produced.

When looking at training of operatives to install concrete kerbs it was found the industry's main training organisation, Construction Industry Training Board (CITB), generally provided training for employees of main contractors whilst the work would be carried out by sub contractors. The Highways and Construction Training Association HCTA had a large local authority base which provided training for it's own workers. Both the CITB and the HCTA concentrated their training on the technical aspects of the kerb installation, included a small element covering manual handling and the use of mechanical lifting clamps.

Figure Two: Vacuum Lifting Equipment

The use of the REBA (Hignett and McAtamney 2000) postural analysis tool confirmed that the manual operations produced heavy postural loading which was reduced with the use of the mechanical grabs and virtually eliminated, for the lifting operation, with the use of vacuum lifting equipment. The shovelling and hammering operations were also of a medium to high risk and were still present when using mechanical grabs or vacuum lifters.

Use of Hierarchal Task Analysis to deconstruct the operation into tasks and sub tasks enabled a tabulated form of the smaller elements to be used to identify risks and begin to find appropriate solutions. Then comparisons of operations were made to see which carried less risk.

Discussion

The visits to sites showed the heavy manual handling to be widespread and very much the accepted way of working. Although visiting a relatively small sample of the total amount of kerbs work this did include manual handling, the use of scissors clamps and vacuum lifters. For each of these operations there were good techniques and bad techniques observed.

Alternatives to the concrete kerb have not always been developed to reduce or eliminate manual handling. Harlow Council have with the assistance of a company who deal with recycled rubber developed a rubber kerb to be used where concrete kerb replacement is high due to impact of heavy vehicles. Although there is little weight advantage with this alternative the number of manual handling operations is reduced because the kerb is replaced less often.

Focus group meetings held to discuss specific issues brought out some of the problems that were not obvious from just observing the work. The manufacturers were content to continue to provide the industry with what was asked for and adding a note in their health and safety sheet for the product to say that manual handling should be avoided.

Figure Three: Slip Form Method

From early investigation the use of lifting equipment was the industry choice rather than using kerbs made of alternative materials or alternative production methods such as slip form construction. The equipment varied considerably and in some cases in the view of the focus group some manual clamps represented a greater risk than manual handling.

When tackling the question of design we were able to gather wide representation for our meeting and discussed the issues in detail. Parallels were drawn with the mobile elevated working platforms. Once they were rare and now they are commonly seen on sites. Also, the alternatives to kerbs were compared to proprietary thin surfaces replacing hot rolled asphalt and the need to provide certification for new products so that designers are supported in their choice in specifying them.

It was felt that designers were not present on sites often enough and this made their part in complying with the Construction (Design and Management) Regulations (CDM) difficult.

The two main training organisations Construction Industry Training Board and the Highways and Construction Training Association provided training for the workers who were then left and not policed to see if they put their training into practice.

The REBA tool was useful in putting numbers to the work observed and thus confirming views of seriousness of the practices through observation in addition it allowed the identification of those parts of the body most heavily loaded. The REBA scores do however only represent a snapshot of the operations and it is still necessary to consider the exposure of the workers to each task and what other work is carried out by them on the days they are not laying kerbs.

Using the findings of this project a prototype assessment tool was constructed to enable the severity of a kerb installation project to be quantified. It is hoped to get feedback from industry partners as to the effectiveness of the tool before tailoring it to be used with other manual handling operations.

Conclusions

The construction industry appears to be lagging behind manufacturing with regard to manual handling operations, ergonomic problems in the manufacture of prefabricated concrete elements were being investigated in the early eighties (Grandjean 1983). This is largely due to the temporary nature of construction sites and a transient workforce being difficult to regulate.

The industry is reacting to enforcement, tackling sections of the problems of manual handling as they become Health and Safety Executive enforcement targets, going from cement bags to heavy blocks and now kerbs. Everyone is looking for guidance in a quick and easy to follow form that can help them comply with a raft of regulations that are more detailed and which they are unlikely to tackle.

It is obvious that change is required as bad techniques using a vacuum lifters are far better than good techniques using manual handling methods.

It is hoped progress can now be made in complying with the manual handling regulations as the industry tackles the targets set at the stakeholders meeting. With this in mind Interpave who represent the manufacturers of the concrete kerbs have asked Loughborough University to work with them to provide guidelines for their products to be used in a safe and healthy manner.

REFERENCES

Buchholz, B. et al. 1996, PATH: A work sampling-based approach to ergonomic job analysis for construction and other non-repetitive work, *Applied Ergonomics*, **27**, 177-187

Hignett, S., McAtamney, L. 2000, Rapid Entire Body Assessment (REBA), *Applied Ergonomics*, **31**, 202-205

Grandjean, E 1983, Occupational Health Aspects of Construction Work, *WHO Euro Reports and Studies 86.* World Health Organisation , Geneva

Niskanen, T. 1992, *Accident risks and preventative measures in materials handling at construction sites.* Helsinki: Helsinki University of Technology; 72pp

Measurements of the Forces needed to Move Patients in Mobile Lifting Hoists and Hospital Beds

David McFarlane

Health and Community Services Team,
Occupational Health and Safety Division,
WorkCover New South Wales

Abstract

Our measurements of the forces needed for manual handling tasks in hospitals and aged care facilities have shown that the forces for pushing wheeled loads are often higher than expected (particularly when the workplaces are not designed for their use). These high force levels were often associated with injuries in cases where only one worker performed the tasks. In one investigation of a series of serious shoulder injuries at a retirement hostel we assessed the task of moving a mobile hoist between a bedroom and an en-suite bathroom task. We found that the injuries were caused by the difficulty of moving the hoist over a ridge in the floor. The practice of moving beds up ramps has also caused a number of injuries in recent years. In one investigation we found that the task of transporting patients in their beds up a ramp in a hospital had caused three compensation cases in the course of a year. The force needed for this task was found to be excessive for one person.

Introduction

As an ergonomist working in a safety inspectorate I am often asked to perform assessments of tasks that have been associated with major accidents and injuries. One very common finding is that the employer has underestimated the level of force needed to perform a task and has permitted one worker to perform the task alone due to staff shortages. Published data on investigations of this type seems to be scarce.

Method

We simulated the tasks while applying the forces needed as pull forces instead of push forces (it is usually not practical to measure the force needed to push a large load over a sizeable distance as it is so difficult to steer it). The forces were measured in kilograms with a Mecmesin advanced force gauge (model AFG 1000N). In some cases the force measurements were instantaneously recorded and graphed (the data was down-loaded to a lap-top computer and stored with Dataplot software). The slope of the ramp at the scene of the incident was measured with an inclinometer. The instrument used was a Macklanburg Duncan "Smarttool" electronic spirit level (an inclinometer that is capable of measuring gradients to the nearest 0.1 degrees).

Results

Investigation 1.
We made a detailed assessment of a task that caused two reported accidents at a retirement hostel. This task was moving a patient on a patient hoist over a lip (a carpet trim strip) between a bedroom and an en-suite bathroom. A member of staff (a 1640 mm tall female) simulated the task to assess the use of a mobile patient lifting hoist (a Pro-Med Pixel) to move an adult female weighing 65 kilograms as a surrogate patient.

The hoist was pulled across the lip after it had been stopped outside the 740 mm wide doorway and re-positioned so that it could pass through without any part of the patient or hoist colliding with the doorframe or the sliding door. The force required to perform the task of pulling the loaded hoist from the bedroom to the bathroom was then measured (in kilograms). A 14.60 kg force was needed to pull the loaded hoist across the lip from the bedroom to the bathroom when the measurement was made at a brisk walking speed. The task was repeated at a slow speed and the hoist was stopped with the castors at the outer edge of the lip. A 44.18 kg force was needed to pull the loaded hoist across the lip between the bedroom and the bathroom at negligible speed after the hoist had been stopped outside the door in order to ensure that it could pass through the door safely.

Graph 1: The force needed to pull patient lifter across door lip at walking speed

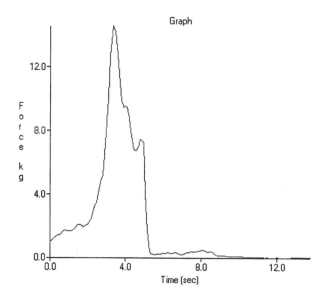

Only the fourth tug was large enough to accomplish the task (as shown in graph 2).

**Graph 2: The force needed to pull patient lifter across the lip of
the bathroom door at negligible speed after stopping**

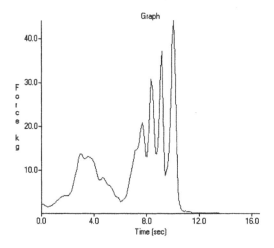

Investigation 2.
The Rehabilitation Coordinator of a public hospital supplied us with a risk analysis
showing that in one year alone three of the largest Workers' Compensation claims made
by orderlies were based on injuries associated with the task of pushing beds up ramps.
The association of the task to the subsequent injuries was traced through hospital incident
reports. One of the claims resulted from a back injury, one resulted from a knee injury
and one resulted from a hernia.

During our inspections we observed that this task was always performed by one worker
working alone (although the management insisted that it was supposed to be performed by
two workers). The maximum slope of the steepest ramp (a ramp adjacent to a Radiology
Department) was 5.5 degrees; this indicated that its design complied with the
requirements of the building code. We measured the forces needed to push beds up this
ramp. A large force was needed to pull a bed (loaded with a surrogate patient weighing
100 kgs) up the ramp; a peak force of 34.4 kgs was recorded and similar high force levels
had to be maintained for about half a minute (see graph 3). We concluded that the force
needed for this task was excessive for one person. It was also tiring. The problem of turning
the load at the head of the ramp after it had acquired considerable momentum could also
have been a contributory factor.

**Graph 3: The Force needed to Pull a Bed with
a Patient weighing 100 kgs up a Ramp**

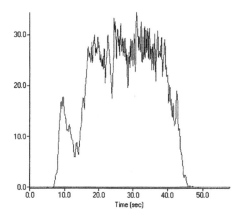

Maximum force levels found in investigations

In the three years 1998 to 2000 we investigated three complaints from workers in the health and aged care industries that related to the application of pushing forces to wheeled loads (beds, patient lifting hoists and trolleys) with both arms at or near shoulder height. In these cases the highest force needed for the tasks varied from 18 to 30 kgs (the tasks required 18, 26 and 30 kg forces respectively); none of these tasks had resulted in any injuries. However, consultations with the workers and simulations of these tasks showed that they were tiring (particularly those where a trolley had to be moved for a considerable distance).

In the same time period we did three investigations of similar manual handling tasks in the health industry (two of them were described above) that had resulted in serious injuries (those where a worker had been absent from work for a week or more). All of these tasks required forces that exceeded 34 kgs (the tasks required peak forces of 34, 44 and 66 kg forces respectively). The forces had to be sustained for less than a minute.

Force Limits for Pushing Tasks associated with Complaints

The maximum force needed for these tasks did not exceed the 30 kg force limit they proposed for two-handed pushing at shoulder height by fit young men. However most of the workers performing these tasks were females. The range of peak forces associated with these complaints were reminiscent of proposed limits for pushing forces for females based on psychophysical criteria (Shoaf et al, 1977).

Force Limits for Pushing Tasks associated with Injuries

The amount of force needed to push a loaded patient hoist from a bedroom to the bathroom across a ridge on the floor (an edging strip between the carpet of the bedroom and the vinyl floor of the bathroom) was estimated to be about 44 kilograms. The force needed to push a bed (with a patient and life support equipment on it) up a ramp varied from 34 to 37 kilograms. The forces required for both of these tasks exceeded the 25 kg force limit for two-handed pushing tasks at chest height for fit young males proposed by Davis and Stubbs (1977) and the Manual handling Research Unit (1980). Both of them also exceeded the 30 kg force limit they proposed for two-handed pushing at shoulder height by fit young men. Moreover many of the injured workers were females.

Discussion

The practice of moving beds through narrow doorways and up ramps has caused a number of shoulder injuries in recent years. Tasks involving moving beds and patient lifting hoists over ridges on the floor and onto the floors of lifts have also caused problems. We noticed that some employers have attempted to estimate the forces on the basis of published data or relied solely on the opinions of their supervisory staff. Consultations with workers are evidently part of the answer but on their own they do not always identify the precise causes of injuries. Calculations of the forces appear to be only partially successful. For instance, Potiki has estimated that the amount of force needed to push or pull a non-deformable wheel (or castor) over a 5 mm high ridge is between 15 and 20 percent of the trolley's weight (Potiki, 1994). Potiki used a formula to calculate the force needed to push a trolley over a vertical ridge but the formula does not include the velocity so any estimate derived solely from it does not take account of the speed factor. His formula predicts that a 30.2 kg force would be needed to pull the hoist over a vertical 6 mm high lip. We found that a 44 kg force was needed. Our measurements suggest that the formula may underestimate the force needed at a very low speed. We conclude that actual force measurements are often necessary.

Conclusions

Our measurements of the forces needed for some manual handling tasks in hospitals have shown that the peak force levels needed to move patients in hoists and beds can be much higher than anticipated if the workplace is not specifically designed for their use.

Having manual handling tasks performed by only one worker appeared to be a risk factor in tasks associated with serious injuries.

Our findings suggest that when assessing manual handling tasks there is a need to both consult the workers to pinpoint the most onerous tasks and make force measurements.

References

Davis P. and Stubbs D., 1977, Safe levels of manual forces for young males II, *Applied Ergonomics*, **8**, 4, Diagram 13, 224.

Manual Handling Research Unit, 1980, "*Force Limits in Manual Work*", (University of Surrey), 24.

Potiki, J., 1994, "*Trial Of A Task/User/Environment Methodology To Develop Ergonomic Guidelines For The Design And Use Of A Manually Handled Hospital Trolley*", a treatise for the degree Master of Safety Science at the University of New South Wales, 225 to 230.

Shoaf C., Genaidy A., Karwowski W., Waters T. and Christensen D., 1997, Comprehensive manual handling limits for lowering, pushing, pulling and carrying activities, *Ergonomics*, **40**, 11, Table 14, 1192.

PROCESS OWNERSHIP AND THE LONG-TERM ASSURANCE OF OCCUPATIONAL SAFETY: CREATING THE FOUNDATIONS FOR A SAFETY CULTURE

C.E. Siemieniuch[1], M.A. Sinclair[2]

[1]*Dept of Systems Engineering,* [2]*Dept of Human Sciences,*
Loughborough University
LE11-3TU

This paper addresses the longer-term (i.e. 10 years or more) assurance of safety, from an organizational perspective. The problem addressed is that of the general, imperceptible trend towards unsafe conduct of company operations, as first enunciated by Rasmussen (2000), and discussed by Amalberti (2001). An example is given to illustrate the difficulties of detecting this before disaster strikes. The paper goes on to discuss one way in which this problem could be addressed; essentially by good corporate governance, knowledge management and ownership of processes. Links are made to the literature on these topics, and a blueprint to help organizations to gain the benefits is outlined. The paper ends by outlining how an organization may be readied for knowledge management, without which the rest of the measures suggested are vitiated.

Introduction

This paper is specifically NOT concerned with day-to-day management of safety, but with the assurance of safety over the longer term – a decade, or longer. The problem of interest has been identified by (Rasmussen 2000); his elegant diagram is shown in Fig 1 below

In this diagram, the central region indicates the area of safe operation for the company's systems. As the organisation drifts towards the boundaries, so business death becomes more likely. But management, in response to changing circumstances in their market, will always be seeking greater efficiency allied to reduced costs – the 'down-sizing' issue. Concomitantly, operators and managers together, although not operating in concert, will be exercising the Principle of Conservation of Energy, whereby short cuts, whether procedural or cognitive, will be discovered and will gradually become standard practice.

The net effect of these two aspects is to create a gradient from right to left, down which the organisation drifts, until it crosses the 'disaster' boundary. The issue is to f8ind countervailing pressures to flatten the gradient, and it is this which we discus in this paper.

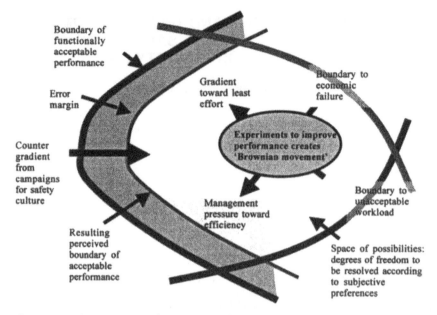

Fig 1: Rasmussen's diagram, redrawn, illustrating how organisations can drift into disaster.

In addition to this, Amalberti (2001) has discussed the paradox that essentially safe systems can be made less safe by the efforts of safety managers, over the longer term (a decade or more).   Amalberti's argument hinges on two observable aspects of organisational behaviour; firstly, managers move on from one job to another in periods of less than a decade; and safety programmes to make safe systems even more safe tend to use as their starting point the elimination of such errors that occur, by procedural changes, allied to hardware changes.   Safety managers make their reputations by demonstrating improvements in safety performance, a 5% improvement will take about 306 years to be statistically significant, whereas a 15% improvement will show in about 32 years. The plausibility of procedural changes, leaning largely on face validity, will serve as proof of a manager's efforts.

But what is to ensure that these safety 'improvements' will be maintained? It should be noted that in 40 years, all of a company's current employees will have left, and new generations will be in place. Theirs will be different interests, probably serving different goals. Clearly, we are discussing a knowledge management issue here, under the title of organisational learning.

Does this happen in practice?   Well, yes; Piper Alpha, Bhopal, and Columbia are egregious examples of this, and there are many more.

Corporate governance and disasters

If there is one consistency among disaster investigations ons, it is that either the necessary systems and procedures (for training, auditing, control, etc.) were not in place, or that they were in place but their operation was of insufficient quality. Both of these issues are issues of corporate governance, and we discuss this briefly.

We use the OECD definition:

> "Corporate governance is the system by which business corporations are directed and controlled. The corporate governance structure specifies the distribution of rights and responsibilities among different participants in the corporation, such as, the board, managers, shareholders and other stakeholders, and spells out the rules and procedures for making decisions on corporate affairs. By doing this, it also provides the structure through which the company objectives are set, and the means of attaining those objectives and monitoring performance" (OECD, 1999)

This definition includes both the internal and external affairs of the organisation; it implies the need for risk management; it embraces the whole organisation; and, although it is well-hidden within this definition, it does imply that the adequacy and quality of processes are significant aspects of good corporate governance .
These aspects are summarised in the following set of questions:
- Are the strategic goals appropriate for the company, given its history and competitive context?
- Has the company assessed adequately the risks associated with these?
- Has the company created appropriate processes, alliances, partnerships, etc. to deliver the goals, bearing in mind the risks?
- Have the processes been engineered as 'best in class'?
- Are the processes operated as 'best in class'?
- Does the company have processes for self-renewal (e.g. process auditing; capability acquisition; change management)
- Does the company audit itself regularly and transparently?

It follows fairly swiftly, from consideration of these points, that the organisation would have to pay co-ordinated attention to all of the following, as components of corporate governance:
- Structure (e.g. allocation of responsibility and authority; autonomy)
- Infrastructure (e.g. IT&T networks; security; access)
- Resources (e.g. time, money, people, knowledge & skills, equipment, and the distribution of these)
- Leadership (e.g. commitment to goals, support, clarity of communications)
- Culture (e.g. trust, willingness to learn, tolerance & retrieval of errors)
- Policies (e.g. resource management, change management, evaluation, suppliers, customers)
- People (e.g. selection, training, appraisal, knowledge, commitment)
- Processes (e.g. maturity, simplicity, metrication, controllability)
- Technology (e.g. maturity, deployment, utilisation, replacement)
- Knowledge (e.g. formal, tacit; organisational configuration; lifecycle)

Considering in particular the 'drift to disaster', it would appear that the 'safety case' approach being adopted in the developed world will have to consider all of these aspects. However, it is not apparent that attention to all of these would prevent a series of decisions by different people, all acting in the best interests of the stakeholders, from precipitating a disaster, as Amalberti (2001) has indicated. What is missing from this list is continuity, and wisdom. Continuity refers to the accumulation of knowledge and experience of a process, so that the decisions in the example above are not made in isolation, but are made in the context of prior decisions, and provide a path into the future. Wisdom is the ultimate goal of knowledge management, and is a blend of experience and knowledge. Given these, it is probable that most disasters could be obviated.

Hence, we now need to consider how continuity and wisdom can be provided within the organisation, and how these can be expressed effectively in control. One way in which this could be accomplished is by the notion of Process Ownership, and we turn to this next.

Process ownership

First, we define a 'Process', for the purposes of this paper:
- A Process has customers
- The Process is made up of partially-ordered sequences of activities
- The activities create value for the customer
- The activities are carried out by combinations of technology (machines, software) and people
- A Process can involve several organisational units, either within a company or across several companies
- A Process is instantiated by the allocation of goals, resources, responsibilities, and authority, and by the acceptance of appropriate metrics for measuring the performance of the process
- Resources typically comprise space, money, machinery, software, communications, people and knowledge

Process Ownership was introduced in the 1990s (e.g. Hammer 1996), with the notion of value chains. However, this early conception saw the role of the Process Owner more as an operational role than as a governance role. If, however, the role is re-defined to place emphasis on the latter, we may have better control over the drift to disaster. The intention here is that the process is 'owned' by a given individual, a 'Process Owner', responsib le for the safety and integrity of the process, and who 'leases' the process to a Process Manager who is responsible for the day-to-day operation of the process and making a profit, etc.. A given process might be owned by a nominated Manufacturing Systems Engineer as a generic process, and instantiations of the process (including one located in Las Palmas, as well as others in China and France) are operated by local managers and operators.

The Process Owner's responsibilities may be defined as:
- documenting the process as 'best current practice'.
- maintaining the integrity of the capability within the process (tools, procedures, skills, and the health and safety of its stakeholders).
- authorising improvements to the process, to ensure it continues to be 'best current practice'.
- ensuring that process changes do not have bad effects on related processes (and *vice versa*).
- Supporting the change process for making process improvements.
- Authorising physical instantiations of the process in a given geographic location, and ensuring that any changes necessary for the process to fit the local context do not harm the integrity of the process
- Ensuring that the process metrics are properly used, and the results are made accessible.

The Process Owner thus becomes distinct from the Process Manager, who is responsible for process performance goals. The Process Owner now maintains corporate governance over the process, and is the repository of process knowledge and process history, both of which are fundamental to continued process safety as discussed above.

The advantages of this approach are:
- it focusses management attention on the prime assets of the organisation – its

knowledge, and the efficient deployment and utilisation of that knowledge.
* it engenders a focus on strategic considerations
* it provides a basis for a thorough understanding of process capabilities with their related safety issues within the enterprise
* it presents a coherent structure for good governance for safety, and for the maintenance of any 'safety cases', as demanded by regulatory authorities;
* it provides a built-in bias against the 'slow drift to danger' identified by Rasmussen and others (though it does not eliminate it).

The disadvantages are:
* Process owners move on, too.
* the role will become ineffective unless the Process Owner is supported and resourced from the highest levels of the organisation, especially with regard to sufficient time to execute the role properly, and to have sufficient authority to stop the process should the Process Owner have cause for alarm about the state of the process in a given instantiation.
* Process Owners will be unable to perform their roles effectively unless the organisation has reached a high level of process maturity
* it divorces direct responsibility for process integrity from process performance.
* there will be differences between the goals for Process Owners and Process Managers/Operators, with the potential for considerable conflict.
* Where a process consists of many sub-processes, each with a process owner, it is possible that conflict will occur between the sub-process owners, leading to delays in innovation and capability acquisition.
* the inevitable creation of an hierarchy of roles for process ownership may be seen to be an exercise in over-staffing, to be resisted fiercely in the interests of profit and efficiency
* the pool of people competent to undertake the role in a given organisation appears to be small (see, for example, the list of components of corporate governance outlined above); the pool of those capable of giving training and acting as experts in process ownership appears to be much, much smaller.

Nevertheless, this approach seems to offer a means of addressing the drift to disaster problem, which is not particularly evident in current organisational scenarios. It also appears that companies are gradually moving towards this approach (unfortunately, this is not the stuff of experiments), and it is hoped that case evidence may follow in the next decade.

Finally, we hope that this paper will help thinking in this domain.

References

Amalberti, R. (2001). "The paradoxes of almost totally safe transportation systems." Safety Science 37, 2-3, 109-126.

Hammer, M. (1996). Beyond re-engineering, Harper Collins.

OECD (1998). Policy implications of ageing societies. Paris, Organisation for Economic Co-operation & Development: OECD Working Papers Vol VI, no 21.

Rasmussen, J. (1997). "Risk management in a dynamic society: a modelling problem." Safety Science 27: 183-213.

Rasmussen, J. (2000). "Human factors in a dynamic information society: where are we heading?" Ergonomics 43(7): 869-879.

Turner, B. and T. Kynaston-Reeves (1968). The concept of Temporal Disjunctive Information. Private Communication.

Turner, B. A. (1978). Man-made disasters, Wykeham Publications.

ROAD TRANSPORT

USABILITY OF PEDESTRIAN CROSSINGS: SOME INITIAL FINDINGS

Sandy Robertson & Roselle Thoreau

Centre for Transport Studies,
University College London,
Gower St.,
London WC1E 6BT. UK
e-mail usaped@transport.ucl.ac.uk

Whilst some people drive vehicles on the roads in the UK virtually everyone uses the road system as a pedestrian at some point. The government is keen to encourage walking as a mode of transport and existing studies have indicated that difficulty in crossing the road can be a barrier to walking. The current study (USAPED project) aims to develop a framework to assess the usability of pedestrian crossing places, and as part of that process is investigating what makes a crossing easy to use. For this study only crossings that are not controlled by traffic signals are being considered (zebra crossings and informal crossing places). The study uses a combination of workshops and observational studies to obtain information on what makes a crossing easy to use and how the different types of road users interact. A description of the methodology and early results for this work will be presented in this paper.

Introduction and background to project

In the Government's White Paper on Integrated Transport (DETR 1998a), quality of life has been identified as being dependent upon transport. Walking has been identified as a key form of transport which is to be encouraged through local transport plans in which strategies to promote walking and cycling are included. Routes that are safe, convenient and comfortable are expected to increase levels of walking. But however successful these routes may be in reducing conflict between pedestrians and other road users, crossing the road will remain for the foreseeable future a key feature in journeys on foot. Hence improving the usability of crossing places to make walking more convenient and comfortable as well as safer will help remove barriers to walking. Innovative ideas for achieving this will give rise to a requirement for tools to assess their effectiveness. The USAPED project funded by EPSRC seeks to address this.

The objectives for the project as a whole are to:

- Review existing knowledge of behaviour at pedestrian priority crossings and at crossing places.
- Identify what is perceived by road users as being important in making a crossing place easy to use.
- Relate the behaviour of road users to the factors perceived by them to be important.
- In terms of these and other factors define the relevant elements of usability of the crossing place.
- Identify and quantify observables that could measure or act as proxies for good or poor usability at a sample of crossing places.
- Develop/demonstrate a practical framework to assess the usability of existing and new designs of crossing places.
- Apply the framework to zebra crossings and at least one type of crossing place and where appropriate derive recommendations concerning guidance for the provision of crossing places.

This paper reports on user workshops that specifically address the second point above.

Use of the road system
The road system in the UK is governed by various laws which are enforced by the police. For the road user, guidance on the use of the road system comes from road safety training and from a document called The Highway Code (DETR 1999). The Highway Code gives guidance as well as identifying legally binding rules of conduct for different road users.

Use of pedestrian crossing places
Formal pedestrian crossings in the UK are traditionally either zebra crossings which in principle provide pedestrians with immediate right of way, or signal-controlled crossings which provide right of way only after a period of delay which depends on traffic conditions. The Highway codes states that at zebra crossings drivers must give way when someone has moved onto a crossing. (DETR 1999), and vehicle drivers are recommended to slow down and let pedestrians cross. Vehicle users are recommended not to wave people across. Pedestrians are reminded that traffic does not have to stop until someone has moved onto the crossing.

Less formal crossing places include infrastructure such as pedestrian refuges and flat topped humps that encourage pedestrians to use them. These types of crossing places are likely to be of increasing importance. In this paper, signal controlled and zebra crossings will be referred to as crossings and the locations where informal facilities are provided to help people cross will be referred to as crossing places.

Guidance for pedestrians crossing the road is given in the Green Cross Code as described in the Highway Code, but there is little guidance for pedestrians or vehicle users specifically on using informal crossing places such as refuges (traffic islands).

User workshops
A series of workshops were arranged so that they could be followed by observational studies at site in the vicinity of where the sample of users was drawn. The observational studies will draw upon the outputs of the user workshops to form a framework for the observational analysis which will quantify issues identified in the user workshops.

As of January 2004 three pairs of user workshops have been undertaken and pilot filming at one pair of sites. This paper will describe initial findings from these workshops.

The workshops and filming are being undertaken in a total of six local authorities (five London boroughs , central, inner and outer London as well as one shire county). While it was originally planned to separate pedestrians from other road users, because of small numbers volunteering to attend, the groups in the initial 3 pairs of workshops had to be combined.

The sample

For each workshop participants were self-selected and were asked to complete a questionnaire prior to the workshop. During the workshop two paper based activities in the form of questionnaires were undertaken and a short film shown various examples of crossings and crossing behaviours was shown. Discussions on issues relating to crossings were held after the paper based activities and after the film had been shown.

Table 1. Sample for workshops

Workshop Number	Borough	Number of participants	Proportion of females	Proportion with drivers licences
1	Richmond	4	0.5	1.0
2	Richmond	6	0.3	0.8
3	Bromley	4	0.8	0.3
4	Wandsworth	3	0.0	0.7
5	Bromley	11	0.6	0.8
6	Wandsworth	3	0.3	0.7

The 31 participants had a mean age of 58.3 years, with 42 percent being over 65 years of age. Fourteen of the participants were female and 17 were male. The participants who had reported some kind of formal training regarding rules of the road accounted for 26, of which 19 had full car drivers licenses whereas the remaining participants had either a motorcycle license or had done a cycling proficiency test.

Pre workshop questionnaires

Each workshop participant was asked to complete a questionnaire prior to attending the workshop. The purpose of the questionnaire was to a) obtain background information on the participants and b) obtain some travel information about the participants. The travel diary section asked about journeys made and the use of crossings on days where the respondents walked. An average of 5.2 trips per person per day were made by the respondents. Of those trips an average of 3.4 are made on foot. The average length of the walking trips was 826.6m and the average daily distance walked by respondents was 2720m. This was considerably higher than data obtained by Ward et al (1994) who found that the average distance walked by a person who walks on an average day was 1154 metres. The average walking journey of the sample took 15.8 minutes and an estimated average journey speed of 0.91m/s. The number of crossings on each walking trip using are shown in Table 2. The values obtained are higher than those found by Ward et al (1994).

What is clear is that for the sample, about one third of crossing are made using a crossing place.

Table 2. Number of times that a pedestrian crosses the road

	Per walking trip (N=105)	Per Day (N=31)
Cross at a crossing place	0.8	2.7
Cross within 50 metres of a crossing place	0.06	0.2
Cross where no crossing place is present	1.5	5.1
Total crossings	2.36	8.0

Early findings from workshops

While it would be premature to make definitive statements about the findings, the following sections describe some of the key points to emerge from an initial analysis of the workshops that have been run:

Use of different infrastructure by different local authorities in different boroughs.
During the course of the workshops and from site selection activities, it became clear that the use of different types of infrastructure (e.g. zebra crossings, refuges, signal-controlled crossings) was different in different local authorities. This appeared to reflect both the situations present in different areas, and also the approach to road system design taken within and between the Local Authorities. The outcome of such differences is that road system users may be faced with unfamiliar infrastructure when travelling away from their local area. As road users have a range of travelling patterns (from people who generally remain in a local area to those who travel widely round the country) the impact of this variation in infrastructure will affect road users in different ways. It was clear that in some workshops, the participants were not familiar with some types of crossing place and were uncertain about the rules governing their use.

Understanding of rules
Road users are unclear about the formal rules for using pedestrian crossings and crossing places. In particular there was frequent misunderstanding of the rules for using a zebra crossing. Although the highway code states that a vehicle does not have to stop for a pedestrian until the pedestrian places a foot on the crossing, many road users indicated that they believed that cars should stop if a pedestrian indicated that wished to cross. There was, however, little consensus about what constituted an indication of intention to cross.

Design of infrastructure
Correct siting of crossings is perceived by road users as important in making them useful for road users and especially pedestrians. Some users perceived refuges as not being big enough to accommodate the needs of users with pushchairs or bicycles.

Strategies for crossings.
During the workshops it emerged that some road users develop their own strategies for

using crossings/crossing places that may differ widely from (and in some cases contradict) the advice in the Highway Code. One of the most extreme examples was of an individual who at a refuge, waved a school bag at cars until one stopped. This strategy appeared to work for that individual and as such was reinforced.

Prediction of behaviour of other road users
Road users feel that being able to predict the behaviour of other road users make crossings more easy to use. This was perceived by both pedestrians and car users. From the perspective of car drivers the issue of being able to a) identify the presence of a crossing and b) being able to identify the desire of pedestrians to cross were seen as being important. For pedestrians the issue of being able to predict if a driver would stop (or give way) was seen as important.

Communication between road users
A number of road users from the perspectives of car drivers and pedestrians report that communication was used to assist in determining the intentions of road users. The use of an acknowledgement (thank you) was perceived as being a positive thing which could encourage a more cooperative approach by road users.

Conclusions

This initial data from the user workshops indicates that there are a number of issues that may contribute to making crossings more easy to use. While there is some indication that infrastructure design (location and visibility of crossings) can play a part, there are also indications that the ways in which the infrastructure is used and the ways in which the road users react to each other may play a big role in the perceived ease of use of crossing places. The latest information about this project may be found at:
http://www.cts.ucl.ac.uk/usaped.html

References

DETR 1999, *The Highway Code*, The Stationery Office (London)
Ward, H., Cave, J., Morrison, A., Allsop, R., Evans, A., Kuiper, C. and Willumsen, L. 1994, *Pedestrian Activity and Accident Risk*, AA Foundation for Road Safety Research, (Basingstoke)

A PARADIGM FOR THE DISPLAY- AND CONTROL-INTENSIVE VEHICLE

Calvin K. L. Or

Mississippi State University,
Mississippi, U.S.A.

An automotive company pioneers the use of by-wire technology in a vehicle design that includes of high-tech but unconventional devices (Birch, 2003). The devices are display- and control-intensive that drivers' visual and spatial resources could be demanding. Driving with those devices may engender different driving behavior and performance than that from a traditional vehicle. The critical question is this: how does interference potentially generated by the use of such devices? This article presents the human factor issues as a paradigm for the vehicle design. It discussed the use of information navigation system, manual control, and visual distraction and workload. In addition, the use of digital human modeling technology and driving simulator in vehicle design and development are illustrated.

Introduction and background

An automotive company intends to develop a future vehicle that includes of high-tech but unconventional devices. In such conceptual vehicle, the center of the yoke has a monitor showing roadway information as well as vehicle data; rear vision is acquired through CCTV screens instead of using traditional rear mirrors; control buttons are mounted in the two columns of the yoke functioning as acceleration and others; push buttons are used for the selection of forward, neutral, or reverse mode; the yoke includes two handgrips for turning and braking the vehicle that substitute the conventional controls. Assumed the design has widespread acceptance in the market, however, it is ambiguous that the drivers could handle the vehicle without detrimental effects on driving performances since they could be required to trigger a sequence of buttons and visually pay attention to the information presented from the monitor whilst they are engaging the primary control at a high speed with heavy traffic. Driving in a vehicle with those unfamiliar controls, add to this, the use of visual and mental demanding device, the critical question is this: how does interference potentially generated by the use of such devices? The advent of the monitoring systems and the unconventional controls is about to transform the driving skills. Already the application of advanced technologies to the domain of ground transportation has made available extensive roadway and vehicular information databases, integrated auditory and visual devices. For an automotive company to develop a safe future vehicle, like the conceptual vehicle, the driver's interaction with the devices and other elements, such as the vehicle-turning handgrips, should well be understood.

Design thoughts

Applying human factors in early stage of vehicle design process could tremendously reduce design and engineering costs (see Chaffin, 2002). Better understanding of good design practices in automobile will manifest considerably higher safety and less likelihood of human errors during the human-vehicle interaction. Good ergonomics is good economics when considering the cost reduction of design processes as well as of the compensation for injuries and fatalities. Although the future vehicle has not yet penetrated the market, it is imperative to identify the potential interferences generated by the use of the display-intensive devices (DID) and the control-intensive devices (CID) whilst driving. The questions concerning safety and the performances are: 1. Does the DID increase drivers' visual attention and workload? 2. Does increasing the number of controls cause driving performance to deteriorate? 3. Does the driver turn and brake safely using the handgrips? 4. How is the skills transfer problem while co-existing both the traditional and the conceptual vehicle? 5. Can the CCTV screen be used effectively for rear vision? 6. How does the dynamic environment, such as traffic and weather, affect driving performances with the use of those devices and controls? 7. What would be the training, acceptance, and satisfaction of the vehicle? Drivers are required to make decisions about steering, accelerating, braking, and performing secondary tasks while they are moving at speed. It is known that the conceptual vehicle has several unfamiliar buttons serving those functions. Regardless of vehicle compartment comfort and driver satisfactions, the design engineers must primarily consider the maintenance of driving performances when driving in the vehicle.

Vehicle and information roadway device

Driving task by itself is already a high visual demanding task in which drivers have to consistently pay attention to entities inside or outside the vehicle. When using the monitor of the future vehicle, alike using an intelligent transportation system (ITS), drivers devote extra visual resources during information retrieval. Since there is affinity of the nature between ITS and the display device (the monitor) of the conceptual vehicle, interferences produced by ITS use were depicted as a paradigm for the monitor design. Visual distraction should be the primary concern of any DID design in automobile. Such system aims to support drivers to make decision of changing different driving strategies. However, evidences have shown that using DIDs cause deleterious driving performance, leads to less safe driving, increases attentional demands and cognitive load (Liu, 2001). Driver's visual attention is always shifted away from road to the device for acquiring information. Tsimhoni and Green (2001) stated that driving performance declined as visual demand increased. It is important to have a "good" interface design for such display. A good interface design must not introduce the increase of drivers' visual and cognitive demand; otherwise, it may help to improve the performances. Effectiveness of DID interface design or the approach of information presentation to the driver has been examined (Green, 2000). Green (2000) suggested that the system should be able to present information that can be readily seen or heard. It is suggested that minimal effort is needed when users perceive system output. The system shall also enhance drivers' decision making without error. A DID with voice control, drivers' visual load could be alleviated if the information flow was well managed, and the potential distraction in such system could be reduced (Brewster, 2003). Drivers' workload cannot be neglected but shall be assessed when considering the usability of the DID. In the context of driving, several methods can be used to evaluate workload when drivers are performing secondary tasks: 1. Driving performances assessment (i.e. maintain proper lane position), 2. Physiological measures (i.e. electrocardiogram), and 3. Subjective workload rating (i.e. NASA TLX). Driver's mental underload (Brookhuis & deWaard, 2001) or excessive overload increases the likelihood of human errors (Sugimoto *et al*, 1997) and decreases the driving performance, such as slower brake reaction time. An in-vehicle DID, for example ITS, requires drivers' visual engagement when acquiring information. Their mental demand is superimposed if the in-vehicle task involves reading of displays and operating manual controls. Driving in a straight road at the same time performing a cognitive demanding task is recognized to have potential risk of crash. It further increases their cognitive burden if they were turning curves (Hancock *et al*, 1990), add together with high traffic density and bad weather condition, other secondary tasks such as conversation with passenger or using a cellular phone, and the use of the non-conventional controls and handgrips, thus risk of crash is magnified. It is imperative to examine how drivers manage many things at once with their limited ability and their visual distraction away from the road whilst driving concurrently using the monitor.

The "Yoke" control

Vehicle acceleration, the selection of the modes, turning, or braking is achieved using push buttons and handgrips. Drivers have to trigger an appropriate button among many choices to activate the corresponding function. The safety issue is this: how could drivers reach the right control among many choices within a safe duration. Degree of eye-hand co-ordination of the driver shall be examined, such as in turning control. Besides, studying the accuracy and time-to-trigger of the activation of a control with certain constraints are important. If the examination did not demonstrate the performance enhancement under a regular condition, it could be problematic if demanded by a safety critical scenario. When driving in the vehicle, the resources dedicated to primary driving task might be diverted to decision making about the controls. If this is so, driving performance could be tremendously degraded. Feyen *et al* (2000) showed the impairment of driving performance and workload with increases in the number of in-vehicle controls. To ease the reaching of a particular control, "frequency of use" principle can be applied to the design. Research findings from Curry and Jaworski (1998) contributed to designers locating the controls for frequently used functions on the steering wheel. Although the entertainment device was examined in their study, such idea can still be mapped to the yoke control design. Burnett (2000) stated that the location of a function or whether a control is positioned near/on the steering wheel is essential to examine. A recommended reaching time to a control is approximately 0.5 to 1.5 seconds so that drivers would feel the minimum temporal and mental pressure (Wierwille, 1993). Again, if the system was used in a safety critical scenario, a more critical reaching time should be considered.

Human-vehicle interaction

Testing the driver-vehicle interacting performance in a real-world context allows designers to exploit the design flaws in which a highest ecological validity is provided. However, the vehicle is still in its developing stage, with respect to safety issues, the examination may not be carried out in real driving dynamic, but could be performed using digital human modeling (DHM) technique or driving simulation. Chaffin (2001) successfully completed several case studies in which DHM was used to analyze and improve the design of a work cell and vehicle. Properly to use DHM in a design analysis could enhance the number and quality of design options and decrease design time. Most DHM software packages support not only the representation of human form or human visualization, but also allow ergonomists to employ the ergonomic analysis tools (i.e. comfort assessment, analyses of human visual field, reaching, and interacting posture) for the evaluation of the interaction of a digital-represented human in a virtual environment. All those are the tools allowing the evaluation of vehicle cockpit regarding driving interaction.   Driving simulator technology grows tremendously and that has been widely used in vehicle-related research over the past decade. It has been proposed that using driving simulator in the research has several advantages: safety, ease of data collection, experimental control, efficiency, and expense. However, such technology does not promise to provide a seamless human factors research tool for performances testing. Fidelity and behavior validity are always the concerns in a research involving driving simulator (Godley *et al*, 2002). Simulators with high fidelity recovery of detailed visual images, sound, and tactile feedbacks such as vibration, and large–scale motion cues are very important (Freeman *et al*, 1994). A driving simulator with low fidelity does not provides the same visual, audio, and tactile sensory input to users that may have an impact on the results. Since there are some pitfalls when using driving simulator for examining the interacting performances, using such technique with a well strategy is needed throughout the testing in order to successfully perform the evaluation.

Discussion and conclusion

The concerns of design were presented in this article with regard to usability, accessibility, and human-vehicle interaction of the conceptual vehicle. The design and manufacturing process becomes more effective and efficient if the design engineers understood all the intrinsic and extrinsic factors that may influence the decision of the design. Influential factors such as environmental variables should be considered. Although age effect has not been discussed in this article, research findings showed that aging could be a contributory factor to the degradation of driving performance and situation awareness (e.g. Imbeau *et al*, 1993). If the DID and CID increases the burdens, for instances visual load and reaction time, of young drivers, so do the old peers. As was noted earlier, visual or cognitive overload is caused by the visual predominance of the information presented. Since DID, such as ITS, has been recognized as a visual and cognitive demanding in-vehicle device, the research in regard to "driving performances and ITS" can be considered as a paradigm that similar phenomena could be applied for testing and understanding the degree of visual and cognitive demand required by the monitor system of the conceptual vehicle. The use of voice control or multimodal interface may facilitate drivers' visual attention on DID. With regard to the intensive control design, keeping in mind that increasing number of controls decreases driving performance and workload. It is valuable to minimize the number of control if possible. The design engineer must have to consider the issues: 1. Minimize the amount of time drivers shift their vision away from the road, 2. Minimize the number of controls, 3. Optimize placement of the display and controls, 4. Use multimodal interface if needed, 5. Consider the timing of messages presentation if auditory interface is used, 6. Augment driver ability in decision-making, 7.

Facilitate best navigation while minimizing distraction by considering the character size or other elements on the display, 8. Improve driver satisfaction and acceptance, 9. Consider the skill-transfer towards the non-conventional vehicle, 10. Provide a user centered design. DHM software provides a "non-hazardous" means for testing the interaction of the designated system and users. If real-road test is not possible, conducting a testing in a simulated environment also allow the examination of system performance and drivers' behaviors. On-road experiment may not possible to examine the non-conventional conceptual vehicle since it is unclear that drivers can control the car safely. The testing may be confined to the means of simulated driving that gives a closest driving scenario to real world. If researcher had a sound plan for simulated driving test, the drawbacks of using such technique will be minimized.

References

Birch, S. 2003, Driving the Hy-wire. *Automotive Engineering Int'l*, SAE Int'l, April 2003, 105-108

Brewster, S.A. 2003, Non-speech auditory output. In J.A. Jacko & A. Sears (Eds.) *The Human-Computer Interaction Handbook*, (Mahwah, NJ: Erlbaum), 220-239

Brookhuis, K.A. and deWaard, D. 2001, Assessment of drivers' workload: Performance and subjective and physiological indexes. *Stress, Workload and Fatigue*, (Mahwah, NJ: Erlbaum), 321-333

Burnett, G.E. 2000, Usable vehicle navigation systems: Are we there yet? *Vehicle Electronic Systems 2000-European conference and exhibition*, ERA Tech. Ltd, ISBN 07008 06954

Chaffin, D.B. 2001, *Digital human modeling for vehicle and workplace design*, (Warrendale, PA: Society of Automotive Engineers)

Chaffin, D.B. 2002, On simulating human reach motions for ergonomics analyses. *Human Factors and Ergonomics in Manufacturing*, 12, 235-247

Curry, D.G. and Jaworski, T. 1998, Frequency of use of automotive stereo controls. In *Proceedings of the HFES 42nd Annual Meeting*, (Santa Monica, CA: HFES), 1210-1214

Feyen, R., Liu, Y., Hoffmeister, D., Zobel, G., Rupp, G. and Bhise, V. 2000, Effects of shared secondary controls and operational modes on performance and perceived workload during a simulated driving task. In *Proceedings of the HFES 44th Annual Meeting*, (San Diego, CA: HFES), 290-293

Freeman, J.S., Watson, G., Papelis, Y.E., Lin, T.C., Tayyab, A., Romano, R.A. and Kuhl, J.G. 1994, The IOWA driving simulator: An implementation and application overview. *SAE Technical Paper #950174*

Green, P. 2000, The human interface for ITS display and control systems: developing international standards to promote safety and usability. *International Workshop on ITS Human Interface*

Godley, A.T., Triggs, T.J. and Fildes, B.N. 2002, Driving simulator validation for speed research. *Accident Analysis and Prevention*, 34, 589-600

Hancock, P.A., Wulf, G., Thom, D. and Fassnacht, P. 1990, Driver workload during differing driving maneuvers. *Accident Analysis and Prevention*, 22, 281-290

Imbeau, D., Wierwille, W.W. and Beauchamp, Y. 1993, Age, display design and driving performance. In B Peacock & W. Karwowski (Eds.), *Automotive Ergonomics*, (Taylor & Francis: London), 339-358

Liu, Y.C. 2001, Comparative study of the effects of auditory, visual and multimodality displays on drivers' performance in advanced traveler information systems. *Ergonomics*, 44, 425-442

Sugimoto, F., Toyabe, T. and Sato, A. 1997, An analysis of human error in skill task and visual recognition under auditory. In *Proceedings of the IEEE international Conference on Systems, Man, and Cybernetics, Computational Cybernetics and Simulation*, 211-216

Tsimhoni, O. and Green, P. 2001, Visual demand of driving and the execution of display-intensive in-vehicle tasks. In *Proceedings of the HFES 45th Annual Meeting*, (Minneapolis, MN: HFES), 1586-1590

Wierwille, W.W. 1993, Visual and manual demands of in-car controls and displays. In B. Peacock & W. Karwowski (Eds.), *Automotive Ergonomics*, (Taylor & Francis: London.), 299-320

In the driving seat – a proactive system for driver and vehicle risk assessments

Wendy Morris[1], Sonja-Louise Schwartz, Mary Sweeney[2], Christine Haslegrave[1]

[1] *Institute for Occupational Ergonomics, University of Nottingham, University Park, Nottingham, NG7 2RD.*
[2] *Anglian Water Services, Occupational Health Dept, Henderson House, Lancaster Way, Huntingdon PE29 6XQ.*

The Occupational Health Department at Anglian Water Services (AWS) have for several years been working with the Institute for Occupational Ergonomics (IOE) to develop a proactive system for the management of work related musculoskeletal disorders. A part of this work has been to develop a system of risk assessment for the provision of company vehicles. The IOE and the Occupational Health Department have therefore worked closely with managers and the Transport Department to develop a user-friendly system that minimises the risk of selecting and procuring an inappropriate vehicle for a driver. Such a system is part of a wider strategy for the management of musculoskeletal disorders including early treatment and rehabilitation programmes. This paper will discuss the process of developing the system and the difficulties encountered.

Introduction

A review of the health and well-being of employees was undertaken by AWS in 1998 and this identified a high proportion of sickness absence due to musculoskeletal disorders. To understand the work related risk factors that may exist within the organisation and to develop measures to eliminate or control these risks, the IOE was invited to develop an ergonomics management programme. The early stages of the programme involved a review of archive data, identification of problem areas, prioritisation of work and initial workplace assessments. Two workplace assessments were undertaken by the IOE that concerned the work of tanker drivers and the work of networks repair operatives. In both assessments, issues concerning the design of vehicles used by employees to undertake their work were raised. The findings from the assessments were reported to the company and the material was used to tailor an ergonomics training programme for company personnel. The purpose of the training programme was to develop an ergonomics team of company personnel to be able to undertake basic ergonomics assessments and to form an 'in-house' resource. Members of the team were drawn from Occupational Health, Safety, Transport, Training and Operational Sectors of the organisation. There have been a number of changes to the personnel of the team for a number of reasons over the course of the programme and further training has been required for new team members. As part of the programme of work with AWS to support the ergonomics team, the IOE has undertaken a number of workplace assessments and reviews of products, and has developed guidance for design engineers.

An issue that has continued to require attention in the management and rehabilitation of many employees who have developed musculoskeletal problems, both work and non-work related has been the suitability of the company vehicle they use. There have been a number of changes to the structure and organisation of the company since the ergonomics management programme first started. Some of these changes have meant that many employees have been relocated from local depots to more distant centralised offices, work areas have changed and as a consequence the mileage and time spent in vehicles travelling have increased for many employees.

There have been a number of reports in the scientific literature of musculoskeletal disorders in high mileage drivers (Porter et al 1992, Porter 1999). Factors that were found to be related to reports of discomfort were long periods of driving, use of a manual gearbox and lack of seat adjustment. The research has tended to focus on car drivers and it is important to note that attention to issues of comfort for van drivers has lagged considerably behind the improvements in the car market. With this in mind the Transport and Occupational Health Departments of AWS have sought to develop a number of tools which if used in the selection of a vehicle for a driver may prevent employees developing musculoskeletal discomfort from driving or can assist in the identification and resolution of problems with existing vehicles. This work has also been supported by health promotion and rehabilitation programmes. The ergonomics programme has therefore had both a proactive and reactive role for the reduction of sickness absence due to musculoskeletal disorders.

Vehicle Ergonomics

A review of the literature concerning vehicle ergonomics found a number of studies that have considered the physical effects of driving on the spine and associated structures (Battie et al 2002, Haslegrave and Mellor 2000), the self reporting of discomfort of drivers (Boshuizen et al 1992, Porter et al 1992), the role of ergonomics in the design of occupant packages within vehicles (Roe 1993, Porter and Porter 1997) and some guidance on driving positions (McIlwraith 1993, Harrison et al 2000,). Later in the programme of work, additional resources were located on the Internet, which considered the assessment of the vehicle for an individual (www.drivingergonomics.com and www.car-seat-data.co.uk).

There are a number of interactions for the driver in a vehicle to consider, such as reach to controls and pedals, the seated posture, clearance for the legs and feet and vision of the external environment, mirrors, controls and displays, which influence the posture that a driver adopts within a vehicle. There are also differences to consider between the reclined driving posture, which is traditional within most cars and the more upright driving position found in vans and lorries. The range of vehicles within a certain price range is generally wide and encompasses a variety of styles, so that an individual purchasing a vehicle for themselves has the opportunity to test a wide range of vehicles and find one that is a 'best fit' for them, considering all these different interactions. A driver being provided with a vehicle for their work by their employer generally does not have the same range of choice. This may be because the company has agreed a fleet deal with a particular manufacturer or has a limited list of manufacturers as suppliers, or because the task has certain requirements for a particular type of vehicle. Such constraints limit choice and may therefore limit the ability to achieve an optimal match between the individual and the vehicle. Finally in some situations the driver may 'inherit' a vehicle selected by a previous employee or a manager or share a vehicle with a number of other employees. The length of contract on a vehicle can vary but within AWS it is between three and five years. It was therefore important to provide appropriate guidance to managers and individuals when selecting their vehicles to avoid costly mistakes. The IOE began to work with both the Transport and Occupational Health Departments of AWS to build such guidance documents for both managers and individuals who may be selecting a vehicle for other employees or themselves, based upon the scientific literature that was available. The guidance required the user (either the driver selecting a vehicle for themselves, or a manager selecting a vehicle for an employee) to first identify the task requirements for the vehicle, then to assess the intended driver in the proposed vehicle. The guidance also provided a flow chart for decision making as to the appropriate action required if an employee reported musculoskeletal discomfort associated with the use of a company vehicle. The guidance forms were reviewed by the Occupational Health Team and subsequently made available to managers and individuals via the company intranet. Awareness of the availability of the guidance was raised through company communications and there was a considerable degree of interest.

Difficulties

As people accessed the guidance forms, the Occupational Health Department received a number of requests to present the guidance document concerning driving posture in a risk assessment format. In response to these requests a driver / vehicle risk assessment form was produced and also made available on the intranet. The risk assessment form was quite widely used by individuals, predominately employees reviewing their current vehicles rather than in the selection of a new vehicle. The Occupational Health Department received a considerable number of enquiries concerning the risk assessment form from both drivers and their managers. In dealing with these enquiries a number of problems with the risk assessment form were identified. A review of this form was therefore undertaken, learning lessons from the process to date, with an aim to develop a more robust system. To support this review a participatory approach was used and a team of interested parties, including representatives of end users of the forms, was bought together to review the difficulties experienced with the current system and consider proposed revisions to the forms.

The original risk assessment form considered various aspects of the seat and driving position as 'hazards' and a number of people had misinterpreted the guidance, reporting that their vehicle was unsuitable when on further review this was not the case. The original form therefore generated a number of 'false positives' that were time consuming to resolve. The process also had not identified who should perform the risk assessment with the driver and it was considered that a lack of training on the part of the assessors (often a work colleague) was generating some of the difficulties of interpretation of the form. There was a need to improve the reliability of the form to reduce the workload of the Occupational Health Team who were asked to give a second opinion when drivers reported that their vehicle was unsuitable. A level of objectivity was also sought to balance the effect of aesthetics of the vehicles. Helander (2003) has noted in a recent study of the ergonomics of chairs that "in the end aesthetics may be more important than ergonomics – at least to the customer who will be guided more by aesthetics than longer-term ergonomic factors". The make and model of a vehicle has social and cultural connections for the driver and may have an influence the outcome of a vehicle assessment. The new process therefore sought to clearly identify the task requirements for the vehicle. Once these had been identified a suitable model from the fleet range was selected which could match these criteria. The prospective driver was then assessed in the proposed vehicle to identify if they were a suitable match.

This work was undertaken against a background of increasing interest by the Health and Safety Executive (HSE) in work-related road safety. Although there was much discussion in safety and fleet managers' publications, clear guidance as to what was required from a driver / vehicle assessment was not available. Only recently have the HSE published a guidance document concerning work-related road safety entitled "Driving at Work" (HSE 2003). This draws attention to the need to consider ergonomics in the provision of vehicles for employees but does not fully indicate what a suitable ergonomic assessment may involve.

Progress to date

Following the review of the existing forms and proposed revisions, the forms were piloted in two areas of the company by a two different groups of users (line managers and occupational health advisers). Two meetings were held during this trial period where aspects of the forms were discussed and updated to account for questions or problems with the forms that had arisen from their use. The feedback from the users from the pilot studies was very positive and the new system appeared to eliminate the difficulties that had previously been reported. The forms have had a final revision to enable these to sit within the intranet system of the organisation and to link closely with the work of the Transport Department so that all the advice and information provided to drivers and their managers is consistent. The final stage is the preparation of the forms and process for presentation to the company's management board and this work is in hand. It is anticipated that once their approval has been gained, a period of communication and training for line managers will be required to minimise any misunderstandings. Information about the new system will also be communicated to employees through various company communications.

Conclusions

AWS has taken a proactive approach to the reduction of work-related musculoskeletal disorders through a programme of education, risk assessment and early rehabilitation. As part of this programme the IOE has supported the work of the Transport and Occupational Health Departments in the selection and provision of suitable vehicles for employees who are required to drive as part of their work. There is a growing interest in the area of work-related road safety but guidance as to how companies must manage this risk has only recently been made available. The IOE therefore undertook a review of existing guidance and research on vehicle ergonomics to support the development of driver advice and driver / vehicle risk assessments. The process of development has been iterative in nature, learning lessons along the way and recognising the benefits of a participatory approach in the development of the proposed system. The successful implementation of the new system will depend on the effective communication to line managers of the benefits for them and their employees from following the system. The participatory team of end users have also been involved in the development of the training and communication programme.

References

Battie M C, Videman T., Gibbons L E., Manninen H., Gill K., Pope M., Kaprio J. (2002) Occupational driving and lumbar disc degeneration: a case-control study. *The Lancet*, 2002, Vol 360, November 2, 1369-1374.

Boshuizen H.C., Bongers P. M., Hulshof C. T. J., (1992) Self-reported back pain in fork-lift truck and freight-container tractor drivers exposed to whole body vibration. *Spine* 1992, Vol 17, No 1, 59- 65.

Harrison D, D,. Harrison S, O,. Croft A, C,. Harrison D, E,. Troyanovich S, J,. (2000) Sitting Biomechanics, Part II: Optimal Car Driver's Seat and Optimal Driver's Spinal Model, *Journal of Manipulative and Physiological Therapeutics*, Vol 23, No 1, 37-47.

Haslegrave C. M. H., Mellor M. A., (2000) Stadiometer measurements of driver's spinal response to steering force and vibration. In *Proceedings of the IEA 2000 / HFES 2000 Congress, San Diego California.* HFES Santa Monica Vol. 5, 485-487.

Helander M., (2003) Forget about ergonomics in chair design? Focus on aesthetics and comfort! *Ergonomics* Vol 46, No 13, 1306-1319.

HSE (2003) *Driving at work, managing work-related road safety.* INDG382, HSE.

McIlwraith B., (1993) An analysis of the driving position· in the modern car. *British Osteopathic Journal* 1993, Vol XI, 27-34

Porter J. M., (1999) Driving and musculoskeletal health, The Safety and Health Practitioner Supplement July 1999, 8-11.

Porter C. S., Porter J. M., (1997) An 'inside-out' approach to automotive design. In Seppala P, Luopajarvi T, Nygaard CH, Mattila M (Eds) *From Experience to Innovation – IEA '97. Proceedings of the 13th Triennial Congress of the International Ergonomics Association, Tampere Finland, June 29–July4 1997,* Helsinki, Finnish Institute for Occupational Health, Volume 2, 90-95

Porter J. M., Porter C. S., Lee V. J. A., (1992) A survey of driver discomfort. In Lovesey E. J., (ed) *Contemporary Ergonomics 1992,* London, Taylor & Francis 262-267.

Roe R. W., (1993) Occupant packaging. In Peacock B., Karwowski W., (eds) *Automotive Ergonomics*, London, Taylor & Francis

RAIL

PILOTING A METHOD TO INVESTIGATE THE THOUGHT PROCESSES BEHIND TRAIN DRIVER VISUAL STRATEGIES

N Brook-Carter[1], A Parkes[1] and A Mills[2]

[1]TRL, Old Wokingham Road, Crowthorne, Berkshire, RG45 6AU, UK
Email: nbrook-carter@trl.co.uk Tel: 01344 770305

[2]Rail Safety and Standards Board

Train drivers are required to monitor the dynamic scene visually, both outside and inside the train cab. Poor performance on this visual task may lead to errors, such as signals passed at danger (SPADs). It is therefore important to understand the visual strategies that train drivers' employ when monitoring and searching the visual scene for key items, such as signals, and to gain an insight into the thought processes behind these strategies.

A pilot study was carried out by TRL on behalf of the Rail Safety and Standards Board, in which train drivers drove in-service trains wearing a state of the art eye tracking system and were later interviewed to try and understand the thought processes behind the visual strategies they used.

Findings from the pilot in relation to the effectiveness of the verbal protocols used to investigate driver thought processes are presented in this paper. The problems encountered when collecting the verbal protocol data and the initial trends identified from this data are discussed.

Introduction

As a result of a number of recent incidents on the railways, which have been attributed to errors in human performance an increased effort has been focussed on conducting research investigating Human Factors issues associated with train driving. One of these areas of research concerns the visual behaviour and performance of train drivers.

Train driving is primarily a visual task. Train drivers are required to monitor the dynamic scene visually, both outside and inside the train cab. Poor performance on this visual task may lead to errors, such as signals passed at danger (SPADs). For this reason it is important to understand the visual behaviour of train drivers, the strategies they employ when monitoring and searching the visual scene and the cognitive processes associated with these visual strategies.

This paper discusses a pilot study carried out by TRL on behalf of the Rail Safety and Standards Board and presents the methodology used and some of the findings, particularly relating to the mental processes underlying those visual strategies.

During the pilot train drivers drove in-service trains wearing a state-of-the-art eye tracking system to record eye fixations. These data were supplemented by verbal reports by the driver to build a more detailed picture of the underlying strategies used by the driver. The pilot was supported by the London Lines Train Operating Company C2C and took place on a line running from Shoeburyness to Laindon in Essex.

The work follows a previous study (Groeger et al, 2001) and extends the methods used. A larger set of data on train drivers' visual behaviour was required to provide a greater understanding of the strategies adopted, such as the visual cues from route knowledge and the relevance or irrelevance of trackside information. The methodology piloted here is therefore intended for use in a series of larger scale trials.

The aim of the pilot was primarily to test the methodology and for this reason, only 5 drivers were involved. Therefore, no firm inferences can be made from the results.

This paper presents the methodology applied during the pilot, specifically considering the method for obtaining data on the drivers' cognitive processes. The limitations and advantages of the verbal protocol method used are discussed and some initial trends are presented.

Methodology

Procedure
The basic procedure adopted during the pilot study was:
- fitting and calibrating the eye tracker to a driver in a station office
- the driver boarding and driving an in-service train
- conducting re-calibrations at station stops when appropriate
- carrying out a verbal protocol on video footage of the route following the drive

Eye tracker
The light-weight eye tracker uses a dichroic mirror to reflect the eye-image onto a highly sensitive head-mounted video camera and a separate scene camera.

The tracking technique used by the system is a 120Hz pupil and corneal dark-eye tracker. The eye tracker automatically tracks point-of-regard and the correlation of the raw eye position to the precise position on the scene, in real time. The image being viewed by the participant is identified by crosshairs on the recorded video footage. The eye tracker headband was connected to eye tracking equipment by a long cable and the equipment carried on the trains was portable and could be fitted into a small bag.

Calibration
Calibration is necessary to match the driver's point-of-gaze to the equipment output co-ordinates. Initial calibration was carried out in the office before boarding the train and involves five registration points. The advantages of calibration at the station were the ability to calibrate with fixed points and the reduced time pressure under which the calibration takes place. Further calibrations were recorded whilst on the train. They were conducted by asking the participant to look at five specified points in the train cab.

Drive
The route chosen for the pilot ran from Shoeburyness station to Laindon station. This route was chosen as it is a simple route and does not contain any complex signals or gantry signals. The train was a 357, which has in-cab equipment including AWS

(Automatic Warning System), DRA (Driver Reminder Appliance) and TPWS (Train Protection Warning System).

Participants

Five participants were involved in this pilot study. The driving experience of the train drivers involved ranged from 9 weeks to 26 years. Participant age ranged from 30 to 55.

Verbal protocols

Verbal protocols were conducted following the drive. Drivers were shown a video clip of a section of the route they had just driven. The sections chosen were judged by a local Managers as having a relatively high workload in comparison to the rest of the route.

The first set of video clips did not depict a crosshair indicating the drivers' direction of gaze. Whilst watching the clips the drivers were asked to imagine that they were driving the train and to talk through what visual behaviour and strategies they think they would adopt. This involved describing what they believed they would be looking at and why. If the drivers experienced difficulties in carrying out verbal reports and only provided brief verbalisations, the experimenter prompted the driver with probe questions such as: *'What are you looking at here?'* or *'Why are you focusing on that particular feature?'*

The second stage of the interview involved the drivers watching a set of video clips of the same 'events', but taken from the actual journey just carried out. These recordings were played via the eye tracker software and therefore displayed the crosshair indicating the driver's point of regard on the visual scene. Whilst watching the clips the drivers were asked to talk through their visual behaviour and strategies with reference to the position of the crosshair on the scene. The drivers were asked to describe what they were looking at and why they believe they were adopting this particular visual behaviour.

Results

The small size of the sample in the pilot study limited the amount of analysis that could be conducted on the verbal report data. However, the detail of the verbal reports was extremely useful in suggesting the reasons behind drivers' visual strategies. The following are examples of some of the common explanations provided by drivers for looking at different aspects of the scene:

Signals:

- Drivers reported looking at the signals ahead in order to determine their next action, e.g. braking. The preceding signal aspect was also very important in this respect; drivers reporting that they particularly looked out for the next signal if they were running on cautionary signals but could look at the scenery if they were running on greens.
- Drivers reported paying particular attention to certain signals such as those that protect a crossover or have been passed at danger in the past.
- Drivers reported they kept looking at the signal as they approached it, even on green, because they had been known to change suddenly in the past.

General Hazards:

- All drivers reported looking at the station platform for passengers that were too close to the edge, or were behaving unusually. Several drivers mentioned that they looked out for children in particular.

- Several drivers indicated that trackside workers could be expected in certain places and that they looked out for them in these areas particularly.
- Several drivers reported that they looked at other trains to see if they had any defects and to pass a signal to other drivers that all is well.
- Drivers reported checking overheads to make sure nothing was hanging from them.
- Several drivers reported looking at the track either for damp spots or to check the points are set in the right direction
- Several drivers reported looking out for the neutral section where power could not be drawn

Other visual behaviour:
- If stopping, drivers reported looking for the stopping board on the station platform
- Drivers reported using the following visual cues as braking points: grey boxes, bridge, signal after footbridge and next signal
- One driver indicated looking for the church and bridge to see where he was on the route

Discussion

Effectiveness of the data collection procedure
The verbal report obtained information relating to the drivers thought processes and strategies or reasons for looking at certain aspects of the visual scene, which could not have been obtained solely from the eye tracking data.

Some problems were encountered when conducting the verbal reports. Verbal reports were particularly problematic on sunny days when the quality of the scene camera output was limited and the participant was not always able to distinguish what he was looking at.

Participants also varied in their ability to verbalise their visual behaviour and in some cases experimenters were required to use several prompts. Experimenters ensured that prompts were neutral and not leading, such as 'what were you looking at here and why?'

Both the video clip footage that did not contain a crosshair and the eye tracker footage with a crosshair superimposed were useful during the verbal reports. During the video clip without a crosshair participants were able to consider what they would normally look at and try to explain why. The eye tracker footage with a cross hair acted as an additional prompt or a re-inforcer, where drivers could describe why they looked at different aspects in the scene.

Recognising the limitations of the verbal protocol
The aim of the verbal protocol is to provide information relating to the cognitive processes of the train drivers. The verbal protocol requires introspection and verbalisation, and the train drivers' ability to do this depends on the level of consciousness they have regarding their behaviour, and will be influenced by their understanding of the trial requirements.

Nisbett and Wilson (1977) have argued that such introspection is almost worthless, as people are generally unaware of the processes influencing their behaviour, particularly behaviour dependent upon a practised skill. They claim that when trying to explain their behaviour, people apply or generate theories rather than remembering the actual cognitive processes. Further, as Hannigan and Parkes (1988) point out, the data collected will be a function of both the participant's ability to verbalise cognitive experiences and the degree to which the participant tailors the verbalisation to meet the expectancies and understanding of the investigator.

Despite findings to support the arguments of Nisbett and Wilson, verbal protocols can identify those processes which people are consciously aware of. Ericsson and Simon (1984) found that participants are more likely to provide accurate introspection when asked to describe what they are attending to or thinking about, than when required to interpret a situation or their own thought process. Further, as Bainbridge (1974) had asserted, whilst a protocol does not give complete, or necessarily reliable, data on the participant's thoughts, it is a source of information which can not be obtained other ways.

Ericsson and Simon stated that the information less likely to be reported by verbal protocol are the cues that allow people to recognise stimuli. This should be taken into account when analysing the verbal protocols, in that the train drivers may well be unaware of a number of the visual cues or 'passing points' which they frequently use.

It is important to note the limitations of verbal protocols. The data obtained can not be assumed to be the whole picture with regards to the underlying thought processes, strategies and tactical decisions relating to the train drivers behaviour, but rather an indirect view of the underlying mental process. In sum, during verbal protocol, the train drivers are expected to present a rationalisation based on their own internal model of train driving, which will not completely mirror the actual cognitive processes underlying the visual behaviour, but which will provide a valuable interpretation of events.

Initial trends

Since the verbal protocols were only conducted by five drivers on two station approaches, , no firm assumptions can be made relating to train driver visual strategies and their underlying cognitive processes from this limited data. However, the initial trends which emerged from the results of the verbal protocols indicate that the train driver's visual strategies are dependant on a number of factors, including position on the route, preceding signal aspects and expectations about hazards. Drivers reported looking out for specific hazards which they expect to occur at particular positions on the route. Drivers visual behaviour appears to be affected by the preceding signal aspect and more attention was reported being paid to the next signal when running on cautionary signals. Drivers appear to use some visual items which are not part of the rail infrastructure to identify their position on the route, such as churches and bridges. Finally a number of different visual cues were used as braking points by the drivers. Some of these cues were part of the rail-infrastructure and others, such as bridges, were not.

References

Bainbridge, (1974). Analysis of verbal protocols from a process control task. A summary of the cognitive processes of operators controlling the electricity supply to electric-arc steel-making furnaces. In Edwards, E. and Lees, F.P. (eds.) *The Human Operator in Process Control*, Taylor and Francis Ltd, London, pp. 146-158.

Ericsson, K.A. and Simon, H.A. (1984). *Protocol analysis*. Cambridge, MA: MIT press.

Groeger, J.A. , Bradshaw, M.F., Everatt, J., Merat, N., & Field, D. (2001). Pilot of train drivers' eye-movements, University of Surrey.

Hannigan, S. and Parkes, A. (1988). Critical incident driver task analysis. In A. G. Gale et al (Eds.) *Vision in Vehicles II*, Elsevier Science Publishers B. V., North Holland.

Nisbett, R.E. and Wilson, T.D. (1977). Telling more than we can know: verbal reports on mental processes. *Psychological Review*, **84**, 231-259.

CIRAS: AN EVOLVING CONFIDENTIAL REPORTING SYSTEM

Mary Miller, Suzy Broadbent, Zoë Mack, Julie Bell

CIRAS,
Suite 13,Emerson Business Centre, Regent House,
Heaton Lane, Stockport, SK4 1BS.

CIRAS is the UK rail industry's confidential reporting system. It was implemented nationally in 2000, following the Ladbroke Grove incident, as a method for railway staff to report concerns regarding safety in confidence. It was hoped that self-reporting of errors and violations would be encouraged due to the confidential nature of the system; however such reports have been relatively low in number. This paper discusses the different types of reports processed by CIRAS and examines how the emphasis might be altered to promote the reporting of human factors type reports. CIRAS could then focus analysis on those human-performance issues that the industry may not be aware of. Such analyses could provide companies with valuable information that could be used to address human errors and violations for staff across the network.

Overview of CIRAS

CIRAS stands for Confidential Incident Reporting and Analysis System. CIRAS offers the UK railway industry an alternative channel through which staff can make safety reports, in complete confidence. CIRAS is completely independent allowing railway staff to report safety concerns without having to go through their line managers or other company channels to make a report. However, the system is not intended to replace the normal reporting channels (e.g. defect reporting, incident reporting) already operated by railway companies.

Development of CIRAS

CIRAS was first founded in 1996 as a pilot study carried out by Scotrail who hoped to collect greater quality information about incidents when they occurred. This would enable the company to find out more about why incidents occurred, in particular, more about the human errors and performance shaping factors which contributed to adverse events.

After the pilot study, more railway companies voluntarily joined CIRAS. Between 1996 and 2000, Railtrack Scotland Zone, GNER, Virgin Trains, First Engineering and GTRM joined. Following the investigation into the Ladbroke Grove rail crash in 1999, it was mandated for the railway industry to implement a national confidential reporting system. It was decided to expand the existing CIRAS system and the national system was rolled out from 1st June 2000. Additionally, some companies have voluntarily joined the system.

How CIRAS operates

CIRAS[1] operates by allowing reporters to make reports by phone or by sending a report form to CIRAS offices. Reports are usually followed up by an interview, normally over the phone. CIRAS interviewers are trained to probe about incidents, human behaviour and influences, to find out about human error and performance along with the various performance shaping factors. All reports are fed to the Core Facility that maintains a national database and analyses the data. Analysis of the data can reveal trends of issues that concern reporters and provide in-depth analysis of risk and safety issues. Companies can then use this information, in conjunction with information from other sources in the industry, to consider which areas should be prioritised as safety and risk concerns. It is hoped that, a greater number of reports about human error in the future could be particularly useful to companies, because such information is hard to predict or discover by other methods. CIRAS is ideal for such reports, and can process them in a way that does not compromise confidentiality of the reporter, a factor that has been claimed to be important in encouraging self reporting (O'Leary & Chappell, 1997 cited in Reason, 1997).

Reports received by CIRAS

Since June 2000 over 3,000 reports have been received. Throughout this time, some consistent patterns have been noted.

Type of issue reported

Figure 1 shows that the most frequently reported issues to CIRAS are rostering problems and training/ briefing problems. This pattern has been evident since CIRAS started to collect reports nationally in 2000.

---

1 CIRAS is operated by a number of independent organisations

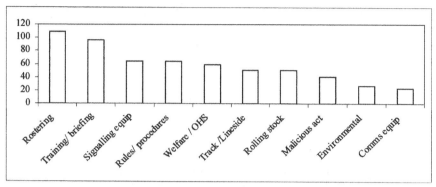

Figure 1 – Most frequently reported issues

Reporting rate over time

It can be seen from Figure 2 that the reporting rate increased over the first year of national operation until a plateau was reached. Currently the reporting rate is fairly stable with between 200 and 250 reports received per quarter. This pattern of reporting rate is similar to that experienced with other confidential reporting systems in other industries in both the UK and abroad[2].

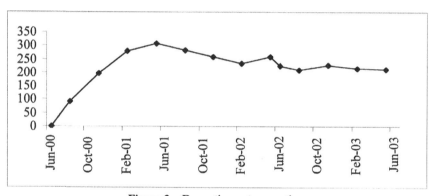

Figure 2 – Reporting rate over time

Use of company channels in addition to CIRAS

At its inception CIRAS was established as a complementary method to company safety reporting channels, intended only to be used when company channels are not considered appropriate by staff (e.g. for sensitive issues where disciplinary procedures are feared as a consequence of reporting). However, based on the usage profile of CIRAS since national implementation, it is evident that a large proportion of reports submitted to CIRAS (around 70% of reports) have already been reported via company channels.

---

2 This was confirmed by participants at a recent international forum on confidential reporting at which there was representation from systems operating in a range of industries and cultures.

This 'dual' reporting is explained by a variety of reasons. A perceived lack of action by the company and the reporter's desire 'to get something done' were the foremost reasons given. A further group of reporters had been prompted to use CIRAS because they believed the situation was so dangerous that 'duplicating' reports was seen as necessary to increase the chances of the problem being addressed. Another motivation for using CIRAS was the fact that reporters are able to obtain feedback in relation to any actions taken or an explanation as to why no action is considered appropriate (this feedback is not always available when using other reporting channels).On the basis of this information, it is possible that improvements to company reporting channels could result in a lower rate of 'dual' reporting, coupled with fewer non-sensitive CIRAS reports. In the short term, however, CIRAS serves a valuable role in allowing safety information to be captured and passed to the appropriate company for action.

Company acceptance of CIRAS

Over the last few years, there has been some resistance to CIRAS from some railway companies who are members of CIRAS. This has been attributed primarily to:

- the financial and time cost of CIRAS to companies
- the fact that CIRAS is mandated, not voluntary
- attitudes of senior management in some companies
- lack of specific information in some CIRAS reports due to confidentiality reasons
- CIRAS being perceived to undermine internal company reporting channels.

Although in some companies there have been good levels of acceptance, support and participation in relation to CIRAS, problems have been experienced with a minority which have been a barrier to the development and expansion of CIRAS. CIRAS has worked with companies to overcome these problems and gradually, through increased promotion such as staff briefings and brain-storming meetings with managers and boards, progress has been made in developing more positive relationships. The use of Liaison Committee Meetings (LCMs) has been an important step in forging such relationships. LCMs are held quarterly and allow company representatives to discuss and peer-review responses to CIRAS reports while also allowing company representatives to air their views on the system. Attendance at LCMs has increased since their inception suggesting that CIRAS has gradually become more accepted by companies. Indeed, a recent questionnaire designed to gather feedback from company representatives with regards to the LCMs found that many company representatives found them a useful method of achieving best practice. It is hoped that, in the future, these positive relationships with companies will encourage the use of CIRAS, and assist in developing a better reporting culture across the industry.

Furthermore, it is hoped that recent changes in outputs to the industry will help to promote better working relationships with companies. Until recently, national analysis reports tended to be lengthy and rather academic in approach. Recent changes, made in response to feedback from companies, have made these more industry focused. In particular, specific reports for individual sectors of the industry provide targeted analysis of a relevant key issue chosen by the industry. Detailed information has also been given on the use of company channels in conjunction to reporting to CIRAS. This could be used to help companies improve internal reporting channels.

What CIRAS would like to achieve in the future

Future development of CIRAS will see it trying to further emphasise to companies and reporters the importance of human error and near miss reporting, enabling CIRAS to build a database of such reports for analysis purposes. CIRAS will try to encourage reports about near misses and human error by taking a number of steps:

- Actively encouraging the reporting of human error and near-miss information by promoting human factors in quarterly journals and highlighting the importance of human error reporting in briefings with frontline staff and managers.
- Increased emphasis on analysis of human factors issues combined with increased education on the value of human factors in understanding and preventing human error.
- Providing information to industry about the use of company reporting channels in combination with CIRAS, to help companies improve reporting systems and reporting culture.

Other confidential reporting systems have noted that it takes 5-10 years for a reporting system to mature fully[3]. This infers that CIRAS is still a relatively young system, and still in the process of being accepted. In this way, CIRAS may still be receiving reports by people who are "testing the system". It is expected that, a larger number of reports of a human factors nature may be received once the system is more widely trusted and accepted. The 1st International Confidential Reporting Systems Forum (2003) established that it is rare to get good self-reports of error early in the development of the system and this is thought to be especially true for a system which is mandated. A further point raised at this event is that success of a system is hard to measure, and that different parties involved have different criteria for success. It is hoped that, with increased promotion and understanding of human factors and its importance, and the acceptance of the system which will come with time, CIRAS will generate a higher number of human factors reports, and that this will lead to improved human factors safety and risk information for the railway industry.

References

Reason, J. (1997). *Managing the Risks of Organisational Accidents*. (Ashgate, Aldershot)

---

3 This was confirmed by participants at a recent international forum on confidential reporting at which there was representation from systems operating in a range of industries and cultures.

TRAIN CAB ERGONOMICS – FROM A DRIVER'S PERSPECTIVE

Zoë Mack, Suzy Broadbent, Mary Miller and Julie Bell

CIRAS,
Suite 13, Emerson Business Centre, Regent House,
Heaton Lane, Stockport, SK4 1BS

CIRAS is the UK rail industry's confidential safety reporting system. Since 2000 it has collected information on safety related issues from drivers, signallers, maintenance staff and other railway personnel. Within the national CIRAS database, issues related to the ergonomics of train cabs have been reported with a focus on poor cab layout, cab temperature, ambient noise and alarm presentation. This is despite some involvement of ergonomists in the assessment and modification of the cab design and layout. These reports mainly come from drivers, some of whom are concerned about the increased risk of error due to fatigue or distraction. New trains with better cabs are now being introduced but constraints such as the need to retrofit older trains mean long standing problems are difficult to rectify. The issues reported reflect the experience of some train drivers and are not necessarily representative of driving conditions in the rail industry.

Introduction

CIRAS stands for Confidential Incident Reporting and Analysis System. It provides a confidential system to the UK rail industry that allows railway staff to report their safety concerns without having to report to their line managers, thus removing the potential for ramifications or recriminations from managers and colleagues. However, the system is not intended to replace normal reporting channels, only to provide an alternative to those who feel they need a confidential route. CIRAS was originally set up for the train operator Scotrail in 1996. Interest in the system spread and companies such as GNER, Virgin, Jarvis, First Engineering and Network Rail Scotland became enrolled on a voluntary basis. Following the Ladbroke Grove rail disaster the government mandated that all UK Railway Group Members should implement a confidential reporting system. This resulted in CIRAS becoming a national system in June 2000. Currently, over 80 companies are enrolled from various sectors of the railway industry.

Reports received

Since the system was introduced nationally CIRAS has received 105 reports regarding train cabs. Table 1 gives details of the types of train cab problems raised through CIRAS. Some reports related to more than one of these categories, for example one report was made regarding cab temperature, the poor design of cab seats and a loud rattling noise from the windows in high speeds.

Table 1. Breakdown of Train Cab Reports

Issue	Number of reports
Poor cab design/layout	43
Temperature problem in train cab	36
Poor cab environmental conditions (other)	28
Noise problem in train cab	20
Windscreen poorly designed	10
Window defect	8
Passenger communication system poorly designed	6
Alarms poorly designed	6

The main issues raised by reporters were poor cab layout or design, temperature and noise problems and other environmental problems. The issues reported reflect the experiences of some train drivers and are not necessarily representative of driving conditions in the rail industry. Some of the issues raised are discussed below. All quotes given are from reporters who cannot be identified, for reasons of confidentiality.

Temperature problems

Temperature problems were reported to CIRAS with the cab being too hot in summer and too cold in winter. Many of these were associated with 15 series cabs. Most reporters were concerned about the distraction caused by their discomfort. It is known that distraction effects can occur due to both hot and cold environments and this in turn can degrade performance (Parsons, 1995). Therefore the potential for a driver error is increased. A SPAD (Signal Passed at Danger) was considered by reporters to be the main safety consequence that could occur, *"Well I mean you might SPAD or miss a station or anything potentially of that nature if you're not concentrating on your driving and you're more concentrating on keeping warm"*.

Fatigue has been found to be one of the side effects of excessively high temperatures (Gafafer, 1964 cited in Matthews *et al*, 2001) and this association was often highlighted by reporters. *"You're wringing wet with sweat and we've only got one type of uniform and that's a winter style uniform and it is 120° in the cab. It's very difficult, you feel like falling asleep and it's constantly like that but you've got to concentrate all the time."*

A combination of factors was often cited as contributing to driver discomfort, for example inadequate ventilation, hot cab environment and drinks not permitted in the cabs. This could lead to dehydration and therefore distraction. When some drivers tried to rectify the problem by opening the windows further problems occurred; the noise level increased and it became difficult to hear the vigilance device *"... you've got air-conditioned units for the passengers but there seems to be nothing for the drivers other than opening the window and then as I said you can't hear the vigilance when you're opening the window because of the noise off the rails, so there's really nothing for air conditioning at all."*

The opposite problem manifests in winter. There have been reports of draughts in the cabs and heaters that do not work. Drivers have reported using newspapers and masking tape to try and block draughts.

Noise

Alarm presentation within the cab is important (Sanders & McCormick, 1992), especially in an emergency. Reports were received stating that background noise in the cab was masking the AWS (Automatic Warning System) making it difficult to hear. *"If you've got the wind with the gangway doors, you're talking about a high-pitched whistle coming in, you could miss the AWS, I mean that's how loud it is, it's really, really loud and annoying."* However, many of the noise related reports were to do with the level of noise produced by the AWS in other cabs. One reporter stated that some drivers covered up the amplifier to reduce the volume of the warning horn as they felt it was too loud causing distraction and discomfort. Whilst people do vary greatly in their response to noise (Haslegrave, 1995), it has been found that performance can be adversely affected by noise, especially discontinuous or unexpected high levels which make it difficult to concentrate (Kroemer & Grandjean, 1997). Therefore an excessively loud AWS could cause as many problems as one that is not heard enough, as both could lead to a driver error.

Design problems

A range of cab design problems were reported and most of the noise and temperature related safety issues were ultimately due to design problems. The visibility of the speedometer was a problem at night for a reporter who had to use a torch to read it. *"It's about the speedometers and the desk lights on the unit, none of us can read them, especially in tunnels or night time. There's a plate which blocks the lights coming through to the actual speedometer and of course the engines are getting old now, they're not being cleaned, and there's only a little slot for the actual light to come through, which makes it very, very hard to see how fast you're going"*

Problems with chairs were also reported including one with a fixed backrest that could not be adjusted *"It's just the position because the back rest of the seat is fixed to the wall, it's got a very limited range of movement, so most of the time you're sat on a seat that's sort of set at a right angle"*. Research shows that increasing a seat angle beyond 90 degrees reduces intervertebral disc pressure and muscle strain (Kroemer & Grandjean, 1999). Consequently, drivers may benefit from having an adjustable chair, as have been fitted in more modern train cabs. Another report concerned a driver who repeatedly hit his knees on the desk when rough riding. Discomfort is known to cause distraction (Bridger, 2001) which is obviously undesirable in a safety critical role.

The design of specific controls has also been reported as being problematic. For example the forward motion of the controller used to operate the throttle in one cab seemed to differ to the other cabs that the driver had used "*I mean it's the only section we've got that you actually open the throttle pushing it forward, most traction is in reverse so if you do fall forward you're actually shutting the controller but on this type you're actually opening it up or pushing it forward*". This could cause confusion as the control may not behave in the same way as a driver's mental model of the system formed from past experience (Wilson & Rajan, 1995) or from stereotyped expectations (Kroemer & Grandjean, 1999). Therefore drivers might make an error by accelerating instead of braking.

Another reporter had an issue with the delay inflicted by the TPWS (Train Protection Warning System). When changing ends of a train the brake could not be taken off before a certain length of time had expired. This meant that some drivers took shortcuts and violated procedure by driving from the wrong end of the train when travelling a few yards in sidings or speeding (to make up time) which could increase the potential for error. "*But of course it's not a good practice so it's been discouraged. And of course it is a disciplinary offence if you're caught driving from the wrong end. But it's encouraging people to do that because of this delay.*" This type of report highlights the advantage of confidential reporting systems such as CIRAS as drivers may not report incidents where they are aware of incorrect behaviour (violations) if CIRAS did not exist.

Discussion

The reports received by CIRAS suggest there are a large variety of ergonomic problems associated with the train cab design of some, particularly older, units. Most of these problems could lead to driver distraction and fatigue but there were also cases of drivers using more dangerous procedures (violations) to save time or effort. These factors can affect driver performance and could result in a driver error along with adverse consequences, such as a station overrun or failure to notice track workers or obstructions on the track.

Interestingly, responses given by companies to some of the reports detailed here stated that ergonomists had been involved in the design, modification and assessment of the cabs. In some cases drivers were also involved. It may therefore be surprising that these problems continue to exist. One possible explanation is that ergonomists may have been involved in the initial design but not once the stock has gone into service. This suggests that designs may not have been tested in the full range of driving conditions. There were indeed a number of reporters who cited the difficulty of recreating the problems that drivers were experiencing. Often the problem only occurred in certain conditions for example, cab temperature problems occurring in extreme weather conditions or visibility problems occurring when it was raining at night or in bright sunshine. Therefore some problems would be left unrecognised if the ergonomist was only involved with the cabs in the initial stages of design. In addition, although there was some user involvement with cab design this may have been insufficient. In one case a driver claimed the finished design was different to the suggestions he had made initially.

There have been constraints on the design of cabs as there is a need to retrofit new equipment into old cabs such as TPWS. This has been happening for many years as reporters recognise, "*As the units have got older, I mean they've gone through refits but nothing's been done about the seats, so it seems to have been going on for quite a while.*"

Although ergonomists were involved in some initial designs, much of the older rolling stock was designed at a time when there was not such a wide awareness of ergonomics. It's difficult for ergonomists to have a large impact in retrospect on the cab design as a whole and consequently some problems are difficult to rectify and have remained. It seems there needs to be some compromise with keeping original cabs in use and the understanding that drivers' comfort and concentration are necessary to the safe running of trains. If train cabs are evolving, the role of the ergonomist should be evolving alongside this. Involving an ergonomist and indeed drivers during the initial service period, may verify design adequacy and identify problems experienced in a variety of conditions and so ultimately improve safety.

These reports show that drivers are a valuable source for identifying ergonomics problems that may not have been picked up by designers, managers or even ergonomists. This reemphasises the importance of frontline staff in the successful design of train cabs and indeed any human machine interface.

Conclusion

The types of cab design problems reported to CIRAS by drivers were varied and are not necessarily representative of cab conditions across the rail industry. The range of problems reflects the range of conditions and requirements that some drivers are exposed to during their work. CIRAS may highlight areas for ergonomists to improve their impact on train cab ergonomics in the future. The fact that problems persist even after ergonomic input show that the evolving nature of cab designs requires continuous involvement of ergonomists. The importance of user centred design or participative design is also reemphasised. This paper found that CIRAS can be a valuable resource for the railway industry and for ergonomists. In particular self reporting of errors and violations, which is unlikely to occur through other channels, can provide useful insight into drivers' behaviour helping to improve train cab design in the future.

References

Bridger, R.S. 2001, *Introduction to Ergonomics*. (McGraw-Hill, Singapore)
Haslegrave, C.M. 1995, Auditory Environment and Noise Assessment. In J.R. Wilson & N Corlett (eds.) *Evaluation of Human Work. A Practical Ergonomics Methodology.* (Taylor and Francis, London), p506-540
Kroemer, K.H.E. & Grandjean, E. 1999, *Fitting the Task to the Human: A Textbook of Occupational Ergonomics*. Fifth Edition. (Taylor and Francis, London)
Matthews, G., Davies, D.R., Westerman, S.J. & Stammers, R.B. 2000, *Human Performance: Cognition, Stress and Individual Differences*. (Psychology Press, East Sussex)
Parsons, K.C. 1995, Ergonomics Assessment of Thermal Environments. In J.R. Wilson & N Corlett (eds.) *Evaluation of Human Work. A Practical Ergonomics Methodology.* (Taylor and Francis, London), p483-505
Sanders, M.S. & McCormick, E.J. 1992, Human Factors in Engineering and Design. 7th Edition.   (McGraw-Hill, New York)
Wilson, J.R. & Rajan, J.A. 1995, Human Machine Interfaces for Systems Control. In J.R. Wilson & N Corlett (eds.) *Evaluation of Human Work. A Practical Ergonomics Methodology.* (Taylor and Francis, London), p357-405

Early Human Factors Assessment of the new European Standard for train driver signalling information.

Joanna Foulkes, Nick Colford, Jenny Boughton

Human Engineering Limited
Shore house, Westbury-on-Trym,
Bristol, BS9 3AA,

The European Rail Traffic Management System (ERTMS) is the arriving standard for railway interoperability in Europe. Part of this standard describes the European Train Control System (ETCS) future standard for in-cab signalling for trains in Europe. The driver's interface was produced several years ago following an ergonomics study of the task of train driving. Each national railway organisation has the responsibility of fitting this European Standard to the needs, practices and traditions of the national railway.

This paper covers the successful application of the methods and tools used in an early human factors assessment of the impact of adopting ETCS on UK train driving, and the results obtained. In addition it recounts some of the practical lessons learned and some of the pleasures and pitfalls of a truly "early" human factors assessment.

Introduction

The work reported is the early human factors assessment of the compatibility between the ETCS standard and UK national railway practice. Separate detailed task analyses were produced of train driving in the UK and train driving using ETCS. From this, a combined task analysis for hypothetical future UK-ETCS train driving was produced. This formed the basis for predictive numerical workload assessments of train-driving scenarios, which were then compared with the expert workload assessment of experienced train drivers.

The SHERPA (Systematic Human Error Reduction and Prediction Approach, Kirwan, 1994) method was applied to the task analysis and the results of human error and workload assessments were combined. On the basis of this systematic analysis human factors issues have been identified fed into the system specification and design process at a very early stage.

Production of a Predictive Hierarchical Task Analysis (HTA)

Data Collection

Data were collected from a range of UK Train Operating Companies. Interviews with Drivers and Training Managers were completed along with footplate rides where drivers' actions and eye view video footage were recorded. Video footage was taken of all the external information available to the driver to enable the cognitive elements of the HTA to be identified. The use of different safety and legacy systems in use across the network were recorded and the train driving Rule Book studied.

Data regarding the ETCS system were drawn mainly from the CENELEC (European Committee for Electrotechnical Standardisation) ETCS standard and other documentation specifying other ETCS applications. Several trips to Switzerland were made to observe a working ETCS-fitted train and track section, where drivers' actions and eye view footage were recorded.

Hierarchical Task Analysis

The data were used to produce a whole task HTA for UK driving and a part-task HTA for ETCS driving (only tasks related to the ETCS system). Included within the HTAs were cognitive tasks involved in processing the acquired information and planning for the journey ahead. It was necessary to structure the HTAs down to a very low level of detail. This was required to complete predictive modelling as it allowed predictive task times to be calculated. Once complete, these task analyses were then combined to produce a predictive UK-ETCS task analysis.

Workload and Error Assessment

Workload Assessment Tool

The workload assessment tool ATLAS[1] was used to complete the workload assessment of various UK-ETCS driving scenarios. There are several steps to the workload assessment using ATLAS; these include taxonomy generation, timeline prediction, scenario building and workload assessment and interpretation.

Taxonomy generation

ATLAS requires the action descriptions described within the HTA to be described at a behavioural level. This allows a predictive time for each task to be calculated using PerfCalc[2], which is based on established cognitive theories. ATLAS also enables a conflict analysis to be completed on those tasks that overlap during the scenarios to establish whether the tasks can be completed in parallel according to cognitive processing theory.

Timeline prediction

In order to build a representative scenario it is necessary to construct a time line of events to which tasks can be anchored. The modelling was completed using a representative piece of track, onto which scenarios of driver behaviour were mapped through realistic journey speed

---

1 © ATLAS is copyright by Human Engineering Limited 1995-2004
2 © PerfCalc is copyright by Human Engineering Limited 1995-2004

profiles. This involved the creation of a model stretch of UK railway line (based on real track geometry). This provided a geographical representation of the features that the driver would encounter on the track. The distances between these were then converted into timings by overlaying a driving speed profile to which the driver would adhere. The timings provided by the speed profile were integrated with the features on the track, enabling predictive scenarios to be built using the tasks in the predictive UK-ETCS HTA.

Scenario building

For the purposes of modelling, the train driving task consists of four elements: maintaining vigilance out of the window; acquiring information; formulating a driving plan; and controlling speed. These four elements are included in the scenario in between the events on the timeline; the plans from the task analysis were followed in response to each event. Once the sequence of tasks was completed, groups of functionally linked tasks are produced and then placed in the sequence.

Workload assessment and interpretation

This assessment consisted of the numerical analysis of workload, and its interpretation in the context of the skilled train driver's behaviour, to assess the manageability of the workload. Each workload conflict was individually and contextually evaluated and categorised according to the predicted operational impact of the workload. This process drew upon Subject Matter Expert (SME) opinion gathered during baseline workshops where subjective data were collected. The resulting description of workload was then compared across the different systems. This comparison led to the final results of the workload analysis from which the conclusions and recommendations were drawn.

Human Error Assessment

The Human error assessment was a relatively straightforward exercise of applying the SHERPA methodology to the taxonomy of tasks used in the workload assessment. The work performed in preparing the HTA and the workload assessment provided enough detail to enable a SHERPA assessment despite the early stage of the project. Error reduction mechanisms were produced for those errors that resulted in a safety consequence or had a "high" probability (according to SHERPA practice).

There is an established link between high workload and the likelihood of human error. The preceding workload assessment identified groups of tasks that may present workload problems for drivers. As a precaution therefore, errors in the performance of these tasks were revisited and the probability values for these tasks were increased by one level in each case (low to medium and medium to high). As a result, a number of additional errors became high priority and they were subjected to the error reduction exercise.

Results

The final output of the work (in addition to reports documenting the methods and results at each stage of the work) was a detailed human factors issues register. The register was organised not only to log and describe the issues themselves but also to provide immediate recommendations, guidance for further work, and categorisation of the issues within the overall project systems engineering issues database.

The issues register was set up early in the work programme and maintained throughout. It enabled the documentation of issues that did not strictly fall into the scope of each stage of the work, such as detailed design features of the driver-machine interface. As each stage of work was completed, any significant issues resulting from the specific analyses – the workload issues and the human error issues – were added to the issues register.

Three fields in the ERTMS systems engineering issues database were of particular relevance to the human factors issues raised. These consisted of:

- A **rationale** for the issue to provide the engineers with the underlying principle for the concern arising from the human factors analysis and the basis for the guidance that followed.
- **Guidance** on resolution of the issue; either direct advice on changes that need to be made to the existing system or an indication of the sort of changes that might be of benefit together with an indication of the further work needed to produce definitive guidance.
- A major **system area** for the issue

The major system areas required by the ERTMS systems engineers to allocate responsibility for the further tracking and development of the issues were:

System Design: The design requirements for components (hardware and software) of the system;

Application Rules: The rules on where and how to deploy .those components across the railway infrastructure; and

Operations: The operational rules, skills requirements, training etc. required to enable people to work together consistently and safely using the system.

By following this system, even the incidental human factors issues raised during the project did not get lost but were documented alongside the findings of each specific assessment. Inevitably, the amount of detail in the rationale and guidance fields was less than for the issues raised directly by the results of the workload and human error assessments. Nevertheless it resulted in many issues being logged and the start of an approach to dealing with them being provided for the systems engineers.

Lessons Learned

The practical lessons learned about performing a detailed early human factors assessment on a major engineering project are summarised below.

Writing Reports for System engineers
The subject matter of this assessment was technically very complex therefore technical sections of reports were accompanied by a summary for non-human factors professionals. All findings and issues were reported in terms directly applicable to the overall systems engineering data management system.

HTA

In order to satisfy the requirement of the HTA plans, it was necessary to have a fundamental understanding of what strategies drivers would use to respond to situations, and what rule/ skill based behaviour they would use to achieve their goals. Driver strategy was developed through interviews with drivers; review of the train driving rule book; and, where necessary assumptions were made that were agreed with SMEs.

Workload

The *Timeline Prediction* work was costly and time consuming and not strictly human factors analysis. The work fell to the human factors team because it was only applicable to the human factors assessment. The need for this sort of activity, in support of the human factors assessment should be considered as part of the human factors work.

Human Error

The SHERPA methodology requires detailed descriptions of tasks and system features that are often lacking, early in a project. A detailed system baseline solely for the human factors assessments was produced in collaboration with the systems engineers and user representatives. On the basis of that baseline it was possible to provide detailed human factors recommendations before the official systems engineering baseline was established.

Conclusion

As reported here it has been possible to conduct, on the basis of task analysis, a very detailed early assessment of a complex system early within a project. The effort required, and in particular the amount of non-human factors work involved, was substantial. However, the sight of detailed human factors recommendations finding their way into a requirements management system before many major design decisions are made is a rare satisfaction.

Acknowledgements

The work reported here was performed under contract from Rail Safety and Standards Board (RSSB). The authors extend special thanks to Dr. Mark Young, formerly of RSSB, and Andy Baker of Davis Associates Ltd and Human Factors representative for the National ERTMS programme for their contributions to the human factors work reported here.

References

Kirwan, B.,1994, *A Guide to Practical Human Reliability Assessment.* Taylor and Francis, London.

AIR TRAFFIC CONTROL

Review of Task Analysis Techniques for use with Human Error Assessment Techniques within the ATC Domain

Katie Callan[1], Carys Siemieniuch[1], Murray Sinclair[2], Laurence Rognin[3], Barry Kirwan[4] and Rachael Gordon[4]

[1] *Department of Systems Engineering, Loughborough University, LE11 3TU, UK*
[2] *Department of Human Sciences, Loughborough University, LE11 3TU, UK*
[3] *STERIA for EUROCONTROL Research Centre, Bretigny sur Orge, FRANCE*
[4] *EUROCONTROL Research Centre, Bretigny sur Orge, FRANCE*

In order to deal with the increased traffic levels, the Air Traffic Management (ATM) system must take advantages of new technologies and procedures which will aid the Air Traffic Control Operators (ATCOs) in their complex job. New allocation of spacing tasks between controller and flight crew is envisaged as one possible option to improve air traffic management. Safety assurance requires full analysis of the possible consequences of new procedures; one method for achieving this is through Human Error Analysis (HEA). In order to carry out such a HEA, the spacing task had to be captured within a Task Analysis (TA) model. This paper covers the evaluation of three TA techniques that can be used as a precursor to HEA analysis – Hierarchical Task Analysis (HTA), STATEMATE and Integration Definition for Function Modelling (IDEF$_0$).

Spacing Tasks

New allocation of spacing tasks between the controller and flight crew is envisaged as one possible option to improve ATM. The motivation is neither to 'transfer problems' nor to 'give more freedom' to the flight crew. In essence, the purpose is to identify a more effective distribution of tasks which will be beneficial to all parties (Grimaud et al. 2001).

The allocation of spacing tasks to the flight crew – denoted airborne spacing – is expected to increase controller availability and to improve safety. This in turn could enable better efficiency and/or, depending on airspace constraints, more capacity. Additionally, it is expected that flight crew would gain in awareness and anticipation by taking on active part in the management of their situation. Airborne spacing assumes new surveillance capabilities (e.g. Automatic Dependent Surveillance-Broadcast) with new airborne functions such as Airborne Separation Assistance System (ASAS).

Human Error Analysis (HEA)

The major goals of HEAs are to assess the impact of human errors on system safety, to suggest ways of reducing the impact of human error and/or reduce the frequency at which it occurs. There are a number of different methods available, but these must essentially provide the answers to the following questions (Kirwan, 1990):

- ° What can go wrong? (Human error identification)
- ° How often will a human error occur? (Human error quantification)
- ° How can a human error be prevented from occurring or its impact on the system be reduced? (Human error reduction)

Air traffic control is a complex system and as such the controllers (and pilots) cannot afford to make certain types of errors. HEA offers a means of investigating the consequences of different human errors that may occur due to the implementation of new procedures such as spacing assurance.

The HEA method which was used by EUROCONTROL was named TRACEr – the technique for the retrospective and predictive analysis of cognitive error in air traffic control. For details of the method see Shorrock and Kirwan (2000).

In order to carry out a HEA assessment, the task which is being evaluated must be analysed in a way that allows all interactions and possible activities involved to be represented through a TA model.

Task Analysis

In order to select three TA methods for detailed analysis, a literature review took place. Each method was reviewed against a set of choice criteria, the results of which are given in Table 1 (more details of how and why the criteria were chosen can be found in Callan, 2003a).

From this choice matrix three methods were investigated further: HTA, STATEMATE and IDEF$_0$. This selection was made on the basis of these methods meeting the majority of the selection criteria. Each method was initially evaluated as being capable of capturing, specifically in the ATC domain, the human-machine interactions along with the various decisions that have to be made by the controller.

The resulting graphical representations were relatively concise in nature whilst allowing a detailed level of description of the tasks. Additionally, the methods were deemed capable of illustrating the error paths that may occur; a requirement for input to TRACEr.

The main areas of concern with the three methods, particularly the IDEF$_0$ and STATEMATE approaches, included the cost implications of purchasing the software, the readability and clarity of the end product. In addition the level of training required to become proficient in the use of the tools, the repeatability and consistency of the methods were issues to be considered.

Initial Task Analysis

As a means of comparing the three TA methods and of presenting the techniques to the stakeholders at EUROCONTROL, a small portion of the spacing task was analysed. From this initial application of the three methods, to the same portion of the task, it was concluded that the STATEMATE tool was too complicated and too resource-intensive for the analyses in which EUROCONTROL were interested. The tool is primarily used for software development and as such there would be a large portion of redundant functionality with the (expensive) tool. Additionally a large training budget would be required to enable people to use the technique.

Table 1: Choice Matrix for Reviewed Task Analysis Methods

Assessed Task Analysis Techniques	Integrated View	Capture Communications Flow	Capture Organisation Aspects fo Delegation	Capable of Capturing Possible Errors	Software Supported	Transparent, Consistent & Self Descriptive	Concise Representation	Hierarchical in Nature	Cost and Time Effective for the Delegation Application	Easily Understood by Non-Experts of the Tool
Hierarchical Task Analysis (HTA)	☑	☑	☑	☑	☑		☑	☑	☑	☑
Statecharts (STATEMATE)	☑	☑	☑	☑	☑		☑	☑		
Goals, Operators, Methods and Selection Rules (GOMS)	☑		☑		☑	☑	☑	☑		☑
Cognitive Task Analysis		☑	☑	☑					N/A	
IDEF (Integration DEFinition language)	☑	☑	☑	☑	☑	☑	☑	☑		
Fault Tree Analysis and Event Trees			☑	☑	☑	☑		☑		☑
Table top analysis	☑	☑	☑	☑		☑			N/A	
Charting and Networks Methods	☑		☑	☑		☑			N/A	☑
Link Analysis		☑				☑	☑		N/A	☑

Detailed Application

The first stage of the detailed analysis was to use the procedural documents to produce a flowchart of events. This enabled different stakeholders and subject matter experts to view the information and validate the content of the models; it was established that the task shown along a timeline was easily understood by all stakeholders and no semantics required explanation.

The information was interpreted and translated into an $IDEF_0$ model and a HTA model, the main semantics of which are shown in Figures 1 and 2 respectively. Due to the nature of the tasks being modelled there was a need to show different actors involved with different activities. For the $IDEF_0$ model this was achieved by using colour coded controlling arrows, and for the HTA model colour coded activity boxes were adopted.

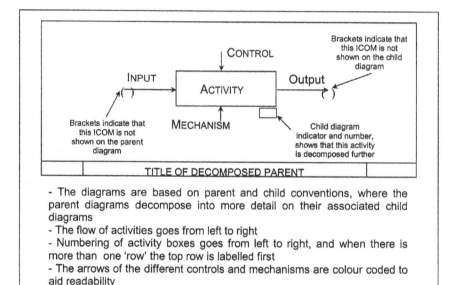

Figure 1: Diagram highlighting the main semantics of IDEF$_0$

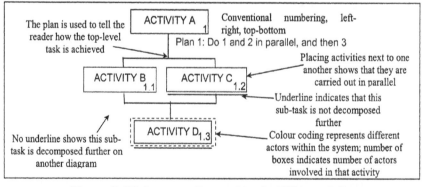

Figure 2: Main semantics used in the HTA models

From the detailed analysis of the spacing tasks using the two TA methods it was found that the IDEF$_0$ model allows a more concise model to be built up for the task in question. The diagrams showed a much more detailed picture of the tasks due to the Input, Control, Output and Mechanism (ICOM) arrows. The IDEF$_0$ models included information flows between the different sub-tasks and allows the reader of the diagrams to be aware of the information that the controllers and pilots have available to them in order to carry out the activities.

Due to the way in which the graphical representation of the $IDEF_0$ method is built up it is more intuitive to follow the flow of activities when compared to the HTA equivalent models. In order to understand the HTA content fully, the user must first read and understand the plans associated with the diagrams, and then interrogate the activity boxes. It is felt that the $IDEF_0$ model offers a more dynamic representation of the task, and the visual representation is easily grasped when the semantics of the method have been explained.

Discussion

$IDEF_0$ provides a more rich picture of the modelled task, which may make the models easier to use by other members of a team, with different knowledge. This is due to the additional information that is held within $IDEF_0$, the users are provided with more information about the problem context and the task in question, which may put them in a better position to carry out their subsequent work from the TA model.

From the diagrams that were produced for this work it is felt that the $IDEF_0$ models would be of more use when discussing the design and implementation of a system with designers and engineers. The method allows both the human and physical hardware/software aspects of the system to be represented and shown more easily. There is also the advantage of having the mechanism information included in the diagrams, this shows what is required (i.e. equipment) for the activity to take place. This means that the effects of failure of that equipment can be traced through the numerous activities.

Alternatively, the HTA diagrams are not as cluttered as the equivalent $IDEF_0$. HTA does not provide information regarding the inputs, outputs or mechanisms, but this could be overcome by supplementing the traditional diagrams with a tabular task analysis which contained the additional information. The major drawback of this would be that the user would have to cross-reference the different sources of information and there would be an increase in the number of diagrams and tables needed to represent the task. The HTA models could have notes included against each activity detailing the missing information, but this is likely to reduce the readability of the diagrams dramatically.

The major advantage that HTA has over the $IDEF_0$ method is the fact it is widely used and understood in the field of human factors, hence the cost of training in the technique is minimal. However, once the semantics of $IDEF_0$ have been grasped the reading of the diagrams is fairly straightforward and the richer picture has many benefits.

The overall results are are under review by the client organisation – EUROCONTROL.

References
Callan, K (2003a). *Investigation of Task Anlaysis Techniques for the Delegation of Spacing Assurance, Loughborough University and EUROCONTROL*
Grimaud, I., Hoffman, E., Rognin, L. and Zeghal, K. (2001). *Delegating upstream – Mapping where it happens. Internation ATM R7D Seminar, Santa Fe, USA*
Kirwan, B (1990). *Chapter 28 Evaluation of Human Work, Taylor and Francis Ltd*
Kirwan, B and Ainsworth, L. K (1992). *A guide to Task Analysis, Taylor and Francis Ltd*
Shorrock, S. T. and Kirwan, B (2000). *Development and application of a human error identification tool for air traffic control, Applied Ergonomics Vol 33 (2002)*

THE EFFECT OF HUMAN INTERVENTION
ON SIMULATED AIR TRAFFIC

Hugh David

R+D Hastings,
7 High Wickham, Hastings,
TN35 5PB, UK

A set of 52 records of students attempting to control the same initial traffic provided the opportunity to develop metrics of the differences due to their actions and of the similarities in the evolving situations.
Potential conflicts, conflict resolution orders and differences in subsequent traffic generated are investigated.
Samples begin to differ about three minutes after the start, but are not completely different until about 30 minutes after the start.

Introduction

Real-Time simulation originated as a purely empirical technique, devoid of any theoretical foundation, driven by the need to develop methods of exploring the behaviour of systems that were under consideration, before spending scarce resources on possibly ineffective systems. The engineers involved were aware that human beings varied, but had no time, knowledge or wish to take more than a minimal account of this variation in their explorations.

Over the years, knowledge of the behaviour of individuals and groups in working environments, and reasonable expectations for their capacity have accumulated in those military and civil environments where it has been possible to carry out systematic trials and observations. However, the high cost (until recently) of simulators, and the continuing high cost of diverting expert operators from their actual tasks has limited the possibilities of exploring large samples of simulated behaviour.

In the course of a project, reported elsewhere, (David and Bastien 2003) on the development of a future, more human-operator compatible, Air Traffic Control system, a body of data became available on the treatment of the same sample of air traffic by 26 postgraduate students, each student repeating the exercise twice.

This paper describes the speed of divergence of these 52 parallel universes, as it can be deduced from the available records.

Experimental Procedure
Full details of the experimental equipment and the procedures involved are provided in David and Bisseret (2003). The analysis of data recorded is discussed in David (2003). To understand this paper, it is only necessary to know that the operators were presented only with potential future conflicts and provided with means to solve them. For brevity,

'conflict' will be used in place of 'potential future conflict'. These 'conflicts' occur when they are first detected by the system, not when separation is actually lost, and cease when action is taken that solves the conflict.

The first conflict occurred at 47 seconds into the simulation, and was identical in all 52 exercises. This conflict was solved in each exercise, using generally similar orders. The differences in these orders have no effect on the exit point of the aircraft involved, since the system automatically returns aircraft to track and varies speed accordingly. They may, however, involve different paths to that exit point, which may render a candidate aircraft unacceptable. The second substitute aircraft generated will have the same entry time as the first, but will have a different exit time. The traffic records for each exercise will therefore differ from the entry time of the first aircraft to have a different exit time. Because the random-number generator employed will no longer be 'in step' all subsequent aircraft will differ, and the exercise will follow a different course.

It is, of course, possible that differences in the response to the first conflict will not cause the rejection of subsequent candidate aircraft, and that two simulations may continue in an essentially identical fashion, until they are differentiated by responses to other conflicts.

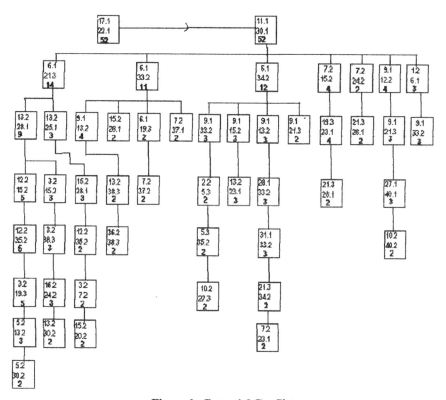

Figure 1 - Potential Conflicts

Analysis

Three possible measures of divergence, 'conflicts', orders and aircraft entry/exit times have been explored.

Conflicts

The data recorded for each exercise included potential future conflicts displayed whenever an order was accepted, or a routine screen update was performed. By recording the first and last occurrences of a particular potential conflict, it is possible to determine for what time a particular conflict was present. There is no rigid connection to the order accepted, since one order may solve several conflicts. Technically, it was possible for an order to create another conflict, although this never in fact happened. The time at which the conflict would start, if no action were taken, was also recorded.

Figure 1 shows the differentiation of the exercises. (To permit a visible diagram at this scale, all unique sequences are omitted.) Each box specifies the two aircraft involved in the conflict and in bold type the number of runs for which this conflict occurs. Although the mean start times are available for each conflict, they cannot be displayed in this form. The first two conflicts are identical in all 52 exercises, then at successive stages there are 7, 14, 10, 7, 5, 4 and 1 sequences of conflicts that occur in two or more runs.

The mean time of the second conflict was 69.3 seconds after the start of the simulation.
The mean time of the last shared conflict in a sequence of nine successive conflicts was 203 seconds.

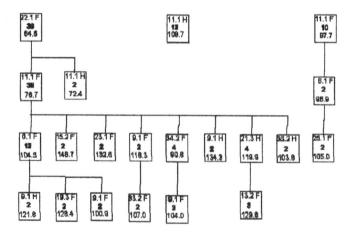

Figure 2 - Conflict Resolution Orders

Conflict Resolutions

A similar approach can be made through the study of the orders given in each run. Orders can be described some detail, (David 2003), but, for this purpose, can be classified by the type of manoeuvre involved,(change of Flight Level, Heading, or Speed) and the aircraft involved. Although there is necessarily variation in the size of manoeuvre and the time it is maintained before returning to track, these differ relatively little in their effects on the airspace 'occupied' by the aircraft affected.

Figure 2 shows the sequences of aircraft and orders for the 52 runs. As before, single runs are not shown, to provide a viable diagram. Each box specifies the aircraft to which the order is given, the mean time at which the order was given, and, in bold type, the number of runs for which the sequence applied.

There were three different first orders, then 3, 9 and 6 alternatives at successive levels. At and after the fifth order all sequences were unique.

The mean times for the three first orders was between 64 and 108 seconds, and the mean time for the longest common run of four orders was 137 seconds.

Table 1 - Entry times of first aircraft with differing exit times

Sample	Number	1st. Diff.	Last Diff.	Ar. Mean	Std. Devn.
A1 Vs A2	26	189.7	1856.6	898.0	413.9
A1 Vs B1	325	196.0	1859.0	888.2	402.0
A2 Vs B2	325	153.3	1922.1	907.5	424.6
A1 Vs B2	650	169.8	1891.4	887.6	414.9
Statistical	Significance	$p<0.001$	$p>0.05$	$p>0.05$	$p<0.01$
Overall	1326	172.4	1890.3	892.7	414.1

Aircraft Generated

The simulator in use employed a stream of pseudo-random numbers to generate random distributions within pre-set limits for the flight characteristics of simulated aircraft at their planned entry time. All simulations began with defined entry times for the first forty aircraft entering the area, and were identical until the first potential conflict was detected. The control applied was sufficient to ensure, with a few exceptions, (5 in 10,000 aircraft) that aircraft left at the time, height and place originally planned. When a conflict was 'solved', the trajectory of at least one aircraft was altered. Each new aircraft, whether from the initial 40 or replacing departing aircraft, was generated at its entry time. If the aircraft initially proposed did not meet the criteria for acceptance, (absence of a conflict within 120 sec of entry or exit) a further aircraft was generated in its place until a suitable candidate was found. Differences between simulation records could be expected to develop from the first difference in aircraft accepted, which could be identified by the entry time of the first aircraft to have different exit times in the records being compared. Subsequent aircraft could be expected to have different exit times, since the pseudo-random number generator would be at a different point in its cycle in different runs. There is no 'ideal' traffic, from which exercises differ, so comparisons can only be made between pairs of exercises. Figure 1 shows that there are 26 pairs of exercises where the first and second exercises for the same participant are compared, 325 where first exercises are compared for two different participants, 325 where second exercises are compared for two different participants and 650 exercises where exercises for different participants and different runs are compared. For each pair of exercises, the exit times were compared for all two hundred-odd aircraft, storing the entry time for the first aircraft carrying that code which had different exit times. These entry times were then sorted into order, and the first, last, mean and standard deviation of these values (40 in each case) were recorded, with the type of comparison (coded A1A2, A1B1, A2B2 and A1B2 respectively). Table 1 lists the arithmetic means of these parameters for each sample and the overall arithmetic means, with the estimated statistical significance of the differences, as estimated by analysis of variance ($F_{3,1321}$). The significant results suggest that second runs start to differ earlier than first runs, and that their rates of divergence are more variable.

The mean time for the first divergence, before which the traffic in the two runs was essentially identical was 172 seconds, and for the last, after which the two traffic samples had nothing in common, 1890 seconds.

Discussion

The mean time of the second conflict was 69.3 seconds after the start of the simulation. The mean time of the last shared conflict in a sequence of nine successive conflicts was 203 seconds.

The mean times for the three first orders was between 64 and 108 seconds, and the mean time for the longest common run of four orders was 137 seconds.

The mean time for the first divergence, before which the traffic in the two runs was essentially identical was 172 seconds, and for the last, after which the two traffic samples had nothing in common, 1890 seconds.

These figures differ because they measure different aspects of the problem. They are in general consistent, suggesting that conflict resolution orders follow conflict detection, and that changes to the traffic will follow the conflict resolutions.

Conclusion

It appears that the 52 runs start to differ about three minutes after the start, and that differentiation is completed only thirty minutes after the start.

References

David, H., 2003, What Can You Get from a Dribble File? In Paul T. McCabe (ed.) *Contemporary Ergonomics 2003*, (Taylor and Francis, London) 277-282

David, H. and Bastien, J-M. C. 2003, Second Evaluation of a Radically Revised En-Route ATC System. In Paul T. McCabe (ed.) *Contemporary Ergonomics 2003*, (Taylor and Francis, London) 305-310

MEASURES OF PERFORMANCE IN AIR TRAFFIC CONTROL

Hugh David

R+D Hastings,
7 High Wickham, Hastings,
TN35 5PB, UK

It has long been a problem in the scientific analysis of Air Traffic Control (ATC) that there is no adequate measure of the efficiency with which the traffic is controlled. This paper considers criteria that such a measure should meet, reviews some widely used measures, and proposes a relatively simple on-line measure that meets the criteria proposed.

Introduction

Air Traffic Control, in its current form, is a technically complex system involving groups of human operators, assisted by a variety of technical devices. Most of these devices have been adopted from other fields, at different times and different stages of development. There has never been any systematic attempt to design the Air Traffic system. Problems, when they appear, are solved by pragmatic methods. Problems are rarely anticipated, and little is known about the safety or efficiency of the system in quantitative terms.

This paper considers measures of performance, rather than of capacity. Even with this restriction, there are problems. In many domains, measures of speed and accuracy can be used - separately or in combination. In Air Traffic Control, provided tasks are completed in time, the precise time is not important. Equally, although 'conflicts' (infringements of minimum separation limits) should and 'collisions' (physical contact between aircraft) must be avoided, the actual separation is relatively unimportant. Neither conflicts nor collisions are practically useful as measures, since in any normal system they are so rare that most comparisons of performance over meaningful periods tend to be 'goalless draws'.

Another approach is to estimate 'cost'. In most systems, controllers provide very wide margins of safety (Lafon-Millet, 1978) since this 'costs' nothing in terms of the effort required, and may reduce the effort required to monitor the system. Controllers are not aware of the relative economic costs of different types of solution to potential future conflict problems - information that is not generally available, in any case. Although estimates have suggested, for example, that changes in Flight Level are more economic than changes in speed or aircraft heading, even this generalisation may be invalidated by special factors, such as the need for aircraft to observe landing curfews, to connect with other flights, to comply with noise reduction procedures or to comply with restrictions on aircrew working hours.

Characteristics of a Measure

Any proposed measure of ATC performance should satisfy a number of conditions.

- It should be reproducible.
- It should be sufficiently sensitive that it can distinguish relatively subtle differences in the Air Traffic control activity.
- It should not be intrusive.
- It should be quantitative.
- It should be 'automatable', preferably within the system, real or simulated. (The expense and unreliability of manual data collection and the technical difficulties of synchronising external systems make internally generated measures practically necessary.)

Previous Studies

There have been many attempts to construct methods for estimating sector capacity. Many of these are essentially similar. One sector estimation and one allied method are described here.

Messerschmit-Bolkow-Blohm (MBB) Method

Although the MBB method is well known in the ATC world, it does not appear to have been widely discussed in the professional literature. Brauser (1975) is the earliest reference available.

This method assigns units of work (UW) required according to characteristics of flights through the sector (Table 1). The method assumes, on the basis of empirical studies, that control capacity will be saturated when speech loading reaches 45% of available time in 'en-route' sectors, and 60-80 % in approach sectors. The sector capacity is defined as 5.5 UW in a six-minute interval, or 50 UW per hour. (One UW corresponds to 64 seconds of speech.) The limit can be expressed as a nominal number of aircraft by dividing the capacity by the average difficulty per aircraft. If the average difficulty is 1.5 then the capacity of an 'en-route' sector would be 5.5/1.5 = 3.67 aircraft in any 6 minute period, or 33.3 aircraft per hour. Brauser mentions that the effects of sector configuration and time of flight in the sector are not considered in the MBB method, as then defined, although studies were under way to determine these effects.   Since the average difficulty will depend on the average frequency of such activities as weather reporting, or of "Pop-ups" (Aircraft entering a sector without having filed a flight plan.), the capacity determination will require a prolonged period of observation to determine how frequent such events -particularly conflict resolution, actually are in the sector. The assumption that the relative frequency of such events is independent of the actual traffic level may be questioned. Conflicts and holding are more probable in heavy traffic conditions, while communication trouble would appear to be inherently unpredictable. Over the years since 1975, procedures have altered, so that, for instance, "Pop-ups" are now very rare.

Table 1 - MBB COEFFICIENTS

Coefficient	Definition	Value (UW)
Basic	Basic unit score	1.0
S0	Airline flight	+ 0.1
S1	Charter flight	+ 0.2
S2	Military flight	+ 0.2
S3	Climbing/Descending	+ 0.24
S4	Vertical Handover/Take-over	+ 0.26
S5	Handover to/from TMA	+ 0.38
S6	Vectoring (e.g. Approach)	+ 0.3
S7	'Pop-Up'	+ 1.3
S8	Conflict Solution	+ 1.4
S9	Weather Report	+ 0.3
S10	Target Warning	+ 0.2
S11	Holding	+ 0.6
S12	Communications Trouble	+ 0.6

Thirteen years later, Van Elst (1988) reported an essentially unchanged MBB method, (the attempts to incorporate sector configuration and flight time in sector having apparently not born fruit) together with a 'simplified method' used by the French ATC service, which derived two different figures for executive (radar) control and procedural control. The first derived an average workload per aircraft by dividing the aircraft into cruising flights, climbing or descending flights within the sector, and flights entering or leaving an upper sector. For each group the total number of minutes spent in the sector was multiplied by a coefficient (1, 1.24 and 1.62 respectively) corresponding to the MBB parameters (Basic, +S0 + S3, +S5), from Table 1. The executive capacity was determined by dividing the frequency-weighted average into a nominal capacity of 500 UW per hour. The procedural capacity was estimated simply as 15 active flights present or within ten minutes of entry to the sector, whatever the sector geometry or traffic flow characteristics.

Versions of the MBB model have been realised at the EUROCONTROL Experimental Centre (EEC) for a Fast-Time (arithmetical) simulation model (SIMMOD) and for the EEC Real Time (Humans-in-the-loop) simulator. In real-time simulation no satisfactory method exists for determining which activities by controllers represented conflict resolution so certain control orders were assumed to represent this activity.

Problems were encountered with shorter time intervals, in place of a standard hour, in deciding to which period coefficients should be allocated - for instance, should the coefficient (S4) corresponding to entering or leaving a lower sector be applied throughout the flight or only at the relevant time?

TRACON Method

The TRACON method, developed by Wesson Incorporated, and employed in the ATC simulation game TRACON II (Wesson International, 1990) operates in a slightly different context, being used to provide feedback to the participant of his performance after a simulation run. The TRACON model provides a performance score after an exercise using the elements provided in Table 2. It does not provide an on-line score. Points are allocated on the successful completion of each flight. The initial allocation is based on the flight profile, plus a quantity depending on the type of flight. Points are subtracted for any commands, and for delays in takeoff. Larger numbers of points are subtracted for failed handoffs, missed approaches, allowing aircraft to fly off-screen, or for infringements of separation. An actual collision causes the simulation to stop, and erases all data. TRACON also provides a 'criterion score', which is an estimate of the best score achievable, and expresses the achieved score as a percentage of the 'criterion'.

Table 2. TRACON Scoring (Simplified)

Type of Flight	Points	Actions / Errors	Points
En-route	500	Any Command	-10
Departure	500	Delay per Minute	-20
Arrival	800	Missed Handoff	-500
Twin-Prop	+100	Missed Approach	-250
Turbojet	+200	A/c Off Radar	-1500
Business Jet	+300	Less than 3 NM (per 15 sec)	-1000
Airliner	+400	Less than 1 NM (Per 15 sec)	-5000
Military	+500	Collision	STOP

Proposed Measure of performance

Following the TRACON method, we allocate points for successful flights, but on a continuous basis rather than at completion. This enables us to obtain an on-line measure of performance.

We assume, arbitrarily, that acceptance from an adjacent sector or airport earns 100 points, as does handover to an adjacent centre. Correct line-up for approach, however, earns 400 points. We allow 20 points per minute for normal flight, so that the TRACON value corresponds to 15 minutes flight time from entry to exit - a generally realistic estimate. We penalise delays at the same rate, whether they represent delayed departure, sector acceptance or sector handover.

Orders are costed at 10 points, as by TRACON, but flights are allowed one order each for entry and exit. Departures are allowed one order when they enter airspace (in addition to the one allowed for take-off). Arrivals, which must be lined-up to intercept the glide path at the correct height and speed, are allowed five orders, corresponding to three vectors, a speed change and a height change. Rather than estimate appropriate times for these orders, the 'scorer' simply does not penalise the first two orders for 'en-route', three orders for departures and seven orders for arrivals.

Other penalties foreseen by the TRACON system are maintained, except that an arbitrary 10,000 point penalty is applied for a collision, rather than stopping the simulation (and losing the current data).

Finally, the coefficients for type are applied to all transactions involving the aircraft, as multiplying factors of 1 for single-engined aircraft, 1.2 for twins, 1.4 for turboprops, 1.6 for business jets, 1.8 for airliners and 2.0 for military aircraft, rather than as a simple addition.

Table 3 (below) illustrates the evolution of this index for a notional exercise in which two aircraft pass through a sector, one 'en-route' and one arrival, with a missed approach which required a 'go-around' and second attempt to land. In actual use, the criterion and total values and their ratio would be calculated at each renewal of the display image, and an analog image of the ratio would be provided if immediate feedback were needed. More aircraft produce less violent variation of the performance ratio.

Space does not permit full discussion of the underlying methodological problems of performance indices, of adaptation to more modern systems or of determining reasonable points allocations.

Table 3 - On-line Performance Index

Minute	Aircraft A	Aircraft B (Criterion)	Aircraft B (Actual)	Total (Criterion)	Total (Actual)	Ratio (%)
1	Airliner	Business		0	0	100
2	(x 1.8)	Jet (x 1.6)		0	0	100
3	180 (O) E			180	180	100
4	36			216	216	100
5	36			252	252	100
6	36	160 (O) E	160 (O) E	444	444	100
7	36	32	32	512	512	100
8	36	32 (O)	32 (O)	580	580	100
9	36	32 (O)	32 (O)	648	648	100
10	36	32 (O,O)	32 (O,O)	716	716	100
11	36	32	32	784	784	100
12	36	640 (O) A	-400 (MA)	1470	420	28.6
13	36	(arrival)	-32	1506	424	28.2
14	36		-32 (O)	1542	428	27.7
15	36		-32	1578	432	27.4
16	36		-32	1614	436	27.0
17	36	E= Entry	-32	1650	440	26.7
18	180 (O) X	X= Exit	-32	1830	578	31.6
19	(en-route)	A= Arrival	-32	1830	546	29.8
20		O= Order	-48 O	1830	504	27.5
21		(O) Allowed	-32	1830	472	25.8
22		MA =	-32	1830	440	24.0
23		Missed	-32	1830	408	22.3
24		Approach	+630 O X	1830	1038	57.2

References

Brauser, K., 1975, Methods for the determination of the Control Capacity of ATC services. Annex 6 to the Minutes of Session 3/75 of the Aviation Commission of the FRG Institute of Navigation, Frankfurt, 14 Nov 1975.

Lafon-Millet, M-T., 1978, *Observations en trafic réel de la résolution des conflits entre avions évolutives*, Raport INRIA No CO/R/55, (Rocquencourt, France:INRIA)

Van Elst, J. (1988) *Summary of Methods and Models used for estimating Sector Capacity*. Directorate of Operations, EUROCONTROL, Brussels.

Wesson International Inc., 1990, *The TRACON II Multi-player ATC Simulator*, (Wesson International, Dallas, Texas)

DEFENCE

Battlespace Information and Knowledge -
A Strategic Research Programme for BAE SYSTEMS

Laird Evans

BAE SYSTEMS
Advanced Technology Centre
Human Factors Department

The Human Factors Department of BAE SYSTEMS' Advanced Technology Centre is one of the longest established industrial human factors groups in the UK. This paper describes its origins, and its current capabilities, before focussing on one major project that aims to address some of the key human issues relating to Network Enabled Capability (NEC).

Origins & History

The BAE SYSTEMS corporate research centre sits within the Shared Services business group of the Company and is known as the Advanced Technology Centre (ATC). The ATC Human Factors Department is one of the longest established industrial human factors groups in the UK. The origins go back some forty years or so, to the mid 1960s and the development of guided weapon systems, such as Martel, Swingfire and Rapier. In those days the Department was part of the British Aircraft Corporation's (BAC) Guided Weapons Division.

In 1983 the Department became part of the newly formed Sowerby Research Centre (SRC), which was named after James McGregor Sowerby, a former research director of British Aerospace (Dynamics Group), who had a vision of a single, corporate research centre for the Company. This vision was realised in 1987 when the SRC became the corporate research centre for British Aerospace (BAe). The Department remained within the SRC until 2000, when the ATC was formed following the merger of BAe and Marconi Electronic Systems.

Current Capabilities

At the time of writing this paper[1], the Human Factors Department comprises 22 members of staff. Most members of staff have a first degree in Psychology and/or Human Factors and many have second degrees in related disciplines.

The Department's capabilities have changed significantly in the forty years or so since it came into being. This is partly in response to the increasing complexity of aerospace and defence technologies that we have witnessed over that time period, and partly in response to the changing requirements of BAE SYSTEMS' Business Units (BUs) and Joint Ventures (JVs).

Currently, the Department can claim to have enviable capabilities, in some cases world class, in the following fields:
-   Human-Centred Design
-   Vision Modelling
-   Display Design & Optimisation
-   Mental Workload Assessment
-   Situational Awareness Assessment
-   Human Factors of Training
-   Environmental Factors Affecting Human Performance
-   Virtual Reality
-   Cognitive Modelling
-   Virtual Reality
-   Novel Control Technologies
-   Distributed Team Working

As suggested above, the Department's principal raison d'être is to undertake human factors research of the highest quality in support of the BUs and JVs of BAE SYSTEMS. In addition, however, the Department plays an important role in facilitating co-ordination and synergy throughout BAE SYSTEMS and in identifying and promoting human factors best practice. This is achieved through participation in a number of cross-Company groups:
-   Human Factors Integration Special Interest Group (HFI-SIG)[2]
-   Training, Simulation & Virtual Reality Forum (TSVR)[3]
-   Human Factors Advanced Technology Group (HF ATG)
-   Systems Engineering Council R&T Sub-Group

---

[1] December 2003
[2] Founded and organised by the ATC Human Factors Department
[3] Founded and organised by the ATC Human Factors Department

Battlespace Information & Knowledge

It was stated previously that the Department's capabilities have changed over the years to meet the changing requirements of BAE SYSTEMS. This is well illustrated by the ATC's response to the emphasis that is now being placed on achieving Network Enabled Capability (NEC), not only from within BAE SYSTEMS but also from the UK Ministry of Defence (MoD).

In 2002 the ATC created a strategic research project known as *Battlespace Information & Knowledge (BIK)*. BIK is currently planned as a 5-year programme of research. It is led by the ATC's Human Factors Department but involves three of the other departments within the ATC: Advanced Information Processing, Communications & Signal Processing and Sensor Systems. The principal aim of BIK is to provide a focus for some of the key human issues associated with NEC. To achieve this, a human-centric approach is being adopted within all work elements of BIK. By this it is meant that all work packages (WPs) are driven by an understanding of the military end-users' requirements for information and knowledge, and technologies are being developed to support those requirements.

There are four key work elements within BIK, namely:

- Human-Centric Information System Modelling
- Information Fusion
- Image Exploitation
- HCI Technologies for Enhanced Situational Awareness

Human-Centric Information System Model (HCISM)
The aim is to develop a modelling approach that will allow the information flows within a particular aspect of the battlespace to be modelled, with particular reference to the attainment of situational awareness by one or more of the key players within that part of the battlespace. Information flows within and around the Intelligence Cell of Brigade HQ have been taken as a point of focus during the first two years of BIK.

To date[4], a modelling notation and methodology have been developed based on UML[5]. Models are developed in two stages. In the first stage, a conceptual model is developed. Essentially, this allows information flows within military organisations to be represented in a number of diagrams, namely:
- Connectivity Diagrams, which show the principal entities in a system, or a part of it, and indicate the flow of communications throughout the system;
- Interaction Diagrams, which show specific interactions between entities; and

---

[4] December 2003
[5] Unified Modelling Language

- Activity Diagrams, which show the dynamic behaviour of system entities, modelled as finite state machines.

In the second stage, these diagrams are used to develop a software representation of the information flows. It is intended that the software implementation will allow 'what ifs' to be analysed. For example, what if a particular function that is currently performed by a human be replaced by some new technology? How will that affect the information flows within the part of the battlespace under scrutiny and how will that affect the attainment of SA by the key human player(s) in the battlespace?

The HCISM approach appears promising, although validation of the approach is required. In particular, any instantiation of an HCISM will include a representation of at least one human operator. To date, little effort has been placed on this aspect of the model, which is, arguably, the most challenging. It is acknowledged that the current representation is simplistic and that this aspect of the model must be improved. This is one of the major challenges facing the BIK team in 2004.

Information Fusion

Technological advances in military C4I systems now make it possible for vast quantities of data and information to be presented to human decision-makers. As a result, information fusion has been identified by the BIK team as a crucial but immature component technology that will be essential to the efficient functioning of future military information systems. It has also been noted that there appears to be a dearth of research into information fusion. In this context, it should be noted that *information fusion* should not be confused with *data fusion,* for which there is an abundance of research. Within BIK, information fusion refers to the fusion of non-numerical data, such as symbols and military text messages.

The BIK team has selected the fusion of separate land, sea and air pictures to form a single, common operating picture, as its application of interest. Fusion algorithms based on logical reasoning have been developed and at the time of writing a technology demonstrator, based of VR-Forces and Matlab is planned for delivery at the end of December 2003.

A key point to note is that, in keeping with BIK's human-centric approach, the focus of this WP is information fusion to support the human decision-maker, not to replace him.

Image Exploitation

The interest within this WP is the interactions between human image analysts and the outputs from various forms of machine assistance. Currently, automatic target recognition (ATR) systems are being examined and Synthetic Aperture Radar (SAR) imagery has been selected as the imagery of interest, as it is known that this form of imagery is extremely difficult to interpret, at least at the level of vehicle identification.

In 2002/03 an Image Exploitation Demonstrator and experimental test-bed was developed. At the time of writing, this facility is about to be used to conduct human factors experiments to investigate the interactions between human observers and the outputs from ATR systems. The aim of this element of research is to specify the necessary performance characteristics of ATR systems but to do so from a human perspective.

HCI Technologies for Enhanced Situational Awareness
The focus of this WP is battlespace visualisation. It is believed that for certain levels of military command, volumetric displays are the displays of the future. These displays present 3D images, similar to holograms, which can be viewed without the aid of special glasses. In addition, volumetric displays can potentially aid collaborative working, as more than a single person can view the display at any one time and get different perspectives on the image being viewed, depending on where they are standing. It has been hypothesised that volumetric displays will enhance shared situational awareness, although this has yet to be proven.

Although it is acknowledged that volumetric displays may not become commercially viable for military applications for some years, this strand of BIK is aiming to undertake fundamental human factors experiments now, to address visualisation of, and interaction with, battlespace entities with 3-dimensional volumetric displays. To enable these experiments, an existing 3D stereobench is being developed as a *model* of a future volumetric display. It is anticipated that by undertaking these experiments now, BAE SYSTEMS will develop knowledge that will enable it to place intelligent specifications on volumetric display manufacturers when these displays become commercially viable. To date, an experimental facility has been developed and experiments will begin in 2004.

Conclusion

BIK represents a multi-year commitment by BAE SYSTEMS to fund research into some of the key human issues associated with NEC. All major work packages within BIK are adopting a human-centric approach, by which it is meant that all are driven by an understanding of the military end-users' requirements for information and knowledge, and are aiming to develop technologies to support those requirements.

STUDYING THE USE OF COLOUR FOR MILITARY COCKPIT DISPLAYS

Harrie G.M. Bohnen

National Aerospace Laboratory NLR
P.O. Box 90502
1006 BM Amsterdam, The Netherlands

Three successive experiments with increasing task fidelity were conducted to study the effect of colour coding on military cockpit displays. The experiments were performed using a PC-based tabletop arrangement, a part-task flight simulator and a full-task flight simulator.

The single-task experiment showed that subjects could better detect and identify display objects in multi-chrome display versions than in a monochrome display version. Significant performance differences were also found among the multi-chrome display versions. The experiments with higher task fidelity showed smaller (objective) performance effects.

More degrees of freedom in behaviour with an increase in task fidelity could be an explanation for the latter phenomenon. Therewith behaviour is more difficult to assess. Furthermore, subtle differences in detection performance could get bogged down as many tasks were performed simultaneously.

Introduction

Due to the large amount of information needed by a fighter pilot to perform his task, military cockpit displays are receptive to clutter. Colour capability is an option to accommodate high symbol densities. Grouping information by colour, such as red for threats and green for friendly aircraft helps the pilot to sort the information (Murch and Taylor, 1986). Colour also increases the rate of information transfer to the pilot. As a consequence search times for targets decrease, resulting in more accurate target detection.

Multi-chrome displays are developed for tracking radar display pages and navigational display pages (Fig. 1) present on two separate multifunction displays. Head-down in the cockpit these pages provide a large quantity of tactical and navigational information. Colour coding is applied as a display feature redundant to symbol shape, especially for display objects that have a variable position on the displays. The tracking radar display provides radar, friend or foe identification and data link information. For air-to-air missions the display presents information for single or multiple targets. The navigational display presents tactical and navigational data using a lateral view. The tactical overview consists of data link communication and friend or foe information. The navigational overview provides routes and waypoint information with a 360-degree coverage area around the aircraft.

The starting point of the design of the multi-chrome display versions is the ability of a pilot to quickly detect and search for tactical information such as the presence and position of friendly, unknown and hostile aircraft symbols and data link information. The design of two multi-chrome displays is primarily based on subject matter expertise. These display versions are tailored to the design of the aircraft. For the design of three other multi-chrome displays, theoretical considerations for the use of colour are applied as well as a model of an operator in his working environment. In addition, standards and regulations regarding the use of colour on cockpit displays are taken into consideration. Each version emphasises some of the design rules that often conflict.

Three successive experiments with an increasing level of task fidelity are performed. The rationale is an evolving understanding of the effects of colour design choices. First, the colour effects are studied in a simple task environment followed by tests in increasingly realistic task environments. Changing the task of the subject from the simple, PC-based search and identification task to a realistic fighter pilot task increases fidelity. The major aim is to assess whether colour shortens target search times and whether that effect is greater at higher display densities. The second experiment is performed in a part-task flight simulator. Several missions are flown by fighter pilots. Pilot task related performance parameters such as head-out time and time-to-a-kill and mental workload are used to study the effect of colour coding. The final experiment is conducted in a full–task flight simulator to validate the findings of the earlier experiments. The same means as described for the second experiment are used to assess pilot performance.

Methods

Subjects
Six subjects participated in each experiment. University students performed the tabletop experiment. The part-task and full-task flight experiments were performed by (former) fighter pilots. All subjects had normal vision. None of the subjects participated in more than one experiment.

Apparatus and display versions
The tabletop arrangement consisted of a PC plus screen. The part-task flight experiment made use of a fighter pilot station. The multifunction displays were equipped with CRT's whereas an elementary outside view was generated on which a virtual head-up display was projected. The full-task flight experiment made use of a real cockpit equipped with LCD's. A dome was used to project the out-of-the-window view.

The monochrome version of the displays consisted of three intensities of green. The use of green (for border text and some symbols) and yellow and white for other symbols characterised the first expert-based display version (I). The use of white (text and some symbols) and blue and yellow (symbols) the second one (II). In one rule-based display version priority was given to stereotype ideas people had about the meaning of colours. Furthermore, text and several symbols were displayed in white (III). Another rule-based version concentrated on consistency in the use of colour for text (green) with the use of colour on other cockpit displays (IV). In addition, yellow, blue and white were used for The last rule-based version (V) took advantage of the design principles of display versions III and IV and text was displayed in green. All five multi-chrome display versions contained intensity modulated colours.

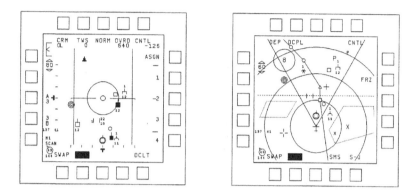

Figure 1. Tracking radar display (left) and navigational display (right)

Experimental procedures
Subjects were extensively trained in the use of the apparatus (PC-configuration and/or flight simulators) and studied one by one the colour display versions. Pilots also studied the meaning of all display objects. Flights to get used to each colour display version preceded the experimental flights. All three experiments used a within-subjects design with colour coding (monochrome and multi-chrome display versions) and display density (number of targets) as factor. The number of colour display versions ranged from five in the single-task experiment and four in the part-task flight experiment (version V excluded) to two in the full-task flight experiment (version II, IV and V excluded).

Results

Single-task experiment
Search times for targets were higher for the monochrome display (1757 ms) when compared to all multi-chrome display versions (1304 ms). The maximum difference of 517 ms was found for version III. Among the different multi-chrome displays significant differences were found too. One of the subject matter expert-based display versions (II) and one of the rule-based display versions (V) showed the worst performance. No effect of display density was found. Evidence was present for the fact that search times increased for targets with a colour identical to fixed-position display objects. This was most clear for targets coded green in cases where text was coded green too (I/IV and V). Counting targets in an exhaustive search measurement took about 4000 ms for the monochrome and about 3000 ms for the multi-chrome display versions.

Part-task flight experiment
The percentage of killed aircraft was statistically not different for the various display versions. However, a difference of 5% in favour of the multi-chrome displays was present. The time interval from target pop-up to target hit was slightly longer for the

multi-chrome display version designed for consistency (IV) than for the other display versions (138 s versus 130s). Colour coding did neither influence the amount of viewing time spent by the pilot at one of the displays nor the frequency with which the pilot looked at one of the cockpit displays or outside the cockpit nor mental workload. The monochrome display was rated less useful than the multi-chrome display versions. This was independent of the tasks of the pilot. The display that applied colour stereotypes and used the colour white for text (III) was rated most useful. Likely because grouping was applied properly. The coding of hostile objects was rated less useful in the expert-based display versions (I/II) than in the rule-based display versions (III/IV). The multi-chromatic display that applied stereotypes (III) was preferred to the monochrome display version.

Full-task flight experiment

Colour coding did not affect any of the assessed performance parameters, neither the pilot's viewing behaviour nor mental workload. Both multi-chrome display versions (I/III) were rated as significantly more useful than the monochrome display and both were rated equally useful. No differences were found between both multi-chrome display versions. Especially the coding of target objects (5 in total), mainly done by allotting yellow was rated in favour of the multi-chrome display versions. Coding of cursors by white (III) is more useful than coding by green such as applied in the monochrome display version. All pilots preferred the multi-chrome display versions to the monochrome display version. Half of them preferred version I the other half version III.

Discussion

The results of the single-task experiment clearly show that search times for 'target' objects are shorter when colour coding is applied (about 25%). This effect was irrespective of display density, which confirms the parallel processing theory of colour. Also in line with earlier findings are the search times for 'target' objects that are longer in the presence of display objects with the same colour.

At the $p<0.05$ statistical significance level, colour coding did not influence the behaviour of the pilots in the part-task experiment. However, colour coding has influence on behaviour when considering $p<0.30$ (Wickens, 1998). With colour coding, pilots spend slightly more time looking outside and less on those displays. This indicates that it is easier to retrieve information. Colour coding according to stereotypes (e.g. friendly aircraft in green) together with text in a neutral colour (white) makes that targets can be distinguished more easily. Colour coded displays are sampled less frequently at higher symbol densities, which suggests that search times are shortened (also see Backs and Walrath, 1992). Disregarding, even when considering $p<0.30$ colour coding has no effect on pilot workload. The less strict p-level also does not change the earlier conclusion regarding no effect at all of colour coding on performance in the full-task experiment.

The approach to perform three successive experiments with an increase in task fidelity did not result in convergence of objective evidence. The opposite is true as performance effects decrease with an increase in task fidelity. The pilots that performed the part-task experiment also conducted the single-task experiment for training, and their results are in line with those of the students. This brings us to the question why only some small performance effects were found in the part-task experiment. A theoretical explanation is that colour is not always the best-recalled coding format under high task load. The utility of colour is task specific, being suitable for segmentation, but not identification. Segmentation may be viewed as a primarily perceptual use of colour. Identification is dependent on central cognitive processing. According to Dudfield and Hughes (1993) colour serves a purpose in tasks concerned with searching for or detecting information at unknown spatial locations in complex and cluttered displays. On the other hand, colour coding is often of less value in monitoring tasks or identification tasks, in which the spatial location of the target is cued to the operator, as is partly the case in the part-task experiment. An important difference between the single-task and the multiple-task experiments is of course that the first requires only the performance of one single task whereas the latter require simultaneous performance on multiple tasks. This explanation is in agreement with suggestions that the use of colour is well suited for a predominantly perceptual task -as the single-task experiment - but may not confer an advantage for concurrent tasks. Furthermore, the beneficial effects of colour coding on central processing may be limited in high-density displays under time-constrained conditions (Backs and Walrath, 1992) as present in the multiple-task experiments.

In both multiple-task experiments pilots favoured the use of colour, most likely due to esthetical appeal. Its likely that colour also positively influences the pilots' perception of the overall system quality.

Reising and Aretz (1984) state that displays that are multifaceted in nature will benefit from the use of colour and that one key to potential benefits of colour lies in time-constrained conditions. They conducted a high-fidelity flight experiment to study colour on cockpit displays and found no differences in performance. One of their explanations is that colour had only a subtle effect on the display they studied. This explanation does not hold for the present study as the tracking radar and navigational displays are far more complex than their display was. Reising and Aretz state that these types of studies should apply more than one metric to tease out subtle performance differences. With the increase in fidelity from single to multiple-task environment the means to assess performance in this study increased in number and complexity. In the part-task experiment few performance benefits were found. In the full-task experiment no effects were present even though performance was assessed in all its dimensions. Therefore, the main relevance of gathering performance data in the multiple-task experiments is maybe that colour did not prove to be counter-productive for the tasks of the pilots. Together with the positive effects found in the single-task experiment and the appreciation of the pilots, this may be sufficient reason to introduce colour in the fighter cockpit.

References

Backs R.W. and Walrath L.C. 1992, Eye movement and pupillary response indices of mental workload during visual search of symbolic displays, *Applied Ergonomics*, **23**, 243-254

Dudfield H.J. and Hughes P.K. 1993, *The use of redundant colour coding as an alerting mechanism in a fast jet head-up display*. Technical report: DRA/AS/FS/TR 93018/1.

Murch G.M. and Taylor J.M. 1986, *Effective use of colour on avionics displays*. Human Factor Research Textronix, Inc.

Reising J.M. and Aretz A.J. 1984. Colour coding in fighter cockpits: it isn't black and white. *Proceedings on the second Symposium on Aviation Psychology*, 1-7

Wickens C.D. 1998, Common sense Statistics, *Ergonomics in Design*, **6** (Santa Monica), 18-22

HUMAN RELIABILITY

BREAST SCREENING: EXPERIENCE VERSUS EXPERTISE

Alastair Gale, Hazel Scott and David Wooding

Institute of Behavioural Sciences, University of Derby, Kingsway House, Kingsway, Derby DE22 3HL

In the UK fewer radiologists are now specialising in breast cancer screening. Consequently, a number of radiographers have been specially trained to interpret mammograms so as to double-read with existing radiologists. Each year some 97% of these film-readers read a set of difficult cases ('PERFORMS') as a means of self-assessing their skills. We investigated whether these trained radiographers performed as well as breast-screening radiologists on this difficult set of cases. Overall, radiographers who have not read the same volume of cases as radiologists did not perform as well on this particular task. These results support the current UK practice of radiographers double reading with radiologists.

Introduction

Some 15 years ago the UK introduced the national breast cancer screening programme, since when all women between the ages of 50 and 64 are invited every three years to attend for screening. Currently some 1.4 million women a year (Patnick, 2003) are screened at over 100 screening centres. The number of women being screened is expected to increase markedly over the next few years for various reasons, including possible extensions of the screening age range. Screening involves taking two mammographic X-ray views of each breast and then these mammograms are visually examined by specialists for any radiographic signs which can indicate early malignant disease presence. In the past these specialists have typically been experienced consultant screening radiologists.

Currently some 560 individuals are involved nationally in interpreting these mammographic images. This large number of personnel then enables a woman's mammograms to be read by more than one person (usually two people – this is known as double reading) where they independently examine the same case and then their decisions are compared. This procedure has been shown to increase the potential of detecting any early signs of cancer.

Unfortunately, in recent years fewer radiologists are entering this particular radiological domain for various reasons, which then produces a possible skill shortfall. Such a potential loss in experienced personnel could impact upon the ability of the screening programme to meet demand and so training schemes have been established to train up some skilled radiographers to interpret mammograms. Once these individuals have successfully completed their training then they are able to double read cases with radiologists. This then raises the interesting question of how good are these radiographers as they will not have had the same experience of reading mammographic cases as the radiologists have had. Typically radiologists will have read at least 5,000 screening cases a year.

Comparisons between radiologists and radiographers based on real life measures could prove to be inconsistent for many different reasons, not least the high variability in the appearances of the cases which individuals examine. However, data from a national self-assessment scheme where individuals all examine the same cases should lend itself to such examination.

Method

Every year in the UK all individuals who interpret breast screening mammograms are offered the opportunity to participate voluntarily in a confidential self-assessment scheme (PERFORMS) in which they examine a series of difficult recent breast screening cases sampled from around the UK (Gale & Walker, 1991; Cowley and Gale 1996). The cases comprise malignant (where the pathology of each case is known), benign (based on pathology) and normal (based on a three year follow up of the woman where no cancer has subsequently been found) appearances. Annually, two sets of 60 cases are interpreted by participants in the scheme, where a case comprises the four mammograms of a woman representing two X-ray views of each breast. Immediate and confidential feedback on their own mammographic interpretation skills, as compared against the opinions of a group of expert radiologists and also as against known pathology, is given.

Subsequently, when all participants have completed all 120 cases then they receive very detailed information on how well they performed. In this feedback the 'national radiological opinions' about each case are first determined from the overall data and then each individual's decisions concerning the radiological appearance of every case are calculated against this national opinion. Data are subsequently fed back to individuals on how well they have performed as compared to the anonymous data of their peers. Whilst the scheme is voluntary in nature some 97% of applicable screening personnel currently take part.

In screening, a decision is made about whether to recall (where a woman's mammograms suggest the need for further investigations) or not recall (where the woman's mammograms are normal in appearance and suggest no indication of potential disease presence) each woman. Although the PERFORMS scheme utilises recent screening cases the data from the scheme is not necessarily directly equivalent to that produced when an individual is actually screening. This disparity is for various reasons, which include; how an individual psychologically approaches participation in the scheme, the information which they have to report for each case examined (which is much more than that required in actual screening) and the high number of cancers in the self-assessment scheme. This difference is readily evident in the fact that it generally takes someone at least twice as long to read the same number of cases in PERFORMS as in real life screening.

There are various potential performance measures which are employed in the PERFORMS scheme and these data include information such as: the number of correct recall decisions; correct return to screen decisions; cancers correctly detected, radiographic features correctly identified and located appropriately, and Receiver Operating Characteristic measures such as d' prime (a measure of the detectability skill in identifying cancer).

When individuals take part in the scheme they do so in similar environmental conditions (e.g. a quiet dark room with the mammograms presented on a radiological multiviewer). It is therefore possible to compare meaningfully the data of different groups of individuals. Consequently, recent anonymous data from the PERFORMS scheme were examined for some 250 consultant radiologists and 102 specially trained radiographers who had taken part.

In medical imaging studies an individual's responses are generally considered as a 2 x 2 matrix of true (TP) and false positive (FP) responses and true (TN) and false (FN) negative decisions (table 1) as judged against a 'gold standard' opinion. In the present study the individual's decisions of 'recall' or 'not recall' are judged against the overall mean decision for all participants about whether or not a case should be recalled.

Table 1. Matrix of responses

		Mean opinion	
		Recall	Not recall
Individual	Recall	TP	FP
radiologist	Not recall	FN	TN

Commonly used indices of imaging performance include measures of sensitivity and specificity. Sensitivity is a percentage figure indicating the individual's ability to identify correctly those cases which should be recalled (TP/(TP+FN)) and specificity is the figure indicating the ability to identify correctly those cases which are normal (TN/(TN+FP)). Here we use the measure 'correct recall' percentage as an equivalent to sensitivity and 'not recall' or 'correct return to screen' percentage as a specificity measure.

The data used here concerned how individuals had classified each case. Three measures were examined: correct return to screen; correct recall (both measures as compared against the overall participants' opinion about whether or not to recall each case), and the number of cancers correctly detected (judged against the known pathology of each case).

Results

One-way ANOVA was carried out on the recent data from two self-assessment sets of 60 cases. In terms of the number of cancers correctly identified it was found that the radiographers performed very well (mean value: 93% of cancers correctly detected). However, they performed consistently slightly lower than the radiologists (mean value: 96% of cancers correctly detected), this difference is significant [$F_{(1, 307)} = 25.416$, $p < .001$].

The radiographers' correct recall percentage (mean = 89.65%) was significantly below that of the radiologists (mean = 92.37%), [$F_{(1, 307)} = 17.086$, $p < .001$]. There were no differences (p = n.s) for the overall correct return to screen percentages and both radiologists (mean % correct = 90.91) and radiographers (mean % correct = 90.76) performed at a similar level for this measure (Figure 1). This suggests that the radiographers' level of specificity was comparable to that of the radiologists, although their levels of sensitivity were less accurate. A lower sensitivity score would also result in fewer cancers being correctly identified.

The performance (mean score) of the two groups was then further analysed by the mammographic case type of normal, benign or malignant (Figure 2). A two-way ANOVA with one within group factor (classification of case) and one between group factor (occupational group) showed that there were significant overall group differences [$F_{(1, 42234)} = 20.26$, $p < .001$] and also a significant effect of classification type [$F_{(2, 42234)} =$

368.36, p < .001]. A priori t-test analysis revealed that for all case classifications radiologists were better (in terms of mean score) than the radiographers (p < .05).

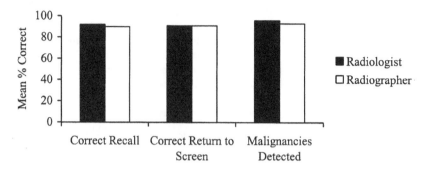

Figure 1. Overall performance of the radiologists and radiographers

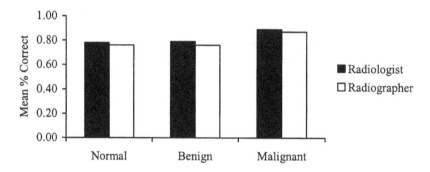

Figure 2. Performance by case classification

Discussion and Conclusion

The purpose of this investigation was to examine some of the data from the PERFORMS scheme to see how well the trained radiographers had fared. It is evident that they performed less well than the radiologists on this particular task, although these differences were very small they were still significant. Whilst the radiographers' specificity was similar to that of the radiologists, their sensitivity was less. There are various possible reasons for this, such as the difficulty of the individual cases and the type of mammographic features present in these particular cases (Scott *et al.*, 2003). In general the radiographers had had much less experience of interpreting cases than had the radiologists, although they did have previous experience of mammography.

However, it must be considered that these data emanate from a self-assessment scheme and not from real life screening data. Thus some caution must be used in interpreting the results. It may be that the radiographers approached participation in the scheme differently (e.g. with more care) than do the radiologists, which may possible inflate the data in favour of the former group. In the PERFORMS scheme individuals are presented with more cancers than they would see in a comparable number of cases in everyday screening (the cancer incidence rate is circa 6.6%). Therefore in the scheme, although the individual cases are screening cases, individuals have to adopt a different overall reporting bias in their decisions about whether to recall a case or not.

Further research is required to compare *directly* the radiographers' and radiologists' real-life performance based on this, as well as the number of years and volume of cases read. Although there were significant differences between the groups for individual test characteristics, the radiographers still scored very highly on all measures and their cancer detection level was only a little over 3% behind the radiologists.

Clearly these radiographers were well able to identify those cases which should be recalled or not, and they were quite expert at this task - really without the experience of the radiologists. This finding supports the current practice of them double reading in the national screening programme alongside radiologists. Their performance data also suggest that they may well even be able to function as single readers without recourse to a radiologist.

References

Cowley H. and Gale A.G., 1996, Minimising human error in the detection of breast cancer. In: S. A. Robertson (Ed). *Contemporary Ergonomics 1996,*(Taylor and Francis, London).

Gale A.G. & Walker G.E. 1991. Design for performance: quality assessment in a national breast screening programme. In E. Lovesay (Ed*.) Ergonomics - design for performance 1991* (Taylor & Francis, London).

Patnick, J. (Ed.) 2003. *Annual review 2003: Saving women for 15 years*, NHS Cancer Screening Programmes, Citigate Communications, London

Scott H.J., Gale A.G., & Wooding D.S., 2003. Comparing breast screening performance of Radiologists and trained Technologists using test sets of mammograms; the relationship between performance, volume of cases read and case difficulty. Presentation at the Radiological Society of North America Conference.

Acknowledgement

This work is supported by the NHS Cancer Screening Programmes

TXT-ING : LINGUISTIC DEGRADATION LEADS TO INCREASED TRANSCRIPTION ERRORS.

Neil Morris, Jenna Hadley, Katie Mackintosh, & Debbie Middleton.

University of Wolverhampton, School of Applied Sciences, Psychology, Millennium Building, Wolverhampton WV1 1SB, UK.

The widespread use of text messages (SMS —short message service) transmitted using mobile phones has resulted in the adoption of a truncated English not dissimilar to that employed in telegrams. Text messages are a maximum of 160 characters and thus create limitations in the content of messages that are not evident in other communication media such as e-mail. This txt dialect is linguistically impoverished and we report data showing that transcription into conventional English from this format has an error rate comparable to transcribing random words. Errors of both intrusion and omission were much greater than for grammatically correct messages. The major implication of this is that although text format is a quick and inexpensive mode of communication it should not be used to transmit information when accuracy is of great import. The development of a standardised glossary and syntax may help to ameliorate this problem.

Introduction

One of the most useful features of the modern mobile telephone is the SMS (short message service) or text facility. This facility allows typed messages of up to 160 characters to be transmitted from one mobile telephone to another (or in some systems to an e-mail account). Katz and Aakhus (2002) have argued that the mobile phone will have an enormous impact on all aspects of life because, in effect, it allows for perpetual contact. One need never be *incommunicado*. However it is not always convenient or desirable to be always available to converse with others. SMS allows one to be perpetually contactable but allows one to orchestrate this so that such contact does not have to occur in real time and the response delay introduced by text messaging allows one to consider ones reply. As a result this creates a quite different style of communication to that which occurs when speaking on the phone (Ling and Yttri, 2002). Text messaging is likely therefore to increase in frequency and it may well be used to transmit important information that needs the immediacy of mobile phone contact but must provide a clear text record of the message. Thus it is important to examine ergonomic considerations relevant to SMS usage.

The 160 character limitation on text messages creates problems similar to those associated with telegrams. If a large amount of information is to be transmitted it must either be sent as more than one message or the message must be terse and spaces (which count in the 160 characters available) must be minimised. Words like text are abbreviated to txt to save a character space and for you becomes 4U. These are fairly well established conventions but there are no such widely accepted conventions for most of the things that one might want to say. Indeed Skog (2002) has argued that groups of teenage users have developed clique-based SMS languages and these can be used as camouflage to exclude outsiders from understanding group communication (Ling and Yuri, 2002). Thus private languages may develop to maintain secrecy but for widespread communication it is important that the linguistic contractions employed are either widely learned or are self evident.

Knowledge of the rules of a language reduces the amount of processing capacity that must be dedicated to extracting meaning from it (Crystal, 1987). Shannon (1950) demonstrated, quantitatively, that English is massively redundant, that is, there are constraints, within an English sentence, on the ordering of words. In Chomsky s (1968) terminology this is a learned surface structure but he also posits a deep structure —a hard wired universal grammar that, if this is a valid concept, would constrain considerably the structure of SMS grammar. Grammar and syntax allow a reader to rapidly parse highly predictable parts of a sentence (Fry, 1970). Such redundancy reduces the information load but at the same time disambiguates sentences (Crystal, 2001). Morris and Jones (1995) adapted Shannon s (1948) system of grading approximations to English to create texts that violated English grammar to varying degrees. They removed every seventh word from a body of text and then randomly re-inserted it elsewhere in the same text. In other versions every sixth, third or every other word were shuffled in this manner. Participants were required to transcribe these modified texts and an unmodified version. Error analysis showed poorer performance with degraded texts. Another study, Morris and Jones (1991), demonstrated that the transcription task burdened verbal working memory and that transcribing the degraded text used up more capacity.

In the present study we use cursive transcription as a measure of distortion arising from using different formats of text. The transmission of information from one medium to another with a human being as an intermediary usually results in information loss and distortion (see, for example, Allport and Postman, 1947 on rumours) but the format of presentation is crucial to the severity of this. In this study we consider the possibility that text messaging may be so linguistically degraded as to resemble moderately scrambled English text. To test this we compare grammatical English text transcription with transcription of the same text degraded in the same manner as in Morris and Jones (1991:1995) and with the same material translated into text message format.

Method

Participants

60 Wolverhampton University undergraduates, 32 female and 28 male, aged 18 — 50 (mean age 28) volunteered to participate in this study. Levelof SMS use was very varied but 58 students reported using SMS at least once per week and the two remaining students did not possess mobile phones.

Materials & Procedure

Participants were each allocated randomly to one of three conditions. 20 were allocated to transcribe grammatical English texts (grammatical group), 20 were allocated to transcribe the same texts except that some words had been removed and then randomly re-inserted (Random group) and the third group of 20 participants transcribed the same text but after it had been translated into Text format (the Text group). The texts referred to everyday matters such as arranging to meet friends.

The grammatical text consisted of twenty sentences (each with between 157 and 160 characters, including spaces). To create the Random condition text the above texts had every fifth word removed and then randomly re-inserted into the sentence from which it was drawn. For the Text version the original Grammatical version was translated into text so that, for example, tomorrow became 2moro . As a result of contracting the texts in this way the Text versions were between 140 and 150 characters. The sentences were presented on a classroom projector screen and generated using PowerPoint2002. Following a pilot study to establish the time required to transcribe the material each sentence was presented individually for one minute. Participants were given a note pad on which they were to write, verbatim in the Text and Random conditions, the sentences presented on the screen. Randomised texts had to be transcribed with the position of the transposed words restored to the position necessary to restore a grammatical structure. In the Text condition they were required to write the Text message in grammatical English. Each sentence had to be written on a separate page.

Group testing was used but the three conditions were run separately. Phase 1 of the study involved collecting baseline data. All three groups were presented with five sentences in grammatical English that they were required to transcribe. This was to establish that the three groups did not differ in transcription ability *per se*. Following this they were given condition specific instructions. They were then asked to transcribe as accurately as possible the messages that were flashed on the screen. Subsequently, a scorer who did not know which condition the transcripts were derived from scored the transcripts in terms of intrusions (things written down that were not presented at that point in the message) and omissions (words missed out). These were converted into percent correct scores to allow comparisons across conditions.

Results

The data were subjected to four one way, independent samples analysis of variance (anova). The first two anova s examined the baseline intrusions and omissions data and revealed that the three groups did not differ in their production of intrusions or omissions (both comparisons - $F(2,57)<1$, $p>0.05$). However once the three groups moved on to phase 2 of the study (where the format of the texts was manipulated) there were marked differences in intrusion error rates ($F(2,57)=28.15$, $p<0.001$). *Post hoc* comparisons revealed that the Grammatical group had fewer intrusions than either the Random group ($p<0.01$) or the Text group ($p<0.01$) but error rates between the Random group and the Text group did not differ ($p>0.05$). In the omissions analysis a similar finding was forthcoming. The anova was significant ($F(2,57)=15.52$, $p<0.001$) and *post hocs* revealed that there were fewer omissions by the Grammatical group than the Random group ($p<0.01$) or the Text group ($p<0.01$) and performance levels did not differ between the Random group and the Text group ($p>0.05$). The means and standard deviations of all these data are shown below in Table 1.

To summarise, the three groups did not differ in initial transcription skill. However the patterns of results for intrusion and omission errors are very similar in the main body of the study. The error rates associated with transcribing text message format and randomised English are not distinguishable, statistically, and both display inferior performance to transcription of grammatical English.

Table 1: Means and standard deviations (in parentheses) for percent transcription errors in a baseline condition (all participants transcribed grammatical English) and when grammatical English, random English and Text format are transcribed.

BASELINE

Condition	Grammatical	Random	Text
Intrusions	3.42	3.10	3.05
	(1.85)	(1.88)	(1.80)
Omissions	8.27	8.96	7.39
	(8.86)	(11.42)	(5.57)

MAIN STUDY

Condition	Grammatical	Random	Text
Intrusions	3.72	10.42	11.37
	(1.74)	(5.33)	(2.37)
Omissions	8.29	27.00	22.12
	(8.15)	(13.43)	(9.59)

Discussion

It cannot be argued, from these data, that the same cognitive processes necessarily underpin the transcription of both randomised English and Text format. However in a meaningful sense they can be said to be approximately equal in difficulty. Error rates in both conditions are well below ceiling allowing for the possibility that one might be demonstrated to be a more onerous task than the other. In both conditions there was a translation stage. In the randomised English condition participants had to transcribe the message with the transposed words replaced in their correct location and in the Text format translation into grammatical English was required at transcription. Since all three groups were matched for transcription skill at baseline this suggests that the increased error rate in transcription is due to a combination of linguistic degradation intrinsic to the messages and the additional burden of translation. These components are not separable in this study.

The process of transcription is not well understood (see Morris and Jones, 1995 for a discussion of this) but it places a burden on working memory even when text remains on screen throughout transcription. Although we tend to think that copying from a screen, verbatim, is a fairly superficial task it is clear that semantic and structural processing is involved. Degraded texts, even when copied in the format presented, are inaccurately transcribed (Morris and Jones, 1995). Thus transcription is a useful measure of semantic processing. It is very likely that verbal working memory is configured to take advantage of the rules of grammar whether they are learned as a specific language is acquired or hard wired as a Chomskyan deep structure (Chomsky, 1968). If this is the case then message formats violating familiar grammatical structures will, in a cognitive sense, be linguistically impoverished. There will be a mis-match between executive schema for processing, say, English and the format in which Text format is initially encoded.

However, as indicated earlier, the advantages of SMS are numerous. It is unlikely that the linguistic impoverishment of this communication medium is insurmountable. With respect to the bulk of SMS interactions, ambiguity or error probably has only trivial consequences. Skog (2002) has shown that user groups can develop conventions that ameliorate this problem, within the realm of their discourse. If SMS is to be used between strangers to transmit non-trivial information then the way forward seems to be to develop conventions that are, as far as possible, universal to a particular SMS national language.

References

Allport, G.W., and Postman, L. (1947) *The Psychology of Rumor.* New York: Holt.

Chomsky, N. (1968) *Language and Mind.* New York: Harcourt, Brace and World.

Crystal, D. (1987) *Cambridge Encyclopedia of Language.* Cambridge: Cambridge University Press.

Crystal, D. (2001) *Language and the internet.* Cambridge: Cambridge University Press.

Fry, D.B. (1970) Speech reception and perception. In J. Lyons (Ed.) *New Horizons in Linguistics.* Harmondsworth: Penguin.

Katz, J.E. and Aakhus, M. (2002) *Perpetual Contact: Mobile Communication, Private Talk, Public Performance.* Cambridge: Cambridge University Press.

Ling, R., and Yttri, B. (2002) Hyper-coordination via mobile phones in Norway. In Katz, J.E. and Aakhus, M. (2002 Eds.) *Perpetual Contact: Mobile Communication, Private Talk, Public Performance.* Cambridge: Cambridge University Press.

Morris, N., and Jones, D.M. (1991) Impaired transcription from VDU s in noisy environments. In E.J. Lovesey (Ed.) *Contemporary Ergonomics 1991.* London: Taylor & Francis.

Morris, N., and Jones, D.M. (1995) Cursive transcription errors using restricted displays. In S.A. Robertson (Ed.) *Contemporary Ergonomics 1995.* London: Taylor & Francis.

Shannon, C.E. (1948) A mathematical theory of communication. *Bell Telephone System, Monograph B-1598,* Technical Publications.

Shannon, C.E. (1950) Prediction and entropy in printed English. *Bell System Technical Journal,* **30,** 50-65.

Skog, B. (2002) Mobiles and the Norwegian teen: identity, gender and class. In Katz, J.E. and Aakhus, M. (2002 Eds.) *Perpetual Contact: Mobile Communication, Private Talk, Public Performance.* Cambridge: Cambridge University Press.

AN APPLICATION OF META-ANALYSIS TO STUDIES OF RULE VIOLATION

Gary Munley[1], Joyce Lindsay[2] & Elaine Ridsdale[2]

[1]Department of Psychology & Speech Pathology
Manchester Metropolitan University
Manchester, M13 0JA

[2]Serco Assurance
Thomson House
Warrington, WA3 6AT,

This paper presents the results of a meta-analysis conducted of studies of rule violation. The study was conducted in order to provide support for Williams' (1996) extension of the Human Error Assessment and Reduction Technique (HEART), proposed as a method for the quantification of violation likelihood. The findings suggest further factors that should be considered when assessing violation likelihood and provide an alternative ordering of the importance of the factors identified by Williams.

Introduction

Whilst techniques developed for the derivation of human error probabilities have reached a state of relative maturity, techniques for the assessment of the likelihood of violation are much less well developed. Violations can be defined as "any deliberate deviations from the rules, procedures, instructions and regulations drawn up for the safe operation and maintenance of a plant or equipment." (HFRG, 1995). Research, e.g. Jacobssen and Svenson (1994) has suggested that violations may be a more common form of failure than the types of human error currently considered by human reliability analysis. Mason (1997) reports that violations of rules and/or procedures are identified as significant contributors to around 70% of accidents in some industries. It is clear, therefore, that considerable effort should be directed toward the development of techniques for identifying and quantifying the likelihood of violation.

To date three approaches have been proposed for assessing violation, these are: Survey of Rule Violation Incentives and Effects, SURVIVE (Holloway, 1990); Improving Compliance with Safety Procedures (HFRG, 1995) and an extension to the HEART (Williams, 1996). The first two of these provide qualitative techniques for the assessment of violation likelihood; the extension to HEART is the only technique currently available for the quantitative assessment of violation potential. The extended HEART uses the structural features of the original HEART and replaces Generic Task Descriptors with Generic Violation Behaviours (GVBs) and Error Producing Conditions with Violation Producing Conditions (VPCs). Each GVB has an associated nominal violation probability and each VPC a weighting value that can be used to increase the violation probability depending upon task conditions. The numeric values associated with GVBs and VPCs were derived from a literature review conducted by Williams, but the means by which the figures were derived is not presented. The research reported here presents the results of a meta-analysis of studies of rule violation, conducted in an attempt to provide some support for the VPC list produced by Williams and also for the importance ascribed to each of the conditions as indicated by the VPC weights.

Meta-analysis was developed as a method of research synthesis in the 1970s in order to extract meaningful conclusions in research areas where individual primary studies appeared to produce contradictory or inconsistent findings (Lipsey & Wilson, 2001). It provides a method for quantitative statistical review of a body of literature in order to identify the importance or size of the relationship between any two variables. The primary metric of meta-analysis is the effect size statistic. Effect size statistics represent the strength and direction of the relationship between any two variables of interest. e.g. frequency of rule violation and gender. This report describes an attempt to derive effect size statistics for each of the VPCs suggested by Williams in the extension to HEART.

Method

Literature Search
A literature search was undertaken to identify a set of studies that could be subject to statistical review. Thirty two keywords derived from existing violation assessment techniques were used as search terms in three electronic databases: Ergonomics Abstracts; PsycINFO and Science Direct to identify papers for possible inclusion. The abstract of each paper was reviewed and compared against inclusion criteria to determine whether the full paper should be obtained and assessed in detail. Those papers selected for detailed review were then read in order to extract the necessary data to derive effect size statistics. The papers used by Williams in producing the extension to HEART were also evaluated and included in the meta-analysis where possible.

Data Analysis
In order to combine effect size statistics from different studies, these need to be converted to a common metric. Wolf (1986) and Rosenthal (2000) suggest the most convenient and commonly used metric is the Pearson Product Moment Correlation r. Thus, each research finding used in this meta-analysis was converted to Pearson's r using the methods prescribed by Wolf (1986) and Glass et al (1981). A further transformation commonly used in meta-analysis, the conversion of Pearson's r to Fisher's Zr, was also adopted. Weighted mean effect sizes were computed using sample size, in the form of an inverse variance weight, to weight each of the effect size statistics provided from individual studies. For reporting purposes the weighted mean Zr is converted back to the original metric of Pearson's r.

The above conventions were used to produce a weighted mean effect size for each of the VPCs identified from the literature search. This weighted mean effect size provides an index of the importance of the variable in relation to violation frequency. Although no formal criteria exist for judging the importance of effect sizes, Cohen (1977) provides a heuristic that has become convention for classifying the size of effect sizes. For correlation effect sizes, r values <0.1 are considered small, r values around 0.25 are considered medium and those > 0.4 are considered large.

As well as establishing the weighted mean effect size for a relationship of interest, it is also necessary to provide an index of the homogeneity of the effect sizes used in deriving the mean. In a homogenous distribution the dispersion of effect sizes around the mean can be accounted for by sampling error. If an effect size distribution is found to be heterogeneous this suggests that there may be significant differences among the studies from which the mean effect size statistic is derived. These differences between effect sizes may be due to fixed factors, such as systematic differences between study characteristics, e.g. differences in sample characteristics or sampling strategy, etc. Alternatively, differences between effect sizes may be due to unidentifiable random factors and methods are available to recalculate the mean effect size, based upon an estimate of the random variance that cannot be attributed to sampling error. These correction procedures were applied in this study to any set of effect sizes that was found to be heterogeneous.

Results

The results are summarised below in rank order based on the strength of the weighted mean effect size for each variable.

Copying
A paper by Wogalter et al (1989) reported two independent studies testing the impact of a rule violating role model on the behaviour of a second person. The weighted mean effect size for the impact of a non-compliant role model on violation was $r = 0.725$, a large effect size using Cohen's (1977) criteria, accounting for 52.6 of the variance in violation. The meta-analysis result suggests that the strongest determinant of violation likelihood is the behaviour of others in the immediate task environment. Copying behaviour was identified as a VPC in Williams' revision of HEART, but was rated as the fourth most important factor.

Convenience/benefit
Convenience or benefit afforded to the individual as a result of committing a violation was also found to have a large relationship with violation frequency. Three studies assessed the relationship between the two variables (Yagil, 1998; Jason & Liotta, 1982 and Wogalter et al,1998) providing five individual effect size statistics. The weighted mean effect size for this relationship was $r = 0.435$ equivalent to 18.9% of the variance in violation. This finding suggests that the benefit or convenience arising from violating is an important determinant of its occurrence, supporting the findings of Williams (1996) and Holloway (1990).

Risk Perception
Risk perception was found to have a medium size relationship with violation; weighted mean effect size $r = -0.29$, equivalent to 8.4% of the variation in violation. Four studies examined the relationship between risk perception and violation; Yagil (1998), Lawton (1998), Adams-Guppy & Adams (1995) and Kanellaidis et al (1995). This finding suggests that if people believe that rules or procedures prevent them from coming to harm they are more likely to be adhered to. Risk perception can be considered to represent an assessment of the potential costs that might arise from committing a violation and is likely to be one of the factors compared against the benefits accrued by violating when deciding upon behaviour. Risk perception is not considered as a specific VPC in the extension to HEART, although it is used both in the HFRG (1995) approach and the SURVIVE method of Holloway (1990).

Time Pressure
Time pressure was also found to have to have a medium effect size relationship with violation; weighted mean effect size $r = 0.28$ accounting for 7.8% of the variability in violation behaviour. The studies used in the derivation of this effect size were those reported by Wogalter et al (1998) and Adams-Guppy & Adams (1995). Time pressure was not included in the extension of HEART to assess violation potential, but is an EPC in the original HEART formulation. It appears logical, however, that if an operator is under time pressure he may take short cuts when completing tasks resulting in the requirements of procedures being violated.

Severity of Sanction Associated with Violation
Three studies, West et al (1975), Ferguson et al (1999) and Friedland et al (1973), were included in the meta–analysis in which the severity of sanction was systematically manipulated or varied naturalistically and its effect on violation noted. From these three studies, four effect size statistics were derived which combined to produce a weighted mean effect size of $r = -0.245$ indicating that rate of violation decreases as severity of sanction increases. Whist the severity of sanction was not included as a VPC in the revision to HEART it is used in assigning the GVBs.

Age
Age has a weighted mean effect size of $r = -0.235$ accounting for 5.5% of the variability in violation. This value was derived from studies by Diaz (2002), Lawton et al (1997), Laujenan et al (1998) and Westermann & Haigney (2000). All studies show that frequency of violation reduces with age. Age is not a factor considered in any of the current approaches to violation assessment. It is considered that the effect of demographic variables such as age are likely to be mediated by other attitudinal variables such as risk perception or perception of benefit.

Experience
Two studies, Lawton et al (1997) and Laujenan et al (1998) examined the relationship between the frequency with which a task was performed and the number of violations committed. The studies demonstrated that as experience of a task increased so did the frequency of violation; weighted mean effect size $r = 0.185$ accounting for 3.4% of the variability in violation. Experience with a task was not considered as a VPC in the HEART extension, neither is it used in the other methodologies currently available. Again it is considered that the effect of experience is likely to be mediated by attitudinal variables as in the case of age.

Presence of an Authority Figure
Two studies, Wogalter et al (1998) and Sigelman & Sigelman (1976) examined the impact of having a person in authority in a situation where it was possible for a violation to occur. The weighted mean effect size $r = 0.17$ equating to 2.9% of variability in violation being explained by this variable. Williams (1996) also identified lack of an authority figure as contributing to violation potential, but placed more weight on this than the present analysis would suggest appropriate.

Gender
Six studies contributed effect size statistics to the assessment of the impact of a person's

gender on violation potential (Rimmo & Aberg, 2000; Diaz, 2002; Lawton et al, 1997; Yagil, 2000; Laujenan et al, 1998; and Westermann & Haigney, 2000). All studies revealed that males were more likely to violate than females. This replicates the finding of Williams (1996), however, the weighted mean effect size associated with gender is rather small, $r = 0.135$ accounting for only 1.8% of the variability in violation behaviour.

Perceived Ability
Two studies (Adams-Guppy & Guppy, 1975; and Laujenan et al, 1998) examined the relationship between a person's perception of their skill in performing a task and the frequency with which they commit violations. These studies revealed that as people feel they are more skilled, their rate of violation reduces, weighted mean effect size $r = -0.131$ accounting for 1.7 % of the variability on violation. Again perceived ability was not considered as a VPC in the HEART extension.

Discussion

The meta-analysis process identified ten factors contributing to violation for which effect size statistics could be derived. Of these ten factors only four were common with the VPCs used in the revision to HEART proposed by Williams (1996) and a further two might be considered to represent factors contained in the assignment of generic violation behaviours. Further, the rank ordering of factors produced by the meta-analysis is markedly different from that derived by Williams (1996). Differences between the findings might be explained by differences in the literature search strategy adopted; the papers included in the respective analyses or the methods used to assess the importance of the variables related to violation.

With respect to his study, a number of weaknesses of the meta-analysis must be acknowledged. The most significant of these relates to the number of individual effects sizes used to calculate the mean effect sizes reported. For five of the ten variables found to be linked to violation, only two individual effect size statistics were available from which to derive the mean effect size statistics. For the remaining five variables the number of effect size statistics contributing to the mean effect size ranges from 3 to 6. Thus the mean effect size statistics reported here are based only on a small number of individual effect sizes. Whilst this suggests that there are only a small number of studies investigating violation behaviour, this in fact is not the case. The small number of effect size statistics derived is due in large part to inadequate reporting of statistical data in primary research studies. This is a common problem in meta-analytic research and is reported widely by those conducting such research.

References
Adams-Guppy, J.R. & Guppy, A. (1995). Speeding in relations to perception of risk, utility and driving style by British company car drivers. *Ergonomics*, **38**, 2525-2535.
Cohen, J. (1977). *Statistical Power Analysis for the Behavioural Sciences*. (Academic Press, New York).
Diaz, E.M. (2002). Theory of Planned Behaviour and pedestrians' intentions to violate traffic regulations. *Transportation Research Part F*, **5**, 167-176.
Ferguson, S.A., Wells, J.K., Williams, A.F. & Feldman, A.F. (1999). Belt use rates among taxicab drivers in a jurisdiction with license points for non-use. *Journal of Safety Research*, **30**, 87-91.
Friedland, N., Thibaut, J. & Walker, L. (1973). Some determinants of the violation of rules.

Journal of Applied Social Psychology, **3**, 103-118.

Glass, G.V., McGaw, B. & Smith, M.L. (1981). *Meta-analysis in Social Research.* (Sage, London).

Holloway, N.J. (1990). "SURVIVE": A safety analysis method for a survey of rule violation incentives and effects. In M.H. Walter & R.F. Cox (Eds.) *Safety and Relaibility in the 90s.* (Elsevier Science, Barking).

Human Factors in Reliability Group (1995). *Improving Compliance with Safety Procedure: Reducing Industrial Violations.* (HSE Books, Sudbury).

Jacobsson, L. & Svenson, O. (1994). Self-reported human errors in control room work. In G.E. Apostolakis & J.S. Wu (Eds.) *Proceedings of PSAM-II An International Conference Devoted to the Advancement of System-based methods for the Design and Operation of Technological Systems and Processes.*

Jason, L.A. & Liotta, R. (1982). Pedestrian jaywalking under facilitating and non-facilitating conditions. *Journal of Applied Behavior Analysis*, **15**, 469-473.

Kanellaidis, G., Golias, J. & Zarifopoulos, K. (1995). A survey of drivers' attitudes toward speed limit violations. *Journal of Safety Research*, **26**, 31-40.

Laujenan, T., Parker, D. & Stradling, S.G. (1998). Dimensions of driver anger, aggressive and Highway Code violations and their mediation by safety orientation in UK drivers. *Transportation Research Part F*, **1**, 107-121.

Lawton, R. (1998). Not working to rule: Understanding procedural violations at work. *Safety Science*, **28**, 77-95.

Lawton, R., Parker, D. & Stradling, S. (1997). Predicting road traffic accidents: The role of social deviance and violations. *British Journal of Psychology*, **88**, 249-262.

Lipsey, M.W.& Wilson, D.B. (2001). Practical Meta-analysis. (Sage, London).

Mason, S. (1997) Procedural violations – causes, costs and cures. In F. Redmill and J. Rajan (Eds.) *Human Factors in Safety Critical Systems.* (Butterworth-Heineman, Barking)

Rimmo, P. & Aberg, L. (1998). On the distinction between violations and errors: Sensation seeking associations. *Transportation Research Part F*, **2**, 151-166.

Rosenthal, R. (2000). Meta-analysis: Recent developments in quantitative methods for literature reviews. *Annual Review of Psychology*, **52** 59-82.

Sigelman, C.K. & Sigelman, L. (1976). Authority and conformity: Violation of a traffic regulation. *The Journal of Social Psychology*, **100**, 35-43.

West, S.G., Gunn, S.P. & Chernicky, P. (1975). Ubiquitous Watergate: An attributional analysis. *Journal of Personality and Social Psychology*, **32**, 55-65.

Westerman, S.J. & Haigney, D. (2000). Individual differences in driver stress, error and violation. *Personality and Individual Differences*, **29**, 981-998.

Williams, J.C. (1996) *Assessing the Likelihood of Violation Behaviour.* Unpublished Dissertation University of Manchester. Department of Psychology.

Wogalter, M.S., Allison, S.T. & McKenna, N.A. (1989) Effects of cost and social influence on warning compliance. *Human Factors*, **31**, 133-140.

Wogalter, M.S., Magurno, A.B., Rashid, R. & Klein, K.W. (1998). The influence of time stress and location on behavioural warning compliance. *Safety Science*, **29**, 143-158.

Wolf. F.M. (1986) Meta-analysis: Quantitative Methods for Research Synthesis. (Sage, London).

Yagil, D. (2000). Beliefs, motives and situational factors related to pedestrians' self-reported behaviour at signal-controlled crossings. *Transportation Research Part F*, **3**, 1-13.

Yagil, D. (1998). Gender and age-related differences in attitudes toward traffic laws and traffic violations. *Transportation Research Part F*, **1**, 123-135.

PATIENT SAFETY

MEDICAL ERROR AND HUMAN FACTORS INTERVENTIONS: TWO CASE STUDIES

Kathleen A. Harder[1] and John R. Bloomfield[1,2]

[1]*University of Minnesota, Minneapolis, USA*
[2]*University of Derby, Derby, UK*

Two on-site case studies using Human Factors principals in an attempt to reduce medical error and improve patient safety were conducted in Midwestern medical facilities. The first study focused on heparin administration procedures, which had become complicated and were causing confusion. Information was obtained from nurses, nurse educators, pharmacists, and hospital administrators. Recommendations for simplifying, streamlining, and standardizing the heparin procedures were made. The second study, precipitated by a recent error, focused on cataract surgery. Cataract surgery was observed and information about surgical procedures was obtained from attending and supervising nurses, and administrators. Recommendations to use an additional patient identifier and to modify procedures were made.

Introduction

In the early 1990s, the Harvard Medical Practice Study investigated the outcomes of more than 30,000 patient cases in 51 randomly selected hospitals in New York State. The results of this study indicated that (1) adverse events occurred in 3.7 percent of the cases (Brennan, Leape *et al*, 1991), and (2) 13.6 percent of the adverse events led to the death of the patient (Leape, Brennan *et al*, 1991). Extrapolation from these results suggests the number of deaths due to medical error may be as high as 98,000 annually in the US. This is more than double the annual fatality rate on US roads—e.g., the number of people who died in crashes in 2002 was 38,309 (FARS, 2003). Not surprisingly, these findings have spurred attempts to reduce medical error and improve patient safety. An important part of these efforts has involved applying Human Factors principals in the investigation of heath care systems. This paper describes two case studies in which Human Factors investigations were conducted at health care facilities in Minnesota. The first focused on the administration of heparin in a large hospital. The second was conducted in the surgical unit of a large eye care facility.

Case Study #1: Heparin Administration Procedures

The first case study focused on heparin administration in a large Midwestern hospital. Heparin is used to prevent blood coagulation, and if a patient is given too small or too large a dose, there can be severe consequences. An attempt had been made to standardize heparin administration using computerized protocols, but errors were still occurring. The hospital convened a Heparin Error Reduction Workgroup who asked us to conduct a Human Factors analysis of heparin administration procedures. We conducted interview sessions with the physician and pharmacist who developed the computerized heparin protocols. We interviewed the staff pharmacists who dispense the drug. We discussed the protocols with nursing administrators and with nurse educators. And, we interviewed nurses at five nursing stations—including cardiovascular nursing stations, where heparin is administered extensively, and medical/surgical nursing stations in which it is used less often.

Potential Sources of Error
We identified the following potential sources of error.

1. Data Entry. When heparin a patient was about to be given heparin, all the possible heparin procedures were presented on the computer screen.

2. Heparin Dosing Terms. The terms 'Therapeutic' and 'Prophylactic' were applied to the strength of the heparin dose. When heparin is prescribed for a patient undergoing 'Therapeutic' treatment, he or she is invariably given a relatively high dose of heparin. When heparin is prescribed for a patient undergoing 'Prophylactic' treatment, he or she usually receives a relatively low dose of heparin. However, the terms 'Therapeutic' and 'High' are not synonymous; and neither are the terms 'Prophylactic' and 'Low'. Sometimes patients who are being treated with heparin 'Prophylactically' require relatively 'High' doses of heparin.

3. Unclear Data Entry Measurements. In some areas of the hospital, the weight and height of patients were recorded in metric units. In other areas, the weight and height of patients were recorded in English units.

4. Physician Order Forms. The physician ordering forms were not structured—as a result, their written orders could be ambiguous.

5. Temporary Assignment of Nurses. Nurses on temporary assignment to nursing stations at which heparin may be administered may not have had the requisite training.

6. Update of Information. The data entry procedure that was used the first time that heparin was administered to a patient in the hospital had to be repeated each day that the patient remained on the drug.

7. Lack of Computers and Printers. Because of a lack of computers, some nursing stations still had to use the previous paper heparin administration procedures. And in other nursing stations although they had enough computers, there were not enough printers and/or the printers were not near enough to the computers.

Recommendations
To address these potential sources of error, we made the following recommendations.

1. New Data Entry Procedure. When heparin a patient was about to be given heparin, we recommended all the possible heparin procedures should not be presented on the computer screen at the same time. Instead, we recommended that the computer should present a series of steps, with no more than three choices per step being presented to the user. In the first main step, the user would be asked whether the patient was 'Male', 'Female (but not pregnant)', or 'Female (and pregnant)'. In the second main step, the user would be asked about the dosing level—i.e., whether it should be a 'High level' or a 'Low level'. If a Low level were indicated, the next step would be to ask whether it should be administered by 'Intravenous infusion' or 'Subcutaneously'; and, if the answer were 'Subcutaneously', the user was asked whether the type of heparin to be used would be 'Lovenox' or 'Not Lovenox'.

In the final main step, information about the patient's weight and height would be entered.

2. Clarification of Heparin Dosing Terms. We recommended clarifying the heparin dosing terms. The terms 'Therapeutic' and 'Prophylactic' should be replaced by 'High' and 'Low', or by the terms 'Therapeutic (High)' and 'Prophylactic (Low)'.

3. Clarification of Data Entry Measurements. The hospital should standardize its measurements and use metric units. However, even if this occurs, since (in the US) the world outside the hospital is unlikely to convert to metric units, we recommended that the computer accept weight and height information in metric and English units. Further, we recommended that feedback should be provided and confirmation required. The procedure for entering weight information should be as follows. After a patient's weight is entered, the system should query whether it was entered in 'kilos' or 'lbs'. If the response is 'kilos', the system should convert 'kilos' to 'lbs', give feedback in 'kilos' and 'lbs', query whether or not the weight is correct, *and* require confirmation before proceeding. Similarly, if the response is 'lbs', the system should convert 'lbs' to 'kilos', give feedback in 'kilos' and 'lbs', query whether or not the weight is correct, *and* require confirmation before proceeding. With regard to height, we recommended that the patient's height could be entered in 'centimeters', in 'inches', and in 'feet and inches'. And, feedback should be given in all three ways. This, in addition to reducing errors related to confusing 'inches' with 'centimeters', would eliminate English unit errors, like '60 inches' for '6 feet'—the type of error that might occur when the user is highly stressed or fatigued (both conditions that occur too frequently in hospitals). Also, we recommended programming upper and lower weight and height barriers to protect against entries that are outside the sensible range.

4. Use of Structured Physician Order Forms. We recommended that physician ordering forms should be modified to reflect the changes recommended for computer data entry, so that the written orders would not be ambiguous. The physician respond to (1) whether the patient is 'Male', or 'Female (but not pregnant)', or 'Female (and pregnant)', (2) whether the heparin dose should be 'Therapeutic (High)' or 'Prophylactic (Low)', (3) if a 'Prophylactic (Low)' dose should be administered as an 'Intravenous Infusion' or 'Subcutaneously', and (4) whether Subcutaneously delivered heparin should be 'Lovenox' or 'Not Lovenox'.

5. Instruction of Temporary Assignment of Nurses. We recommended that nurses on temporary assignment should have the nurse in charge of a nursing station where heparin is administered (1) explain that heparin doses are determined using a computer querying system, (2) explain that the terms 'Therapeutic (High)' and 'Prophylactic (Low)' refer to dosage levels and *not* to treatments, and (3) be told that he or she should *always* seek help if unable to understand any of the steps recommended for the new data entry procedure.

6. Changing the Requirements for Updating Information. The data entry procedure used the first time that heparin was administered to a patient in the hospital should be as outlined above. However, on subsequent days when the heparin dose is modified, only the following entries should be necessary—(1) the patient's name, (2) the patient's hospital ID number, (3) the current dosing level, and (4) the data returned daily from the lab. This information would be sufficient to calculate the modified dosing level. The amount of data-entry work would be reduced. In addition, confusion about whether the patient's weight should be his or her weight on starting the heparin treatment or his or her current weight (which may be considerably different in some cases) would be removed.

7. Adding Computers and Printers. We recommended adding computers and printers.

Case Study #2: Cataract Surgical Procedures

The second case study was conducted in the surgical unit of a Midwestern eye facility. The day after an error was discovered—the wrong intraocular lens (IOL) had been inserted in the wrong patient—we were asked to review the procedures involved in cataract surgery. The facility has eight operating rooms (ORs) which are in constant use. In our Human Factors review of the facility, we observed the various stages that a patient experiences—(1) arrival in the waiting rooms prior to surgery; (2) the surgical procedure in which a cataract is removed and an IOL is inserted; and (3) post-operative procedures in the recovery rooms. In addition, we observed (1) pre-operative ordering of IOLs; (2) selection of IOLs from their storage area; and (3) handling of the IOLs that were to be used in surgical procedures. Also, we interviewed members of the IOL team—including attending nurses and supervising nurses—and facility administrators.

Potential Sources of Error

In this case, we identified the following potential sources of error.

1. Patient Identification. Many identity checks were made throughout the procedure. However, only one identifier—the patient's name—was used. [The error that precipitated our review occurred because two patients with the same name had cataract surgery scheduled, with the same eye surgeon, within a few weeks of each other.]

2. Location of Patient Charts and IOLs and Lens Order Forms. The patient charts, IOLs, and lens order forms for all patients scheduled for surgery in a particular OR were placed in that OR at the beginning of the day. However, the order in which patients arrive in the OR may change. Also, some physicians schedule patients in two ORs, and may switch patients from one to the other if the patient order changes.

3. IOL Check. Just before insertion in the patient's eye, the IOL was checked against the lens order, but it was not checked against the patient's chart.

4. IOL Selection. Some physicians do not request the IOLs they require until bring the day of surgery—this could result in hurried IOL selection and an error.

5. Hard Copy of the Surgery Schedule. A small font size was used for the hard copy version of the surgery schedule, which would be needed if the computers fail.

Recommendations

To address these potential sources of error, we made the following recommendations.

1. Adding a Patient Identifier. We recommended that each time the patient's identity is verified, the nurse and/or physician should ask the patient for his or her name *and* birth date. Birth date was recommended as the second identifier, rather than social security number, because it is easier to remember. Also, some patients may not have a social security number. Using an additional form of patient identification will make it less likely that patients with the same name will be confused.

2. Changing the Location of Patient Charts, IOLs, and Lens Order Forms. We recommended that (1) the patient charts, IOLs, and lens order forms for the patients scheduled for surgery should be (1) kept outside the OR; (2) kept on carts, with one cart for each physician; (3) arranged on each cart in the order in which the patients are scheduled for surgery; (4) go with each patient into the OR after his or her identity has been checked. Also, if, during the surgery, the physician decides to change the power of the IOL, then the member of the IOL team who selects a new IOL should remove the IOL that is no longer needed from the OR. These steps will reduce the possibility of error—if there is only one lens packet in the OR, the possibility of mixing up IOLs in the OR will be eliminated. This recommendation (1) will be particularly important if the personnel assisting the physician change, due to scheduled breaks, before the IOL is inserted, (2) address the potential problem that could occur if the lens packet is placed on the OR counter without any confirming information, and (3) minimize the possibility of an IOL mix-up that could occur as a result of a change in schedule and a change in the OR to which a patient is assigned.

3. Using Patient Charts for IOL Check. We recommended that when the physician confirms the IOL is correct just prior to insertion in the eye, he or she should check the IOL against the patient's chart. This will eliminate the physician's reliance on memory.

4. Selecting of IOLs prior to day of Surgery. Physicians should be discouraged from bringing a list of required lenses on the day of the surgery. Hurried IOL selection on the day of surgery is more likely to lead to an error than if they have been selected prior to that day.

5. Increasing the Font Size of Hard Copy of the Surgery Schedule. The font size for the hard copy of the schedule placed on the door in the OR should be at least 14 pt. Then, if the computers fail, it will be easier to read the schedule, particularly in an OR with dim lighting.

Outcomes

Heparin Administration
The recommendations made for changes the heparin administration procedures were reviewed and implemented by the hospital's Heparin Error Reduction Workgroup. In the two months after their implementation, there was a 15.5% reduction in heparin errors resulting in harm to patients on the cardiovascular nursing stations. In addition, feedback indicates that the upgraded computer interface has been well received by the nurses using it.

Cataract Surgical Procedures
The IOL team (1) added the recommended patient identifier, (2) changed the location of the patient charts, IOLs, and lens order forms, (3) check IOLs against the patient charts, and (4) have increased the font size of the hardcopy of the surgical schedule. At the time this paper was written there had been no more patient identification errors.

References

Brennan, T.A, Leape, L.L., Laird, N.M., Hebert, L., Localio, A.R., Lawthers, A.G., Newhouse, J.P., Weiler, P.C., and Hiatt, H.H., 1991, Incidence of adverse events and negligence in hospitalized patients. Results of the Harvard Medical Practice Study I, *New England Journal of Medicine*, **324**, 370-376.

Leape, L.L., Brennan, T.A, Laird, N.M., Lawthers, A.G., Localio, A.R., Barnes, B.A., Hebert, L., Newhouse, J.P., Weiler, P.C., and Hiatt, H.H., 1991, The nature of adverse events in hospitalized patients. Results of the Harvard Medical Practice Study II, *New England Journal of Medicine*, **324**, 377-384.

FARS, 2003, Fatality Analysis Reporting System, National Highway Traffic Safety Administration.

Identifying and Reducing Errors in the Operating Theatre

Dr Ken Catchpole
Professor Marc de Leval

Cardiothoracic Unit
Great Ormond Street Hospital for Children NHS Trust
London.

The operating theatre can be viewed as the apex of the patient experience, and is one of the most crucial components of the healthcare system. This makes the identification and reduction of errors in the operating theatre an important stage in making wholesale improvements in the reliability of the healthcare system. The current work seeks to identify errors at a behavioural level in surgery, failures at a systemic level before, during and after surgery, and their overall contribution to patient outcome. Our work is taking place at three centres in the UK, and by working closely with professional pilots, we hope to utilise many of the lessons learned from other high reliability organisations.

Introduction

It is now well accepted that a system solely reliant on human performance 'at the sharp end' will ultimately be a system that fails (e.g. Reason, 1990; Helmreich, 2000). Undesirable outcomes occur as a result of series of undesirable events or states – some of which may have arisen months before, or may be endemic in the culture – that form a cascade of errors, ultimately predisposing the single individuals who form the last line of defence to make the final, critical mistake. Systems that have defences in depth – configurations to capture or reduce errors at many different levels, long before they can accumulate to create an adverse event – will be more reliable, and resistant to the inevitable human errors that occur as part of normal system operations. The maintenance of those defences requires a sufficient culture of safety, and central to such a safety culture is the blame-free and systemic consideration of error. After de Leval *et al* (2000), who offered evidence for a connection between non-technical errors in cardiac surgery and eventual patient outcomes, the current work, jointly funded by the Department for Health and the National Patient Safety Agency, seeks to examine and reduce errors in surgery at three very different UK centres. This paper describes our progress to date, with particular focus on our initial work in neonatal cardiac surgical procedures.

Error in Surgery

The operating theatre can be viewed as the apex of the patient experience. It requires an accurate diagnosis methodology, a suitable and willing patient, pre-treatment planning, post-treatment care, specialist equipment, an appropriate workspace, a team of highly skilled individuals, and an infrastructure and culture to sustain and integrate all these components in the best possible way. If surgery is the apex of the healthcare system, then iatrogenic[1] injury during surgery is the nadir of healthcare error. The individuals in the operating theatre may be predisposed to error by poor diagnosis, poor equipment, poor planning, or a range of other technical and non-technical latent failures. In order to reduce adverse patient outcomes, it is vital to understand the wide variety of components that predispose iatrogenic injury, which requires the study of events, behaviours, or components of the system which are not, on their own, iatrogenic, but may cause iatrogenesis in certain configurations. The likelihood of poor patient outcome will increase with the number and frequency of errors in theatre (de Leval *et al*, 2000), but may be mitigated by the skill of the individuals and their ability to work together as a team (Carthey *et al*, 2003). If we are to reduce errors, then we must approach the study definition from a systems perspective; we must examine why surgical teams behave sub-optimally, the sorts of sub-optimal events that occur, the frequency with which those events occur, and their relationship with patient outcome.

Cultural Assessment

In order to understand the team and institutional cultures that individuals in the operating theatre find themselves, a hierarchy of three measures is being adopted (Figure 1). A Checklist of Institutional Resilience (CAIR), completed by two human factors practitioners, provides a score that reflects the key safety features that are encouraged by the practices and policies of the parent institution. The second measure, a questionnaire to be completed by a broad sample of healthcare practitioners in each instituion, examines safety culture from the perspective of incident reporting, and assesses the ability of the individuals in the institution to learn from their mistakes. The final questionnaire will be administered to those involved in surgery, and measures the team culture and attitudes among the staff. This Operating Theatre Team Management Questionnaire (OTTMAQ) has been adapted from a similar instrument

---

[1] Derived from the Greek *Iatros*, meaning physician, and *-genic*, meaning induced by, is commonly used to refer to any disease or complaint caused by the caregiver or caregivers. As we are unaware of any other industry that has a specific label for human error, this not only reflects the accepted frequency with which these undesirable events occur, but also the prevalence of blame culture within the industry.

used extensively in the aviation industry (Helmriech & Merritt, 1998). It is hoped that for each of the three research centres, we will be able to develop a different cultural profile that reflects the level of safety and co-ordination in the surgical team. We will then conduct in-theatre observations to examine how these differing cultural profiles affect the events in theatre.

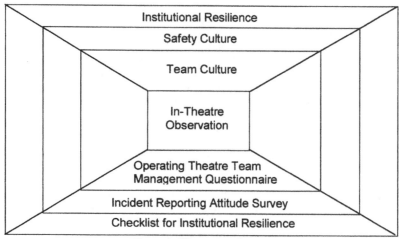

Figure 1: Hierarchy of measures

Surgical Assessment

To complement the broad assessments of safety and teamworking cultures at the three study centres, we will examine the number and types of negative events[2] found in surgery. Having observed over 40 cases prior to the full development of the measurement instrument, our general approach will be to measure three parameters from a suitable observation point in theatre; technical events, non-technical events, and discipline-related events. In order to understand the complex procedures involved, a series of task analyses were conducted to assess the tasks involved in a range of surgical procedures. From these, it was then possible to construct observational checklists both of technical events, and the standard verbal protocols that indicate good practice. As a result of observing these procedures, we have also developed a discipline checklist that allows the measurement of the occurrence and management of breaches in discipline in theatre, such as the management of distractions, or violations of hygiene protocols. The often unpredictable nature of surgical procedures means

---

[2] We have been keen to avoid the word 'error' at this stage, both because it is an emotive label in the medical profession, and because the term must be nested in some concept of outcome, and which we argue is something that cannot necessarily be observed prospectively.

that we also accept that we will not be able to predict all negative incidents that might occur, and for these we have included an area on the checklist for free-observations.

As well as an in-situ observer methodology, we have been deploying a video collection methodology. Utilising two cameras positioned at the head and the foot of the operating table which record the interactions of the anaesthetic team and the theatre staff, respectively, one overhead camera recording the surgical field, and a direct video feed from the anaesthetic workstation, we are able to record, in some detail, the events that occur during one procedure. Though this may well form a useful record of many of the procedures we observe, the primary purpose is to evaluate the utility of video methodologies for in general surgery for error analysis. This will be done by comparing the results of what can be viewed using the video data with the observations in theatre. Initial impressions are that while it will be possible to observe many events in theatre, many of the more subtle behaviours will not be observable. In addition, the acoustics of the operating theatre mean that, despite high quality sound equipment and careful placement of the microphones, much of the voice communication that can be heard using the human ear is lost during the recording.

Patient Outcome

Clearly, events in theatre will have an influential impact on the successful recovery of the patient. We hope to use procedural length and time of stay in hospital as indicators of outcome, but may also examine a range of other outcome variables that are routinely collected by the care teams. However, as we are observing a range of procedures, at a range of centres, we know that many of the events observed in theatre occur in isolation will have no significant impact on the patient, either because the team was able to compensate for the negative event, or because the event itself had no direct effect on the patient. In order to obtain some measure of the former, we will score the performance of each team according to a behavioural marker system adapted from Carthey *at al* (2003) at the end of each operation. In our analysis we will also compensate for factors associated with the patient and the procedure.

Further Work

As part of ongoing efforts to improve the process of the patient handover from theatre to intensive care, additional studies are adopting a similar observational style. The handover is of particular interest because it follows on directly from the surgical procedure, is time critical, has both technical and teamworking challenges, is relatively invariant across a number of different surgical procedures, it often occurs at the end of a long procedure when team members are tired, and because there is currently very little task standardisation or protocol. There is also particular interest in the briefing that is given by the surgical team to the intensive care team, because the information that is communicated during that time is vital in terms of the treatment that the patient will receive for several hours subsequently. Once we have evaluated the handover process, we intend to design and implement a training programme, and re-evaluate the success of the training at a subsequent date.

We also feel attention to pre- and de-briefing surgical teams – something that at the moment is done, if at all, on an ad-hoc basis, rather than as a matter of course – could yield excellent results in terms of team-building over time, in helping teams learn from the events they encounter in surgery, and in preparing teams for the events that they may encounter in the future. This sort of approach is used very effectively in aviation, and we hope to instigate a study of this type in the near future.

Conclusions

In the operating theatre, a wide range of system parameters – both technical and non-technical – contribute to patient outcome. While the complexity and unpredictability of medical error cases make it exceptionally difficult to study prospectively, the work of a number of research groups both in the United States (e.g. Helmreich and Merritt, 1998; Weinger et al. 2003; Xiao et al, 1996;), and in the United Kingdom (e.g. Vincent et al 2001; Flinn et al 2003) is currently seeking to tackle these complex questions. Our investigations hope to add substantially to this research by examining the link between safety culture, teamworking, errors in the operating theatre, and patient outcome.

References

Carthey, J., de Leval, M. R., Wright, D. J., Farewell, V. J., and Reason, J. T. 2003, Behavioural markers of surgical excellence. *Safety Science*, **41**, 409-425.

de Leval, M. R., Carthey, J., Wright, D. J., and Reason, J. T. 2000, Human Factors and Cardiac Surgery: A Multicenter Study. *The Journal of Thoracic and Cardiovascular Surgery*, **119**, 661-672.

Flin, R. H., Fletcher, G. C. L., McGeorge, P., Sutherland, A., and Patey, R. 2003, Anaesthetists' attitudes to teamwork and safety. *Anaesthesia*, **58**, 233-242.

Helmreich, R. L. and Merritt, A. C. 1998, *Culture at Work in Aviation and Medicine*, (Ashgate, Aldershot).

Helmreich, R. L. 2000, On error management: lessons from aviation. *British Medical Journal*, **320**, 781-785.

Reason, J. T. (1990). *Human Error*, (University Press, Cambridge)

Vincent, C., Neale, G., and Woloshynowych, M. 2001, Adverse events in British hospitals: Preliminary retrospective record review. *British Medical Journal*, **322**, 517-519.

Weinger, M. B., Slagle, J., Jain, S., and Ordonez, N. (2003). Retrospective data collection and analytical techniques for patient safety studies. *Journal of Biomedical Informatics*, **36**, 106-119.

Xiao, Y., Hunter, W. A., Mackenzie, C. F., Jefferies, N. J., and Horst, R. L. 1996, Task Complexity in Emergency Medical Care and Its Implications for Team Coordination. *Human Factors*, **38**:4, 636-645.

MEDICATION ERRORS AND THE JUDICIOUS USE OF COLOUR ON THE LABELLING OF MEDICINES

Ruth Filik, Kevin Purdy, & Alastair Gale

Applied Vision Research Unit
University of Derby
Derby DE22 3HL

Medication errors occur as a result of a breakdown in the overall system of prescribing, dispensing, and administration of a drug. A common contributory factor is similarity in packaging, often from the use of nearly identical packaging for two separate items. This paper reviews the debate surrounding the use of colour in drug labelling to aid the identification of drug products, and describes preliminary research investigating the use of colour on the labelling and packaging of medicines.

Introduction

Based on recent estimates, medical error is among the leading causes of death in developed countries, with errors involving medication among the most common of all medical mistakes (Kohn, Corrigan, & Donaldson, 2000). The Department of Health (2000) recommends that serious errors in the use of prescribed drugs be reduced by 40% by 2005. Many factors can contribute to errors being made, including: failed communication; poor drug distribution practices; dose miscalculations; drug and drug-device related problems; incorrect drug administration, and lack of patient education (Cohen, 1999). The Report to the Committee on Safety of Medicines from the Working Group on Labelling and Packaging of Medicines (2001) states that a review of data collected in relation to dispensing errors, or near-misses, points to similarity in drug names (both look-alike and sound-alike similarities) and similarity in drug packaging as being the two main contributory factors. The current paper is concerned with issues of similarity in drug packaging and labelling, specifically relating to the use of colour.

Colour can be used on packaging, labelling, or the drugs themselves either to code or differentiate products. Colour coding is the systematic application of a colour system to identify specific products (e.g. Warfarin 1mg tablets are brown, 3mg are blue). Colour coding may be distinguished from colour differentiation, which entails using colour to distinguish a product, but the colour itself does not mean anything. Colour differentiation can highlight strength differences within a range, or help distinguish different products across ranges.

There is some debate surrounding the use of colour on the medicines labelling. One perspective is that colour can aid the correct identification of medicines, for example, by reducing the similarity in appearance between confusable medicines, and as an additional safeguard to complement the use of appropriate fonts and letter size on packages (Delaney, 1999). Using multiple colours, designs (e.g. blocks of colour in circles, squares, rectangles) could lead to safety advantages such as it would be easier to distinguish packs that had been put in the wrong drawer (Burgess, 1999).

The other side to the debate is that the colouring of patient packs for generic pharmaceuticals can only lead to error, at both pharmacist and patient levels. The main concern being that it can lead to too much reliance on colour, leading to complacency (Mason, 1999). Intuitively, using colour to differentiate strengths of a particular drug within each drug line is a sensible idea and does highlight differences between strengths. However, this may lead to the possibility that drugs are selected incorrectly by association with colour and strength instead of name and strength (Underhill, 2001). One potential problem arises when manufacturers use the same colours across different lines, and make all of their packaging the same size. The result is that entirely different drugs have virtually indistinguishable packaging (Gentle, 1999). This problem is compounded when drugs with similar names that are made by the same manufacturer, have similar coloured packaging, and some strengths that are the same are placed next to each other on the dispensary shelf. Problems may also arise when products with the same colour are commonly co-prescribed (often to elderly patients), possibly leading to an increased risk of the patient taking the wrong drug at home. Therefore, some people argue that it is safer to require that all generic packs are printed in black on a white background, forcing users to read the label, and removing the possibility of duplication of colours (Freedman, 1999).

The Report to the Committee on the Safety of Medicines from the Working Group on Labelling and Packaging of Medicines (2001) evaluated the use of colour on packaging. The report noted that it has for a long time been policy that colour coding is not appropriate on grounds of safety. The major reasons for this are: a limited number of colours are available for use; the incidence of colour-blindness; individual perception of colour; difference in appearance of the same colour in different lighting conditions; no colour coding system could positively differentiate between all 12,000 medicines authorised in the UK, and in the absence of a nationally or internationally agreed colour code any UK system could be perceived as a barrier to trade. The primary problem with the use of colour on the labelling of medicines was considered to be that colour can become a short cut to the identification of the medicine rather than reading the label text itself. The use of colour coding was therefore not supported. However, it was agreed that the judicious use of colour on packaging materials might help to differentiate between products within a company portfolio, particularly between different strengths of the same product. It was agreed that the prime function of a label is to identify the medicine, what it is for and how to use it safely, and that there is no substitute for reading the label. Following this, the Medicines Control Agency (2003) produced a document outlining best practice guidance for the labelling and packaging of medicines. The stance taken on the use of colour was that, "Innovative pack design that may incorporate the judicious use of colour is to be encouraged to ensure accurate identification of the medicine".

Two key empirical questions would appear to emerge from this discussion. Firstly, whether colour can be used to help healthcare professionals and patients distinguish different products (e.g. products form the same range that are different strengths). Secondly, whether this could lead to colour being used as a shortcut to identify a product, resulting in less careful inspection of other information on the label such as the product name. The following sections will describe preliminary research investigating these questions.

Experiment 1

The question investigated in Experiment 1 was whether colour can aid the identification of a product of a particular strength within a range. Participants were shown a target product (presented on a computer screen for a limited amount of time) that they subsequently had to search for amongst an array of products.

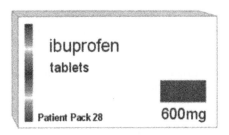

Figure 1: An example pack.

The array consisted of eight different strengths of the target drug. For example, the target might be ibuprofen tablets 600mg, and the array would be ibuprofen tablets 100, 200, 300, 400, 500, 600, 700, and 800mg. Packs were presented in a circle around the centre of the screen. Half of the time a pack with the same strength as the target product was present in the array and half of the time it was not. The participants' task was to indicate whether or not the target pack was present in the array by pressing one of two buttons, marked 'Y' and 'N'. Mock drug packs were designed for the purpose of the experiment (see Figure 1). Packs had a rectangular block above the strength, which in half of the trials was coloured, (producing eight different colours in the array) and in the other half of the trials all of the blocks were grey. Therefore in half of the trials, colour could be used to help distinguish products of different strengths and in half the cases it could not. It was hypothesised that if colour aids identification of products within a range, then participants should make fewer errors when they can use colour as an aid to identification.

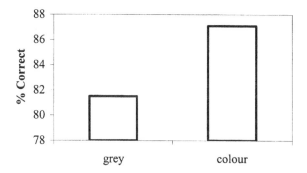

Figure 2: Percentage correct responses for coloured and grey packs.

Results showed that participants made significantly fewer errors when they could use colour as an aid to identify a product in a range (see Figure 2).

Experiment 2

Experiment 2 addressed the question of whether colour can become a short cut to the identification of a medicine, rather than reading the label text. One real-life situation in which this could be a problem would be where a product with a similar name as the target product was also available in the same strength and was the same colour as the target.

The procedure for Experiment 2 was identical to that of Experiment 1. Once again, packs either had a coloured block above the strength, in which case colour could be used as a cue, or a grey block, in which case it could not. In Experiment 2, the target pack was never present in the array, but was instead replaced by a product with a similar name. For example the target would be sulfadiazine tablets 500mg and the array would contain sulfasalazine 50, 100, 200, 300, 400, 500, 600, and 700mg. Therefore it was always a similarly named drug with the same strength and colour as the target that was present in the array, so the correct answer was always 'No, the target is not present'. So that the required response was not always 'No', an equal number of filler trials were added in which the target *was* present in the array. If colour made participants less likely to attend to other information on the label, such as the name, then they should make more errors in which they incorrectly say that the target is present (i.e. respond 'Yes') when the packs are coloured, than when they are not.

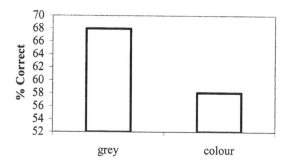

Figure 3: Percentage correct responses for coloured and grey packs.

Participants made significantly more errors when they could use colour as an aid to identify a particular strength of drug (see Figure 3).

Discussion

The current paper investigated the use of colour on drug packaging and labelling as an aid to product identification. Experiment 1 provided support for colour aiding the identification of a particular strength of a product within a range, with participants making fewer errors when they could use colour as a cue to identification. However, Experiment 2 provided evidence for colour being used as a shortcut to identify the medicine, leading to incorrect identification. When a product had a similar name and the same strength as a target product, participants were more likely to incorrectly identify it as being the target drug when they could use colour as a cue.

One limitation of the current research is that the task was fairly artificial; it would be informative to investigate this issue in a more 'real life' setting. However, it is useful to initially investigate such issues in a laboratory environment, where factors can be more tightly controlled, and there is no risk to the patient.

It could be concluded that these results provide empirical support for the unambiguous, or judicious, use of colour in the labelling and packaging of medicines. For example, colour could effectively be used to distinguish different strength products within a range, but could be misleading if similar colours were used across product ranges, particularly those with similar names, which may be situated close together on a shelf.

References

Burgess, P. D. (1999, September 25). More colour please! [Letter to the editor].
   The Pharmaceutical Journal, **263** (7064), p.484.
Cohen, M.R., (ed.) (1999). *Medication Errors*. Washington: American Pharmaceutical
   Association.
Delaney, T. (1999, October 30). Colour has a role [Letter to the editor]. *The Pharmaceutical
   Journal*, **263** (7069), p.708-709.
Department of Health (2000). *An organisation with a Memory*. London: HMSO.
Freedman, N. (1999, September 4). Colour not the answer [Letter to the editor].
   The Pharmaceutical Journal, **263** (7061), p.349.
Gentle, J. (1999, August 14). APS packaging concern [Letter to the editor].
   The Pharmaceutical Journal, **263** (7058), p.240.
Kohn, L. T., Corrigan, J., & Donaldson, M. S. (2002). *To err is human: building a safer
   health system*. Washington DC: Institute of Medicine.
Mason, J. P. (1999, August 21). A vast improvement [Letter to the editor]. *The
   Pharmaceutical Journal*, **263** (7059), p.276.
Medicines Control Agency (2003). *Best practice guidance on the labelling and packaging of
   medicines*. London: Medicines Control Agency.
*Report to the Committee on Safety of Medicines from the Working Group on Labelling
   and Packaging of Medicines* (2001). London: Medicines Control Agency.
Underhill, R. (2001, May 26). Ridiculous and totally unprofessional [Letter to the editor].
   The Pharmaceutical Journal, **266** (7149), p.718 – 722.

WARNINGS

HAZARD COMMUNICATION OF
WARNING SIGNAL WORDS AND SAFETY PICTORIALS

Oguzhan Erdinc

Turkish Air Force Academy Industrial Engineering Dept.
Istanbul, Turkey

In this study, hazard communication of 9 signal words and 7 safety pictorials was rated by Turkish Air Force Academy cadets. A one-way ANOVA showed that main effects of signal words and pictorials on hazard communication were significant. DEADLY and a pictorial of a skull and crossbones were found to communicate significantly the highest level of hazard. Though being recommended in safety standards to convey high, intermediate and low hazard levels respectively, hazard communication levels of DANGER, WARNING and CAUTION were not found to be significantly different. Pictorials with clear themes communicated significantly higher hazard levels than pictorials with abstract themes. This study, conducted with a particular, non-native English speaker group, was expected to contribute new findings to warning literature.

Introduction

An effective visual warning should be able to communicate hazards as required. This goal can be achieved through the proper selection and design of its components. Numerous authors and institutions have reviewed studies conducted on visual warning design; a need to analyze the effects of design components on hazard communication is often propounded in these reviews (Lehto, 1992(a and b), OSHA, 1997, Rogers *et al*, 2000).

Warning signal words and safety pictorials are major components of visual warning design. Use of signal words and safety pictorials in warning design has been strongly recommended in the warning literature. The presence of signal words was found to enhance hazard perception (Wogalter *et al*, 1994). Interaction of signal words with other warning design components such as color or symbols is another direction of research (Braun and Silver, 1995, Chapanis, 1994, Wogalter *et al*, 1994). Along with the rapid evolvement of globalization, pictorials have become attractive means of conveying public or safety information in multi-national scales owing to their language-free nature, ease of implementation and communication capability. Authors have focused mostly on the comprehension of pictorials believing that hazard communication could be realized once the meaning of a pictorial is understood correctly (Davies *et.al*, 1998, Jones, 1978, Wolff and Wogalter, 1998). In the present study, hazard communication of 9 signal words and 7 safety pictorials was evaluated via ratings of undergraduate Turkish Air Force Academy (TUAFA) cadets. Effects of visual warning design components may drastically vary among viewer groups depending on the differences in cognitive and cultural structures. These differences should be elaborated to attain a deeper insight of hazard communication and warning effectiveness in global scales. Thus, performed with a non-native English speaker group of undergraduate TUAFA cadets, this study was expected to contribute new findings to the warning literature.

Signal words rated in this study were, DEADLY, DANGER, WARNING, CAUTION, ATTENTION, STOP, LETHAL, NOTICE and NOTE. Low, intermediate and extreme/fatal levels of hazards are recommended to be communicated by CAUTION, WARNING, and DANGER respectively in safety standards and guidelines (Lehto, 1992(a)). While a consistent set of findings in warning literature reveales that DANGER connotes higher levels of hazard than WARNING and CAUTION, significance of difference between hazard communication levels of WARNING and CAUTION is hard to mention (Chapanis, 1994, Wogalter *et al*, 1994, Wogalter and Silver, 1995).

DEADLY was found to connote higher hazard levels than other recommended signal words do (Wogalter and Silver, 1995). Though being its synonym, LETHAL has less frequent usage than DEADLY does and it was selected to observe familiarity-related differences on hazard communication. ATTENTION, NOTICE and STOP were also included in different studies and expected to convey low levels of hazard (Wogalter and Silver, 1995).

As for safety pictorials, not only comprehension of specific meaning of a pictorial, but also the theme and shape of pictorial were considered to influence hazard communication. For that reason, pictorials of various themes and shapes, with or without a specific meaning were selected for present study. All of these safety pictorials were manipulated in achromatic form to avoid confounding effects of color. Selected pictorials and their associated meanings were presented in Table-1.

Participants

Sixty-seven 3rd year undergraduate TUAFA (Ind. Eng. Dept.) cadets (62 males, 5 females ranging in age between 19-22, with a mean age = 21 years, SD = 0.6352 years), with a sufficient command of English, consented to participate in the study. The study was conducted during an ergonomics class.

Material and Procedure

Hazard communication of selected signal words and safety pictorials were rated seperately. Material consisted of three pages; the first including personal questions about participants (e.g., age, gender), second and third including signal words and safety pictorials respectively. Signal words were written in 16 point Times New Roman font in capitals with an Office-Word program. All pictorials were approximately 2 X 2 cm. and printed in achromatic (i.e., black and white) form. A 5-point rating scale was designed with anchors; (1) None of them, (2) Low danger, (3) High danger, (4) Very high danger, (5) Fatal danger. Fatal danger was accepted to be the highest level of hazard. Participants were given one of the three versions of tests, in which signal words, safety pictorials and anchors were arranged in different orders for randomization. They were instructed to rate the hazard they perceived from each item seperately.

It was emphasized that they were not asked to put the items in rank order and there were no "right" or "wrong" answers. The purpose of the study and the points assigned to anchors were not explained. They were told to mark "None of them" in case they did not know the meaning of the items or they did not perceive any hazard at all. No time limits were imposed on the participants.

Table 1. Signal words, safety pictorials and associated meanings, rating results, rankings and homogeonus subsets

Multiple Comp. Subsets	Signal Words	Mean Rating Scores	Std. Dev.	Multiple Comp. Subsets	Safety Pict. & Meaning	Mean Rating Scores	Std. Dev.
Highest rank	**DEADLY**	4.582	0.9	*Highest rank*	[skull & crossbones] 1	4.731	0.665
					Toxic hazard, poison		
Subset 1	**DANGER**	3.373	0.794	*Subset 1*	[radiation symbol] 2	3.493	1.541
	WARNING	3.239	1.060				
	CAUTION	3.134	1.086		Radiation hazard		
Subset 2	**ATTENTION**	2.343	0.88		[flame symbol] 3	3.164	0.979
	STOP	2.03	0.9		Flammable hazard		
	LETHAL	2	1.403	*Subset 2*	[X symbol] 4	2.209	0.664
Subset 3	**NOTICE**	1.552	0.907		No specific meaning		
	NOTE	1.403	0.552		[! triangle] 5	1.940	0.868
					Hazard in general		
				Subset 3	[prohibition circle] 6	1.576	0.656
					No specific meaning, prohibiton		
					[filled circle] 7	1.507	0.943
					No specific meaning		

Results

Mean rating scores, standard deviations and multiple comparison subsets are presented in Table-1. A one-way ANOVA performed on the mean rating scores showed significant main effects of both signal words (p<0.001) and safety pictorials (p<0.001). Mean rating scores of signal words and safety pictorials ranged from 1.403 to 4.582 and from 1.507 to 4.731 respectively. Further Newman-Keuls range tests demonstrated significant differences between mean scores of signal words (p<0.01) and safety pictorials (p<0.03). DEADLY yielded significantly the highest hazard communication level. The rest of the signal words were grouped into three homogenous subsets. Differences within-subsets were not significant, but differences between-subsets were. The first subset was formed by DANGER, WARNING and CAUTION, second subset by ATTENTION, STOP and LETHAL and third subset by NOTICE and NOTE from high to low mean rating scores.

The 1st pictorial, skull and cross bones was found to communicate the highest level of hazard. Subsets of remaining pictorials were formed in the same manner with signal words. The first subset was formed by 2nd and 3rd, second subset by 4th and 5th and third subset by 7th and 8th pictorials from high to low mean scores.

Discussion

Present study had certain limitations. Signal words and safety pictorials were viewed in list form, which lacked the natural context of warnings. Though relations are proposed to exist between hazard communication and comprehension in warning literature (OSHA, 1997), this study excluded comprehension concerns.

Differences between hazard communication of DANGER, WARNING and CAUTION were not found to be significant in contrast with relevant recommendations (Lehto, 1992(a), Wogalter et al, 1995). This finding, especially regarding to WARNING and CAUTION is consistent with previous warning research (Chapanis, 1994, OSHA, 1997, Wogalter et al, 1994, Wogalter and Silver, 1995). Given the fact that English is the international communication language, to build a reliable 'hazard communication hierarchy' of English signal words is an important requirement to ensure warning effectiveness in global scales. In this respect, current hierarchy of signal words should be reconsidered. Further studies should elaborate effects of signal words in natural context of warnings and should include evaluation of compliance attained by different signal words as well as hazard communication.

Due to its strong hazard communication and explicitness, DEADLY can be employed in warning design to convey fatal hazards effectively. However, potential adverse psychological effects of sensing the threat of death each time it is used could be a drawback for the use of DEADLY. Systems which are not used in public could be more appropriate to employ DEADLY (e.g., industrial or weaponary systems).

Low rating scores of LETHAL indicated that unfamiliarity with signal words substantially reduces hazard communication for non-native English speaker groups. Hence, when warnings are designed for global use, signal words should be readily understandable in order to ensure warning effectiveness.

The 1st pictorial, skull and crossbones conveyed fatality very explicitly. Together with the skull and cross bones, the 2nd and 3rd pictorials, which comprise the first subset, can be classified as pictorials with clear themes. Thus it can be suggested that, clarity of theme enhances hazard communication of pictorials. In order to increase reliability and applicability of safety pictorials, themes regarding to hazards should be symbolized explicitly instead of designing abstract signs.

Multiple comparison subsets showed that, triangle pictorials conveyed higher levels of hazard than did circular pictorials. This finding supported that shape-coding manipulations can be used to control hazard communication of pictorials.

The 5th pictorial, though recommended in guidelines to indicate high, even extreme hazard levels, yielded low rating scores. This could be attributed to its abstractness. The exclamation mark in it indicates importance but it might fail to communicate hazard. Furthermore, this pictorial is extensively used in computer warnings, and the majority of the hazards expressed in computer warning messages can be avoided by a mouse click easily. This could be a factor which diminishes the hazard communication capability of the 5th pictorial. The fact that all participants were young, computer-generation cadets supports this idea. Computers, being the most common medium of life today, might alter attitudes toward warnings and this consideration points out new extensions for future warning research.

References

Braun, C.C. and Silver, N.C. 1995, Interaction of signal word and colour on warning labels: Differences in perceived hazard and behavioral compliance, *Ergonomics*, **38**, 2207 - 2220

Chapanis, A. 1994, Hazar associated with three signal words and four colors on warning signs, *Ergonomics*, **37**, 265-275

Davies, S., Haines, H., Norris, B., and Wilson, J.R. 1998, Safety pictograms : Are they getting the message across? *Applied Ergonomics*, **29**, 15 – 23

Jones, S. 1978, Symbolic representation of abstract concepts, *Ergonomics*, **21**, 573– 577

Lehto, M. R., and Salvendy G. 1995, Warnings : A supplement not a substitute for other approaches to safety, *Ergonomics*, **38**, 2155 – 2163

Lehto, M. R. 1992(a), Designing warning signs and warning labels:Part-1: Guidelines for practitioner, *International Journal of Industrial Ergonomics*, **10**, 105 – 113

Lehto, M. R. 1992(b), Designing warning signs and warning labels:Part-2: Scientific basis for initial guidelines, *International Journal of Industrial Ergonomics*, **10**, 115 – 138

OSHA 1997, Hazard Communication: A rewiev of science underpinning the art of communication for health and safety, *www.osha.gov*

Rogers, W.A., Lamson, N., and Rousseau G. 2000, Warning research : An integrative perspective, *Human Factors*, **42**, 102 – 132

Wolff J.S., and Wogalter M.S. 1998, Comprehension of pictorial symbols: Effects of context and test method, *Human Factors*, **40**, 173 – 186

Wogalter M.S., and Silver N.J. 1995, Warning signal words : Connoted strength and understandibility by children, elders and non-native English speakers, *Ergonomics*, **38**, 2188– 2206

Wogalter M.S., Jarrard S.W., and Simpson N.S. 1994, Influence of warning label signal words on perceived hazard level, *Human Factors*, **36**, 547 – 556.

METHODS

Ergonomics Guidelines – A Help or a Hindrance?

Dave Gregson & Andy Gait

Synergy Consultants Ltd
Yewbarrow
Hampsfell Road,
Grange-over-Sands,
Cumbria, LA11 6BE

There is a large body of ergonomics guidance that has been published to assist in the uptake of ergonomics principles, advice and methods. However, it is the authors' experience that finding the appropriate guidance for a particular application can be time consuming, with much of the information proving to be conflicting, out-of-date or unnecessarily technical. It is often daunting for non-ergonomists to apply the guidance and it is therefore not surprising that ergonomics issues are often overlooked.

This paper will consider the difficulties in providing effective guidance and discuss specific areas where improvements can be made. It will debate the topic of guideline usability for both ergonomists and those without ergonomics training.

Introduction

Ergonomics guidelines in the form of standards, regulations and textbook material can have profound effects upon improving designs in terms of safety, reliability, efficiency and comfort. However, guidelines can be used inappropriately or given scant attention particularly by non-ergonomists. This paper describes the author's experience as relatively newly qualified ergonomists in the use of guidelines material within the Defence and Nuclear Sectors.

Generic and Application Specific Guidance

Generic guidance has a clear advantage in that it can potentially be used repeatedly to improve and assess the use of ergonomics knowledge for wide ranging projects. This repeated use means that a consistent approach can be applied and familiarity with the material can be gained by ergonomists.

However, generic guidance cannot easily define the best ergonomic design because it often depends upon the particular application, its users and other environmental and organizational factors. The most effective guidance will consider the context of the work that is to be carried out, including an understanding of the user's experiences, skills, expectations, workload, working environment and task specifics etc. This depth of knowledge about the context of work is a pre-requisite to the effective use of any guidance and this requires suitable experience of the particular field and may necessitate a task analysis approach.

In most cases application specific guidance is preferred, but given the innumerable applications for ergonomics this cannot always be achieved. Often generic guidance has to be used and this can be applied with various levels of success to specific applications.

A common problem with generic guidance is that it can be difficult to translate into practical design decisions. If designers are to effectively employ ergonomics guidance, then in general, more detailed and application specific advice is needed that does not require further interpretation. Generic guidance often uses more loosely worded terms and therefore there is a potential for different designers to interpret the requirements for compliance with various degrees of stringency. The Display Screen Equipment (DSE) Regulations (Health and Safety Executive, 1992) is an example of generic guidance. It is the authors' opinion that the extent to which this guidance is used successfully is significantly dependent upon the user's interpretation of the guidance.

If there is no application specific guidance available then it will often require an ergonomist to ensure that the generic guidance is suitably tailored to meet the particular requirements of the application.

One of the major issues concerning the effective use of ergonomics guidance is understanding when compliance with the guidance is essential and when there may be some leeway in its adoption. There are trade-offs that have to be made between ergonomics compliance, running to program and basic economics. The difficulty in defining the precise benefits of ergonomics compliance in economic terms often means that guidance is not complied with. It can therefore be useful to detail the relative merits of compliance with the various guidelines, and where possible to define when compliance is essential or when it may be 'desirable'. This is often difficult to define and can be increasingly difficult to determine if the guidance is generic and not application specific.

If there is no indication of any priority in terms of compliance with the ergonomics guidance then the designer or engineer may simply select to comply with the guidance that is cheapest or the least hassle to implement.

Size, Structure and Content of Guidance

The designer or engineer may have access to a large volume of ergonomics guidance but clearly this does not mean that this will necessarily be used. It is easy to be overwhelmed with the size of some of the guidance and many are cumbersome to use. Users of guidance often do not have the time to read the further references and filter out the irrelevant, out-dated or already implemented.

Two commonly used sources of human factors material are NUREG 0700 (US Nuclear Regulatory Commission, 2002), which provides guidance on Human System Interface Design for the nuclear industry and Defence Standard 00-25 (Ministry of Defence, 2000) which covers a comprehensive range of issues for defence related applications. These documents contain vast amounts of guidance that require considerable time and effort to understand and assess designs for compliance. As a result, these documents are particularly daunting for those with no human factors experience.

Some documents can be difficult to use because the guidance is scattered about and excessive cross-referencing is required to understand the issue. An example of this kind of document is the Engineering Equipment and Materials User Association (EEMUA) guidance on Alarm System Design, Management and Procurement, (EEUMA, 1999).

Given the timescale of projects and the poor structuring of documents, guidance is often used in an unsystematic fashion with concentration on an insufficient range of ergonomics issues. Also insufficient guidance may be used because the user finds and persists with the first source of information found which may be inappropriate or out-of-date etc. The designer or engineer may simply not have the time to undertake ergonomics literature searches and contend with any potential conflicts between guidance.

An important element in encouraging compliance with ergonomics guidelines is to present the guidance in a form that is suitable for the intended audience. There appears to be a relatively limited amount of guidance that is usable by non-ergonomists and does not require prior knowledge of ergonomics issues. The following should be considered when designing guidance documents:

- Define the intended audience of the guidance
- Define how the guidelines were determined
- Detail practical applications of the guidance (e.g. use of anthropometric tables)
- Remove jargon and irrelevant technical detail
- Limit the degree of cross referencing required
- Provide definitions, illustrations and common language to support understanding
- Use meaningful headings that allow the novice to locate the specific guidance that is sought

Employment of Ergonomists

Ergonomics guidance material should not be seen as a potential replacement for employing ergonomists who are experts in the field of concern. An ergonomist can be used to identify and present suitable guidance, aid the understanding of guidance and assess ergonomics compliance. Often ergonomics issues may arise that could not have been easily predicted and in these circumstances the presence of on-site ergonomists is valuable. Also an ergonomist with expertise in the chosen field may have more up-to-date knowledge of ergonomics developments than the information contained in routinely used guidance documents.

Contractual Obligations

Defence and other industrial sectors often have contractual obligations to meet particular ergonomics guidelines or regulations. However, some of the guidance can prove to be overly constricting and considerable time and resource can be spent pursuing deviations from inappropriate guidance. The adherence to guidance can prove problematic if the author of the guidance could not have predicted the context of use. Therefore, it is important to involve ergonomists in defining guidance material that should be followed and providing advice on the wording of any compliance statements.

Effective Application of Ergonomics Guidance

The following are suggested means of improving the use of ergonomics guidance:

Understand context of problem
This should ensure that the appropriate guidance is used and adapted where necessary to meet the particular demands of the application. This may require task analysis but will certainly require significant input from experienced personnel in the field.

Dedicate time to understanding which guidance is relevant
This may require consultation with an ergonomist to be aware of the documents that could be useful. There may not be application specific guidance available and therefore significant time may be required to identify the relevant sections of guidance and understand which guidance is the most accurate and appropriate.

Consolidate guidance
Useful guidance may come from a range of sources and it would therefore be beneficial to consolidate the guidance into a singular format to improve accessibility and ease of use.

Consider ergonomics training to understand guidance
Some guidance may be necessarily technical and therefore some training may be required to the non-ergonomist or inexperienced ergonomist. This should prevent ergonomic guidance being ignored because it is simply not understood.

Encourage records of compliance
The application of ergonomics guidance can be improved by encouraging the completion of ergonomics compliance records. The requirement to provide a response to individual guidelines should increase the likelihood of the guidance being read and understood and can promote communication between ergonomists and non-ergonomists.

Employ ergonomist knowledgeable in the field
Ergonomists should be employed to define the appropriate guidance, tailor generic guidance to specific applications, produce application specific guidance and assess for compliance. They can also be used to deal with ergonomics problems that may not be dealt with by the guidance documentation that is provided.

Conclusions

This paper has highlighted some of the various approaches to improving the use of ergonomics guidance. The relevant approach to adopt is dependent upon the time and resources available to the company of concern. Unfortunately, time may not be afforded to in-house ergonomists to develop more specific and consolidated guidance. However, the compilers of guidance material could make some significant improvements by giving greater consideration to the users of the guidance and detailing how the guidance can be practically applied.

References

Engineering Equipment and Materials User Association (1999) Alarm Systems – A Guide to Design, Management and Procurement.

Heath and Safety Executive (1992) Health and Safety (Display Screen Equipment) Regulations.

Ministry of Defence, (2000) UK Defence Standard 00-25, Human Factors for Designers of Equipment DSTAN Issue 7 CD-ROM.

US Nuclear Regulatory Commission (2002) NUREG-0700 Human-System Interface Design Review Guidelines. Revision 2. U.S. Nuclear Regulatory Commission. Washington DC. USA.

Ergonomic Ergonomics?

Dianne Williams

Health Safety & Engineering Consultants Ltd.
70 Tamworth Road
Ashby – de – la – Zouch
Leicestershire
LE65 2PR
01530 412777
Dianne.Williams@hsec.co.uk

Working as an Ergonomist in industry can involve using a variety of assessment tools. Two case studies are explored to illustrate problems encountered when using tools for two very different purposes: expert witness work and assessing fairground rides. In the first example, RULA (Rapid Upper Limb Assessment), the Manual Handling Operations Regulations, and the Manual Handling Assessment Charts (MAC) are all discussed. The second case study looks at tools for assessing restraints in fairground rides. This example looks at the Health & Safety Executive's recommended method and available anthropometric data. The paper concludes that whilst there are many good assessment tools, these are often limited in their use in situations where there are very strict time and resource restraints.

Introduction

There are many assessment tools available to the Ergonomist, such as Heuristic Evaluations, Cognitive Walkthroughs, Task Analysis and Risk Assessments. The effectiveness of applying contemporary ergonomic tools in an industry environment shall be explored in two case studies.

Case Study One: The Expert Witness

Many Ergonomists carry out litigation work, with a common area for claims being for musculoskeletal injuries. This example follows the assessment technique carried out on a case involving an employee of a pottery maker who allegedly sustained a shoulder injury. The job involves lifting a wet plate with its mould from a turntable, and placing it to the left side on one of eight shelves of a dryer. When the plate is dry, the mould is placed back onto the turntable, and the dry plate is placed on one of 6 shelves of a trolley.

As the injury is an upper limb disorder, one possible place to start would be by using RULA (Rapid Upper Limb Assessment). Assessing the task in this way results in an action level of 2 for putting plates on the top shelf of the dryer, and an action level of 3 for putting the dry plate on the bottom shelf of the trolley. According to RULA, action level 3 means that changes are required soon. However, the assessment gives no indication as to whether this type of injury would be foreseeable. It also considers that if a posture is repeated more than 4 times in a minute, a score of 1 should be added to the score. There is no different scoring for varying levels of repetition, nor is there any weighting for repetitive movements by different body parts. This is a particular problem when assessing a task with a particular injury in mind. There is also no indication of any weighting system to show the difference in risk between a wrist pronating a hundred times in a minute, and somebody carrying out a task that involves inclining the trunk forward by 10°, five times in a minute.

To take into account a variety of postures throughout a task, it would be necessary to sample and analyse the postures at appropriate intervals; a luxury a solicitor will usually not allow you. There is an additional problem in that you have to watch the task carried out in its real environment, so when things are machine based production managers will not allow you to slow down or stop the line, and it is very difficult to accurately assess quick changes in angles of deviation. You can, where allowed, take a video of the actions involved – however, where there are large machines and cramped environments it is often impossible to get a correct angle to be able to analyse wrist and finger postures from a video. This highlights the need for a quick form of analysis that can be carried out in situ without impeding on the task or its environment.

The Guidance on Regulations for the Manual Handling Operations Regulations (1992) are commonly used in cases like this, as the allegations state the regulations have been breached. A problem with this is that it is simply a guideline to *recommended* lifting limits, it is not a definitive "if you lift anything over this weight you will get an injury". If a weight does exceed the recommended limit, a further assessment needs to be carried out. The regulations also state that they seek to prevent injury to any part of the body, not just the back. But in terms of *predicting* injury after it has occurred, it is very difficult to apply the guidelines to other parts of the body. This is particularly true in the case of cumulative disorders rather than traumatic injuries, as there is no indication of the effect of continuously lifting weights within the guidelines over a long period of time.

A more detailed manual handling assessment can be carried out, using MAC, a relatively new tool designed by the Health & Safety Executive (HSE). However, this is designed to identify levels of risk associated with different aspects of a task with a view to helping people reduce the risks involved in manual handling operations. It also states that it is for musculoskeletal disorders rather than upper limb disorders (as a shoulder injury can be both, this presumably indicates that it is aimed at only assessing back injuries), so it is not really very relevant in this situation, even though the Claimant's case is that there is a breach of the Manual Handling Operations Regulations.

HS(G)60 looks specifically at Upper Limb Disorders and contains a risk assessment form to fill in regarding the job involved. It provides a comprehensive view of the risk factors, and also gives some good detail of some of the injuries that are 'upper limb disorders'. However, there is still no definition of 'force' or at what rate of repetition a job becomes likely to pose a 'foreseeable risk of injury'.

A possible way to determine stress on the body during dynamic work tasks is through ECG recordings. However, this is expensive, and time consuming. This is the basic limiting factor for an ergonomist carrying out litigation work: time. With unlimited time you may be able to use extensive assessment techniques, but solicitors will often be reluctant to pay the fees for five hours of analysis on top of the time to write the report and carry out an assessment.

Several sources give definitions of force and repetition, but all vary and few have been validated. The issue is complicated further by some medical experts disputing whether an Upper Limb Disorder can be 'Work Related' at all. As there is often competition between experts, and due to the confidentiality within litigation, there is little opportunity to collect the data necessary for a comprehensive assessment tool for specific foreseeable injuries.

Experts being unable to agree on basic risk factors for injuries will not help the reputation of Ergonomics. To move forward in this area will improve the standard of risk assessments and the prevention of injuries. It will benefit the public, people making legitimate claims, companies with high insurance premiums and 'Claim Cultures', and Ergonomics as a profession.

Case Study Two: The Fairground Assessor

Theme parks in the UK must adhere to strict procedures and rigorous testing to ensure high safety standards. There are two main issues when looking at the ergonomics of passenger safety: containment and restraint. Containment involves keeping the passengers in the ride, and is obviously an absolute minimum for rides such as fast roller coasters. Restraint is keeping the passenger secure so they cannot hurt themselves on things either inside or outside of the ride. This constantly grows more difficult, as rides are increasing in their speed, force, and rapid changes of movement. Rides cannot simply be fast anymore: the theme park visitor demands to be hanging, standing upside down, on the longest, fastest, tallest, scariest ride with the most 'air time' ever. At the same time, they want to feel safe in the knowledge that they won't get hurt on the ride. This paradox in requirements means that restraints must be absolutely secure, with no margin for error, whilst imposing as little as possible on the rider.

A slight discrepancy in anthropometric data may not have a huge impact on the design of workstations or chairs, but could have fatal consequences if solely relied upon for fairground assessments. To err on the side of caution would mean searching through all sources to find the largest and smallest possible measurements for each dimension, which would take a lot more time than is available with the short timescales and quick turnaround needed by the leisure industry on these sort of rides.

The HSE provide information on containment and restraint for fairgrounds in a video entitled "Thrills, Not Spills" (HSE, 1996). A form is also included that shows which measurements should be taken of the ride, and which anthropometric dimension the measurement should be assessed against. If you are looking to set a minimum height restriction on a ride, and want to use the HSE's method, they recommend that dimensions that are important for restraint purposes should be identified, such as popliteal height, which is important for passengers to be able to brace themselves. The HSE say that you should measure the relevant dimension (usually the seat pan height), then find the 5th percentile who have the nearest popliteal height, and use the 5th percentile height of this group to set the limit. Using the example of a seat pan height of 348mm gives a height restriction of 1410mm. This all appears logical, except that tall nine year olds can now ride this ride, who have a potential popliteal height of only 300mm. This problem is exacerbated by the fact that many people (including some ergonomists) use the anthropometric tables incorrectly. It is often assumed that a person of 5th percentile stature will also be in the 5th percentile for all other dimensions. However, as all good ergonomists realise, each dimension should be taken individually: for example, there may be very tall but very thin people who cross over several percentiles for different dimensions.

There is also a problem when people incorrectly choose between different age groups of adults. For example, if you were looking at the adult riders you may choose to use the group aged 19-65. However, looking at stature, you are then using a 95th percentile measurement that is smaller than it would be if the age group 19-25 were to be used. So, using the table of 19-25years old should be more reliable in predicting the tallest people's safety. Then you look at measurements such as chest size or hip width, the data for the tallest men is now irrelevant for the largest chest or hip sizes that may be riding. For these measurements it is not only necessary to use females, but also to use the 45-65 age group, as these are larger still.

A further problem arises from the difference in available anthropometric data. Looking at different sources, there are usually discrepancies between all dimensions to some degree. A lot of the problems come from the fact that the measurements are taken of the available population, with adult male data often taken from military groups, so the figures will be biased towards fit, healthy, tall men. In their video "Thrills, not Spills", the HSE recommend using anthropometrics from a British standard, but this has since been withdrawn. This method uses data for the British population, but there should also be consideration that people from other countries are very likely to visit theme parks, who may be different in size to British people. With the 'average' size of a population constantly changing it is important that the data continually adapts.

The anthropometric data available gives a limited amount of dimensions. An important part of successful containment is not only that the rider will not be ejected from their seat during the ride, but also that they cannot deliberately remove themselves from any part of the restraint system. However, this requires detailed data on tissue compression in conjunction with complex joint movements.

An option during containment assessments is to use one of many software applications that use simulated models of humans to aid design. Many of these will let you specify what size adult or child you want. If a 95th percentile 12 year old is the youngest to meet the minimum height restriction for a ride, then it would be reasonable to use this size in the assessment. However, this then means that the software will use all 95th percentile dimensions, whereas in reality it is not unrealistic that a very tall 12 year old could also have a very thin chest depth or hip breadth. This would mean that dimensions of restraints such as lap bars would not be a close enough fit these passengers. Most software also only has certain types of child data available, such as ages 5, 7, 9, and 11. This would mean a large gap in the dimensions needed and will not be suitable for such safety critical assessments as these. You can specify your own measurements on many of these programs, but there is no definitive reference source for anthropometric data. There are several sources available, but all vary in the values they give.

With ride manufacturers having to be more inventive by the day, the ride carriages are becoming increasingly varied. Seats are no longer benches, but instead are moulded to shape. Over the shoulder restraints are now curved to restrain at the abdomen rather than the shoulders. This results in the HSE's recommended ride assessment form not being easily applicable because there are so many more dimensions needed than previously. A further problem with this is that there is likely to be less consistency in measurements as there will be no definitive reference point.

The leisure industry in the UK is highly regulated, with rigorous testing procedures and safety assessments. In the UK the risk of being killed on a fairground is 20 times less likely than in a motor vehicle, 450 times less likely than on a motor cycle and 12 times less likely than walking somewhere (Holloway & Williams, 1990, updated Tilson & Butler, 2001). However, in the ever-developing world of theme park rides, it is essential that the tools we use also progress.

Conclusions

As Ergonomists, our aim is to fit the tools to the user, yet many of the tools we use ourselves do not fit our own requirements! If we find major limitations ergonomic tools ourselves, how can it be realistic to expect designers to use it? There is a need to examine ergonomics to assess the gaps in the tools we have available to us. There are some very effective tools when used in specific situations, yet the whole objective of ergonomics is that we take into account all users, and all their requirements. If progress is not made towards developing 'ultimate' tools, then ergonomics as a profession is unlikely to advance significantly.

References

Holloway, N.J., Williams, R. 1990. An Assessment of Risks at Fairground Rides. HSE report SSG/10/32/3/6-1. Updated Tilson, S.J. & Butler, K.M. 2001.

HSE, 1992. Manual Handling Operations Regulations 1992. Guidance on Regulations L23 (Second Edition) HSE Books 1998 ISBN 0 7176 2415 3

HSE, 1996. Thrills Not Spills: A Guide to the Ergonomics of Passenger Containment, VHS Video.

HSE, 2002. HS(G)60 (rev). Upper Limb Disorders in the Workplace, HSE Books.

A Heuristic Approach to Early Human Factors Evaluation

Stephen Wells[1] and Michael Goom[2]

[1]BAE SYSTEMS
Building 20X, PO Box 5, Filton, Bristol, BS34 7QW

[2]MBDA
FPC 520, PO Box 5, Filton, Bristol, BS34 7QW

Tools which produce results too late to affect anything are a waste of time and money. There is a need for tools which give early answers. These tools may be approximate and may not catch all the issues. However, if they can be used early enough, then they are of value.

This paper takes a lead from a practical approach proposed by Jakob Nielsen and proposes a way to derive heuristics to aid early evaluation.

A case study based on an Office Workspace is presented to illustrate the approach.

Introduction

Many of the tools we use in the world of Human Factors require extensive study and a lot of data. In some cases this is acceptable. If we are studying an existing team in a workspace, which will not change over the duration of the study, then the detail is valuable. On a development project, however, decisions are being made everyday which change the context of the study. The danger is that the study will report after all the decisions have been made.

A study which reports too late to affect anything is a waste of time and money. There is a need, therefore, for tools which give early answers. These tools may be approximate and may not catch all the issues. However, if they can be used early enough, then they are still of value.

The Heuristic Evaluation Process

Heuristic evaluation, as proposed by Jakob Nielsen, has been used very successfully on a number of projects by MBDA Missile Systems and its subcontractors. This type of evaluation of proposed HCI styles has proved to be of great benefit during the early stages of a project, allowing various stakeholders to identify and correct potential problems. It is important that the people undertaking the evaluation are drawn from different backgrounds (e.g. users, systems engineers, ergonomists) to ensure adequate coverage of potential problems. Typically 5/6 evaluators will be able to identify the majority of areas of difficulty very quickly.

The time taken to train people to use the technique is minimal, usually taking less than half an hour. Working either on their own or in pairs the evaluators usually need between 1-2 hours to complete the assessment depending on the complexity of the interface.

Separate the Process from the Heuristics

Jakob Nielsen's ten heuristics for HCI are well known (Nielsen 1994). Nielsen not only describes a set of heuristics for evaluation HCI, he also describes a process for using those heuristics. By separating the process from the heuristics, different sets of heuristics could be used to assess different aspects of a design. Indeed we could imagine a library of sets.

There is one form of heuristic that is not used by Nielsen. This additional heuristic will suggest that a further evaluation should take place using a different set of heuristics. For example, when examining general characteristics of a workspace, a software interface could be evaluated further using Nielsen's HCI heuristics.

Derive New Heuristics

The process we describe in this paper is illustrated in the figure 1.

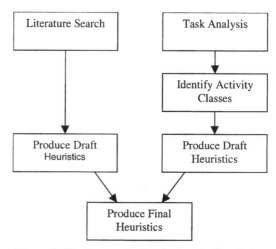

Figure 1: The Process for Producing the Heuristics

A literature search may well produce initial suggestions as to likely heuristics. In the example which follows, a search for information on office workspace design produced several examples of advice from a health and safety perspective. Not all issues concerned with workspace design are concerned with health and safety. In order to ensure that complete coverage of the issues is achieved, a further complementary approach is required.

This complementary approach is based on a Hierarchical Task Analysis. The low-level tasks are compared to identify common features. The term "Activity Class" refers to these common features. They cover basic actions such as "move", "manipulate", "talk" etc.

Each Activity Class becomes associated with a group of tasks. A draft heuristic can then be produced from each Activity Class, being scoped by the associated tasks.

Finally, draft heuristics from several sources are compared to produce a final list.

Case Study: The Office Workspace

Workspaces: Why do we get it Wrong?

The table and chair are so common that we rarely rethink them. There are examples of rethinking. In the days of typewriters, a secretary would often have a second desk at right angles to the main desk. This would be slightly lower and support the typewriter at the correct height for the secretary's hands. This type of consideration is often lacking today.

There is clearly a need to be able to evaluate workspace designs. There are many ways to do this, many of which take specialist equipment and take too much time. What is needed is a practical evaluation method. It need not be perfect. If it identifies the main problems quickly, it will be of value.

Heuristics for Workspace Design

A search for guidance on the Internet produced several sites where Health and Safety guidance is provided for setting up display screens. These do not cover the whole task set, but give a useful cross check on some aspects of it. Typical examples are (The URLs for these sites are given in the references.):

"Chairs should firmly support a comfortable posture, providing support to the lower back region and avoiding pressure on the back of the thighs" (University of California, Berkley)

"If you find that your shoulders ache after a long day at the computer, see how you are holding your shoulders while you are working." (Science Fiction Writers of America)

The advice from these sites is summarised as seven draft heuristics. Examples of the draft heuristics are in table 1.

Table 1: Example Draft Heuristics from Health and Safety Guidance

G4	Is adequate legroom provided?
G5	Is a document holder needed? If so, is one provided?
G6	Are lights sources outside the field of view?

Another useful source of guidance in the UK is the "The Health and Safety (Display Screen Equipment) Regulations 1992". While including some interesting points that do not appear in the advice found on the Internet, much is too general to be useful. For example:

"There shall be adequate space for operators or users to find a comfortable position."

It is difficult to see how such a statement can be tested.

The regulations can be summarised as a set of fifteen draft heuristics. Examples of these draft heuristics are given in table 2.

Table 2: Example Draft Heuristics from Display Screen Regulations

DS5	Is the screen adjustable for tilt and swivel?
DS7	Is space provided for documents?
DS12	Does the lighting allow documents to be read without affecting the ability to read the screen?
DS14	Are noise levels low enough to prevent distraction?

These guidelines concentrate on the Health and Safety aspects of the workspace. While this guidance and regulation is useful it does not provide a complete basis for assessment.

There is a need for an alternative approach which can fill the gaps. A simple Task Analysis can provide this. An informal observation of people in an office lead to a simple list of the tasks. Inspection of the tasks showed that they fall into a number of related types of activity. These "activity classes" cover general activities such as "read", "write", "move" etc. These classes will form the basis of heuristics. Table 3 gives an extract from task analysis showing the activity classes.

Table 3: Extract from the Task Analysis with Activity Classes

Task	SubTask	Sub-Sub Task	Activity Class
Interact with PC	Look at Screen		Read
	Use a Keyboard		Reach, Manipulate
	Use a Mouse		Reach, Manipulate
	Load Software		Retrieve
	Use Software		Read, Manipulate
Interact with Paper based medium	Find Paper/Book		Retrieve, Reach
	Read Paper/Book		Read
	Find Pen/Pencil		Retrieve, Reach
	Write		Write, Manipulate
	Store Paper/Book		Store
Interact with People	Use Telephone	Dial Telephone	Manipulate
		Converse	Listen, Talk
		Make Notes	Read, Write, Manipulate
	Converse		Listen, Talk

The next step is to "invert" this list in order to group the tasks that involve particular activities. So, for example, the activity class "Read" is related to five tasks:

Table 4: Activity Class "Read"

Activity Class	Task
Read	Look at Screen
	Use Software
	Read Paper/Book
	Look at Pin Board
	Make Notes

From this inspection, two draft heuristics could be created (TA8 and TA10 in Table 5). By applying this technique to each of the activity classes in turn, fourteen draft heuristics were generated.

Table 5: Example Draft Heuristics from Task Analysis

TA4	Items on the desktop must be within reach
TA8	Space must be available on the desktop
TA10	Lighting levels should permit reading
TA11	Lighting levels should permit writing
TA12	A surface should be available on which to write

*If required, perform a further evaluation using a different set of heuristics

Each of the three approaches has led to a different list of potential heuristics. The lists must now be consolidated. There is some overlap between the lists where display screens and chairs are concerned. In other areas, the Task Analysis is the only source.

By inspection, the draft heuristics fell into eleven general areas. For example, four of the draft heuristics (G6, DS12, TA10, TA11) were concerned with Lighting. Each of these general areas can be used to form a final heuristics scoped by the draft heuristics. To these are added the note identified in Table 5. The final list is given in table 6.

Table 6: Example of a Consolidated List of Heuristics

1. The seating position should allow the operator to work comfortably The operator should be able to sit comfortably with shoulders relaxed, elbows bent, and forearms, wrists, and hands approximately parallel to the floor. The top of the display should be at about eye height and be adjustable. The keyboard should be separate and have enough space to allow support for hands and arms
2. The chair should be adjustable The chair adjustable so that the lower back is supported, thighs are horizontal and feet are flat on the floor or a footrest. Chair controls should be within reach and easy to use.
3. The operator should be able to move while seated at the desk Legroom and chair movement should allow the operator to be able to move to prevent cramps etc.
4. The operator should be able to move quickly and easily between different locations in the workplace Operators do not remain at one location. A computer operator will move between a desk, a photocopier, a printer etc.

Note: If required, perform a further evaluation using a different set of heuristics - Human Computer Interfaces should be examined using Neilsen's Heuristics

Conclusion

The heuristic approach to reviewing designs is an easy to implement method which produces results quickly. Once it is recognised that the process can be separated from the heuristics, the way is open for the development of different sets of heuristics to examine different aspects of design.

The paper has shown how a literature search supplemented by a simple task analysis could be used to develop further sets of heuristics for other workspaces. For example, a set might be developed for the operation of industrial machinery.

Some questions are still outstanding:

1. "Activity Classes" to focus the tasks. How big would a useful reference list of activity classes be? Probably less than a dozen, but this has yet to be demonstrated.

2. How could heuristics be derived which took into account disabilities? How would the heuristics for an office workplace need to be changed if they are to be used by someone in a wheelchair, or by someone who suffers from chronic back pain?

References

Nielsen J, 1994. *Heuristic evaluation*. In Nielsen, J., and Mack, R.L. (Eds.), *Usability Inspection Methods*, John Wiley & Sons, New York, NY.
The Health and Safety (Display Screen Equipment) Regulations 1992.

Web Sites Used in the Preparation of this Paper

Jakob Nielsen's web site: http://www.useit.com
University of California, Berkley, University Health Services.
     http://www.uhs.berkeley.edu/Facstaff/Ergonomics/ergdesign.htm
Science Fiction and Fantasy Writers of America. http://www.sfwa.org/ergonomics/work.htm

[For full details of the task analysis and workstation heuristics developed please contact the authors]

GENERAL ERGONOMICS

PIN PADS
FIRST STEP TO A CASHLESS SOCIETY?

Tanita M. Kersloot and Paul Snee

*RNIB Sensory Design Services
Bakewell Road,
Orton Southgate
Peterborough, PE2 6XU*

The rapid increase in card fraud losses over the past years required a response, as cards would not survive if losses were to increase even further. The introduction of new services as well as the development of new devices (PIN pads) is helping the fight against fraud. As a result of this, many new PIN pads are emerging from various manufacturers. Although the development of these systems is exciting and should have many benefits, it also raises a number of concerns, in particular in relation to the accessibility to disabled people. Guidelines and standards for PIN pads are limited and PIN pads emerge on the market without having been evaluated by disabled people (the end user). This results in complaints and expensive after market 'fixes', which could have been prevented through the adoption of an inclusively design methodology in the first place.

Introduction

Card fraud losses are growing and were about £411 million in 2001 on UK issued cards (www.chipandpin.co.uk). To try and prevent the card fraud from growing even further, Chip and PIN will be introduced. The Chip and PIN project has been described as the largest UK project within retail since decimalisation with a cost of over 1.7 billion. In future, credit and debit card transactions will be authorised by the customer keying in a Personal Identification Number (PIN) rather than by signing a paper receipt.

The system of paying benefits by order book at the Post Office is also open to fraud and "fraudsters swindle £80m in lost and stolen order books – one hundred from pensioners each week." (Malcolm Wick, Minister in the Department for Work and Pensions, giving evidence to the Trade and Industry Committee on 4 June 2003). Payment of benefits and pensions is also in the process of changing and has started to be paid directly into bank accounts. Withdrawing this money will be possible through the use of a PIN pad.

PIN pads (terminals that accept the PIN card) will be of various shapes and sizes. Some PIN pads will be portable and cordless, others will have a PIN pad built into a terminal, and some will have a PIN pad connected to the existing terminal by a cable. Figure 1 shows an example of what a PIN pad may look like.

Within the UK, the market segment for older and disabled people is growing. Projections suggest that the number of people aged 65 and over will exceed the number of people aged under 16 by 2014. By 2025 there will be more than 1.6 million more people over the age of 65 than people under 16 (www.statistics.gov.uk). Results from the 1996/7 Survey of Disability showed that an estimated 20 per cent of the adult population had a disability. In general the prevalence of disability increases with age (Grundy *et*

Figure 1: Example of a PIN pad

al, 1999). For example 78 per cent of registered blind people are over the age of 65 (source: SSDA 902). The increase in the number of older and disabled people should have implications, placing greater demands on products and services. Especially as most disabled people do go out, as indicated by the 1996/7 Survey of Disability that showed that most had made outings of various kinds in the past four weeks (Grundy et al, 1999). It is essential that PIN pads, for use by the general public, are designed to be accessible by as many people as is reasonably possible including the aging and disabled population.

The attitude toward disabled people has changed over the years. From keeping disabled people isolated from society and treating them in separate institutes, attitudes have moved towards integration within society. As a result, the Disability Discrimination Act (DDA) was introduced in 1995 making it illegal to discriminate against disabled people in some areas.

Disability Discrimination Act 1995

The main message that the Disability Discrimination Act (DDA) transmits, is not to treat disabled people less favourable than others. The DDA covers a wide range from employment, goods facilities and services, premises, transport to education, and is implemented over a number of years. The DDA part III deals with the duties placed on those providing goods, facilities or services to the public. The Act makes it unlawful for service providers to discriminate against disabled people in certain circumstances.

The term 'service provider' is defined as those "concerned with the provision, in the United Kingdom, of services to the public or to a section of the public, with or without payment (s.19(2)), including goods and facilities". There are very few services that are not covered within the act. The only exception is clubs with a genuine selection procedure (such as private members clubs), gyms and fitness centres where you have to join will not be exempted (Casserley, 2003).

Disability is defined as "physical or mental impairment, which has a substantial and long-term adverse effect on the ability to carry out normal day to day activities" (Casserley, 2003).

The DDA part III takes a pro-active approach; it is not sufficient to wait till a disabled person requires a service, but service providers have to be ready for when they do. They have the duty to make reasonable adjustments to practices policies or procedures, physical features and auxiliary aids and services. For example, where an auxiliary aid or service (such as information in Braille, or a sign language interpreter) would facilitate the use of a service, the service provider must take reasonable steps to provide it. This part of the Act comes into force on the 1st October 2004.

A disabled person who believes that a service provider has unlawfully discriminated against him or her may bring civil proceedings. The Disability Rights Commission (DRC) supports disabled people in securing their rights under the (DDA). They help solve problems and achieving solutions, often without going to a court and they also support legal cases to set new precedents and test the limits of the law (www.drc.org.uk). If successful, a disabled person could be awarded compensation for any financial loss, including injury to feelings. The disabled person may also seek an injunction to prevent the service provider repeating any discriminatory act in the future.

The manufacture and design of products are not in themselves covered by Part III of the Act, because they are not involved in the provision of services directly to the public. Nothing in the Act requires manufacturers or designers to make changes to their products, packaging or instructions. However, it makes good business sense for manufacturers and designers to make their goods (and user information) more accessible to disabled customers. The service provider buying the product from manufacturers would want the product to be 'compliant' with the DDA in order to avoid committing an act of unlawful discrimination. At the moment service providers may think that they can take legal action against the manufacturer if they commit an act of discrimination. Manufacturers, however, are either unaware of the DDA act or think it does not apply to them. This issue will not be resolved till after the act is in force in October 2004.

User needs

Manufacturers of PIN pads see their customers as the shops and other service providers buying their equipment. They would design a PIN pad to please their customer and, for example, adopt the customer's corporate colour scheme, rather than using good contrasting colours which would make the PIN pad more user friendly. Within the shop, it may be the headquarters responsible for purchasing equipment, rather than the smaller local branches who would have direct contact with disabled customers. The distance between the PIN pad manufacturer and the disabled end user is large, and could therefore easily been disregarded by the manufacturer. The design could ultimately be the decision of the bank i.e. Barclays colour scheme not necessarily the service providers (e.g. the Post Office where the cash point is located).

The solutions to some of the problems experienced by people with disabilities seem trivial to people without a disability, but they can have a major effect on the usability or accessibility of PIN pads.

People experience problems in reading and understanding the instructions (if available), inserting the card, reading the screen, using the keypad, listening to audible output (if applicable), reading the printed output and retrieving the card. For example, blind people often press the incorrect numeric key due to poor tactile information on the 5 key (normally a raised dot) and the addition of extra keys, such as F1 to F4 which are there for future reference rather than being used at present.

For many disabled and elderly users, the most important aspect of usability of products and services is consistency in the design. Different layouts for PIN pads require different skills for different machines and people may not be able to apply their learning from one PIN pad to another. A prime example of inconsistency is the lack of standards and guidelines and the discrepancies between the guidance that does exist.

Standards and Guidelines

All guidelines for PIN pads (and ATMs) related to the layout of the numerals are consistent. They should all be according to the telephone layout with the numbers one to nine laid out in three rows of three with one to three at the top. However, the consistency in respect to the positioning of the function keys is completely different. The British Standard on key pads (BS EN 1332-3:1999) shows three different layouts in respect to the positioning of the function keys; one vertically arranged to the right of the numeric keys and two horizontally arranged underneath the numeric keys (see Figure 2).

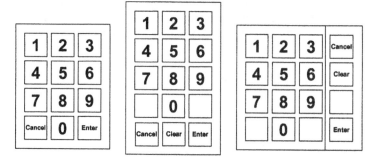

Figure 2: Keypad layouts as recommended by BS EN 1332-3:1999

The ATM guidelines (Centre for Accessible Environments, 1999) mention that function keys are best located to the side of the keypad rather than below and does not recommend either the right or the left side. Dr John Gill (1998), chief scientist at Royal National Institute of the Blind, recommends the right side of the numeric keys. It is understandable that if guidelines and standards are vague or inconsistent, so will the PIN pad itself. This example may seem trivial to people without a disability, but imagine being blind and wanting to press the cancel button; would it be underneath the 7, diagonal to the left of the 0 or in the top right hand corner next to the 3?

Information for designers may be limited or very general, as PIN pads are a fairly new development and with all systems in the early stages of specification, it is difficult to be precise about all requirements. It is time consuming to produce guidelines and standards and manufacturers often do not see direct commercial benefits. The result is that products get rushed through the development stages and the evaluation stage with disabled people is overlooked. The DDA should promote the development of standards and guidelines in respect to designing for disabled people, but it is difficult to interpret the DDA into specifications for PIN pads (and many other customer facing pieces of equipment such as information kiosks). A prime example is the Post Office (not part of the Chip and Pin project) who were the first service provider who purchased PIN pads for all their nation-wide branches. They ignored advice they were given about the inaccessibility of the PIN pad and installed PIN pads on all their counters. Once in use they received complaints from disabled people who were not able to use it independently.

Conclusion

PIN pads may be a first step to a cashless society, but if they are not designed with all end users needs in mind, they will not be accessible to all (or most) people. Therefore, a proportion of the population, in particular the 20 per cent with disabilities will not be able to use the PIN pads and the equipment will not comply with the DDA act. The DDA Code of Practice currently lists a number of steps which are likely, if taken, to result in compliance with the Act. These steps are:

- Training
- Disability audits
- Making adjustments and ensuring that staff and disabled customers are aware of them
- Monitoring and reviewing effectiveness
- Consultation with disabled people and organisations

If these steps and good design practice are adopted by manufacturers and the evaluation stage with disabled people is included, then it is hoped that the benefits of Chip and Pin can be realised, without excluding people with disabilities.

References

Casserley, C. 2003, *Removing Barriers to Access: The Legal Context – DDA Part III* (unpublished).

BS EN 1332-3:1999, *Identification card systems – Man-machine interface – Part 3: Key pads,* (British Standards)

Centre for Accessible Environments, 1999, *Access to ATMs – UK design guidelines*

Gill, J. 1998, *Access Prohibited? Information for Designers of Public Access Terminals,* (Royal National Institute for the Blind on behalf of Include)

Grundy, E., Ahlburg, D., Ali, M., Breeze E. and Sloggett, A. 1999. *Disability in Great Britain,* (Department of Social Security, Leeds)

www.chipandpin.co.uk

www.drc.org.uk

www.statistics.gov.uk

Designers In Schools – Workshop

Meg Galley[1], F. Erg. S
Mike Tainsh[2], F Erg S
Andree Woodcock[3], M Erg S

Ergonomics and Safety Research Institute, Loughborough University and President of the Ergonomics Society[1]; QinetiQ, Ltd, Farnborough, Hants[2]; University of Coventry, Coventry, West Midlands[3].

During 2003, the Ergonomics Society contributed to the Designers in Schools Programme which was a government initiative managed by the Design Council. The workshop will both describe and discuss the Society's contribution. It will be Chaired by Dr M A Tainsh, supported by Ms M Galley (President) and Dr A Woodcock. All three were major players in the contribution of the Ergonomics Society to the Designers in Schools programme. This work included going into schools and teaching.

1. Introduction

The work reported here is part of the activities of the 'A' level Working Group, which was initiated in late 2000. It is supported by Council and Chaired by Mike Tainsh. There was an intention to influence syllabuses within appropriate institutions, influence the content of exams and encourage high quality applications to undergraduate courses in tertiary educational establishments.

Two activities have preoccupied the members of the Working Group over the past three years: the first is the development of Clive Andrews seminal work for secondary students and teachers (Andrews and Kornas, 1982) into material for a website (www.erdonomics4schools.com), and the second is the establishment of an advisory service, accessed from the website, for secondary students where questions can be posted and brief answers provided by experienced members of the Society.

As a result, when an enquiry arrived at Meg Galley's door for the Designers in Schools Initiative which was managed by the Design Council on behalf of the UK government, the Ergonomics Society was in a position to respond. It is important to remember that none of the contributors had any appropriate training or experience. The workshop aims to share this experience and arrive at some conclusions which can be debated, with a view to encourage the teaching of ergonomics in secondary schools and particularly at 'A' Level.

2. Risks

It is essential to understand that at the start of work on this initiative there were considered to be some clear risks:

- Ergonomics is often viewed as a subject more suitable to graduate and postgraduate teaching and research, than to secondary school activities. There is a risk that the students would be confused and hence reject the subject
- There was a desire to involve the students in the subject. This could include laboratory, practical or measurement work. It was considered that this would help encourage the point of view that ergonomics is a science and there is a problem of the 'image' of science amongst younger people.
- There was a lack of precedents, so it was unclear how to approach any presentation or workshop. There is a risk here of failing to gain the students' attention and interest.
- The volunteers' clear lack of classroom experience might result in poor social conditions or interactions with the class and this might hamper the teaching or workshop.

3. The Three Reports

3.1 Andree Woodcock: Bluecoat School Nottingham

The Design Council linked Woodcock with a secondary school in Nottingham, where with the assistance of the Head of Design and Technology she produced four lesson ideas. The visits were scheduled to occur within the Designers and Schools week. The class teacher assisted in each lesson.

The main aim was to introduce children to the scope of ergonomics, by looking at its origins and the application of ergonomics to real world, design activities. At the end of the week the Subject Head was provided with all teaching material for future use by his department. Activities were designed to involve periods of reflection, individual and group activity, and to practice modelling, craft and design skills. Some of the work undertaken will be shown in the workshop. The teaching approach chosen was one of activity-based learning. The children listened to a short PowerPoint presentation on ergonomics as an introduction, leading into a more focussed, project based session. Four subject areas were selected – designing an ergonomically sound chair, a child friendly car, book bag and toothbrush to appeal to teenagers. More details of the lessons are outlined below:

Ergonomically sound chair. This was an all day lesson, for handpicked 14 year olds. The children worked in groups to design, construct and make a model chair based on anthropometric measurements, and also assess a range of chairs around the school for comfort using a short questionnaire. Child friendly car. This was a double period Design and Technology lesson for 14 year olds. Design of school bags. This was chosen because it is an area of concern for teachers and ergonomists. Pupils were firstly asked to assess their own bags (the term ruck sack was dropped as being unsuitable) using a prepared questionnaire as input into group work. Secondly, in groups they were asked to write or draw a storyboard of 'the day in the life of' a school bag – in particular considering how the bag is used, how it was stored and carried at various points of the day. Thirdly, they were required to produce a group poster of their agreed ideal bag, outlining some of the issues they had considered. Design of a toothbrush for teenagers. This was a double period lesson with 12 year olds. A range of toothbrushes was brought in as stimulus material. The aim of the lesson was to consider anthropometry, design and usage – so pupils were asked to consider hand and mouth sizes, the tooth brushing task, and features that would appeal to teenagers. Each pupil was required to produce six rough sketches and a final annotated best sketch.

3.2 Mike Tainsh: Woking High School, Department of Design and Technology

I offered my details to the Design Council who in turn put me in contact with the Head of Department (Michelle Abbs-Rowe) who expressed the wish to have a lesson that could be given to the whole of her year 10/11, and this was agreed.

The choice of subject stemmed from the belief that it was essential to teach the students about ergonomics and that good design involved the principle of matching human characteristics to task characteristics. Attention was focused on sitting in the classroom as this offered the opportunity for the students to measure their own characteristics and the characteristics of the objects in the physical environment immediately around them. The approach to learning and maintaining maximum interest and attention was developed around a concept of a high degree of involvement in a laboratory situation. The main bulk of the teaching material was lifted directly from the "ergonomics4schools" website. The Learning Zone subjects of seating and anthropometry were used.

The lesson was planned to take one hour. There were three lessons with approximately sixty students in total. The introduction to the lesson would be carried out with a PowerPoint presentation and the information presented simply in procedural terms so that the students could understand how to execute the measurement work. The measurement work would be carried out with tape measures and the final analysis was to be "on the board" with calculations being done "on the hoof" by myself and recorded so that all could see, with participation from the students.

The lesson divided into three equal parts: introduction, measurement in groups, reporting and analysis of grouped measurements. All year 10/11 participated, permanent staff were present to ensure good discipline (which was always perfect). I was introduced at the start and then launched into the introduction. The measurement stage proceeded when each student took seven basic measures of themselves (see website), while working in groups, and took appropriate measures of a chair and stool. Finally each group reported on the measures which were plotted on a distribution by myself on a white board. The distributions were then examined to obtain measures of central tendency which were compared to the characteristics of a laboratory chair and stool. As a class we then decided which items of furniture supported the task of sitting better.

3.3 Meg Galley: Burleigh Community College, Loughborough

For various reasons I was not able to join in the activities in the week in June that was set aside by the Design Council. Personnel changes in the school I was allocated, and personal commitments, meant that the session was postponed until the Autumn Term. Subsequent discussions with the teachers indicate that they see this as much better timing than the summer term.

My plan was therefore fitted in around what the students were already doing in class. I was to join two classes from three separate years (11, 12 and 13) and work with them on their projects which were:

- Year 11 – a mechanical toy
- Year 12 – a computer mouse for children aged 7 to 9
- Year 13 – work on individual projects on a tutorial basis.

As background to ergonomics4schools I prepared a short overhead presentation that stressed what ergonomics was, and how they website could support them. I talked about the importance of not designing for the 'average' and showed them how to select appropriate dimensions from data tables. After this short presentation, during which I asked some questions, I moved about the class, which varied in size from about 20 in the younger groups to 4 in the older ones. All lessons were of 45 minutes duration, and the first 10 was taken up with the short presentation and the last 5 with packing up and clearing away, leaving only 30 minutes for the main teaching period.

For the younger groups the pace of work was quite slow with a good deal of 'messing about while the teacher wasn't looking'. The groups working on toys were still in the process of deciding on the final design for their mechanical toy and most of their attention was on materials and how the mechanical linkages would work. We all struggled to see how much ergonomics input there could be, but we did talk about safety, finger trappings and little children putting small toys in their mouths. For the group designing the computer mouse for young children it was easier for them to understand how ergonomics would make a difference and we actually used a photocopier to shrink an adult hand down to the size of a small 7 year old girl. This 'hand' was obviously far too small to use a standard mouse. To me this illustrates the need to demonstrate what might be 'blindingly obvious' to us by the use of simple visual aids. For the year 13s, who were working on very advanced work, I was able to offer detailed advice. When I knew the topic of their project I took them specific information including some evaluation work I had done on ironing boards (for a new ironing board), a copy of a paper on the ergonomics of sanitary ware (for a portable bidet) and the website of a report we had done on cycle carriers for the DTI (for a car mounted cycle carrier). I found this aspect of the work particularly rewarding and think the students also did.

4. Conclusion

It is difficult to draw any major conclusions from the work yet but a few appear clear:

- It is possible to handle ergonomics in classroom situations starting at years 10 and eleven, if the conditions are right and this includes a supportive and resourceful teaching staff.
- The students appear to enjoy the subject.
- There is a shortage of established teaching material – the material.
- It is most beneficial if the ergonomics input relates specifically to something they are already doing as part of their course work
- Lessons need to be kept short to maintain attention and interest.
- Simple, self-evident visual aids need to be available to illustrate points such as the difference between a large man's hand and a small girl's hand.
- For older pupils working on specific projects it is useful to have some advance warnings of their subjects so the presenter can be better prepared.

It is planned to make some of the teaching material available on the website. The discussion and main business of the workshop will follow the presentation of these reports.

5. Reference

Andrews, C J A and Kornas, B 1982, *Ergonomics Fundamentals for Senior Pupils.* Napier College, ISNB 0 902703 14 5.

Acknowledgements

Grateful thanks are expressed to the staff and students of Bluecoats School in Nottingham, the staff and students of the Design and Technology Department at Woking High School, and the staff and students of the Design and Technology Department at Burleigh Community College.

Disclaimer

The opinions expressed here should not be taken to represent those of the authors' affiliations, the Ergonomics Society, the Design Council or its sponsors, or those of the schools.

ACCIDENTS RELATED TO ACCESS PATH USE OF MOBILE MACHINERY IN FINNISH AGRICULTURE

Tiina Mattila[1], Marja Lehto[1], Timo Leskinen[2], Jouni Lehtelä[2], Juha Suutarinen[1], Janne Väänänen[3], Pekka Plaketti[2], Pekka Olkinuora[1]

[1] *MTT Agrifood Research Finland, Agricultural Engineering Research (Vakola), Va-
kolantie 55, FIN-03400 Vihti, Finland*
[2] *Finnish Institute of Occupational Health, Department of occupational safety, Topeliuksenkatu 41 a A, FIN-00250 Helsinki, Finland*
[3] *The Local Government Pensions Institution, Albertinkatu 34, FIN-00101 Helsinki*

The access system means "the system provided on a machine for entrance to
and exit from an operator, inspection or maintenance platform from and to
the ground" (SFS-EN ISO 2867:1998). According to earlier studies the frequency of access path accidents is high in agricultural machinery work. In
this research a case study was done followed by a statistical study. According to the results, an access path accident typically occurs in crop production
and the average duration of absence due to access path accidents was 36.4
days. In a family farm it probably means a lot of difficulties in everyday life
and work, for example delays in sowing and harvesting. Better usability of
access paths would probably prevent accidents in agricultural machinery
work.

Introduction

Getting out of or into the cabin and coupling or uncoupling implements seem to be the
most accidental work phases when working with farm tractors. Statistical analysis of the
accidents in Finnish agriculture 1983–1993 pointed out that about 35% of the tractor accidents occured when mounting or dismounting the cabin (Suutarinen, 1997). Leskinen et
al. (2002) found out that in 1997 about 30% of farmers' injuries associated with the use
of tractors were access path accidents, despite the fact that only a small part of all working time with machines is spent on the access paths. The high relative risk of accidents on
access paths requires further study on causality and prevention.

Statistical analysis

Material and methods
The accident statistics database of the Farmers´ Social Insurance Institution of Finland
(Mela) was utilized. Mela attends the social security of all farmers, fishermen and rein-
deer breeders in Finland (altogether 98 400 farmers and family members in 2002) and the

insurances provided by Mela are compulsory by law. In 2002 Mela paid compensations for 6812 new accidents.

The aim of the statistical study was to analyse the most usual types of access path accidents, the phase of work, the circumstances and consequences and the time of the accident in order to understand better the phenomenon and point out the problem areas. The variables of the database were selected to detect problems related to access path usage. Accident data of 8 years, 1990–1997, were used. The only way to acquire valid data in terms of actual access path accidents was to read the descriptions of all the accidents in the database. About 40 000 descriptions were read, out of which 2212 accidents were discovered to be accidents in work during ingress or egress of a mobile machinery cabin. Female cases were so rare, 110 accidents (5%), that they were eliminated to reduce errors. The final data included 2102 cases.

Results

Most of the accidents occurred in crop production (55%). The most typical types of accidents were falling when stepping, slipping and tripping (50%) and falling from a height (29%). The most typical consequences were strains or sprains (71%), bruise and crush (15%) and bone fracture (11%). In 70% of access path accidents the injured part of the body was lower limbs and in 15% of cases it was body or back (Fig. 1). Uneven and slippery terrain was the most common cause (56%) and the second was machines (35%).

The number of days compensated was used as an indicator of the severity of the accident. The average duration of absence due to access path accidents was 36.4 days. The proportion of serious accidents (the period of disability is more than 30 days) of all access path accidents was 21%. More than a half of bone fractures were serious and most of sprains and strains caused disability of one to two weeks (Fig. 2).

A typical access path accident in Finnish agriculture occurs in crop production, on Monday afternoon in September when a male person falls, slips or trips and his leg is sprained or strained. The injury is caused by uneven or slippery terrain, and the period of disability is 7 to 14 days.

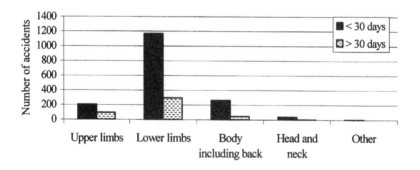

Figure 1. The number of access path accidents grouped by the injured body part and the severity of the accident (duration of disability less than or equal to 30 days and over 30 days)

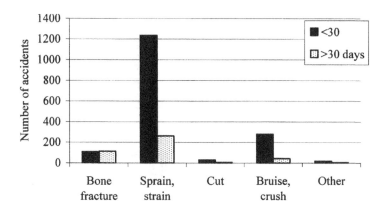

Figure 2. The number of accidents grouped by the type of injury and the severity of the accident (duration of disability less than or equal to 30 days and over 30 days)

A Case study

One access path accident was analysed using a systemic model made for the analysis of accidents (Suutarinen et al., 2002). The model has been considered a useful method for systematic analysis of work related accidents. (Tuominen & Saari, 1982). The aim of the case study was to identify the chain of critical events and circumstances that led to an accident. Especially, the focus was in the division to human related and engineering related accident factors, although only one case was analyzed.

Description of the accident
A farmer was ploughing a stubble field with a three-furrow reversible plough, which was coupled to the tractor with a traditional three-point linkage having a mechanical, manually adjustable and operated top link. The tractor was not normally used in ploughing and the plough was borrowed. The farmer had to adjust the top link several times in the beginning of the ploughing. The tractor was a little bit tilted because of an old furrow. The step of the access path of the tractor was dry and clean. The weather conditions were good and the farmer was working in the daylight. The farmer was wearing normal working clothes and running shoes. The soles of the shoes were smooth and probably slippery. The farmer slipped and fell down when he was entering the cabin for the fifth time during a short period. He was in a hurry and felt himself a little bit frustrated by technical breakdowns with the plough. Figure 3 presents the normal action (in the rectangle with rounded corners), the contributing factors of the accident (in skewed rectangles), the action of the injured person (in rectangles), the situation where the farmer and the causes of the accident got together (in the double lined rectangle) and injuring (in the oval).

The tractor was in good condition and the access path was constructed in accordance with the relevant access path standard (SFS-EN 1553:1998). In spite of this, the access path had many shortcomings. There were no slip–preventive structures on the step sur-

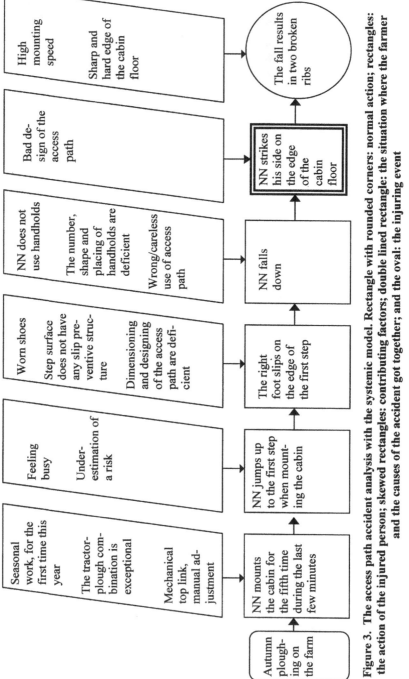

Figure 3. The access path accident analysis with the systemic model. Rectangle with rounded corners: normal action; rectangles: the action of the injured person; skewed rectangles: contributing factors; double lined rectangle: the situation where the farmer and the causes of the accident got together; and the oval: the injuring event

faces and the step structure was not closed from behind, which would prevent the foot from sliding through the gap between steps. The edge of the step was a little bit worn. The access path was not in the middle of the door opening and it was not straight. The steps were partly hard to see. The handholds were difficult to use, because of the deficient shape and placing, and there were too few of them.

In the analysed accident case the main elements leading to the injury were the repeated need of ingress and egress of the cabin because of an old technology (manually adjusted top link instead of a hydraulic one), working frustrated under time pressure, underestimation of the risk, insufficient friction leading to slipping, deficient handholds, and finally a sharp edge of the cabin floor at the door opening. In conclusion, it can be said that interaction of the user's behaviour and deficient design of the access path caused the accident.

Discussion

Access path accidents related to the use of mobile machinery in agriculture are assumed to be similar enough to the access path accidents of other mobile machinery so that these results can be generalized, with due caution, on the use of most mobile machinery. An access path should be designed so that it is obvious for the operator how the path is used safely. Three-point contact should be easy to maintain all the way when mounting or dismounting a cabin. Three-point contact means simultaneous use of two hands and a foot or of two feet and a hand when mounting or dismounting a machine.

The standard EN 1553 should be developed, e.g. by adding the requirement of a structure to prevent foot from sliding through between the steps and clearer requirements for slip prevention and handhold placement.

One way to prevent accidents is to minimize the need for exit of and entry into the cabin by using modern technology, for example implement couplers between the tractor and implement. The implement coupler enables the driver to couple and uncouple implements without leaving the cabin.

References

Leskinen, T., Suutarinen, J., Väänänen, J., Lehtelä, J., Haapala, H. and Plaketti, P. 2002, A pilot study on safety of movement practices on access paths of mobile machinery. *Safety Science* **40**, 675–687.

SFS-EN 1553:1998. Agricultural machinery. Agricultural self-propelled, mounted, semi-mounted and trailed machines. Common safety requirements. 53 p.

SFS-EN ISO 2867:1999. Earth-moving machinery. Access system. 27 p.

Suutarinen, J. 1997, Non-Fatal Tractor Accidents and Their Prevention. *Journal of Agromedicine* **4**, 313–324.

Suutarinen, J., Väänänen, J., Mattila, T., Leskinen, T., Lehtelä, J., Plaketti, P. & Olkinuora, P. 2002. Ajettavien työkoneiden kulkuteiden turvallisuus II (in Finnish with English summary). *Maa- ja elintarviketalous* **18**, (MTT) 69 p.

Tuominen, R. and Saari, J. 1982, A model for analysis of accidents and its application. *Journal of Occupational Accidents* **4**, 263–273.

LISTENING TO THE NAVIGATOR – HOW WE CAN INTEGRATE THE OFFICER OF THE WATCH INTO THE SHIPS BRIDGE

Katharine V. Baylis

8 Broadacres, Fourstones
Hexham, Northumberland
NE47 5LW

Over the last 15 years the manner in which ships are navigated has seen substantial change. Current systems integrate numerous sensors and technologies in order to present the Officer of the Watch (OOW) with a concise appraisal of the real time status of the vessel. Although statistically the occurrence of incidents is decreasing, the criticality of the interaction between the OOW and navigational systems is increasing and continues to account for around 80% of navigation related incidents at sea. In order to reduce this figure and ensure the continued enhancement of the safety of navigation, a greater understanding of the interaction between the OOW and navigational technology is required. This paper investigates what the maritime industry is doing to improve the interface between the ships officer and navigational systems incorporated within the ships' bridge.

Introduction

Shipping is stated as "perhaps the most international of the world's great industries and one of the most dangerous." (www.imo.org/home.asp). Although sharing similarities other industries such as rail, road and aviation the maritime industry is unique in many aspects of the way in which it operates. Teams of seafarers (be they on the bridge, in the engine room or on deck) operate the vital components of the vessel to warrant a safe passage and successful delivery of the cargo to the relevant port. Engineering officers ensure that the engine and associated machinery operates reliably and satisfactorily at all times, being on call 24 hours a day in case a major problem develops. Similarly, on the bridge, systems are monitored round the clock and there is always at least one officer present to conduct the safe navigation of the vessel.

Until relatively recently the maritime industry has concentrated on providing technical solutions to problems affecting ships and the International forum (or UN body) for the maritime industry, the International Maritime Organisation (IMO), has primarily been associated with developing these requirements. Over the last few years however there has been a growing awareness that human involvement is a major contributory factor to incidents and accidents. The human element, human error, or human factors (hereinafter referred to as the human element) contribute to around 80% of incidents occurring at sea today and reducing this figure is considered critical when striving to make improvements to the safety of navigation (Bryant, 1991). The IMO therefore have begun to take steps in order to remedy this situation, one of which being the introduction of legislation and guidelines aimed at incorporating human element considerations into the design and operation of ships.

The Current Status of the Ships' Bridge

Bridge watchkeeping is regarded as the single most important activity conducted at sea. "Upon the Watchkeeper's diligence, expertise, training and efficiency rests the security of the vessel, cargo and all on board" (Woods, 2002). It is therefore essential that the OOW is able to carry out this process as effectively, efficiently and professionally as possible utilising equipment to enhance tasks and ensure the safety of the vessel (Woods, 2002). Historically, the role of the navigator has consisted of direct involvement in the navigation of the vessel and additionally the manual control of a limited number of single processes (Breda, 2001). Modern vessels have however increased dramatically in both size and complexity; at the same time manning arrangements have been reduced (Woods, 2002). In addition to this the duties of watchkeeping today have changed beyond recognition and Breda (2001) states "...on board automated control systems are applied on a large scale, which has induced a considerable change in operator tasks." The role of the navigator today is therefore primarily supervisory, monitoring a variety of processes which are controlled by a range of semi-intelligent sub-systems (Breda, 2001).

The OOW is now also being presented with an ever-expanding array of information, made increasingly possible by rapid developments in technology (Cobley, 2002). In principle the more information the navigator is presented with the more comprehensive and complete an understanding of the situation should be available (Cobley, 2002). This enables more informed decisions to be made in potential close quarters situations (Cobley, 2002). However, it is being realised that the manner in which this information is presented to the OOW can have repercussions regarding its usefulness. Integrated Bridge Systems (IBS) and Integrated Navigation Systems (INS) theoretically enable large quantities of information to be presented to the OOW in a succinct display which have any non-essential items of information filtered out (Wentzell, 2002). The technical specifications and performance standards for IBS and INS were adopted by IMO in 1996. An IBS can perform any two of the following operations; Passage execution, Communications, Machinery control, Loading, discharging and cargo control, Safety and security. An INS however has three different levels; INS A - Sensory inputs –heading & speed etc; INS B - Decision making tools –radar and electronic chart equipment; INS C - Control functions –heading or track control functions and performance monitors (Wentzell, 2002).

It has long been recognised that it is hard to separate the human from the technical elements contained within a total system (Quinn and Scott, 1982). Consequently the human being is regarded as just one system component amongst many others and the human element to be an example of component failure (Quinn and Scott, 1982). The implications of this within the maritime industry have been seen with the occurrence of several high profile incidents that have identified problems at the human-technology interface as causal factors. It is becoming increasingly apparent that these incidents result from problems that are much more deeply rooted than the mere 'end user.' A classic example of this was with the grounding of the passenger vessel The Royal Majesty (see NTSB accident report). Although this incident occurred in 1995, many issues that it raised are still relevant today.

What are the Key Issues Currently Challenging the Maritime Industry?

The key issues that the maritime industry have identified as requiring addressing in order to improve the usability of navigation equipment and in turn successfully reduce the number of human element related incidents occurring at sea are: 1. Complete systems cannot be fully developed within the constraints of existing international regulations, 2. The systems being produced are beyond the capabilities of the operators who use them 3. The sea staff change and so do the pilots, 4. Needs must be adequately addressed, 5 Current training provisions are unsatisfactory with the style, content, duration and frequency, 6. Greater standardisation of equipment is required (i.e. a common user display interface); manufacturers should design systems around specific end users and provide the necessary training for each piece of equipment, 7. The ergonomics of the ships bridge including the design, layout and usability of systems is a critical factor; electronic systems should be designed to complement the manual tasks undertaken by the OOW, and the design parameters must involve mariner input from the outset 8. Greater industry co-operation is required between different sectors to ensure regulations prove as effective as possible 9. The regulations themselves require modification to ensure that they complement the requirements of the industry they have been designed for, 10. Comprehensive integration of the systems utilised within the wheelhouse is a necessity in order to facilitate fast and intuitive decision making of watch officers irrespective of the situation they may confront.

How is the Maritime Industry Addressing these Issues?

The maritime industry has recognised the importance of ensuring a harmonious working relationship between the OOW and the technology they operate, however disseminating this understanding throughout the industry in its entirety is proving to be a lengthy process. Several sectors in particular have invested substantial resources into identifying and understanding relevant human element or ergonomic issues.

As aforementioned, IMO acts as the international forum for the maritime industry; its purpose is to provide the machinery for co-operation among governments and practices relating to technical matters of all kinds affecting shipping engaged in international trade (www.imo.org/home.asp). IMO encourages and facilitates the general adoption of the highest possible standards in matters concerning maritime safety, efficiency of navigation and the prevention and control of marine pollution from ships. It drafts international conventions and treaties which governments then ratify and adopt. The areas of IMO work which are of particular interest are:

SOLAS Chapter V – Regarded as the most important maritime convention, the Safety of Life at Sea Convention has recently had a major revision. This revision has led to the inclusion of more user friendly design criteria to be made a mandatory requirement imposed upon shipowners and operators. These regulations have been drawn from the guidelines produced by IMO (see, for example, https://mcanet.mcga.gov.uk/public).

Guidelines on Ergonomic Criteria for Bridge Equipment and Layout – In December 2000 the IMO produced a set of guidelines for the maritime industry pertaining to ergonomically sound design criteria for vessels.   The guidelines are comprehensive incorporating bridge layout, the work environment (i.e. climate control and vibration), the layout of the workstation (consoles and display arrangements), alarm management, information display, and interactive control. The major shortfall with them is that they are merely guidelines and any operator not wishing to incorporate them into the design of future bridges or indeed in improving existing ones, does not have to (www.seamanship.co.uk).

The International Safety Management (ISM) Code – The introduction of the ISM code was a turning point in the way that maritime administrations saw the future of safety standards. It encourages ship owners to take a more active role in understanding human element issues that are present within their organisation, while concurrently certifying that the responsibility for maintaining high safety standards and operating environments lies with all employed by the organisation (http://www.imo.org/HumanElement).

The View from the Wheelhouse

Although some collaboration between industry sectors led to the development of bridge systems, ships officers did not achieve much in the way of input. As the requirements of the end-user are critical when striving towards a successful design, a questionnaire was distributed to seafarers serving on P&O Nedlloyd vessels. The questionnaire was designed to provide feedback on how mariners perceive the bridge systems they must work with; to identify any recurrent areas of concern and to establish whether training with equipment is of an acceptable standard.

The questionnaire established the following: 1. There have been substantial changes to navigational technology over the last ten years, 2. The overall level of training was stated as satisfactory with (what was regarded as) the ten most important items of navigational equipment (Radar, ARPA, sextant, echo-sounder, ECDIS, GMDSS, GPS, paper charts, VHF, Loran-C), 3. Officers were generally able to familiarise themselves with equipment before standing watch. However, on the occasions when familiarisation was not possible this was due to constraints imposed by lack of time or training 4. Almost every ship the respondents ad sailed on had either a different make or model of navigational equipment, which again had implications for familiarisation 5. All respondents stated that standardisation of navigational equipment would almost, if not always, assist in enhancing watchkeeping practices 6. IBS have been received with a mixed response throughout the industry, 7. Bridge simulator training courses are regarded as a useful tool in assisting watchkeepers to handle close quarters situations that might develop whilst on watch, 8. Officers were, overall, happy to report near misses that may occur whilst on watch if this can be done confidentially, 9. All respondents have at least occasionally felt overloaded with information whilst on watch and adding more technology into the wheelhouse will only add to this overload and confusion, 10. Officers stated that commercial pressures sometimes compromise their application of the collision regulations.

Conclusions

The introduction of new technology into the bridge has forever altered the manner in which ships are navigated and manned. Frequently, newly introduced technology is not compatible with existing equipment, leading to the development of a fragmented system with which officers have to navigate the vessel safely. The maritime industry has begun to take steps to address issues arising from problems at the human-technology interface, however more work is required. The main areas of particular concern are: greater industry co-operation; improvement of regulations; inclusion of new technology into current carriage requirements; coherent identification of the perceived problems; equipment standardisation; system integration; officer training. A greater level of industry-wide co-operation is beginning to emerge, with different sectors meeting for a second time at an 'Integrated Bridge Systems and the Human Element' conference, part of the International Maritime Convention for 2003 (www.seatrade-london.com). However, the sector that has the greatest influence on the purchasing of sound navigational equipment, the shipowner, was not able to be represented.

In order to enhance the working environment for the OOW within the ships bridge, the maritime industry must reconsider the current 'system' in place for designing and fitting out bridges on both new builds and existing vessels. New equipment should have carefully considered technical specifications allowing it to be seamlessly integrated with existing navigation bridge equipment and training. This requires improvements in our understanding of both the context and activities that comprise the ships bridge. Further work will study watchkeepers activities utilising voyage data recorder (VDR) data that provides a complete real time picture of the complete bridge system to be reviewed. Analysis of this data should provide greater insight into the demands of the OOW and hence potentially provide a comprehensive specification for future equipment design.

References

Breda, L. *Capability Prediction: An Effective Way to Improve Navigational Performance*. The Journal of Navigation, Vol 53.

Bryant, D.T., The Human Element in Shipping Casualties. Department of Transport, Marine Directorate. 1991.

Cobley, C. 2002, "Review of Technology Affecting Modern Bridge Systems", *Integrated Bridge Navigation Systems – User Enhanced Design Conference November 2002*. London: The Nau Institute.

http://www.imo.org/home.asp [accessed; 01/02/2003]

https://mcanet.mcga.gov.uk/public/c4/regulations/safetyofnavigation/index.htm   [accessed; 17/10/2003]

http://www.seamanship.co.uk/M_Notice%20Updates/safetyofnavigation/pdf/msc982-ERG.pdf [accessed; 19/11/2003]

http://www.imo.org/HumanElement/mainframe.asp?topic_id=287 [accessed; 12/10/2003]

http://www.seatrade-london.com [accessed; 02/09/2003]

Quinn, P.T & Scott, S.M. 1982. The Human Element in Shipping Casualties – Analysis of Human Factors in Casualties. The Tavistock Institute.

Wentzell, H.F. 2002. "Developing Integrated Bridge and Navigation System: Manufacturers' Viewpoint", *Integrated Bridge and Navigation Systems – User Enhanced Design Conference November 2002*. London: The Nautical Institute.

Woods, P. 2002. "Tasks and Tools – A Shipmasters View", *Integrated Bridge and Navigation Systems – User Enhanced Design Conference November 2002*. London: The Nautical Institute

COMPUTATIONAL ERGONOMICS – A POSSIBLE EXTENSION OF COMPUTATIONAL NEUROSCIENCE? DEFINITIONS, POTENTIAL BENEFITS, AND A CASE STUDY ON CYBERSICKNESS

Richard H.Y. So* and Felix W.K. Lor

Computational Ergonomics Research Team
Department of Industrial Engineering and Engineering Management
The Hong Kong University of Science and Technology
Clear Water Bay, Kowloon, Hong Kong SAR
Email: rhyso@ust.hk, Felix.W.K.Lor@ust.hk

*All correspondence should be sent to Dr. Richard H.Y. So

This paper proposes a sub-discipline called 'Computational Ergonomics' and explains how it will support the ultimate goal of 'studying human by building a human through quantitative modeling'. This idea is not new and has been successfully implemented in the sub-discipline of Computational Neuro-Sciences (CNS) for more than 20 years. In fact, quantitative models developed by CNS researchers are becoming comprehensive enough to explain simple voluntary human behavior already. We believe the timing is right for ergonomics researchers to make use of the many open-source quantitative CNS modeling algorithms as fundamental building blocks of quantitative human performance models. This move is also consistent with the new sub-discipline of 'neuroergonomics' (Spring 2003 issue of Theoretical Issues in Ergonomics Sciences) and the call for more quantitative formal models of human performance (Spring 2003 issue of Human Factors). In addition to explaining the essential elements of the proposed 'Computational Ergonomics', this paper presents a case study to illustrate the benefits of the proposed changes. In particular, how the authors' research on simulator sickness with virtual reality systems has benefited from the proposed changes. This paper intends to raise stimulating and controversial arguments to be discussed during the conference presentation.

Introduction

After more than 5 decades of collecting human performance data through empirical experiments, most current ergonomics research is still focused on running empirical experiments rather than developing quantitative models based on previously collected data. This, to a large extent, might have been due to the habit of not publishing the raw data. The lack of published well-documented raw data prohibits the existence of 'pure' model developers. In other words, ergonomics researchers must collect a lot of empirical data by themselves before they can start developing a model and this posts a very high entry barrier for most researchers and will certainly drive away those computational scientists and mathematicians whose only interests are in data modeling. A second reason for the lack of quantitative human performance models is that most ergonomics problems are very complicated and in the absence of open-source smaller scaled models as building blocks, the task of developing a large scaled quantitative human performance model is too difficult for most researchers. The third reason is that even if we pool all the developed quantitative human performance models together, they may not be compatible with each other. This paper explains how the essential elements within the proposed sub-discipline can address each of the three reasons. Examples of elements include (i) a suggested requirement of publishing open-copyrighted raw ergonomics data for third-party model developers to work on; (ii) a suggested requirement for model developers to define the structures of their models according to the known anatomical fact of human biology – an important element behind the success of Computational Neuro-Sciences (CNS); and (iii) a suggested requirement for model developers to publish their models in the form of open-source computational code library for others to use. The significance of above elements and others are explained in the following sections.

Challenges on developing ergonomics models

Figure 1 illustrates the common modeling approaches in ergonomics. In the following subsections, the challenges related to these modeling approaches will be discussed using the authors' own research on cybersickness.

High entry barriers
Cybersickness is defined as the sickness generated after a user is exposed to a virtual reality (VR) simulation. The authors have reviewed that there have been at least 63 empirical studies on cybersickness examining the effects of 17 independent variables such as types of scene background and scene movement, exposure duration, types of VR display, display's field-of-view, image delays, use of stereoscopic, inter-pupillary-distance mismatch, method of navigation, postures during simulation, amounts of head movement during simulation, age, gender, pre-exposure posture stability, habituation, menstrual phase, and drug treatment. With so many published empirical studies, there should be plenty of raw data that can support the development of empirical regression models, operator control models, or even production-rule models on cybersickness (Figure 1). However, the reality is that although results of statistical analyses on the effects of 17 independent variables are well documented, the raw data collected in the 63 studies were not published. This makes it difficult for any "outsider" to develop quantitative models to predict levels of cybersickness and prohibits researchers of other disciplines such as computer sciences and computational neuroscuences to contribute their expertise. The authors were able to develop models to predict levels of cybersickness (So et al., 2001; Yeun et al., 2002; So et al., 2004) because the authors have themselves conducted many empirical experiments on cybersickness (e.g., So, 1994; So and Lo, 1998; So et al., 2000; Lo and So, 2001; So et al., 2002). A review of literature indicates that the habit of not publishing the raw data is fairly common in the field of ergonomics.

The compatibility challenges and the absence of model building blocks
Because of the high entry barrier discussed in the last sub-section, it is not surprising to find that there are very few attempts to model the cybersickness generation process. As a good ergonomics practice, it is always worth reviewing related studies. A review of literature on quantitative models on opto-kinetic induced motion sickness, seasickness, and space sickness

has proven to be productive because there have been published biological conceptual models on vection-induced motion sickness (work by Prof. Robert Stern, Pennsylvania State University and Prof. Ebenholtz at Schnurmacher Institute for Vision Research), empirical modeling to predict seasickness (work by Prof. M.J. Griffin, Institute of Sound and Vibration Research), and operator control models to explain space and motion sickness (work by Prof. Charles M. Oman, MIT, Prof. Telban, State University of New York). However, the three clusters of research used very different approaches (empirical regression: Griffin, 1990; Kalman filter model: Oman, 1982; adaptive filter: Telban *et al.*, 2001, and biological conceptual model: Stern *et al.*, 1995, Ebenholtz *et al.*, 1994) and shared little common building blocks. Although the authors have benefited a lot from their research methodologies, approaches of modeling, and concepts on vomit generation, reusable model building blocks could not be found. The main reason is that these models have different structures.

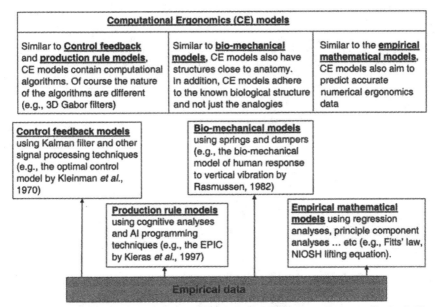

Figure 1 An illustration of the common quantitative model approaches used in the field of ergonomics and their similarity with 'Computational Ergonomics (CE)' models. The differences and the uniqueness of CE models are explained in the text.

Potential Benefits of Computational Ergonomics Approach of Modeling

The authors humbly admit that the use of computational algorithms to model and predict ergonomics behavior is not new. An example related to motion sickness is the motion sickness dose value (MSDV, Griffin, 1990) which takes in the time histories of vertical ship motion accelerations and output the predicted percentages of passengers that would vomit. Another example is the herristic mathematical model to predict levels of motion sickness by Oman (1982). The proposed 'Computational Ergonomics (CE)' approach to modeling is not just the application of computational models, it consists of four essential elements: (i) a CE model structure must adhere to the human <u>anatomical</u> structure, (ii) the willingness to share the source codes of CE model <u>building-blocks</u>, (iii) the use of <u>computational</u> algorithms, consistent to the true biological process, to construct the model building-blocks, and (iv) the sharing of raw <u>data</u>. In short, these four elements are referred to the "A,B,C,D" (<u>A</u>natomical structure, open-source <u>B</u>uilding-blocks, <u>C</u>omputational algorithms, and sharing of raw <u>D</u>ata) of Computational Ergonomics.

Anatomical structure

To solve the compatibility challenges of different ergonomics models, it is proposed that all computational ergonomics models should be based upon the known biological structures. Applying this requirement to the study of cybersickness, a biological plausible model to predict individual vection (illusion of self-motion) ratings while exposed to a VR simulation is developed (Figure 2). From the extensive study of anatomical fact of human biology together with the finding of a biomechanical pathway of vection, a sensation of vection is suggested to be generated at the parieto-insular vestibular cortex (PIVC) (Brandt, 1999). Our eyes first capture a sequence of motion pictures. Then visual signals are transferred via optic nerves to the visual cortex. At the region of primary visual cortex (V1), the spatial frequency of motion pictures is extracted. Efferent signals from V1 are received by the middle temporal cortex (V5). Inside this region, local optical flow vectors are evaluated. On the vestibular signal pathway, physical head movement is measured by semi-circular canals. The head movement stimuli are then fed into vestibular nuclei (VN) and are routed to PIVC where visual and vestibular signals are integrated and hence generate vection. Because this model involves the essential biological process in vection generation, it can be used as a model building block for future models related to vection.

Open-source building-blocks and computational algorithms

The bottom part of Figure 2 illustrates the flow of the computational quantitative model with respect to the biological structured model on vection. Visual and vestibular stimuli are major factors generating vection. Our current model consists of two modules to calculate the spatial frequency and perceived velocity of visual stimuli. Fourier analyzer acts as V1 to extract spatial frequency. Similar to V5, scene velocity detector employs optical flow to estimate the perceived velocity. For vestibular stimuli, we use a head-tracking device as a physical motion detector to measure the head movement. Integrating these three inputs together, we can determine how much vection is generated. Based on this simple biological plausible computational model with the analysis of ergonomics data, we invented a metric called Cybersickness Dose Value (CSDV) that can accurately predict levels of cybersickness (So et al, 2000). Currently, the Fourier analyzer building-block has been made available on www.cybersickness.org. User can upload a snapshot of a VR simulation to the website and the server will automatically calculate its spatial frequency.

In the future, other open-source libraries of useful neuro-computing algorithms in our site will be included in the site. With the recent advances of computational neuroscience in understanding higher brain functions (Dayan and Abbott, 2001), there will be many open-source computational building blocks that are freely available to ergonomics researchers.

Sharing of raw data

In order to solve the high entry barrier challenges, it is proposed for all computational ergonomics researchers to share their raw empirical data in a standard format so that other ergonomics researchers can analyze the data with their own methods. These raw data can be open-copyrighted and require all users to acknowledge the original data owners in their work.

Applying this to our study of cybersickness, a sharing platform has been set-up on www.cybersickness.org for other researchers to download our open-copyrighted raw data collected from two of our recent experiments. We hope that this approach would encourage third-party model developers such as computational scientists and mathematicians to analyze our data.

A biological illustration of possible pathways of cybersickness generation

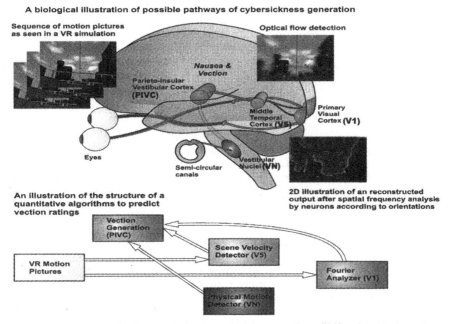

Figure 2 An illustration of a possible biological pathway of vection generation at PIVC region using inputs from V5 and V1 regions. The bottom part illustrates the structure of a quantitative model to predict vection levels.

Conclusions and Future Research Directions

Computational Ergonomics is a new paradigm to link physical and neuro-ergonomics with the help of computational neuroscience. Figure 1 illustrates its relationship with the existing common modeling approaches in the field of ergonomics. This emerging field can encourage researchers with different disciplines to develop quantitative ergonomics models. A well-developed biological computational model with open-source codes can form fundamental building blocks for others to use and the sharing of raw empirical experimental data with a standard format can allow theorists to easily study ergonomics problems. We strongly believe computational ergonomics will be one of the focus areas leading towards the total elucidation of complicated human behaviors.

Acknowledgement and References

This study is funded by the HK Research Grant Council through DAG02/03.EG47 and HKUST6076/99E. Due to page limit, the list of references is not included. Readers can send an email to the authors to request the list of references.

DEVELOPMENT OF A CROWD STRESS INDEX (CSI) FOR USE IN RISK ASSESSMENT

Ken Parsons and Nor Diana Mohd Mahudin

*Department of Human Sciences,
Loughborough University,
Loughborough, Leicestershire, LE11 3TU, UK*

A Crowd Stress Index for use in risk assessment was developed in four parts: a literature review to identify relevant factors and understand the terms 'crowd' and 'crowding'; a study on an underground train; personal interviews; and a climatic chamber experiment in which physiological and subjective measures were taken from people in a crowd. Eight components were identified: physical, social, personal, crowd characteristics, crowd dynamics, crowd behaviour, location and psychological. Ninety-two questions in a risk assessment questionnaire provide the data for scores on 42 themes, leading to scores on the eight components in terms of contribution to crowd stress and finally a single index value ranging from 'no crowd stress' to 'extremely high crowd stress'. The index appears sufficiently well developed for validation in practical applications.

Introduction

A person in a crowd can be exposed to stress due to the proximity of other people. This stress can cause both physiological and psychological strain that can range from discomfort and dissatisfaction to concern, panic and death. There may be positive reactions to being in a crowd. A feeling of purpose, belonging, excitement or simply warmth. This paper presents the development of a Crowd Stress Index. It identifies factors that can contribute to crowd stress and proposes a method that integrates them into a single number that relates to the magnitude of that stress, and hence likely strain, on people in the crowd. A detailed description is provided in Mahudin (2003).

Method

An appropriate structure for the Crowd Stress Index and factors that can contribute to it were not obvious at the outset and hence information was collected in four parts:

1. *A literature review* to determine the nature of a stress index of this type, identify how terms such as 'crowd' and 'crowding' have been used and to review studies of peoples' reaction to stress caused by crowds.

2. *A study on an underground train* to determine the physiological and subjective response of one person in context and in a range of 'crowd' levels.
3. *Personal interviews* with a range of people to identify their perceptions of crowds and their responses to being in a crowd.
4. *A climatic chamber experiment* to identify physiological and subjective responses of people while being exposed to a 'crowd' under controlled conditions.

Part 1 : Literature Review

Le Bon (1895) prophesied that we were entering the era of crowds. They represent a primary datum of social existence. A crowd is more than a mass of people, each crowd is unique and has a life of its own. Dickie (1995) stated that ebullient crowds with inadequate management have within them their own seeds of disaster. Berlonghi (1995) emphasised that those involved in crowd management must foresee the nature of the crowd that will be in attendance, be able to observe the behaviour of a crowd whilst an event is taking place and make timely decisions for effective action.

Mahudin (2003) identified four areas of research: animal research; experimental research on people; conceptual research; and human behaviour research. She considered definitions of 'crowd' and 'crowding' from different perspectives (demographic, phenomenological, social) and crowd stress as 'the feeling of having insufficient space due to the proximity of other people'.

A Crowd Stress Index will therefore predict the extent to which a person in a crowd will feel that they have insufficient space due to the proximity of other people. A high Crowd Stress Index (CSI) value will therefore predict high levels of strain with consequences such as panic, injury and death. A low CSI value will predict low levels of strain and no unacceptable consequences. A particular aspect of crowd stress is strain caused by heat. Griffitt and Veitch (1971) found that people were less friendly to each other on hot days and that even moderate temperature combined with crowding led to aggressive behaviour. Parsons (2003) suggested that examples of aggressive behaviour (e.g. security 'forces' providing a confrontational social context) may encourage aggressive behaviour. Braun and Parsons (1991) conducted a laboratory study into crowding and found that the inability to disperse metabolic heat, including restrictive evaporative loss due to sweating, can cause significant strain even in moderate temperatures and low density crowds.

Part 2 : Crowd stress on an underground train

To obtain information about strain caused by crowding in a practical context, skin temperatures, heart rate and subjective responses were measured on a standing passenger while travelling on an underground train. In addition, information on crowd characteristics was gathered by an observer who was seated nearby. The male passenger was a 24 year old undergraduate of Asian origin. The experiment was conducted from 16.15 hrs to 18.30 hrs in summer with outside (above ground) temperature of 28°C and relative humidity of 47%. Conditions inside the carriage increased over the recording period from 26.6°C and 44%rh minimum to 30.1°C and 52%rh maximum. Mean skin temperature (Ramanathan, 1964) rose from 31°C to 33°C over the session and heart rate increased slightly from 61bpm to around 80 bpm. Heart rate measurement was however unreliable as the recording instrument appeared to receive interference from train operation. Results showed that in general the subject was under little strain. He generally reported his experience in the crowds as pleasant and not tense, not irritable, tolerant, not sticky and had little difficulty in moving around. Crowd density was however generally light. In more dense crowds higher skin temperatures were recorded and the subject perceived the situation as 'very crowded', squashed, warm and slightly uncomfortable.

Part 3 : Personal interviews

The results of the literature review and preliminary study on an underground train indicated that more detailed personal interviews were required to identify a fuller breadth of factors that would influence crowd stress. Individual interviews were therefore carried out with a selected range of ten male and ten female subjects (aged 20 to 45 years). They included students, university staff and the general public. One male subject was a person in a wheelchair. The interview was semi-structured and in six sections. A. Definitions of crowd and crowded; B. General experience in crowds; C. Discussion of scenarios (trapped in a lift, going in and out of a stadium and standing close in a cash dispenser queue); D. Describe feeling if in a crowd presented in photographs (swimming pool, football match, concert); E. Discussion of similarities and differences between pictures of crowds; F. Final comments. Interviews were private, recorded and lasted 30-40 minutes. The results provided seven components and a total of 48 factors within the components (see Figure 1).

Part 4 : Climatic chamber experiment

To complement the studies presented in Parts 1, 2 and 3, a laboratory experiment under controlled conditions was conducted to determine detailed responses of people while in a crowd. One male subject (23 years, 1.70m, 63kg) had subjective and physiological (skin temperature and heart rate) responses recorded while surrounded by nine male subjects in a climatic chamber set to 30°C and 50%rh and 0.15ms^{-1} air velocity. All ten participants and a thermal manikin stood in a space 1.0m x 1.5m representing a tightly packed crowd. Subjects were of Asian origin and did not appear to be disturbed by the crowd. Over the one hour session mean skin temperature rose from 33.5°C to 34.7°C and heart rate from 71 bpm to 93 bpm. The subject initially felt very crowded, uncomfortable, slightly irritable and slightly intolerant. As he knew the others in the crowd well he did not feel that he had to avoid interaction. The 'crowd' around the subject generally felt very crowded, squashed and uncomfortable. All agreed that they needed more space. Behavioural measures indicated that there was a tendency for individuals to reduce crowd density by subtle movements away from those around them. Interviews with crowd members suggested that factors such as density level, space satisfaction, heat, air, lighting, adaptation, room volume, duration of exposure, focus of attention, relationships and social atmosphere all affect crowd stress levels. In particular, the expectation about the crowd, especially knowledge of when they could leave the crowd, influenced strain experienced.

A Crowd Stress index

Figure 1 shows the structure of the Crowd Stress Index that was developed from Parts 1, 2, 3 and 4. Forty-two factors contributed to eight components that lead to a final Crowd Stress Index for use in risk assessment.

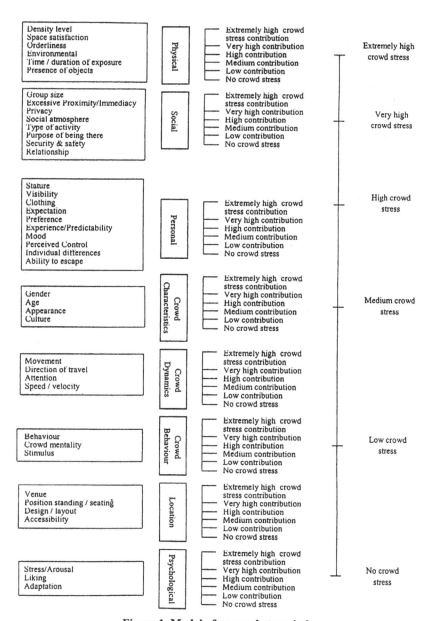

Figure 1. Model of a crowd stress index

Conclusions

A four-part method investigating crowd stress from different perspectives has allowed the development of a Crowd Stress Index. The index is a simple tool that could be used for risk assessment as well as in the management of crowd control. It is sufficiently well developed to allow validation over a range of practical applications.

References

Berlonghi, A.E. 1995, Understanding and planning for different spectator crowds, *Safety Science,* **18,** 239-247

Braun, T.L. and Parsons, K.C. 1991, Human thermal responses in crowds. *Contemporary Ergonomics. Proceedings of the Ergonomics Society's 1991 Annual Conference, 16-19 April* (Taylor and Francis, London) 190-195

Dickie, J.F. 1995, Major crowd catastrophe, *Safety Science,* **18,** 309-320

Griffitt, W. and Veitch, R. 1971, Hot and crowded: Influences of population density and temperature on interpersonal affective behavior. *Journal of Personality and Social Psychology,* **17**(1) 92-98

Le Bon, G. 1895, *La Psychologie des foules* Translated : The crowd. (Unwin, London, 1903)

Mahudin, N.D. Mohd. 2003, The development of a Crowd Stress Index, *MSc project report,* Loughborough University

Parsons, K.C. 2003, *Human Thermal Environments,* Second edition, (Taylor and Francis, London) ISBN0-415-23792-0

Ramanathan, N.L. 1964, A new weighting system for mean surface temperature of the human body, *Journal of Applied Physiology,* **19,** 531-533

ERGONOMICS OF LEFT BRAIN / RIGHT BRAIN IMBLANCE

Bernard Masters

College of Chiropractors
106 London Street,
Reading,
Berkshire RG1 4SJ

Modern assessment of the electrical activities of the brain now allows a better understanding of the two hemispheres' individual functions. The two halves should balance each other. The analytical left pursues order, abhors change and interprets the formal rather than the emotional aspects of speech. In contrast, the right interprets the informal aspects of speech (sees the punch-line of a joke), it is creative, synergistic and seeks novel approaches. When this delicate balance is upset for some reason, there is a one-sided disproportionate rise in pain sensitivity and loss of motor performance. With reference to this cerebral dominance, this paper explores the ergonomics of fitting the task to the circumstances of the individual's brain type - laterality. Examples show that motor/pain problems can be induced in individuals whose work environment does not suit their functional brain type.

Introduction

There appears to be a relationship between physical health and imbalance in the brain's two hemispheres. One cortex becomes dominant (lateralization) and allows the other to express an increase of pain or reduced motor performance on that side of the body. The resultant imbalance is further influenced by emotional and physical environmental stress. An approach to redressing this lateralization concentrates on the function rather than the performance of each neural system. By assaying levels of competence of each system, diagnosis of the longitudinal level of the dysfunction is made and a treatment regime is devised based on the responses of the cortex, cerebellum, midbrain, pons, medulla, spinal cord and 'motor end-effector' to promote neural plasticity and reduce nerve degeneration.

Lateralization - Left-right asymmetry in the brain

The study of neuroscience has grown considerably since the Victorian era and the growth has been exponential in the last two decades (Siegal, 1999). *'The cerebral cortex is the*

RIGHT CORTEX	left cortex
FLEXIBILITY	control
RESPONDS TO LARGE LETTERS	responds to small letters
SYNERGISTIC	analytical
GLOBAL PICTURE	counting individual items
NEWNESS & UNKNOWN	familiar & known
UNFAMILIAR IMAGES	photographs of family
RESPONDS TO EMOTIONAL WORDS	responds to word analysis
PROSODY & PRAGMATISM	vocalisation
IMAGES FROM THE LEFT FIELD	images from the right
MEMORY FOR SHAPES & COLOURS	good memory for details
ARTISTIC DRAWING	technical drawing

Figure 1. Functions of the cerebral hemispheres, (see Carpenter 2003, p. 397)

part of the body that makes us truly human' (FitzGerald, 1994), but this structure is enormously complex with a multiplicity of interconnections. The left brain governs the right side of the body and as most people are right handed, (Kandel et al, 2000), examples in the right to left brain shift in left-handers can result in vague disabilities such as stuttering, apraxia or agnosia (Carpenter, 2003).

The cortex has been labelled numerically relating to function. Wenicker's or area 22 contributes to language processing and is enlarged on the left side (Carpenter, 2003). Next to this language centre in the left hemisphere is the control area of the right hand. 95% of the population is right handed and being more developed in the left brain, this area was interpreted as being the dominant hemisphere (Guyton, 1991, de Meyer, 1994).

The auditory, visual, and somatic interpretive areas all feed into the *general interpretive area or knowing area* (Wernicker' area). Motor expression – speech, resides in Broca's, area 44 & 45 (FitzGerald). During conversation there is increased and matched blood flow in both hemispheres, but the right side is concerned with melodic nuances. Part of the temporal lobe is concerned with sensory function of hearing and reading. If this area in both hemispheres is equal in size, then language is distributed evenly. If one part is underdeveloped this may contribute to dyslexia.

Information processing
Information is processed differently in the two hemispheres. The left, being more analytical, is superior for processing visual, auditory and tactile information. The right hemisphere handles shapes, spatial relationships and music, as this side is synergistic and holistic (FitzGerald, figure 1). Lateralization can be described as an under-performance, or *escape*, of one half of the brain, suggesting that a boost in the non-functioning half is needed in order to rebalance the relationship. In writer's block - difficulty in starting (Bond Solon, 2003), left lateralization subjects should concentrate on logic and order and the right should attempt a non-linear approach, (figure 1).

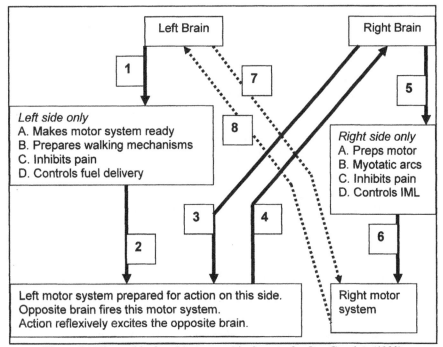

Figure 2. Hierarchical and heteroarchical control, after Carrick (1992)

The left brain [1] prepares the left motor system on the left half of the body [2] so that right brain [3] can produce movement on the left side. Feedback from left sided movement [4] excites the right brain [5] and in turn, prepares the right half of the body for action [6]. The left brain the produces movement in the right body [7], and movement in the right body then excites the left brain [8], which in turn excites the left motor system [1], and the cycle continues. Interrupting this cycle causes neural demise or 'wind-down'.

Modelling human neural control systems.
Control of neural systems appears to be both heteroarchical and hierarchical (figure 2) the brain controls the body, but reflexively, the body controls the brain, (Carrick 1992). In the neuronal doctrine (de Myer, 1996) proposed by Cajal in the late 1800's (Kandel et al, 2000), all nerve cells are related (Moore, 1982) and the six tenets (de Myer) are summarised by 'receiving & transmitting' for survival. Interrupting this relationship causes 'wind-down', but inducing plasticity promotes neural 'wind-up'.

Case studies. No. 1: Right brain physiological problem.

History
Mrs. A, a fifty year old deputy headmistress, presented with an acute back pain of a week's duration, complicated with long term tingling in the right side of her face and

arm.   Coughing and straining gave pain [spinal coverings were irritated].   Activity worsened the pain [exceeded systems' capacity], but rest eased her condition.   She was tired all the time [systems' fatigue] and apart from her school duties, she was studying for a professional doctorate.

Physiological assessment
Measures functional symmetry in neuronal systems, and asymmetry was seen throughout the body, e.g. shorter left leg length (by 10mm) and uneven standing weight distribution [left weight scale 35k & right 45k], increased reflexes in the left upper limb, right enlarged pupil [see figure 2, D].   Finger-to-nose test was inaccurate - she did not know where she was in space.   Functional diagnosis suggested a loss of integration of control between the right cortex and left cerebellum.

Background to stress
In contrast to her right problem, her headmaster appeared to have a left lateralization.   He brooked no argument and used forceful tactics (ironically he had attended a course that focussed on right/left brain dominance).   Right cortical dominant people do not like to be micro-controlled but prefer to work creatively and synergistically with others.   Stress loads the emotional centres to produce natural painkillers, and these inhibit the autoimmune system and infections occur.   Becoming accustomed to the load, but relieving the strain, physical breakdowns occur as a rebound phenomenon.

Case No. 2. Left hemispheric physiological non-ablative lesion.

Presenting symptoms
Mr. B, a production line worker, was referred by a consultant orthopaedic surgeon because of an escalating burning sensation with back pain of seven weeks duration. His occupation involved reaching over a production line. Coughing did **not** provoke the pain [i.e. there was no spinal irritation].

Surprisingly, squatting and bending **did** relieve the pain, whereas, walking or general physical activity exacerbated the problem [this suggested problems with blood flow to the spinal when loaded, see figure 2, 5D – Fuel delivery control by the IML i.e. the intro-medio-lateral column of cells in the spinal cord].

History
As a child he had right sided paralysis when he broke his collar bone. Since his back pain started, he could not stand for long, nor sleep in spite of prescribed pain killers.

Examination
He stood on two scales with weight asymmetry (25k - 35k). There was loss in torso movement. Reflexes were uneven and pendulous [loss of cerebellar control] and he fell over when he walked the heel-toe test with shut eyes. There was an increase in both blind-spots, but the right was bigger suggesting a left cortical problem (figure 2 A). Other tests showed neural asymmetry in the body's systems.

Functional diagnosis
Loss of integration between the left cortex and other systems, (figure 2 C & D).

Background to laterality induced by stressing the systems

He enjoyed working on a particular production line and to avoid being moved, took on the burden of charge-hand. This was not a solution as one worker would not take orders from him. Returning to the shop floor, again he was constantly moved. He worried continually and continuously (perseverated) and could not sleep.

Treatment for laterality right/left

There is a need to understanding that pain is made worse by continuous stress/strain [IML, figure 3, C & D] and that this is a physiological problem. Rehabilitation of the left cortex is made by employing familiar concepts (favourite melodies or a regular beat of music and counting). Management should realise that this is a subtle industrial injury and their assistance will be needed in rehabilitating this worker.

Seeking new challenges will develop the right cortex but these people become easily accustomed to a new situation and need to seek further change by exploring the unknown and listening to new types of music with irregular rhythms.

Conclusion

Meggison et al, (2000) in *Human Resource Development*, suggest a psychological contract with the work force, taking into account the person's profile, the work environment, training and development. Similarly, Harrison (2000) lays out the individual's responsibilities for self-development. Understanding the individual's laterality of brain function could make a contract more productive. Ignoring the worker's brain dominance type appears to have costly effects for these two people. Any (human) system that is eccentrically laden or poorly controlled will not perform at its optimum. Further investigation is needed into the ergonomics of cortical dominance related to individuals and their response to the ambience of the workplace.

References

Bond Solon Training, 2003, *Excellence in Report Writing, Delegate's Workbook* – General Expert Witness Report, (London, privately published), 12

Carpenter R.H.S., 2003, *Neurophysiology*, Fourth Edition, (Arnold, London) 392-397

Carrick F.R. 1992, Clinical & functional neuro-physiology, post-graduate lectures.

De Meyer W.E., 1994 *Technique of the neurological examination,* (McGraw Hill, New York) 27-28, 401

FitzGerald, M.T.J., 1996, *Neuroanatomy*, (W.B. Saunders Company, London) 228-239

Guyton A.C. 1991, *Basic Neuroscience*, (W. B. Saunders Company, London) 244

Harrison R., 2000, *Employee Development*, (Chartered Institute of Personnel and Development, London) 174

Kandel E.R., Schwartz J.H., and Jessell T.M., 2000, *Principles of Neural Science*, Fourth Edition, (McGraw Hill, New York) 1174, 10

Meggision D., Banfield P., & Joy-Matthews J., 2000, *Human Resource and Development*, (Kogan Page, London) 34-42

Moore K.L., 1982, *The Developing Human, Clinically Orientated Embryology*, Third Edition, (W.B. Saunders Co. Philadelphia) 60

Siegal, A. & H. 1999, *Neuroscience,* PreTest, Third Edition, (McGraw Hill, New York) ix

EFFECTS OF PROCUREMENT TYPE
ON HUMAN FACTORS ACTIVITIES

Daniel L. Welch

Advanced Management Technology, Inc.
Arlington, VA
USA 22209

The US Federal Aviation Administration (FAA) is currently developing two satellite-based navigation systems, the Wide Area Augmentation System (WAAS) and the Local Area Augmentation System (LAAS). This paper describes the differing nature of the two procurements and explores the resulting differences in the tasks, tools, and procedures of the associated Human Factors (HF) programs.

Introduction

Human factors is a system engineering discipline and an essential part of the effective acquisition (design and development) of complex systems. It typically employs a series of tasks, tools and procedures to achieve its ultimate goal of maximized system performance through optimized human performance. Precisely which tasks, tools and procedures are employed, however, is highly dependent on the nature of the system being acquired and the nature of the acquisition.

The US FAA is currently developing two satellite-based navigation systems, the WAAS and the LAAS. These acquisitions are on-going and in different stages of development. They have also taken different and somewhat unusual acquisition strategies in order to bring innovative systems into operation as quickly as possible. This paper examines the differing natures of the acquisition strategies and the impact of those differences on the associated HF efforts.

The Wide Area Augmentation System (WAAS)

The WAAS is a differential global positioning system (GPS) augmentation which improves the accuracy, availability, and integrity of GPS, eventually allowing GPS to be used as a primary navigation source from takeoff through Category I precision approach.

Unlike traditional ground-based aids, WAAS will cover a much more extensive service area.

The system currently consists of 25 precisely surveyed Wide-Area Reference Stations (WRSs) with associated antennas, which identify errors in the received GPS signal and transfer that data to two WAAS Master Stations (WMSs). The WMS computes correction information for specific geographical areas and sends the correction to four Ground Uplink Stations (GUSs). The GUS uplinks the correction message to one of two geostationary communications satellites (GEOs) and the GEOs transmit the correction message on the same frequency as GPS (L1, 1575.42 MHz) to GPS receivers enabled for WAAS.

The entire ground-based WAAS system is operated and monitored through two Operations and Maintenance (O&M) consoles, one on each US coast. The O&M console constitutes the major human interface for the ground-based WAAS, along with the maintenance of the WMSs, the WRSs, and the GUSs.

Acquisition Strategy

The WAAS development contract was awarded to Wilcox Electric Inc. in August 1995. That contract was terminated for the convenience of the government in April 1996. The effort was transferred to Hughes Aircraft Company in May 1996. Shortly thereafter Hughes was acquired by Raytheon Systems Company and a rebaselined contract was established in June 1999. By January 2000, safety and technical integrity problems necessitated the formation of the WAAS Integrity Performance Panel (WIPP) to address those issues. In August 2000, an independent review board (IRB) was chartered to review work products developed by the WIPP and assess their technical merit and validity. Based on the work of the WIPP and the IRB, a second rebaselined contract was established with Raytheon in November of 2001.

It is reasonable to state that during this entire period the WAAS Program experienced a number of contractor, scheduling, funding profile, and operations and maintenance concept difficulties. Individually and collectively, these difficulties negatively impacted HF efforts and participation in system design. Initially, the Wilcox effort included a full-up HF effort. When the Hughes contract was initiated in 1996, HF was de-emphasized due to a life-cycle concept which would use contractor personnel in all operations and maintenance positions. (That concept was eventually reversed and FAA personnel are currently operating and maintaining WAAS.) By August 2001, it was decided to place HF "on the back burner" in order to direct full energy towards solving integrity and safety problems. While an unfortunate necessity, these were in fact realistic decisions since external pressure from congress and the airlines impelled the earliest available.fielding of the WAAS.

Impact on Human Factors Efforts

Rather than integrating HF efforts early in the design and development process, the WAAS effort deliberately deferred HF activities until the system was already fielded. As a result, the WAAS as it exists today contains a number of HF issues which need to be addressed. The process of doing HF for the WAAS, therefore, now consists of identifying and correcting those issues, rather than "designing them out" from the beginning. In essence, HF for WAAS is now an exercise in "change control."

Given the acquisition strategy for the entire system, this is actually a reasonable and effective approach. The WAAS was commissioned in July 2003 and is currently transitioning from Initial Operational Capability (IOC) to Full Operational Capability

(FOC). This transition involves extensive hardware and software improvements to the initial system and HF efforts are being effectively incorporated into that change process. The contract Statement of Work (SOW) for the FOC phase details a number of tasks which will ensure maximum improvement to the WAAS through HF.

A critical task analysis (based on a complete task inventory) is being conducted to identify those operator/maintainer tasks essential for safe and efficient WAAS operation. This analysis will form the basis for the identification of critical design considerations (CDCs), defined as elements of the WAAS which will have a significant impact on human performance (and hence system performance) if not properly and adequately designed. Attempts at HF improvements to the WAAS will be focused on the improvement of CDCs.

At the same time, a WAAS HF Action Item List (WHAIL) is being established to enumerate and track individual items of concern; i.e., HF discrepancies, problems, user issues, etc. The WHAIL will contain and define all items under consideration for improvement. WHAIL items must be linked to CDCs and, hence, to Critical Tasks.

WHAIL items will be discerned through a number of efforts. Raytheon is undertaking a program of studies, analyses, and test and evaluation to assess system equipment, software, human-computer interfaces, and procedures. This effort will employ mockups, rapid prototyping and simulations to identify and evaluate WHAIL items, for which alternate design solutions will be developed and appraised. Also, two Users Groups are being established (Operations and Maintenance) to provide subject matter expertise, provide preference and usability data, and review, assess and validate WHAIL items. The User Groups will be a major source of input to the WHAIL.

Note, however, that the User Groups will not act as decision making bodies themselves – rather they will form one data source among others for the government HF Working Group. This body will examine HF issues from the WHAIL, develop solutions, evaluate solutions in light of all competing system engineering requirements, finalize and prioritize a solution package, and proffer a vetted resolution to the WAAS Configuration Control Board (WCCB). The WCCB, in turn, makes final cost, schedule and technical analyses and forwards accepted changes to the contractor for implementation.

Thus, at this point HF for WAAS is very similar to efforts within the US nuclear industry immediately after Three-Mile Island. Through a combination of activities, HF discrepancies are being identified, evaluated, and corrected. The important issue, however, is that through the use of critical task analysis and CDC data, HF energy, time, and money (along with recommended improvements) are being focused on important aspects of system design shown to directly impact human performance and hence, overall system performance.

The Local Area Augmentation System (LAAS)

The LAAS also augments GPS to provide an all-weather (Category I, II, and III) complex approach, landing and surface navigation capability. It is composed of a space segment (the GPS satellites), an air segment (the LAAS-enabled GPS receiver in the aircraft), and a ground segment, known as the LAAS Ground Facility (LGF). This discussion is limited to the LGF.

The LGF is composed of four precisely surveyed reference receivers (RRs) at the airport, which receive GPS information and transfer that information to the primary processor. The processor calculates correction information and transmits that data (along with integrity and approach-related data) via a Very High Frequency (VHF) Data Broadcast (VDB) antenna to aircraft in the vicinity of the airport in the VHF band (108 – 118MHz).

There are two primary human-system interfaces. A Maintenance Data Terminal (MDT) controls the LGF and enables maintenance functions, while an Air Traffic Control Unit (ATCU) provides LGF operational status information to the Air Traffic Controller and permits individual runways to be LAAS-activated or -deactivated.

Acquisition Strategy

In April 1997 the FAA initiated two separate cost-sharing partnerships with industry teams, one led by Raytheon, the other by Honeywell. This original FAA concept, known as the Government-Industry Partnership (GIP), was intended to develop a CAT I LAAS system with industry funds while the FAA prepared operational documentation, provided technical support, and conducted Type Acceptance or certification of the systems.

However, changing FAA requirements for LAAS required a change in acquisition strategy. In April 2002, a Request for Proposal (RFP) was released for a LAAS developmental program. During the source selection process, changing requirements resulted in again redefining the program to a phased approach. Currently, Phase I is a design effort to take the LAAS concept to Critical Design Review (CDR). A successful CDR will enable entry into Phase II, a Limited Rate of Production (LRIP) effort to install Category I LAASs at six operational sites as a proof of concept. Success at the LRIP sites will enable entry into Phase III, which will transition LAAS to Category II/III complex approaches and install the system in additional airports. A contract for Phase I effort was awarded to Honeywell in April 2003. The current acquisition strategy, therefore, aims to leverage the GIP efforts and results in order to phase-in Category I LAAS, then transition to Cat II/III and complex approaches.

Impacts on Human Factors Efforts

Like the WAAS program, changes in the acquisition strategy and other realities have impacted how HF is being conducted for LAAS. During the GIP, the industry partners developed prototypes of the MDT and the ATCU with little government input (beyond requirements). Generally, the GIP partners had developed functional prototypes of the subsystems independent of any formal HF effort, and then subjected the resultant designs to an HF "review" employing the FAA *Human Factors Design Guide* (HFDG) (DOT/FAA/CT-96/1). These designs were presented at formal "demonstrations" where HF considerations were described. The role of the government in the demonstration was to validate that the GIP partner took CDCs into account in the design. Based on the pre-established evaluation criteria, the question of whether or not a proper HF process was employed was *not* an issue.

Changing acquisition and political realities necessitated the change from a "type accepted" LAAS to a full developmental effort. For the initial LAAS developmental program, HF provided requirements-linked and -traceable input to the FAA acquisition documentation (Acquisition Strategy Paper (ASP), Integrated Program Plan (IPP), Request for Proposal (RFP), etc.), and evaluated offeror proposals against HF criteria described in Section L of the RFP. This is standard HF participation for a normal acquisition process.

For the first phase of LAAS development, a fairly conventional HF program is being conducted. System analysis is accomplished through function flow diagram analysis, a task inventory is being developed, and tasks are subjected to a criticality analysis. Tasks identified as critical will be subjected to a more in-depth analysis leading to the identification of CDCs for LAAS hardware and software. Operator and maintainer user groups are employed to provide system, user preference, and usability performance data, while government and contractor HF Working Groups are providing coordination and integration into system engineering efforts.

This is all standard HF process, employing typical HF tools and techniques.

"Fluid" requirements, however, continue to impact the HF efforts. For example, the requirement for control of LAAS-enabled runway ends via the ATCU is currently being reconsidered, due to new FAA restrictions on additional displays in the tower cab. This may obviate earlier work done to develop a user interface for the ATCU and may necessitate an entirely new concept for displaying LAAS data in the tower cab (i.e., a small "idiot light" panel vs. a graphical user interface on a 15" touch-screen monitor). This would require considerable re-evaluation and further HF effort.

In addition, the initial maintenance monitoring concept relied on the National Airspace System (NAS) Infrastructure Monitoring System (NIMS) to alert air facilities personnel to problems with the LAAS. The possible non-availability of NIMS requires a reconsideration of alerting techniques and may necessitate a re-design to provide remote status panels at appropriate locations throughout the airport site. This reconsideration, like that for the changed ATCU requirement, will have to be accomplished "on the move" and require great flexibility in terms of HF activities.

Thus, for LAAS, while a traditional HF developmental effort is being attempted, realities outside the control of the program office and changing requirements are strongly impacting what needs to be done to provide for optimized human performance. The tasks, tools, techniques and procedures employed by HF are therefore changing, in real-time, to respond to the changing demands of the acquisition system.

Summary

Text books, graduate courses, and academic papers provide a theoretical ideal of how HF should be integrated into the system development process. HF practitioners know it never happens that way. The reality is that forces outside of our control constantly impact what we do, how and when we do it, and, most importantly of all, why we do it. We can not simply rely on established, "tried-and-true" HF processes, techniques, tasks, tools, and procedures. Given the reality of an individual program at a particular point in the acquisition cycle, what "normally" would be done, or what is the "commonly accepted" process, may not be the most appropriate for achieving the ultimate HF goal – maximizing system performance by optimizing human performance.

As things change (often chaotically) around us, keeping that goal in mind will help the HF practitioner re-evaluate what *needs to be done* in light of the current actuality, rather than what *normally ought to be done*.

As we wait to participate in that "perfect," or at least "normal," acquisition, we can use the vagaries of the acquisition sequence to predict how best to integrate HF into the program and what tasks, tools, and procedures to employ.

POSITIVE ERGONOMICS: IMPROVING MOOD BEFORE THE WORKING DAY BEGINS.

Orsolina I. Martino and Neil Morris

University of Wolverhampton, School of Applied Sciences, Psychology, Millennium Building, Wolverhampton WV1 1SB, UK

This review takes the form of an ergonomic 'walk' from the moment of waking, through the first cup of coffee, to leaving for work. It considers a regimen for improving mood prior to beginning the working day. Positive mood has been linked to optimal arousal levels and increased somatic comfort, both of which are likely to aid work performance. Empirical findings on the effects of caffeine intake, bathing and breakfast consumption are discussed, taking into account the nature and duration of these effects and how they might be used to the maximum advantage. It is concluded that one's habitual practices before the working day begins have important implications for work performance.

Introduction

It is clear that workplaces can be stressful in a number of ways and a large literature on workplace stress has accumulated. Recently a previously undervalued perspective in psychology, positive psychology, has become influential (Seligman, 2000). One facet of this approach is the idea that psychology should concern itself with preventing psychological dysfunction, that is, the maintenance of psychological well-being. Now ergonomics is traditionally very much concerned with preventative measures rather than *post hoc* repairs so one might argue that there is a long tradition of positive ergonomics. However many areas of interest to ergonomists, for example, designing equipment suited to the physical and mental capacities of workers, are concerned with activities that commence when the worker arrives at the workplace. Stress, however, may be present both inside and outside of the workplace. It may be taken home and rehearsed. If this is the case then the place to begin tackling stress may be at waking and before one goes to work. If one arrives at work already stressed then the overall level of stress experienced at work is likely to be more severe than if one arrived with a calmer 'baseline'. There are various ways in which stress in the workplace can be tackled and these may include, for example, counselling and/or meditation (Morris and Raabe, 2001). However in addition to these interventions the worker may be able to enhance, to a modest extent, their psychological well-being before the working day begins by simply adjusting their behaviour in the hour or so before they set off for work. This paper examines that hour before work in light of the literature on activities that typically occur before the working day begins. We start from the moment of waking.

Caffeine

The first priority at waking is to re-hydrate as significant fluid losses occur during sleep. Although re-hydration should be a main priority upon awakening, the fact that many individuals prefer a cup of tea or coffee to a glass of water suggests an alternative reason behind their choice of early-morning beverage. Caffeine ingestion is usually associated with tense-energy, with many people quite aware that the desired increases in attention and alertness are often accompanied by less pleasant feelings of anxiety, tension and nervousness (e.g. Gilliland & Andress, 1981; Gilliland & Bullock, 1983/84). According to Thayer (1996) this indicates that people use caffeine consciously as self-medication, i.e. to alleviate tiredness, and are willing to tolerate the more negative effects even though calm-energy would be the ideal state to achieve. Thayer argues that it is possible to achieve calm-energy with smaller amounts of caffeine, with individual mood responses dependent on individual differences in tolerance. By limiting one's caffeine intake before leaving for work, it may be possible to benefit from its energising and even relaxing properties without the unwanted side-effects. Following re-hydration the next priority is usually to cleanse the body.

Ablutions

When we consider our morning ablutions there may be a tendency to opt for a shower to save time. However there may be benefits to allotting perhaps 10 minutes to quiet reflection in a warm bath. Such 'downtime' may provide the only quiet period prior to the commencement of the working day. As such it may help to set the calmness level at the beginning of the day a little lower and thus provide an 'innoculation' against the coming stresses of the day. Our research in this area is ongoing. Some of our recent research suggests that a bath may be psychologically efficacious and that the addition of aromatherapy oils to the bath may augment the psychological benefits. Morris (2002) instigated a series of trials that required participants, none of whom had any mental health problems requiring professional psychological treatment, to bathe once per day in a bath containing either grapeseed oil (a placebo) or grapeseed oil and lavender oil. Aromatherapists frequently cite lavender oil as having psychologically beneficial properties (Lawless, 1994). In the first study mood was assessed using the UWIST mood adjective checklist to measure three different mood dimensions (Matthews *et al.*, 1990) before and after the two week bathing regimen. In study two optimism and pessimism about future events were measured before and after the sequence of baths. Mood improved, on all dimensions, even when no lavender was added suggesting that the baths *per se* improved mood. In the second study the level of optimism remained unchanged but pessimism was reduced following lavender baths but not after baths with no essential oil added. These results suggest that mood may be elevated by the relaxing properties of bathing and that matters viewed in a pessimistic light may be viewed a little more positively after self-administered aromatherapy. After a relaxing bath a nutritious breakfast is in order.

Breakfast

The circadian blood sugar rhythm closely follows that of mood, with higher levels observed in the morning than in the afternoon (e.g.Troisi, *et al.*, 2000). This positive relationship between blood sugar levels and mood has been demonstrated in various studies. Benton and Owens (1993) showed correlations between increased blood sugar and feelings of energy with reduced tension. Similarly, Gold et al. (1995) found that inducing hypoglycaemia using insulin infusions raised tension and lowered energy. More recently Martino & Morris's (2003) study revealed that raising blood sugar levels improved mood on three dimensions as measured by the UWIST Mood Adjective Checklist (Matthews *et al.*, 1990) - not only was energy increased and tension reduced, supporting previous reports, but there was also a marked improvement in hedonic tone (a dimension related to the pleasantness of one's internal state). Thus the overall implication is that raising blood sugar enhances both arousal and the actual pleasantness of mood. Higher blood sugar levels have also been shown to improve performance on cognitive tasks (Lapp, 1981; Benton and Owens, 1993; Benton et al., 1994; Martin & Benton, 1999), possibly as a consequence of mood and arousal enhancement.

Despite numerous findings highlighting the positive link between blood sugar levels, mood and performance, the view that raising blood sugar *per se* will be sufficient to bring about such improvements should be treated with a degree of caution. Benton (2001) points out the importance of dietary habits, considering the role of breakfast consumption in subsequent mood and performance. The constituents of a 'healthy breakfast' are discussed in Benton (1996). Benton (2001) found that eating breakfast was associated with improved mood, particularly when it provided high glucose levels later in the morning. Morris and Sarll (2001) showed that a glucose drink improved listening span in students who had missed breakfast, but not in those who did not usually eat breakfast. Likewise Benton and Parker (1998) found that a glucose drink improved memory by reversing the effects of missing breakfast. What these findings suggest is that increasing already moderate blood sugar levels is unlikely to influence cognitive performance; rather glucose can be taken to ameliorate negative effects of low blood sugar on mood. Martino and Morris (2003) showed that a glucose drink significantly improved mood by raising blood sugar to moderate levels. Mood itself was increased to a moderate degree, which should lead to better work performance.

Beginning the day with a nutritious breakfast is therefore beneficial in the sense that one begins the day with an elevated rather than low blood sugar level, helping to establish a state of calm-energy prior to leaving for work. Eating breakfast habitually means that if it is occasionally missed, a convenient form of glucose can be taken to reverse any resulting deficits. However, due to the 'rebound' effect of sugar snacking (see Thayer, 1989), whereby the initial energising and tension-reducing effects can induce later fatigue and tension, this is not recommended as regular practice. Thus the worker is now re-hydrated, relaxed and alert, and fuelled up for the day's work.

Conclusions and recommendations

It is possible to achieve a state of calm-energy before beginning the working day by making simple changes to one's morning routine. As a result stress that arises in the workplace is likely to be more manageable than if it occurred in addition to a less relaxed 'baseline', helping to keep mood and arousal at optimal levels for good performance. It is recommended that:

(1) Caffeine is taken in relatively small amounts to bring about an increase in energy without the increase in tension that accompanies greater amounts.

(2) 10 minutes or so are set aside for a relaxing bath, helping to reduce tension by providing a valuable 'quiet' period prior to beginning the working day. The psychological benefits may be augmented using aromatherapy.

(3) A nutritious breakfast is taken so that the mood-enhancing effects of raised blood sugar levels continue throughout the morning.

References

Benton, D. (1996). *Food for Thought*. Harmondsworth: Penguin.

Benton, D. (2001). The impact of the supply of glucose to the brain on mood and memory. *Nutrition Reviews*, **59**, S20-S21.

Benton, D. & Owens, D. (1993). Is raised blood glucose associated with the relief of tension? *Journal of Psychosomatic Research*, **37**, 723-735.

Benton, D., Owens, D.S. & Parker, P.Y. (1994). Blood glucose influences memory and attention in young adults. *Neuropsychologia*, **32**, 595-607.

Benton, D. & Parker, P.Y. (1998). Breakfast, blood glucose and cognition. *American Journal of Clinical Nutrition*, **67**, 772S-778S.

Gilliland, K. & Andress, D. (1981). Ad lib caffeine consumption, symptoms of caffeinism and academic performance. *American Journal of Psychiatry*, **138**, 512-514.

Gilliland, K. & Bullock, W. (1983/84). Caffeine: A potential drug of abuse. *Advances in Alcohol and Substance Abuse*, **3**, 53-73.

Gold, A.C., MacLeod, K.M., Frier, B.M. & Deary, I. (1995). Changes in mood during acute hypoglycaemia in healthy subjects. *Journal of Personality and Social Psychology*, **68**, 498-504.

Lapp, J.E. (1981). Effects of glycaemic alterations and noun imagery on the learning of paired associates'. *Journal of Learning Disorders*, **14**, 35-38.

Lawless, J. (1994). *Lavender oil*. London: Harper/Collins.

Martin, P.Y. & Benton, D. (1999). The influence of a glucose drink on a demanding working memory task. *Physiology and Behavior*, **67**, 69-74.

Martino, O.I. & Morris, N. (2003). Drinking glucose improves mood at the beginning of the working day. In P. McCabe (Ed.), *Contemporary Ergonomics 2003*. London: Taylor & Francis. 226-231.

Matthews, G., Jones, D. & Chamberlain, G. (1990). Refining the measurement of mood: the UWIST Mood-Adjective Checklist. *British Journal of Psychology*, **81**, 17–42.

Morris, N. (2002). The effects of Lavender (*Lavendula angustifolium*) baths on psychological well-being: two exploratory randomised control trials. *Complementary Therapies in Medicine*, **16**, 223-228.

Morris, N. and Raabe, B. (2001). Psychological Stress in the Workplace: Cognitive Ergonomics or Cognitive Therapy? In M. Hanson (Ed.) *Contemporary Ergonomics 2001*. London: Taylor & Francis. 257-262.

Morris, N. & Sarll, P. (2001). Drinking glucose improves listening span in students who miss breakfast. *Educational Research,* **43**, 201-207.

Seligman, M. (2000). Positive Psychology. In Gillham, J. (Ed.) *The science of optimism and hope*. Philadelphia: Templeton Foundation Press. 415-429.

Thayer, R.E. (1989). *The Biopsychology of Mood and Arousal*. Oxford: Oxford University Press.

Thayer, R.E. (1996). *The Origin of Everyday Moods: Managing Energy, Tension and Stress*. New York: Oxford University Press.

Troisi, R.J., Cowie, C.C. & Harris, M.I. (2000). Diurnal variation in fasting plasma glucose: Implications for diagnosis of diabetes in patients examined in the afternoon. *JAMA, The Journal of the American Medical Association*, **284**, 3157-3159.

ERGONOMICS INTEGRATION IN THE DESIGN OF A ROTARY HAMMER

Eilís J. Carey, Joachim Vedder and Gerold Fritz

Hilti Corporation,
FL-9494 Schaan
Principality of Liechtenstein

This study reports the integration of ergonomics principles in the design of a rotary hammer, which is used in construction trades for drilling into concrete and other base materials. The tool in question was a battery operated tool, and differed fundamentally in design in comparison with traditional rotary hammers. Ergonomics support was provided in developing the design concept, and during the development phase, in the form of design checks. At the prototype stage, a full ergonomics investigation was conducted, comparing the new tool design with two other models. Eight subjects used the tools in four different drilling directions for concrete with and without reinforcement bars. The perceived exertion and subjective assessment of features were recorded for each case, and subjects gave their overall tool preference. One conventional tool gave significantly greater levels of perceived exertion than the others. The use of the tools in concrete with rebar caused greater levels of perceived exertion than in concrete without rebar, and the drilling direction had a significant effect on perceived exertion. The results of the study were used to understand the strengths and weaknesses of the different tool designs, as input for development and marketing.

Introduction

Drilling into different base materials is a common activity in construction trades. For example, in the mechanical and electrical trades holes are frequently drilled into concrete in order to form a fastening point for building services, normally using hand-held rotary hammers or, less frequently, hammer drills. Previous research (Müller et al., 2003) found that 12% of the time of mechanical services workers was spent drilling, and 50% of this was drilling overhead. Rotary hammering work presents several ergonomics challenges, as there is very often a combination of high force exertion, vibration, dust, noise and awkward postures, especially when drilling overhead. This can result in the development of pain and musculoskeletal disorders (Rosecrance et al., 1996).

The first rotary hammer in the world was produced in 1932, and the first cordless rotary hammer was introduced in 1984, yet the design of these tools from the handling point of view has remained fundamentally unchanged, despite evidence of high muscle forces and joint moments (Anton et al., 2001). Rotary hammers have a handle at the rear, which the user typically grasps with the dominant hand, and a side handle which may be grasped with the non-dominant hand, though many construction workers hold the tool with the dominant hand only, particularly for overhead work. The side handle is usually adjustable for left handed people. The tools can be alternated between hammering with rotary action and pure rotary action. The tool bit exchange mechanisms differ across manufacturers. The tools are switched on with a finger-operated trigger switch, located close to the main handle. The tools considered in this study were rechargeable battery powered tools, with battery exchange mechanisms.

Ergonomics support was provided during the development stage of a new rotary hammer, taking the principles of hand tool design, such as those defined by Sanders and McCormick (1992), and the knowledge of the specific applications of the users, into account. In particular, the handling of the tool was a focus of the ergonomics and development work. In order to check the users reaction to the tool, and to evaluate the ergonomics of the tool, an evaluation was made at the prototype stage, taking the new tool and two tools of conventional design. The objective was to identify the strengths and weaknesses of the new design in comparison with the more conventional designs, as input for development and marketing.

Method

Three factors were considered in the study, as shown in Table 1. The combination of these factors resulted in 24 combinations, and each one was performed by each subject. The "Up" and "Down" drilling directions were vertical, and the other drilling directions were horizontal. The drilling depth was 60mm and the drill bit diameter was 8mm for all cases, and the applications were drilling in concrete without reinforcement bars and drilling in concrete with reinforcement bars.

Table 1. Factors for the experiment

Factor	Levels	Notation
Tool	New tool	T1
	Conventional tool A	T2
	Conventional tool B	T3
Drilling direction	Up	D1
	Side at head level	D2
	Side at waist level	D3
	Down	D4
Application	8mm drill concrete	A1
	8mm drill concrete with rebar	A2

The dependent variables were the Borg Rating of Perceived Exertion (RPE) and the subjective response to a questionnaire, distributed after working with each tool. The questionnaire asked for a rating from very good to very bad for features of the tool including weight, vibration, handling, main and mode switches, tool balance, battery exchange, drill bit exchange and ease of use in different drilling directions. Eight male subjects participated in the study, with a mean age of 29.9 years (standard deviation of 3.4 years), a mean mass of 84kg and a mean height of 1.84m. The order of the combinations was balanced over the subjects. The mean grip strength of the subjects was 554.3N.

The force level of a previous study from the literature (Anton et al. 2000) was set at 22.3N, but in this study the subjects regulated the force level themselves. Safety glasses were worn during the work, but gloves were not worn, as these are generally not used on the construction site in the application of the analysed tools.

Subjects familiarised themselves with the tools and the analysis scales before the experiment. Then each tool was used, in the order defined in the experimental design. Two holes were drilled for each of the 24 conditions. After each condition, the Borg RPE scale was completed by the subject. For the case of drilling into rebar, the drilling was stopped after the rebar was hit. Video footage was taken for the entire experiment. The questionnaire of tool features was administered to the subjects after completing the four drilling directions and two application levels for each tool.

Results

The mean perceived exertion for the tools in the four different drilling directions is shown in Figure 1, and ranged from 10.0 (between very light and light on the scale) to 13.7 (between somewhat hard and hard). In each case, the perceived exertion was highest for drilling upwards, followed by drilling horizontally at head level, drilling horizontally at waist level and drilling vertically downwards. Paired two-sample students t-tests of the drilling direction showed a significant difference between all drilling directions ($p<0.01$) except for between the downwards drilling and waist level drilling directions ($p>0.05$). There was a significant difference in the RPE for drilling with and without rebar ($p<0.01$).

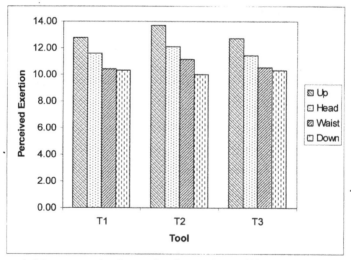

Figure 1. Mean perceived exertion for each tool/direction combination

Students t-tests of the RPE of the tools showed significant differences ($p<0.01$) between Tool 2 and both other tools, as shown in Table 2. Tool 1 did not differ significantly from Tool 3 in terms of RPE.

Table 2. Significance levels of RPE for paired students t-tests of the tools

Tool	T1	T2	T3
T1	-	-	-
T2	**	-	-
T3	NS	**	-

** p < 0.01

NS: not significant

The mean subjective response for different features of the tool are shown in Figure 2. Tool 1 received a higher mean subjective rating than the other tools for weight, vibration, normal handling, tool balance, and ease of use in the upwards and head level directions. The rating for weight and tool balance was significantly higher than for the other tools (student t-test; $p<0.05$). Tool 3 received the highest subjective rating for handling on hitting reinforcement bar and ease of use in the waist level and downwards directions.

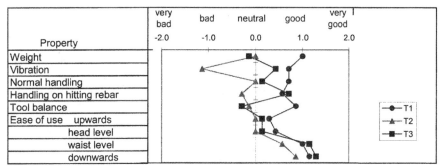

Figure 2. Mean subjective response for several features of each tool

Discussion

The significant difference in RPE for drilling in rebar compared to without rebar is interesting, as the subjects were allowed to stop the drilling once the rebar was found, and so no difference in RPE was expected. The difference found may be due to psychological tension or apprehension that the subjects felt, as a tool may spin when rebar is met, and this can in extreme cases cause injury to the user. Most drilling on construction sites is performed in concrete with reinforcement bars, and the fact that the user almost never knows if the bars will be hit may cause psychological stress.

The drilling direction with the highest perceived exertion was drilling in the upwards direction, which accounts for around 50% of the drilling of mechanical trade workers. Anton et al. (2001) showed that the shoulder moment and upper limb EMG activity increase with reach distance for overhead drilling. Measures to avoid drilling overhead include the use of inset anchors which are cast into the concrete, and rig-mounted tools which can be manipulated from ground level.

The designed tool had a lower mass than the conventional tools, and this may explain the lower subjective rating that it received for the ease of drilling in the downwards direction. The user would have to apply slightly more force on the tool for downward drilling with a lighter tool.

The subjects drilled just two holes for each of the 24 experimental conditions, which may not be representative of the actual use on the construction site, although these tools target the lower end of professional user, who is not using the tool constantly. Nevertheless, a more realistic scenario on the construction site would probably lead to higher RPE levels, and may identify other factors which affect exertion levels, which were not taken into account in this study.

The force which the users applied to the tools was not controlled in this experiment, and several factors are known to affect this force, including the task, experience and strength of the user. Björing et al. (2002) defined several factors which influence grip force, including the experience and strength of the user, the weight of the tool, the size, softness and surface characteristics of the handle, the use of gloves, the presence of liquids, the presence of vibrations, the wrist position as well as the reaction forces of the tool. Further studies recording the overall force applied to the tool and the grip force for different handle configurations will provide additional useful information to developers of these tools.

Conclusions

–   Drilling upwards lead to a significantly higher RPE than any other drilling direction, and was rated as being most difficult by the subject, for each tool.
–   Drilling in concrete with reinforcement bars caused a higher RPE than in concrete without reinforcement bars, even when the reinforcement bars were not hit.
–   The new tool was better in terms of the RPE than one other tool, and showed no significant difference with the other, and was better than both other tools in subjectively rated weight and tool balance.
–   The ergonomics investigation provided important input to the design process, and showed the ergonomics benefits of the new tool. Different configurations of the hands of the user were possible with the new tool, addressing the requirement for drilling in different directions. The integration of ergonomics in the design of this rotary hammer was effective and timely, and resulted in a more ergonomic product reaching the market.

References

Anton, D., Shibley, L.D., Fethke, N.B., Hess, J., Cook, T.M. and Rosecrance, J. 2001, The effect of overhead drilling position on shoulder moment and electromyography, *Ergonomics*, **44**, 489-501

Björing, G., Johansson, L. and Hägg, G.M. 2002, Surface pressure in the hand when holding a drilling machine under different drilling conditions, *International Journal or Industrial Ergonomics*, **29**, 255-261

Müller, S., Carey E.J. and Vedder J. 2003, Physical and psychological work load of construction workers in the mechanical and electrical trades, XVth Triennial Congress of the International Ergonomics Association, August 24-29, Seoul, Korea

Rosecrance, J., Cook, T. and Zimmermann, C. 1996, Work-related musculoskeletal symptoms among construction workers in the pipe trades, *Work*, **7**, 13-20

Sanders, M.S. and McCormick, E.J. 1992, *Human Factors in Engineering and Design*, Seventh Edition, (McGraw-Hill, Singapore)

The effect of intensity of 'side on' simulated solar radiation on the thermal comfort of seated subjects

Sam Vaughan, Jenny Martensen and Ken Parsons

Human Thermal Environments Laboratory
Loughborough University, Loughborough, Leicestershire, LE11 3TU, UK

This report presents the results of an investigation into the thermal comfort of subjects seated 'side on' to simulated solar radiation. Subjective and objective data were collected over a 30 min period while 8 healthy male subjects were exposed to radiation intensities of 0, 200, 400 and 600 Wm^{-2}. The experiment was conducted in a thermal chamber where all other environmental conditions were kept constant. Physiological and psychological measures were taken using two methods. Firstly, local skin temperature was determined using thermistors placed at 11 locations on the subjects' bodies. Secondly a questionnaire was administered every 5 mins during the exposure to ascertain the subject's perceived thermal sensation, preference, comfort and stickiness. Significant increases in skin temperature and sensation were found at all radiation intensities and a significant asymmetry was present across the subjects' bodies. Preference, discomfort and stickiness increased significantly under the 600 Wm^{-2} condition. A discomfort curve was developed to modify existing models to incorporate direct 'side on' simulated solar radiation.

Introduction

This investigation looks at the effects of 'side on' thermal radiation on human thermal comfort. A large amount of research has been aimed at building environments but only recently has it been aimed at vehicles. Madson *et al* (1992) reported that determining thermal comfort in vehicles was more complex than for buildings. The transient non-uniform characteristics of vehicle environments makes it a lot more difficult to measure and predict human thermal responses. Hodder and Parsons (2001) conducted an investigation into the relationship between 'front on' simulated solar radiation and thermal comfort as part of the Brite Euram Automotive Glazing project. They found a strong relationship between the intensity of the radiation and the thermal comfort of the subjects exposed.

One application for this work is when considering train passenger environments, where the passengers would be side on to windows. This situation is different from 'front on' exposure as other areas of the body are exposed. This investigation developed a discomfort curve to allow for incorporation of direct solar radiation into current thermal comfort models. The Predicted Mean Vote thermal comfort index (PMV), (Fanger, 1970; ISO 7730, 1994) takes into account radiation in the form of mean radiant temperature but doesn't consider the effects of direct solar radiation. This investigation looked to further the work of Hodder and Parsons (2001) by considering the effects when subjects were seated 'side on' to the radiation.

Method

Experimental Method

A repeated measures within-subject design was used. Eight healthy male subjects carried out four experimental conditions on separate days, but at the same time of day. They were seated in a Fiat Punto front car seat 'side on' to the simulated radiation and exposed to 4 intensities, 0 (control), 200, 400 and 600 Wm^{-2}. The environment was controlled to be thermally neutral without the direct radiation. The exposure order of the subjects was randomised using two 4x4 Latin squares to reduce presentation order effects. The subjects wore a white cotton/polyester (65/35%) shirt and beige cotton/polyester (65/35%) trousers and their own underwear and shoes. The subjects sleeves were rolled up above the elbow and the total estimated clo value including the seat was 0.7 with an estimated metabolic rate of 70 Wm^{-2}. The average height of the subjects was 1794mm with a standard deviation of 30.17mm, the average age was 24.75 years with a standard deviation of 5.63 years.

The subjects were seated 'side on' to a vertical window with the radiation source at 45° to that window. They wore protective eyewear throughout exposure.

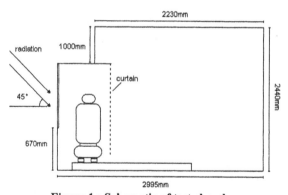

Figure 1 : Schematic of test chamber

The chamber is a purpose built insulated room that was constructed to provide a thermally neutral environment. There is a pane of 1m x 1m x 3.5mm clear monolithic glass mounted vertically at one end of the chamber. A bank of 4 simulated solar radiation lamps (1000W metal halide CSI lamp manufactured by GE lighting, Beeson (1978)) were fixed to a frame. The intensity of the radiation was altered by varying the distance of the lamps from the subjects. This intensity was monitored during the investigation using a Skye pyronometer SP1110 to ensure uniformity across the subjects.

Experimental Measures

Environmental conditions were monitored and recorded every ten seconds throughout each experiment with an Eltek/Grant squirrel data logger. Thermistors were positioned around the chamber to record air temperature, t_a. Also measured were mean radiant temperature, t_r, via a black globe thermometer, relative humidity, rh, with a capacitance hydrometer, and air velocity, v, with a hot wire anemometer.

The subjects' skin temperature was measured using thermistors positioned at 11 points on the body, right and left head, arm, forearm, thigh and calf, and one positioned centrally on the chest. These were recorded every ten seconds using another squirrel data logger.

The subjects completed a subjective questionnaire every five minutes throughout the experiment, rating their thermal sensation, thermal comfort, stickiness and preference, see below

4 Very Hot	
3 Hot	4 Very Uncomfortable / Sticky
2 Warm	3 Uncomfortable / Sticky
1 Slightly Warm	2 Slightly Uncomfortable / Sticky
0 Neutral	1 Not Uncomfortable / Sticky
-1 Slightly Cool	
-2 Cool	

Figure 2 : Example of the modified 7 point ISO scale and the 4 point scale used for comfort and stickiness

Experimental Procedure

The subjects were asked to arrive 30 minutes prior to the experiment and were sat in a thermally neutral preparation room adjacent to the chamber. The thermistors were attached using 3M transpore tape, and the subjects dressed in the standard clothing. The subjects remained in the preparation room until they felt they were thermally neutral. They were then taken into the chamber and seated on the car seat protected from the radiation by the curtain. They were asked to complete a questionnaire to ensure they were still thermally neutral. Once this was completed the subject was slid into the radiation and the experiment commenced. Questionnaires were administered from this moment, and every five minutes, until the experiment finished 30 minutes later.

Results

Using a modified Ramanathan (1964) method for weighting skin temperatures, (average across body sides) figure 3 shows a comparison of the mean skin temperatures of the 8 subjects for the 4 conditions. Using a standard t-test it was found that there were significant differences between all conditions except between 200 and 400 Wm^{-2}, P < 0.01. There were also significant differences between the mean left and right side temperatures for each radiation intensity. The temperature gradients across the body ranged up to 7°C at the maximum radiation intensity and several subjects stated that as a source of discomfort.

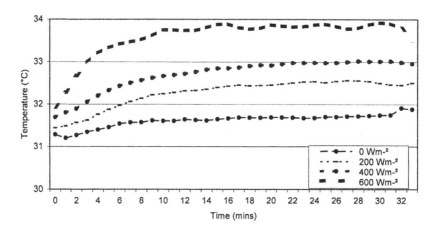

Figure 3 : Comparison of mean skin temperatures for 8 subjects exposed to 4 different intensities of radiation

Subjective Results

Figure 4 shows the subjects sensation vote for the four differing intensities of radiation. A significant difference was found between all conditions and a neutral vote was maintained by the subjects during the control. When considering the end vote, after 30 minutes of experimentation, at the highest radiation intensity it can be seen that the exposure has modified the subjects' vote by 2.4, ie from neutral to between warm and hot.

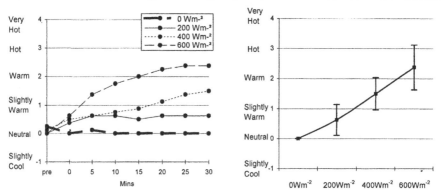

Figure 4 : Mean thermal sensation graph for 4 different solar radiation intensities

Figure 5 : Thermal sensation graph for subject mean end vote

Figure 5 shows the mean end vote, after 30 minutes of experimentation of the 8 subjects. This curve shows a distinct and significant increase in the thermal sensation experienced by the subjects corresponding to an increase in radiation intensity.

Discussion

All of the mean skin temperatures were found to be significantly different from those found during the control condition, P < 0.01. This supports the findings of Hodder and Parsons (2001) although the effects were less pronounced. This may well be due to the area of the subjects exposed as a 'side on' profile gives a smaller surface area for radiation to strike. It can be seen that the temperature of the skin increases immediately on exposure to the radiation and levels out at approximately 10-12 minutes, then stays constant for the remainder of the experiment.

Figure 4 shows the increase in thermal sensation experienced by the subjects, this increase was significant, P < 0.01, in all cases. This shows that the subjects could perceive a difference between all the radiation intensities and supports the increase in skin temperature. While the subjects could perceive the radiation increase the only intensity that caused significant discomfort was 600 Wm^{-2}. Several of the subjects felt that the lower radiation levels were pleasant and actually increased their thermal comfort, although not to a significant level.

Conclusion

The relationship between skin temperature, thermal sensation and radiation intensity has been established for subjects seated 'side on' to the radiation source. Also a discomfort curve for thermal responses to 'side on' simulated solar radiation has been found.

References

Beeson, E.J.G. 1978, The CSI lamp as a source of radiation for solar simulation, *Lighting Research and Technology,* Vol. 10, No 3, 164 – 166

Fanger, P.O. 1970, Thermal Comfort, (Danish Technical Press, Copenhagen)

Gameiro da Silva M.C. 2002, Measurement of comfort in vehicles, *Measurement science and technology,* Vol. 13, R41 – R60

Hodder S. and Parsons K.C. 2001, The effect of the intensity of simulated solar radiation on human thermal comfort, *Automotive Glazing, Brite Euram Project,* BE97-3020 2.4/001

ISO Standard 7730 1994, Moderate Thermal Environments – Determination of the PMV and PPD indices and specification of the conditions for thermal comfort

Madsen T. Olesen B. and Reid K. 1992, New methods for evaluation of the thermal environment in automotive vehicles, *ASHRAE Transactions,* Vol. 92, part 1

Parsons K.C. 1993, Human Thermal Environments, Second Edition, (Taylor and Francis, London)

Ramanathan, N.L. 1964, A new weighting system for mean surface temperature of the human body, *Journal of applied physiology,* Vol. 19, 531-533

ARE YOU COOKING COMFORTABLY? ERGONOMICS IN THE RESTAURANT KITCHEN

Andrea Eagles & Alex W. Stedmon*

*Institute for Occupational Ergonomics (IOE)
School of 4M
University of Nottingham
Nottingham NG7 2RD

By their very nature, kitchens are busy working environments, often with many people performing skilled tasks under severe time pressure to complete dishes to a high standard and to do this is a safe and efficient manner. Using a case-study example, the main aspects of environmental ergonomics, considered in this paper, include: temperature, noise, lighting, ventilation and humidity. Taking each in turn, the issues are highlighted in relation to general ergonomics principles, findings from the case-study example and potential solutions. With a better appreciation of best practise ergonomics principles in the initial design of such environments, or in making recommendations to existing environments, the workplace can support the worker rather than the worker having to adapt to the workplace.

Ergonomics in the restaurant kitchen

Restaurant kitchens are busy working environments with periods of low and high activity, often with many people moving and working in cramped conditions. Such environments can be inherently dangerous due to sharp objects such as knives; various mechanical devices such as cookers and food-processors; slippery surfaces and floors; and very hot liquids such as water in pans and oil in fryers.

Whilst there are strict guidelines for general Health & Safety at Work (HMSO, 1974), there do not appear to be many strict ergonomic guidelines, outside of general workplace design, to support the safe design and use of kitchens. Ergonomics is an important factor in product and workplace design in order to optimise performance and enhance usability, and the design of physical factors in a workspace environment should reflect the importance of ergonomics in order to optimise employee comfort and productivity at work (Wilson & Corlett, 1995).

Using a case-study example of a high-class vegetarian restaurant in London's Primrose Hill area, the main aspects of environmental ergonomics, considered in this paper, include: temperature, noise, lighting, ventilation and humidity. Information for the case-study example was collected during typically busy lunchtime session in March 2003. The case-study example is typical of many restaurant kitchen environments, serving up to 50 customers at any one time, where staff work different shifts, and where food is prepared throughout the day with lunch and evening dinner menus. Taking each of the main environmental aspects in turn, the issues are highlighted in relation to general ergonomics principles, findings from the case-study example and potential solutions.

Temperature

Thermal comfort is defined as a state of mind which expresses satisfaction with the thermal environment, and is influenced by individual differences in mood, personality, individual and social factors (Nicol *et al*, 1995). As the internal core temperature of humans varies only slightly between 36.1°C and 37.2°C, it is often the subjective perception of the thermal environment that is most influential in determining overall comfort (Oborne, 1995). British legislation does not set a maximum working temperature, although the World Health Organisation (WHO) recommends a maximum air temperature of 75°F/24°C for comfortable work. The Workplace (Health, Safety & Welfare) Regulations, state that a reasonable temperature should be maintained during working hours and the Approved Code of Practice to these regulations states, 'that all reasonable steps should be taken to achieve a comfortable temperature' (HMSO, 1992). Indeed, when air temperature increases, workers are more likely to report dissatisfaction over an increase in general humidity, illustrating that this is an important factor in workplace design and the way workers perceive their environment (Palonen *et al*, 1993).

When environmental temperature increases people will sweat and blood will be re-directed to the surface of the skin in order to maintain homeostasis. As such, fatigue and de-hydration can occur sooner than it would in a cooler environment with alertness and mental ability suffering as a consequence (Wilson & Corlett, 1995). Indeed, 'the frequency of accidents, in general, appears to be higher in hot environments than in more moderate environmental conditions. One reason is that working in a hot environment lowers the mental alertness and physical performance of an individual' (NIOSH, 1992).

The temperature of the case-study restaurant kitchen was recorded at 22.8°C. This was a typical busy lunch period with all six gas burners and the oven in constant use for several hours. This general temperature is close to the WHO maximum recommendation. Whilst extractor fans are fitted to remove some of the heat produced by cooking, these do not adequately reduce the temperature especially in summer when the temperature can be much higher. Cheap and simple methods of reducing the heat in a kitchen include: opening the windows, installing more extractor fans and ventilation hoods. Another remedy would be to allow more breaks for the workers to cool off elsewhere, however this is not always possible in a busy restaurant environment.

Noise

In everyday situations, noise can be defined as 'any unwanted sound' that is annoying at a particular moment, and is particularly appropriate when applied to noise in the workplace (Kroemer & Grandjean, 1997). According to Wilson & Corlett (1995), four levels of noise can be calculated for ergonomic assessments:

- sustained noise level– the average level of sound over a given time period.
- median noise level – the level that is exceeded for 50% of a given time period.
- background noise level – the noise level that is not exceeded for 90% of the time.
- peak noise level – the level that is exceeded for 10% of the time.

In general the sustained noise level is often measured. The Noise at Work Regulations (HMSO, 1989), state that if noise levels exceed 85dB, workers need to be informed about risks to their hearing and be provided with ear defenders, and if noise levels exceed 90dB, employers need to control exposure by doing all that is reasonably practicable to reduce it, such as making machinery quieter.

Several electrical appliances are operated in the case-study kitchen environment, often at the same time. Such appliances include an industrial dishwasher, food processors, hand-held blenders and spice grinders. There is also noise from extractor fans and ventilation hoods as well as orders being shouted, a radio and the activity of cooking itself. In common with many restaurants, the customers are protected from extraneous kitchen noise by keeping doors to the kitchen closed, however this can sometimes only serve to trap the noise in the kitchen to the annoyance of the workers. Annoyance can be classified in a number of ways according to the temporal variability, information content, signal to noise ratio, controllability, predictability, individual attitudes, experienced quality, reduceability and functionality (Sailer & Hassenzahl, 2000).

The sound levels in the kitchen were measured during a typical afternoon when several items of equipment were being operated. The general sound level in the kitchen with the radio on was 70dB, and with the radio off it was 60dB. The noise level next to the dishwasher when it was being operated was 72dB. The sound level coming from the vents over the cooker was 60dB. None of the levels of sound measured in the kitchen were excessively high, although over prolonged periods of work it might cause annoyance.

Lighting
Recommendations for lighting levels are often specified in terms of units of illuminance (lux) that refer to the amount of light falling onto a surface. Illuminance will depend on various factors such as the number of light sources in the environment, their distance and angle from the surface and reflections from other surfaces (Kroemer & Grandjean, 1997). For visual comfort and good optical performance, a number of conditions should be met: a suitable level of luminance is appropriate for the task in hand; spatial balance of surface luminances is important for both visual comfort and visibility; temporal uniformity of lighting avoiding fluctuating brightness levels; and avoidance of glare where inappropriate lighting can be a source of visual discomfort (Kroemer & Grandjean, 1997). The Food Safety Act (HMSO, 1990) applies to lighting in catering kitchens. It states that no part of any light fittings should be able to fall into foodstuffs and they should be capable of being cleaned on a regular basis. In a kitchen environment, it is recommended that the standard illuminance should be 500 lux for food preparation and cooking; 300 lux for serveries, vegetable preparation and washing up areas; and 150 lux for food stores and cellars (CIBSE, 1994).

The case-study kitchen has a small skylight, which provides a low level of natural light and there are no other windows. The kitchen door opens out into the dining area but provides little natural light as it positioned far from the main front door and windows of the restaurant. The restaurant, like most professional dining environments, is lit with candles and general low lighting to enhance the mood of the room and support the dining experience. The main source of illumination in the kitchen, therefore, is artificial lighting via a number of fluorescent tubes attached to the ceiling. The artificial lighting levels were measured at various points in the kitchen:

- preparation areas – 180 to 286 lux
- next to oven – 225 lux
- dishwasher area – 185 lux

These readings are well below recommended levels and so extra lighting should be incorporated into the kitchen, especially in preparation areas where employees might be doing delicate tasks with sharp knives. Whilst natural light is minimal, an advantage of artificial light is that it provides a constant light source and so there are fewer fluctuations than with natural light. Also, the effects of glare can be controlled more readily by strategically placing lights so that any negative effects are minimised.

Ventilation & humidity

Adequate ventilation and the rate of air movement are important factors in the thermal comfort of the workplace, and the general principle is that 'effective and suitable provision shall be made to ensure that every enclosed workplace is ventilated by a sufficient quantity of fresh or purified air' (HMSO, 1992).

Adequate ventilation is essential to control health and safety, food hygiene and food safety in kitchens. There are several objectives that ventilation should achieve, which include the following:

- the ventilation through the kitchen must bring in sufficient cool, clean air and remove excess hot air to allow a comfortable working environment.
- the ventilation must dilute and remove products of combustion from gas-fired appliances.
- the ventilation must dilute and remove cooking odours, vapours and steam.
- the ventilation must be kept clean from fat residues to avoid fire risks.
- the system must be quiet and vibration free, with clean incoming air that is not too hot or cold.

Steam, smells and grease in the air caused by cooking can cause condensation in the kitchen. In order to address this, ventilation fans are usually fitted in windows, outside walls, roofs or ceilings. Cooker hoods either extract or re-circulate air and by ventilating air from the kitchen, lower the internal air pressure slightly which helps to keep cooking smells from spreading to other rooms. This is important in preventing any unfavourable smells reaching the dining areas of the restaurant. Extraction units should be correctly sized for the room they are in, with between 10-15 air changes per hour. The Health and Safety in Catering Liaison Committee considers the lack of adequate ventilation in kitchens to be a major problem in catering.

Humidity is also an important factor in the kitchen environment. This is the concentration of water vapour in the air. The ideal relative humidity of air for comfort is in the range 55 - 65 per cent, though the range of 40 - 70 per cent is usually considered acceptable (HMSO, 1993). Low humidity, with a dry atmosphere, will have an effect on the mucous membranes of the nose, eyes and throat of the workers, causing discomfort and irritation.

The case-study kitchen has an extraction fan in the ceiling next to the dishwashing area, and a cooker ventilation hood above the gas oven and hob, which removes cooking odours and heat. There is also one small skylight window that allows very little natural ventilation into the kitchen. A fan has also been installed to improve air circulation but this is not sufficient to adequately cool the kitchen, especially in summer months when the temperature can reach an uncomfortable temperature. The humidity of the kitchen during a busy part of the afternoon, with a dishwasher being operated at regular intervals, was found to be 38%. In this instance, both ventilation is poor and the humidity level is below recommended levels.

Discussion

The physical environment has a great impact on the performance of people in the workplace and can affect health, comfort and overall job satisfaction. All environments can affect a workers health either positively or negatively, and by evaluating the working conditions and the environmental ergonomic factors, positive changes can be made if necessary to the workplace and employee well-being.

The evaluation of this restaurant kitchen reveals areas where improvements can be made to enhance worker health, safety and performance. The temperature in the kitchen was higher than the recommended levels. This could be remedied by the use of a more efficient extractor fan system, which would also improve ventilation and help to lower temperatures. The humidity level was slightly lower than recommended, although not so low as to cause extreme dryness to the nose and eyes. The levels of noise in the kitchen were below those that would cause damage to the hearing or necessitate the wearing of ear protectors, even with several items of electrical equipment, including a dishwasher, being operated. The lighting levels in the kitchen were much lower than recommended levels. Brighter general lighting could be fitted over work preparation areas to improve the levels or supplementary lighting in the form of spotlights could be fitted to provide more light in localised areas.

Conclusion

By their very nature, kitchens are busy working environments, often with many people performing skilled tasks under severe time pressure to complete dishes to a high standard and to do this is a safe and efficient manner. With a better appreciation of best practise ergonomics principles in the initial design of such environments, or in making recommendations to existing environments, the workplace can support the worker rather than the worker having to adapt to the workplace.

References

CIBSE. (1994). *Code for Interior Lighting.* London.

HMSO. (1974). *Health and Safety at Work Act.* London. Her Majesty's Stationary Office.

HMSO. (1989). *Noise at Work Regulations.* London. Her Majesty's Stationary Office.

HMSO. (1990). *Food Safety Act.* London. Her Majesty's Stationary Office.

HMSO. (1992). *Workplace (Health, Safety & Welfare) Regulations.* London. Her Majesty's Stationary Office.

HMSO. (1993). *Workplace (Health, Safety & Welfare) Regulations – Approved Code of Practise.* London. Her Majesty's Stationary Office.

Kroemer, K.H.E. & Grandjean, E. (1997) *Fitting the Task to the Human.* London, Taylor & Francis Ltd.

Nicol, F., Humphreys, M., Sykes, O. & Roef, S. (1995) *Standards for Thermal Comfort:* London, Chapman & Hall.

NIOSH. (1992). *Working in Hot Environments.* NIOSH Publication No. 86-112

Oborne, D.J. (1995). *Ergonomics at Work – 3rd Edition.* Chichester, Wiley.

Palonen, J., Seppanen, O., & Jaakola, J.J.K. (1993). The Effect of Air Temperature and Relative Humidity on Thermal Comfort in the Office Environment. *Indoor Air,* 3, 391-397.

Sailor, U. & Hassenzahl, M. (2000). Assessing Noise Annoyance: An Improvement-orientated approach. Ergonomics, 43, 1920-1938.

Wilson, J. R. & Corlett, E. N. (1995) *Evaluation of Human Work - 2nd Edition.* London, Taylor & Francis Ltd.

DESIGNING-OUT THE CELL OF EXCLUSION

John Mitchell[1] and Robert Chesters[2]

1 Director, Ergonova
2 Design Manager, AHEAD Project, Rowley, Regis and Tipton PCT

Disabled and elderly people are excluded from mainstream life when they try to use products, facilities, services or systems that demand, unimpaired movements, sight, hearing, manual dexterity and intellectual capacity.

The effects are comparable to living in a 'cell of exclusion' that has been inadvertently constructed by society and by disabled people themselves. The cell's four walls represent non-inclusive design development, ignorance about barriers, ignorance about the penalties of exclusion and the absence of disability expertise from design/development processes. The cell's glass ceiling represents society's low expectations of disabled people and its marshy floor represents the low expectations of disabled people themselves. It is hoped that, when the human and economic costs of exclusion are fully exposed, they will act as a powerful driver towards 'designing-out' barriers and achieving full inclusion.

Introduction

Disabled and elderly people are excluded from mainstream life when they try to use products, facilities, services or systems that demand unimpaired movements, sight, hearing, manual dexterity and intellectual capacity (Bennington, 2000).

In effect, non-inclusive products carry an implicit performance specification for their users, such as that set out for mobile phones in Table 1 below.

Table 1 Performance specification from your mobile phone

I require that my users are able to:

- see my display
- hear my speech outputs
- remember, understand and follow my operating procedures
- use my keypad with speed and dexterity

The Penalties of Non-Inclusive Products and Facilities

Exclusion penalizes not only disabled people and their families, but also a wide range of providers and revenue agencies (Mitchell, 2000a). Non-inclusive mainstream products and facilities penalize disabled people by making them less productive and competitive. They may find that function becomes impossible, slow, more erratic or more hazardous. They may have to depend on other people, pay for assistance, pay for special equipment or have to use prominent, unsightly and stigmatizing products.

For example, many blind people who cannot see the displays on mobile phones have difficulty in finding phone boxes as an alternative. The search can take considerable time, involve dependence and possible embarrassment as well.

Manufacturers penalize themselves by reducing the size of their potential markets when they develop non-inclusive products. For example, Martin (1988), estimated that blind people form 5% of the total population. In view of this and the fact that sound technology is already well developed, it is surprising that few manufacturers have incorporated auditory as well as visual displays into mobile phones.

Exclusion penalizes support services because it increases the demand for formal and informal carers. Informal carers must redirect their energies from their existing lives and responsibilities when they provide support. The Treasury and revenue agencies are penalized when disabled people and their carers become less socially and economically active. Their contributions to taxation and to local and national economies are likely to decrease whilst their need for various types of benefits is likely to grow.

Exclusion and modern technology

Older technologies, such as building and printing, could not have been inclusive from the outset because social, material and technological options were very limited.   Modern technology makes it relatively easy to 'design-out' barriers and to produce fully inclusive products and facilities. For example, a fully inclusive mobile phone could well have auditory feedback for blind people, 'speech to text' linkage for deaf people, voice command for non keypad users, and simple operating procedures to ease the mental load for everyone.

Nevertheless, the advent of modern technology does not appear to eliminate the creation of fresh barriers. 27 people with impaired movement, sensation or thinking in Sheffield reported that non-inclusive homes and neighborhoods restricted their everyday choice and function. Few of these problems were due to inadequate technological development. Most could have been avoided by providing evidence to underpin informed choice and by developing good, inclusive practice for the design, layout, installation and refurbishment of homes and neighborhoods (Mitchell, in press).

Few stakeholders will be willing to change long-standing products and practices unless they anticipate substantial benefits as a result.

The Cell of Exclusion

Mitchell (2002) has set out these factors in a systems diagram in which each block represents an ingredient in the exclusion of disabled people, see Figure 1.

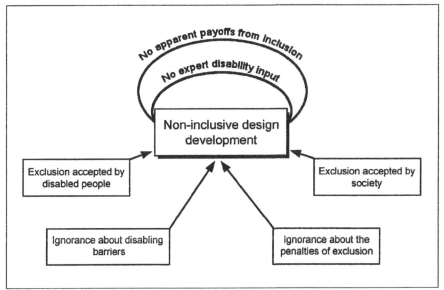

Figure 1. Ingredients in the exclusion of disabled people

The evidential and motivating factors illustrated in the diagram can also be thought of as being wrapped around disabled people to represent the 'cell of exclusion' in which they live, see Figure 2.

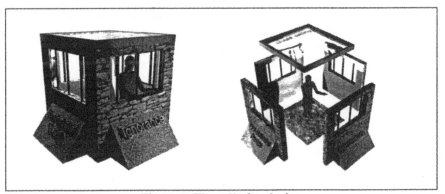

Figure 2. The cell of exclusion

The cell has:

A 'glass ceiling', representing society's low expectations not only of women but also of disabled and elderly people as well. The cell's marshy floor represents disabled and elderly peoples' own low   expectations about themselves and their contribution to society. The low ceiling and marshy floor cramp and restrict the activities of the cell's inmates.

The first wall represents the non-inclusive design and development system. Its performance depends on the instructions it receives from investors. At present, full inclusion is not mandatory. If this was to change in the future, then developers would undoubtedly do their best to meet the new brief. Reprogramming this wall to inclusion would interrupt the development of new barriers. As the newly built world became inclusive, the only remaining barriers would have been inherited from the past.)

Wall Two represents ignorance about the nature and causes of disabling barriers.
Though disabled people understand them well, there are few systematic routes for their feedback (Mitchell, 2000b). Buildings, system and transport facilities are almost never evaluated by their users. Accordingly, developers get very little valid information about the usability of their designs or the new barriers they have created (Mitchell, 1995).

Wall Three represents ignorance about the penalties of exclusion.
Martin (1987) estimated that nearly 14% of adults in the UK are disabled. Population Trends (1988) also estimated that   more than 20% of the population are over 60.   Their combined purchasing power is probably between 20 and 30% of the total market.

Barriers stop disabled people from working, looking after themselves, running their homes, doing their shopping or looking after other members of their families.   They become less productive and more dependent. The consequent penalties may prove to be severe and long-lasting. At present the human and economic costs of exclusion are uncollected and therefore invisible (Mitchell, 2000b).   However, when they are fully exposed it is likely that they will prove to be a major motivator towards full inclusion.

Wall 4 represents the exclusion of disabled expertise from design and planning. Disabled people live with barriers, they understand them and they can help the design/development system to get rid of them.

Conclusions

Each wall of the cell is controlled by a different stakeholder. Inclusion cannot be achieved unless the stakeholders work together and share evidence more effectively.

Currently, designers and developers are not required to use their resources and expertise to produce inclusive outputs. The authors believe that, when the penalties of exclusion are exposed they will provide the motivating force to redirect these efforts towards achieving full inclusion.

References

Bennington, J, Mitchell, J and McClenahan, J. 2000, *Autonomy for Disabled People:Mutual Problems, Mutual Solutions, Report to the King's Fund*, (Kingsfund, London)

Martin, J, Meltzer, H. and Elliott, V. 1988, *The Prevalence of Disability Among Adults*, (OPCS, London)

Mitchell, J, Chadwick A, and Hodges B, (In preparation), *A Home with Nothing Wrong with It, like a Blank Sheet of Paper – feedback and analysis of housing barriers for people with different impairments, Report to Sheffield Housing Department*, (Inclusive Living, Sheffield)

Mitchell, J. 2002, *Shattering the Glass Cell of Exclusion, Invited presentation to Maturity Matters, the 6th International Conference on Aging, Perth, Western Australia*

Mitchell, J and Bennington, J, 2000, *Revealing and responding to the needs of wheelchair consumers, Contemporary Ergonomics 2000, pp 385-390, Ed McCabe, PT, Hanson, E and Robertson, SA* , (Taylor and Francis, London)

Mitchell, J, Bennington, J and McClenahan, J. 2000, *Autonomy for Disabled People: The need for systemic choice and innovation, Contemporary Ergonomics 2000, pp 375-380, Ed McCabe, PT, Hanson, E and Robertson, SA*, (Taylor and Francis, London)

Mitchell, J. 1995, *Do Planning Regulations ensure that New Buildings can be used by all? Contemporary Ergonomics, pp 223-218, Ed McCabe, PT, Hanson, E and Robertson, SA*, (Taylor and Francis, London)

Office of Population Censuses and Surveys 1988, *Population Trends*, (HMSO, London)

CONSUMERS, EVIDENCE AND INNOVATION

Robert A. Chesters[1] and John Mitchell[2]

1 Design manager for Medilink West Midlands
2 Director of Ergonova

Consumers require good quality products to undertake their daily activities effectively, safely and pleasurably. Products that demand unimpaired physical, sensory or cognitive function exclude many of their potential users, or make function more difficult, time-consuming or dangerous. For example, mobile phones that do not provide auditory feedback for blind users. Non-inclusive products cast a long shadow. They restrict market size for manufacturers and retailers, they reduce the productivity and economic viability of disabled people and the dependence they create places a heavy load on support services.

The AHEAD/IDA project in Sandwell sets out to overcome such problems by establishing a West Midlands partnership to underpin inclusive mainstream and assistive design. Rowley, Regis and Tipton PCT have appointed a design manager to lead the interdisciplinary project team. Holistic, consumer-centred methods are used to reveal consumers' priorities and problems, such as their need for easier access and greater safety during hygiene. Their requirements are compared with the performance of existing products and design is concentrated on any emerging mismatches. New product development involves firms as well as larger manufacturers whose spare capacity could offer advantageous quality, flexibility and pricing.

Introduction

It is widely recognised that user centred design can improve quality of life and restore autonomy to disadvantaged and socially excluded individuals. This was proven in part by achievements made by the AHEaD project, a health action zone initiative based in Sandwell in the West Midlands. The AHEaD project was a partnership between Rowley Regis and Tipton primary care trust and Sandwell council. The projects brief was to look at preventing social exclusion in Sandwell through developing three innovative strands, Inclusive design, 1Time banking and 2Social enterprise.

This paper will look in detail at the Inclusive design strand and its approach to user centred design.

---

1 Time banking, a community development concept which encourages individuals to trade in time as oppose to money. Here people trade in skills with no comparative value attached, just an hour for an hour.

2 Social enterprise, developing businesses with social rather than purely commercial drivers.

In the widest context the project set out to do two things, firstly identify new innovation which could help reduce the barriers and stigmas elderly and disabled people face in day to day life. Secondly to develop new products which could be of benefit to local industry, ultimately supporting business in the area. To achieve this, a partnership was established between users and the NHS. A design manager was appointed to co-ordinate the process and develop the Sandwell model for inclusive design. This was a unique arrangement and for the first time the NHS directly employed a product designer to deliver the inclusive design part of the project.

By placing the research and design responsibility within the primary care trust it enabled the project to have a status on 'honest broker'. The design manager was able to occupy this unique status and bring together users, providers and industry which traditionally have very different priorities and ideals but are essential stakeholders in the design process.

Stakeholders in the design process

The key stakeholders are quite generic terms and could be identified for almost any product, see figure 1. Users are characterised as the group identified as needing the product. Providers are those who supply the product to the end user, and industry represents the manufacturers. For example, a motorcar is designed for a specific user group identified through market research; it is provided through dealerships and is manufactured by the automotive industry.

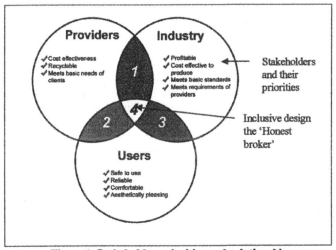

Figure 1. Stakeholder priorities and relationships

The projects earlier work looked at the design of assistive products provided by health and social services. Community equipment services provide assistive products to elderly and disabled people for use in the home, these range from modified cutlery to seating and bathing aids. This product group largely consists of retro-fitted equipment developed to aid people with disabilities in the home. Initial research was conducted with Sandwell community equipment store to identify opportunities for inclusive design in this area. Several equipment types were identified as having huge potential for improvement, the key problems concerned poor usability, no consideration for visual aesthetics, poor safety and quality. Subsequent research identified that in many cases users were so dissatisfied and embarrassed by the equipment they were given that they simply hid it away. Two products were identified as having the most potential for inclusive design.

Cot sides: Used to prevent bed bound users from falling out when sleeping or confused. Current metal frame sides have recently been identified as posing significant entrapment risks to users, in the worst cases can get stuck between the bars and suffocate. The design and materials used in the construction are unsympathetic to the user and are cold to the touch and have a distinct hospital look. Current variants fail to fit securely to all types of bed meaning that they can fail to drop in an emergency. Other interesting results came out of user research. Adults found the term 'Cot side' insulting as it made reference to products used for children.

Toilet frames: Used to give extra support to people with reduced mobility when using the toilet. These were found to be visually ugly and cause problems for other members of the household when using the toilet. These problems add stigma to the product which resulted in a proportion of users hiding them away. Both of these product types were deemed as having a negative affect on the user, worsening their situation by making the whole experience degrading and stigmatising. This was backed up by research conducted with users of such equipment.

User centred design methodology

Throughout the research and design process users were consulted. At first this was simply a process of interviewing individuals with disabilities and asking some simple but vital questions. The questions centered on those used in the Lifemaps model of assessment which poses holistic questions relating to what the disabled user wants from their equipment (Mitchell and Bennington, 2002). The questions address personal needs in line with the users' aspirations. This process allows designers to walk through a day in the life of an individual and identify possible opportunities for innovation. These interviews were held with around 30 individuals with a range of disabilities, the aim was to find and characterise generic problems associated with daily living. The results of this were complex as each individual has unique needs and specific tacit knowledge. However several common conclusions were made.

- All interviewees felt that they were stigmatised by their equipment
- Most interviewed recognised that assistive equipment was essential and that there was plenty of room for improvement
- All interviewees were able to site situations where they had been injured or seriously incapacitated by equipment failure
- Most interviewed said that they have never been asked or involved in the design of equipment
- Many individuals had made modifications to their equipment to meet their own personal needs

Once the priorities and needs of users and providers were clearly identified, the design process could begin. Throughout the project design groups were set up to test new ideas and prototypes with users and providers. Showing designs, models and test rigs to the groups provided the essential feedback needed to guide the design development. Individuals with specific disabilities had insights which would normally be overlooked by designers and engineers, these were often crucial to the developing design.

Out of this process came two innovative products with significant added value in terms of usability, flexibility and visual aesthetics. The resulting toilet frame design has height adjustable arms which can be easily moved out of the way for wheelchair transfer, or to allow other householders to use the toilet without obstruction. Visually the design is interesting and more sympathetic to current trends in home furnishings. It can also fit around any toilet without the need for alteration or permanent fixing.

The new 'Safety sides' have been designed to fit any type of domestic bed securely. Fabric sections span the side of the bed and pose no entrapment risk to the user. Fabrics are more sympathetic, are warm to the touch and can be visually very attractive.

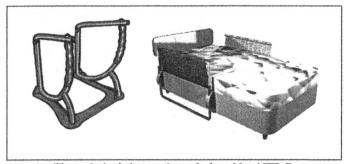

Figure 2. Assistive products designed by AHEaD

This work has proven that inclusive design methodologies have a place in the design of assistive products. A business plan is currently being developed to commercialise these products. The business case for such products is becoming more apparent to industry.

"New markets for products and services will open up in the UK as the number of people who are 60 and over rises by more than a half to 19m in 2030." (Foresight aging population panel, 2000). Figures like this cannot be ignored by industry for much longer, the commercial viability of inclusive design is becoming an economic fact.

Inclusive design in other settings

The project also looked at inclusive design in other areas of health and in the wider consumer world. Medical products used by patients can be enhanced by inclusive design to improve the experience of medical treatment. Here the AHEaD project worked with a regional hospital to develop a product which enabled patients to move more freely when fitted with a catheter. This new innovation gives patients more independence and greater freedom of movement.
AHEaD also worked with a Birmingham based jewelry company on the design of a medical identity tag for children. The companies existing product range consisted of necklaces, watches and bangles which are engraved on the reverse with the Childs medical condition. The company

was experiencing difficulties with products issued for children under ten years old. These existing products were not suitable or durable enough for younger children. As with the early assistive equipment design groups it was necessary to conduct user research with parents and children to find out their particular needs. From this research a range of durable and engaging products was developed. Unlike the traditional jewelry products, these new designs were designed to be indestructible and meet the changing lifestyle needs of children from aged 2 to 10 years old.

In the wider consumer setting AHEaD worked with an innovator on the design of an inclusive bathroom system. This project was an attempt at finding ways to design out many of the traditional barriers users with mobility problems experience in the standard bathroom design. This re-design of the original TRIAD hygiene unit was an attempt to make a more flexible modular system. The function, position and orientation of the toilet, sink, shower and seating surfaces were studied. Designs and full rigs were tested with users to identify the most accessible configuration. The ultimate aim of this project was to design a bathroom which met the needs of able bodied and disabled users alike. During the design process it became clear that the new configurations were taking up less space than standard components, meaning that the new system could be installed in smaller spaces. This makes the design attractive to home owners with limited space but the need for a second bathroom. The final system will be marketed as a compact bathroom without making any reference to ability.

Conclusions

Perhaps the most important conclusion from the AHEaD project work was the realisation that designers don't have the enough information or experience of the needs of disabled people. In order to design more inclusive products designers need to understand disability better. The only real experts here are disabled people themselves who have the insights and tacit knowledge that come from a truly personal perspective. As designers we find it easier to design for ourselves or the 95th percentile man, but this blanket approach widely excludes large groups of society. Starting the design process with ourselves in a very productive strategy, however this must be followed up by involving others with varying abilities.

Ultimately users must be central to the design process if we want to develop truly inclusive products, services and environments.

References

Mitchell, J. and Bennington, J. 2002, *Piloting the wheelchair Lifemap in Sandwell*, (Ergonova, Sheffield)

Foresight Aging Population Panel, 2000, *The age shift, a consultation document*, (Department of Trade and Industry, London)

COGNITIVE PERFORMANCE EVALUATION WITH Δ METHOD AT LIPS

L. BAYSSIE, P. BONNET, L. CHAUDRON, P. LE BLAYE

ONERA/DCSD/PSEV
French Aeronautical and Space Research Establishment
Systems Control and Flight Dynamics Department
Human Performance Simulation and Flight Testing Unit
2 avenue E. Belin, 31055, Toulouse Cedex 4, France
{bayssie,bonnet,chaudron,leblaye}@onera.fr

Abstract

When aiming at improving the safety and the performance of socio-technical systems, the study of the relations between the variations of the operator's cognitive performance and his task is a fundamental issue. In the particular context of human factors studies in aeronautics, appropriate experimental environments and assessment methods are crucial for the evaluation of new interaction systems. We propose a closed-loop methodology, called Δ, for assessing person-system interactions through the causal relations between the variations of the operator's cognitive performance P and the variations of his task T, *i.e.* dP/dT. The Δ method is currently used at the LIPS (Pilot-System Interactions Laboratory).

Our contribution consists in a conceptual and experimental presentation: the Δ model; a first set of aeronautical experiments on noise annoyance and results; the *Eucalepic* conjecture and its application at the Pilot-System Interactions Laboratory (LIPS).

Keywords: Cognitive Performance Assessment (Δ Method), *Eucalepic* Effects, Pilot-System Interactions Laboratory (LIPS).

Performance assessment with Δ

Objective

Controlling complex systems (*e.g.* civil or military aircraft piloting, air traffic control...) generates specific problems in which the quality of the interactions between the operator and his system is critical. Thus, the conjoint improvement of flight safety and performance of aeronautical systems leads to search for some validated modifications of these interaction systems in terms of ergonomic, procedures and training. The main problem is the evaluation of the effective result of an improvement of the interface on the operator's activity. Hence, our approach aims at building an objective methodology called Δ for assessing the effects of evolving interaction systems on the actual performances reached by the operator. The man-machine interaction system evaluation is studied through the causal relations between the variations of the parameters of the task T and the variations of the operator's cognitive performance P, measured through his activity, *i.e.* dP/dT.

Performance evaluation

The operator's activity is classically analysed by two axis: - 1) the estimation of the task's conditions which contributes to the comfort and easiness of the work, - 2) the operator's performances which are actually measurable. Frequently, the only parameters considered for the evaluation of new interaction systems are subjective ones (type 1), *e.g.* inner feeling of workload. Unfortunately, various types of bias appear in the exclusive use of these parameters: - the invasive part of the on-line protocols (e.g. increase of the degree's feeling..) - the self knowledge bias (e.g. rationalisation *a posteriori*..) [1]. As the essential point for the aeronautical conception is to guarantee the results' liability as well as in terms of performances effectively measured as their acceptation and their conscious level by the operators, the operator's performance is defined both in terms of objective and subjective data. This led to establish three classes of parameters :

1) the subject's **Actual-performance** A,
2) the subject's **Feeling** F relative to the task,
3) the **Self-evaluation** S of the operator's own performance.

The global performance is defined as a triple $P = (A, F, S)$ which is analysed combined with the task T (see Table 1).

	Objective	Subjective	Examples
T Task	×		mission, reasoning test ...
A Actual-performance	×		flight parameters, answer vector ...
F Feeling		×	TLX scale, free text ...
S Self-evaluation		×	answer sheet, ... free text ...

Table 1. The quadruple (T, A, F, S).

Pragmatic methodology

The main topic of the Δ method is to analyse the relations between the task (and its variations) and the global performance (and its variations), considering a corpus of experiments. This implies to take into account a set of operators or (ergodic hypothesis) and a set of instances of the operators' performances. The methodology is summarised as follows:

(a) **Representation step**: definition of the quadruple (T, A, F, S),
(b) **Preparation step**: determination of the set of operators: $\{O_i\}_{i \in [1,n]}$ and the controlled variations of the task dT,
(c) **Experimental step**: for all O_i, effective experiments of the tasks T and $T + dT$, recording of (dT, A_i, F_i, S_i) in a database,
(d) **Qualitative analysis step**: the database is formally analysed so as to produce: **clusters** and **rules**.

As far as human factors are concerned, our methodology relies on symbolic data and qualitative analyses. The qualitative analysis tool is the Generalised Formal Analysis technique described in [2]. The mathematical groundings are described in an algebraic frame in which a lattice structure, the Cubes model, is defined for cognitive performance modelling. Rules induction techniques are also used so as to produce causal relations.

Experiments

Introduction

A classical explanation for reasoning errors is a bad ratio of the task's demand and the operator's supply in cognitive resources [3]. Hence, a standard assumption is that a high workload implies a low operator's performance. If we suppose that $t_e <_T t_d$, the task space T, and $p_d <_P p_e$, the operator's performance space P, in which the "e" in t_e (resp. p_e) means "easy" and also "d" = "difficult", are partially ordered spaces, the challenge is to verify if the following standard axiom is always true:

$A_1 = $ *The operator's activity f is a decreasing morphism from* $(T, <_T)$ *onto* $(P, <_P)$.

Aeronautical experiments

Involved in a research program on noise annoyance, a preliminary study was conducted so as to determine the correlations between aircraft's noise parameters and annoyance's feelings. The formal analysis of the experimental data has shown stronger links between annoyance's feeling and the current person's activity than between annoyance and noise's level [4]. In order to get an objective understanding of the noise annoyance's phenomenon viewed as a cognitive performance disorder, we have focused on the study of the relations between the effect of aircraft's noise and the operator's cognitive performance.

We conducted a first experiment – during Paris Air Show 2001 – in which we included a comfort parameter in the environment (aircraft's noise vs calm) of a set of aeronautical relevant cognitive tasks: temporal reasoning in process supervision with the Earth methodology (Artificial Assessment of Human Temporal Reasoning [3]) and deductive reasoning about flight procedures evaluation.

54 voluntary participants – visitors or professionals of the Paris Air Show – have taken part in the experiment. Every participant has passed each task both in a noisy environment (aircraft's noise) and in a normal sound environment.

The performance's measure corresponds to the triple (A, F, S) previously defined : the Actual performance A are quantitative (time response) and qualitative (errors); a questionnaire is distributed to each person in order to express his Feelings F of the task's conditions (real aircraft's noise vs calm) and also his Self-evaluation S of his own performance.

The results are analysed through a qualitative analysis model called "Generalised Formal Analysis" (GFA), which allows qualitative clustering (here, 57 groups of subjects, common (T, A, F, S) features) and symbolic rules induction according to their degree of plausibility. The main result is interesting in so far as for a significant proportion of the subjects (52%), they did improve their performance while they thought they degraded them and also felt annoyed about noisy conditions. Indeed, the functional relation between the partially ordered task space T and the objective/subjective performance P is *not* monotonic.

The *Eucalepic* effects

As we noticed that disturbance may lead to a better performance, A_1 can be refuted in many actual situations. Thus, a second order effect can be described: in particular contexts, function f does increase. This surprising result needs a definition:

the $\boxed{Eucalepic}$ effects occurs when a "more difficult" task leads to a "better" performance of the operator [5] (see Fig. 2).

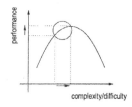

Fig. 2. A eucalepic situation.

The word "Eucalepic" is built from two Greek words:
$\epsilon\upsilon$ = "good" and $\kappa\alpha\lambda\epsilon\pi\sigma\varsigma$ = "difficult".
The working *Eucalepic* situation can be formalised as follows:

$$(ii)\ (t_e <_T t_d) \wedge (p_e <_P p_d)$$

which can be condensed as:

$$\boxed{\text{Eucalepic} = (p_{easy} <_P p_{difficult})}$$

To a certain extent, eucalepic cases are common with any situation in which the relations between stress, workload, complexity... of the task and the performance follows the classical inverted U-shape curve [6], which describes a eucalepic situation. In the same way, eucalepic situations emerged in reasoning activity for which a special model, "Continuous Inference" has been built so as to track the cognitive performance and also to capture the formal neighbourhood of a task T. The model also offers learning capabilities for a human cognitive agent [2].
The current work consists in tuning the Continuous Inference model so as to capture the pivotal mechanism of the Eucalepic effects. The current study is devoted to the analysis of the influence of the hypothetico-deductive content of the instructions on the pilot's activity in the flight simulator LIPS.

Pilot-System Interactions Laboratory

LIPS provides an open and flexible modelling and simulation infrastructure to evaluate new concepts and assistance systems, considering the human operator specific aspects.
The distributed and modular architecture permits real time human-in-the-loop experiments, where the following parameters can be adapted according to the studies: - number of pilots and observers, interaction/command modes for the pilot, number and type of autonomous and controlled mobiles in the general scenario, 3D data bases for the synthetic visualisation of the scene and objects, Head Up Displays (HUD) and control displays, evolution models for the mobiles (civil or military aircraft, helicopters, missiles, terrestrial mobiles)...
Therefore the experimental environment LIPS offers rich functionalities specifically adapted to our new aeronautical experiments.

Discussion

Let us consider the complexity of the information exchanged during a professional dialogue (*e.g.* flight crew activity). Each piece of information is a local task to the listener and can be analysed through its hypothetico-deductive content. Indeed, depending on the context of the mission, the pieces of information will not be provided within the same frame. This tuning seems to be natural [7] but difficult to capture. In the context of a flight mission, one can notice that the time and information constraints are varying with the technical phase of the piloting activity (resp: characteristic time δt): piloting ($\delta t < 1s.$), guidance ($1s. < \delta t < 5s.$), navigation ($5s. < \delta t$). In the same idea, the fine grain analysis of actual crew (and ATC) dialogues have shown that the dialogues could describe the whole spectrum of the basic speech acts [8]: direct ("No windbrakes!"), indirect ("Without windbrakes so as to protect the engine"), or implicit ("If a dysfunction in the calculation device is observed then it could be relevant to avoid any speed reduction device.")

Based on the inferential content of the sentences (negation, implication, disjunction - which are the most critical operators), the compexity of the information is studied through a navigation task, inside which some piloting actions have to be performed. The first pre-experiments, THD_0 on LIPS, suggest that the adequation between the flight phase, the speech acts and the inferential content follow the eucalepic principle. The current complete experimental campaign THD_1 on LIPS (01-02/04) is dedicated to that purpose.

References

[1] J. St. B. T. Evans. *Biais in human reasoning: causes and consequences.* Lawrence Erlbaum Associates, Hove and London (UK), Hillsdale (USA), 1990.

[2] L. Chaudron, N. Maille, and M. Boyer. The CUBE lattice model and its applications. *Applied Artificial Intelligence*, 3(17), March 2003.

[3] J. Pastor. Cognitive performance modeling. Assessing reasoning with the Earth methodology. In *Modelling Human Activity*. Coop'2000, Mai 2000.

[4] M. Boyer. *Induction de régularités dans une base de connaissances, application au phénomène bruit/gêne et ses extensions.* PhD thesis, Supaèro, décembre 2001.

[5] L. Bayssié and L. Chaudron. Delta: a cognitive evaluation of interaction systems. In *CESA'2003Computing Engineering in Systems Applications*, Lille, France, 9-7 July 2003. IEEE Transactions on Systems, Man and Cybernetics.

[6] Robert M. Yerkes and John D. Dodson. The relation of strength of stimulus to rapidity of habit-formation. *Journal of Comparative Neurology and Psychology*, 18:459–482, 1908.

[7] J.B. Van der Henst, L. Carles, and D. Sperber. Truthfulness and relevance in telling the time. *Mind and Language*, pages 457–466, November 2002.

[8] J.R. Searle. Speech Acts. Cambridge University Press, 1969.

CHANGES IN ORGANISATIONAL ROLES WHEN DISASTER STRIKES, WITHIN THE MANUFACTURING DOMAIN

C.E. Siemieniuch[1], M.A. Sinclair[2]

[1]*Dept of Systems Engineering,* [2]*Dept of Human Sciences,*
Loughborough University
LE11-3TU

During a team-working study in a steel rolling mill, a 'disaster' happened; a hot, flat plate undergoing rolling became trapped under a set of rollers. It took 3 days to restore production, at a loss of £1 million per day. The purpose of the study was to predict likely outcomes of a new staffing plan for the steel mill, and this real-life scenario provided some useful information for the exploration and evaluation of this staffing plan. We outline a number of the issues that emerged from this analysis.

Steel mills provide an excellent example of a complex, heavily-automated socio-technical system, with emergent properties. This particular system was designed from a strongly-engineering perspective. The assumption was that 'normal' performance would be the standard state, with a few perturbations from time to time. This was not the case in reality, and frequent human interactions and interventions were necessary for 'normal' performance to be maintained. The implications for human knowledge and communications will be discussed, both from the viewpoint of the operators, and the managers. The effects on this of the proposed teamworking plans are discussed; the likely nature of decision-making; the effects on communications and trust; and the likely effects on the relationship between managers and operatives.

Some general, system-engineering design principles can be drawn from this

Introduction

During a team-working study in a steel rolling mill, a 'disaster' happened; a hot, flat plate became trapped under a set of rollers. It took 3 days to restore production, at a cost of £1 million per day.

The purpose of the team-working study was to predict likely outcomes of a new staffing plan for the steel mill, and this real-life scenario provided useful information and insight for the exploration and evaluation of this staffing plan. We outline some of the issues that emerged from this analysis.

Steel mills are a good example of a complex, heavily-automated socio-technical system, with emergent properties. The mill extended over some 400 metres, in two sections. The rear section was the cool end, where coils of steel plate were sorted and dispatched to customers. This was not part of the study. The front end, the 'Hot Mill' was the focus, where steel slabs were selected according to customer demand, reheated, rolled to near-specification, cooled and rolled to specification, and then cooled, coiled and banded. This is a continuous process where the properties of the product are strongly influenced by the cooling and rolling processes.

This system was designed with cost-efficiency and agility in mind, and from a strongly-engineering perspective; the result was a line controlled by computers having two shop floor managers and six shop floor operators with supervisory control distributed over 300 metres of the hot mill, within a shift of some 30 people. Production was to be 24 hours, 7 days a week, with 21 shifts. There was a 'Day Support Team' with two senior managers, working normal hours each day,

The cultural assumption and managing ethos of the company was that 'normal' performance would be the standard state, with a few perturbations from time to time. Normal performance was visualized as full production with the automation being in control moment by moment, reflecting the organisation's deep understanding of steel making, with the operators adjusting parameters from time to time as required by contingent circumstances, and with planned maintenance of the line. This was not the case in reality as the staffing indicates: frequent human interactions and interventions were necessary for 'normal' performance.

The teams in the study

The current arrangement was that the hot mill shop floor crew comprised one single team, divided loosely into operators and support crew. The plan was to create two semi-autonomous teams in the hot mill in each shift, which would be self-organising. The hot mill would be divided into a 'hot' team, who select the slab, heat it in the reheat furnace, and roll it to near-specification. The 'finishing' team would take the rolled plate, finish-roll it to specification, coil and band it. Support roles e.g. electricians, etc, would be allocated to one or other of these two teams. This was a significant change to the current situation of one, hierarchically-controlled team, with paternalistic management and strong unions. However, because business conditions in this industry had been poor for so long, unions and the management had learnt to work together for their mutual salvation.

As this plan was being put into effect, it became possible to explore the implications of the planned structures.. During the period in which the exploration took place, the 'disaster' happened. When a steel plate is undergoing final finishing, it passes once through a set of 6 rollers, with a guillotine machine at the beginning of the set. .By the time the plate reaches the finishing rollers, it is about 70 metres long, weighs 20 tonnes, is at a temperature of about 900°C and is travelling at 50-70 km per hour. For some reason, the leading edge stopped just after the rollers, causing the rest of the plate to buckle in the rollers, and the trailing edge to wrap around the guillotine. Fortunately, no personnel were near the rollers at the time..

Fixing the disaster

Immediately, management was informed and the line was cleared of all other product. Over the next day, the rollers and guillotine machine were dismantled, the solidified plate removed, replacement rolls were put into the rollers while the old set were re-machined, and the line was reassembled. At the same time, the software was re-analysed to discover if a bug was the cause. By the third day it was possible to restart production. This represents a major effort by maintenance personnel, in the teams

Observed organisational changes

Organisationally, several significant changes occurred. Fig. 1 shows the original organizational set-up, under 'normal' production conditions. The axes of the matrix are 'Degree of discretion in achieving target and executing operations', and 'Degree of responsibility in planning resourcing and scheduling operations to achieve targets'; these form a bi-dimensional map of decision authority, and, by implication, official status. Hence, top right is the acme of power, and bottom left is where nobody should be. The blobs are roles, and arrows are known, well-used communication paths connecting the roles, by which authority is translated into activity. PM is the Plant Manager, top right; FR on right is the Floor Roller, supervising the operation of the mill on the shop floor. Between these two are management and engineering roles; other shop floor operators are all grouped towards bottom left.

While senior management may well be dealing with the external interfaces, their authority (and presence) on the shopfloor is, if anything, diminished

Support staff, particularly fitters, have their roles enhanced (i.e. their blobs have moved upwards and to the right), markedly changing their relationships with other roles compared to normal.

Organisational implications

Utilising other work in the mill on communication patterns, as well as stakeholder interviews, the following conclusions were drawn:

- Under 'normal' operations, the line is still dependent on its older, pre-automation staff for smooth running. But their 'tweaking' skills are being lost due to retirements and loss of use with increasing automation.
- These older workers are also very important in disaster scenarios; their tacit knowledge of the behaviour of the line, gained form manual control days, is very helpful in minimising down time and foreseeing problems.
- Furthermore, the group cohesion built over a long period of time, with the experience gained of each other under different operating scenarios, indicates that the ad-hoc role changes that occur for different scenarios of operation are not likely to cause undue friction and difficulties. This needs to be recognised in the design of roles and teams
- Their understanding related to the behaviour of the whole line , but new teamworking plans envisaged placing a boundary across the line, before the finishing rollers. Either side of this, the teams would have their own performance metrics, goals, and (eventually) loyalties. In effect, an artificial, organizational fence could be created, splitting in two a continuous process whose main operational communication flows run across the fence. It is not obvious how this will assist normal operations, and may interfere with the current distribution of organisational knowledge and decision making structures that evidently swing into place when disaster strikes
- The existing divisions between management ('the suits') and the shopfloor, as indicated by the pattern of communication links, are unlikely to change, given the plans as enunciated. Decision paths and structures are not reflected in communication paths and new team boundaries and roles are likely to increase this dichotomy at the expense of response times and disaster avoidance. 'High power' distances between the two groups will remain.
- The new structure will impair 'situational awareness' since it fails to address the

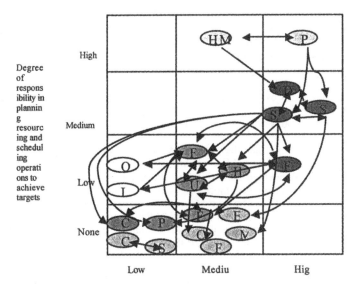

Degree of responsibility in planning resourcing and scheduling operations to achieve targets

Degree of discretion in achieving target and executing operations

Fig. 1 Chart of roles in rolling mill, normal' operations.

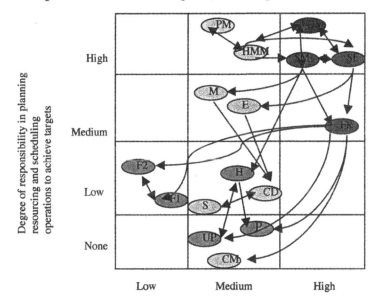

Degree of responsibility in planning resourcing and scheduling operations to achieve targets

Degree of discretion in achieving target and executing operations

Fig 2 Chart of roles for'disaster retrieval'.

need for coherence of communication flows and tacit knowledge sharing (e.g. reports may still be 'massaged' and many key shopfloor roles remain the passive recipients of information with little opportunity to provide real input to the decision making process).

- The intended blurring of roles so that team members can take on more than one role will allow more flexible operation. However, this benefit will depend heavily on management evangelism and trust, allied to quality training over an extended period. Furthermore, such benefits may only be garnered within teams rather than across teams with the consequent negative effects on tacit knowledge transfer.

- Both shopfloor groups; the older workers and the younger workers, expressed concern about the overall loss of skills as the older workers left the business. They both commented that while the steadily increasing level of automation meant that better quality, more consistent steel could be produced, they felt ever less in control of it. The younger workers were definite in saying that in the event of a breakdown of the automation they could not as a team operate the line manually; the older workers were worried that those still in work were losing their skills due to lack of use. Many of them said they no longer trusted themselves to operate the line manually.

- Given the role changes that are indicated above, requiring wider knowledge and authority than under normal operations, it seems evident that simulation of the line in a variety of decision making scenarios would be a very necessary skills building tool. Given the level of knowledge as evinced by the automation, it should not be hard to develop a simulation model of the line on which skills could be dev eloped. In turn this would enhance both disaster avoidance and disaster management skills/roles for all personnel involved.

Acknowledgements

The authors would like to thank the company concerned for providing this field research opportunity, and to recognize the support provided by EPSRC, without which this research would never have happened.

DEVELOPMENT OF A PROCESS FOR HUMAN RELIABILITY ASSESSMENT IN A DEFENCE INDUSTRIES CONTEXT

G.A.L. Ng[1], W.H. Gibson[2], B.A. Kirwan[3], C.E. Siemieniuch[1], M.A. Sinclair[1]

[1]*Loughborough University*
[2]*University of Birmingham*
[3]*Eurocontrol, Bretigny-sur-Orge, France*

BAE SYSTEMS wished to upgrade its in-house human reliability assessment (HRA). It believed that soon its methods and databases would soon cease to be state-of-the-art, and it wished to upgrade both. This paper reports the initial studies defining the problem, and identifying a possible solution. This paper reports the initial studies defining the problem, and identifies a possible solution.

A scoping study was undertaken within BAE SYSTEMS. Because BAE SYSTEMS has a presence in several defence sectors, this exercise produced some differing requirements, reflecting the different characteristics of the clients in the sectors. For example, for the surface ships sector, it is sufficient to use the opinions of Subject Matter Experts from the Navy; for submarines, rigorous, detailed analyses are required conforming to the tenets of the Nuclear Installations Inspectorate.

Other studies into the methods currently available were undertaken. These supported the findings of Dougherty (1990) and Cacciabue (2000) and others: we are still using first-generation techniques, whereas the need is for second-generation, dynamic reliability techniques; the data on which these techniques rely is not consolidated, and have many gaps; at different stages of the design, develop and deploy process there will be a need for different HRA techniques, appropriate to the level of maturity of the product.

Introduction

BAE SYSTEMS has a presence in most sectors of the military domain, dealing with a very wide range of military customers, both national and international. These require various critical subsystems to have a Human Reliability Assessment (HRA) exercise to be performed to qualify and certify the whole system for military use. Unfortunately for BAE SYSTEMS, there is little agreement among the Navy, Army and Airforce of a single military client, let alone across clients, as to what constitutes a satisfactory HRA process. However, because the clients become committed for decades (the life cycle of a defence system tends to be 40 years or more), certain methods become standard practice over the years; it means that different parts of BAE SYSTEMS addressing different clients were using incompatible methods and datasets for this.

BAE SYSTEMS' initiative

BAE SYSTEMS are aware that this state, although standard across all defence industries world-wide, is about to become dated and wished to move ahead of this state. Furthermore, the accepted state of HRA has been criticised for some time. For example, Dougherty (1990) made trenchant criticisms, summarised below:
- Less-than-adequate data
- Less-than-adequate agreement in use of expert-judgement methods, both agreement between judges and accuracy of predictions
- Less-than-adequate calibration of simulator data, because simulators are not calibrated either
- Less-than-adequate proof of accuracy in HRAs, particularly for non-routine tasks
- Less-than-adequate psychological realism in some HRA approaches
- Less-than-adequate treatment of important performance shaping factors (PSFs) such as managerial methods, cultural differences, and interpersonal variability

A consequence of these known deficiencies is that predictions are conservative and probabilities are over-stated. BAE SYSTEMS commissioned a 1-year scoping study, with the aims:
- To determine what are the priorities for generation and use of HRA data within BAE SYSTEMS to meet anticipated requirements
- To document the external availability of current data and methods in the area of HRA from a BAE SYSTEMS perspective
- To establish the reliability and applicability of the Human Reliability Assessment data available to BAE SYSTEMS, both internally and externally
- To improve the quality and maturity of these data where necessary
- To specify a sustainable process to manage the generation, validation and application of HRA data and methods within the Engineering Design Life Cycle in BAE SYSTEMS.

Initially, a number of interviews were conducted with HRA stakeholders. These fell into 4 classes; users, internal clients, Integrated Project Team (IPT) managers, and secondary stakeholders. Some external interviews in other defence companies and academic communities were also held; these corroborated many of the issues that surfaced during the interviews.

Initial findings

- There was little disagreement that a generic process was needed for HRA.
- There was agreement that this process ought to be tailorable to meet the needs of different project teams and the requirements of their particular clients.
- The HRA process would have to be aligned with the major design and lifecycle processes in defence. For instance, the process ought to acknowledge the specifics of

the Capability Maturity Modelling and Integration (CMMI) philosophy, which is becoming widespread in defence circles.

- This HRA process would need to include a toolbox of HRA methods, which could be selected for particular kinds of analyses, and at different stages of the design process.
- These tools would necessarily have to deal with different phases of product maturity; probably concept, detailed design, and design verification, though other phases might well require HRA as well
- Of the many tools explored, only a few seemed to suit the needs of the IPTs with their emphasis on engineering
- There was a dearth of applicable, reliable data. One of the most promising data sets is CORE DATA, maintained by the University of Birmingham, but this too needs extension.
- It appears that only one second-generation tool is applicable to the needs of BAE SYSTEMS; CREAM (Hollnagel 1998)
- Consideration should be given to follow-on projects after this scoping study; firstly exploration of the use of simulator data that already exists as a source for extension of CORE DATA; and secondly, adaptation of existing tools to meet the particular needs of defence is required.
  There appears to be no tried and tested tool which considers teamwork. There is a distinct need for such a tool.

Follow-on work is continuing to refine and extend these findings, and to implement them in a proposed process for executing HRA.

A proposed process for HRA

For the rest of this paper, we discuss initial ideas for a process. Fig 1 below illustrates the basic concept underpinning this process.

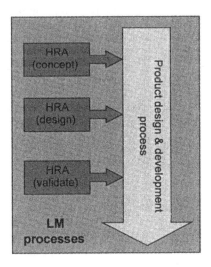

Fig 1 Where HRA tools will be used in the design and certification process.

However, the application of the tools must be managed, so a process is required to do this. Fig 2 illustrates this process.

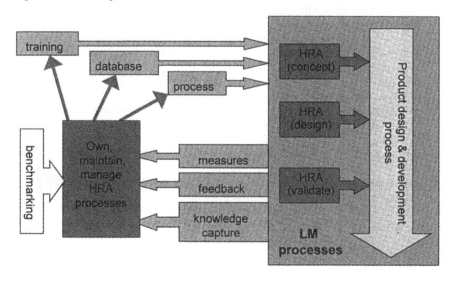

Fig 2 the whole process for managing HRA

At top left are the support requirements for use of the tools. These include maintenance of the tools, upgrading of HRA databases through feedback and external activities, and training of personnel (both initial and ongoing) to ensure there is a ready pool of personnel able to conduct HRA analyses.

In the middle at the bottom are the feedback loops to enable management to happen. This includes feedback on HRA studies carried out, measures of efficiency, and the capture of knowledge about the tools and their performance, and about usefulk data discovered during the studies that could be used in other studies.

At bottom left is the bench-marking activity; this is a slight misnomer, since it includes all outwards-looking activities which are likely to improve the quality and maturity of the HRA process, not just bench-marking per se.

The 'own, maintain and manage' box is where the process is controlled. It is the quality of what goes on in here which will determine whether or not this process is a success, long-term

Risks for this process

The scoping study was also able to identify some of the risks in what is proposed. Three particular risks, relevant to any ergonomics project, are:

Deploying the process only when it is ready. If deployed too early, the process will lose credibility. This means that an extensive pilot phase will probably be needed; this will depend on the availability of resources.

If the HRA process is aligned to another process, if HRA fails then the aligned process will be detrimentally affected also. Again this points to a necessary validation stage before roll-out.

There is a need to prove to the business that there are solutions to three concerns: that the HRA process is not useful, that the HRA process is not credible, that the HRA process doesn't fit with the current processes used. Failure in any of these will result in failure of adopting the process.

Acknowledgements

The authors would like to acknowledge the resources, commitment and enthusiasm that BAE SYSTEMS has provided for this research.

References

Cacciabue, P. C. (2000). "Human factors impact on risk analysis of complex systems." Journal of Hazardous materials **71**: 101-116.

Dougherty, E. M. (1990). "Human reliability analysis - where shouldst thou turn?" Reliability Engineering and System Safety **29**: 283-299.

Hollnagel, E. (1998). Cognitive reliability and error analysis method. London, Elsevier.

TAKING THE WRONG SPUR(R):
STOPPING BEHAVIOUR

John R. Bloomfield[1,2], Kathleen A. Harder[1], and Benjamin J.Chihak[1]

[1]*University of Minnesota, Minnesota, USA*
[2] *University of Derby, Derby, UK*

Lee (1976) used real world stopping behaviour, reported by Spurr (1969), to
support his theoretical work on the visual control of braking. However, as
Kaiser and Phatak (1993) point out, the data used by Lee came from a single
driver who decelerated at approximately the same rate whatever the initial
speed—Lee ignored the stopping data of Spurr's other 14 drivers whose
average deceleration increased as a function of their initial speed. This
selective use of data, may explain why stopping data obtained in a simulator
study, reported by Bloomfield *et al* (2003), do not support Lee's theory. This
paper presents additional analysis of the Bloomfield *et al* data. Between
three and ten discrete stages can be identified in the 98 stopping profiles they
obtained. These stages involve several responses—gradual deceleration,
moderate deceleration, definite deceleration, coasting and, even, acceleration.

Introduction

When developing his influential theoretical work on the visual control of braking, Lee
(1976) was concerned with how drivers avoid colliding with stationary or moving
vehicles. Whenever a driver approaches a vehicle, perceptually the edges of that vehicle
appear to move further from each other. The rate at which the edges appear to move
away from each other can be used to predict time-to-collision (TTC). Lee (1976) used
the term 'tau' to refer to the visual variable specifying TTC when the closing velocity is
kept constant, and suggested that drivers use 'tau dot'—the first temporal derivative of
'tau'—to control the severity of contact. Lee supported his ideas using data reported by
Spurr (1969). Spurr had recorded the deceleration patterns of 15 drivers—six of whom
were professional drivers—as they stopped at bus stop signs and traffic lights. Spurr
reported that, for 14 of his drivers, the "average deceleration increased with increasing
initial speed" (Spurr, 1969, p. 59). In contrast, he also reported (Spurr, 1969, p. 59) that
a 15th driver "used approximately the same average deceleration during stops from all
speeds"

What Lee (1976) failed to report was that the Spurr data he used to support his
theoretical work came from this 15th driver. It was not until much later—after Kaiser
and Phatak (1993) re-examined Spurr's paper—that the fact that Lee used data from a

single driver was pointed out. To say the least, the "nonrepresentative nature of this driver's strategy" (Kaiser and Phatak, p.199) is problematic.

Not knowing that Lee's theoretical ideas were supported by data from a single driver, a number of researchers explored and extended these ideas using experimental paradigms that did not directly involve braking performance. Instead they used passive observers. For example, in one paradigm in which TCC is estimated, the observer approaches a stationary vehicle, at a constant speed but, before he or she reaches the object, it is obscured. The subject continues to move toward the obscured object and indicates the point at which he or she believes a collision would occur. Research using this paradigm is reviewed briefly by Cavallo *et al* (1998) and by Groeger (2000). A second paradigm using passive observers was employed by Kim, Turvey and Carello (1993). They presented a simulated braking pattern, truncated the trajectory, and asked the observers to judge whether the baking pattern would result in a "hard" or "soft" contact.

There has been a considerable amount of work looking at perceptual variables related to stopping behaviour. However, as both Kaiser and Phatak (1993) and Boer *et al* (2000) point out, there has been relatively little experimentation involving active braking. As mentioned above, Spurr (1969) had his drivers stop a car at various places on real roads—though not behind another vehicle. And Boer *et al* (2000) compare stopping performance data obtained both on a closed track and in a driving simulator—reporting considerable differences between the two data collection methods.

The data discussed in this paper are stopping performance data obtained using a driving simulator. As previously reported (Bloomfield *et al*, 2003), these data do not support Lee's theory. Additional analysis of these data, which were collected as drivers arrived at a rural intersection at which there was a stop sign, are presented.

The Experiment

The advanced driving simulator used in this experiment is described in detail in Harder *et al* (2003). The 49 subjects (25 males and 24 females), who participated in the experiment, were licensed drivers between the ages of 18 and 65.

The experiment focused on a particular problem intersection in rural Minnesota. A driving scenario modeled on this intersection was developed for the simulation study. The major road running through this intersection— Minnesota Trunk Highway 58 (MNTH-58)—is a North-South road that passes the East side of Goodhue, a small community with less than 1,000 inhabitants. MNTH-58 is a two-lane undivided highway with a speed limit of 55 mph. It intersects with County Road 9 (CR-9), an East-West road that passes Goodhue on the South side. CR-9 is a stop-controlled, two-lane undivided road, also with a speed limit of 55 mph. There are stop signs at the intersection when it is approached on CR-9, but no stop signs when it is approached on MNTR-58.

There were four drives for each subject—two on MNTH-58 and two on CR-9. The subject drove 8.1 km (5.03 miles) on MNTH-58—4.6 km (2.86 miles) before the intersection and 3.5 km (2.17 miles) after it, when starting from the North; and 3.5 km (2.17 miles) before the intersection and 4.6 km (2.86 miles) after it, when starting from the South. And the subject drove 5.6 km (3.48 miles) on CR-9—2.8 km (1.74 miles) before the intersection and 3.5 km (2.17 miles) after it, when starting from both the

North and the South. The order in which the approaches were driven by each subject was randomized.

The subjects were divided into two groups. Twenty-four subjects in Group 1 drove in a Before Condition in which the intersection was modeled as it currently exists in the real world. The 25 participants who were in Group 2 drove in an After Condition, in which a number of modifications were made in order to increase the saliency of the intersection—the experiment is described in more detail by Harder *et al* (2003). Whether driving in the Before or After Condition, the subject was informed that he or she would be driving on a two-lane undivided highway with a speed limit of 55 mph, and asked to drive as he or she "normally would."

Results and Discussion

In this paper, we are concerned only with the approaches to the intersection on CR-9, from the East or West. [Note: The driving performance of the subjects when they approached the intersection from the North and South on MNTH-58, and did not encounter stop signs is presented by Harder *et al* (2003).]

Distance from Intersection at which Stopping Behaviour Began
On CR-9, before the intersection came into view, the subjects drove at an average speed of 96 km/h (60 mph), whether starting from the East or West. We determined the distance from the intersection at which each subject began to slow down—i.e., when the subject took his or her foot off the accelerator. The mean distance at which this occurred for those subjects driving in the Before Condition was almost identical whether they were approaching from the East or the West. The mean distances were 237.2 meters (778.2 ft) from the East and 236.6 meters (776.2 ft) from the West. Therefore, since the subjects were driving at an average speed of 96 km/h (60 mph), the stopping behaviour began, on average, 8.82 s away from the intersection—a time that is very similar to the 8 s margin value of 'tau' that Lee (1976, p. 443) suggested might be needed "to brake comfortably."

The mean distances at which the stopping behaviour began for the subjects driving in the After Condition were greater than for those who drove in the Before Condition. The mean distances were 263.2 meters (863.5 ft) from the East and 301.4 meters (988.8 ft) from the West. When t-tests were used to compare the distances obtained in the After and Before Conditions, a statistically significantly difference was found for the approach from the West ($p = 0.0357$); while the difference was not significant for the approach from the East.

Given the subjects were driving at an average speed of 96 km/h (60 mph), the mean distances obtained in the After Condition indicate that the subjects began to slow down when they were 9.8 s away from the intersection, when approaching the intersection from the East, and 11.2 s, when approaching it from the West. Therefore, in the After Condition, the times at which stopping behaviour began were somewhat longer than the necessary margin value of 'tau' suggested by Lee.1976.

Deceleration Profiles

Each of the 49 subjects approached the intersection twice on CR-9, once from the East and once from the West. The resultant 98 deceleration profiles were examined. As reported by Bloomfield *et al*, these profiles did not compare well with the theoretical curves expected using 'tau dot'. In fact, a number of different stopping behaviors were observed. Typically, between three and ten discrete stages could be observed in the deceleration profiles. At the extremes, there was one stopping profile with only one stage—in it, deceleration gradually increased and then gradually decreased until the subject stopped—and two profiles in which there were as many as thirteen stages.

The stages that could be identified included the following response categories—(i) gradual deceleration; (ii) moderate deceleration; (iii) definite deceleration; (iv) coasting; (v) accelerating; and (vi) other. Table 1 presents a breakdown, by percentage, of these various types of response for the first three stages of the deceleration profiles.

Table 1. Percent Each Type of Response in First Three Stages

Type of Response	Stage 1	Stage 2	Stage 3
Gradual deceleration	60.0 %	1.4 %	21.1 %
Moderate deceleration	28.4 %	60.5 %	15.5 %
Definite deceleration	3.2 %	19.7 %	19.7 %
Coasting	6.3 %	16.9 %	32.4 %
Acceleration	2.1 %	0.0 %	7.1%
Other	0.0 %	1.4 %	4.2 %

Table 1 shows that, in the first discernable stage, 60.0 % of the profiles began with gradual deceleration. In the second stage, none of these 60.0 % continued with gradual deceleration, but instead with another of the response categories.

Table 1 also shows that, in the first stage 28.4 % of the profiles began with moderate deceleration. In the second stage, none of these 28.4 % continued with moderate deceleration, but instead with another response categories. However, since 71.6 % of the profiles began with something other than moderate deceleration, it striking that 60.5 % of the profiles did have moderate deceleration in the second stage. This means that 88.9 % (28.4 % plus 60.5 %) had moderate deceleration in the first or second stage.

However, by the third stage the response categories were much more spread among the response categories—with as many as 31.4 % showing that the subjects were coasting. And, although Table 1 does not show it, this trend towards spreading out amongst the response categories continued as the number of stages increased from four to ten.

Comments

Boer *et al* (2000) also obtained what they called "multi modal braking profiles" when they explored braking behavour using Nissan's six-degree of freedom simulator. They suggest that this behaviour is uncharacteristic of real world driving performance. However, their comparison was made between simulator driving and driving on a track at the Hokkaido proving ground—and given the conditions of their experiment, it seems likely that their subjects were somewhat familiar with the driving scenario they drove.

The stopping behaviour analyzed here was obtained with subjects who were unfamiliar with the particular roads on which they drove. It seems likely that this is one reason why the braking profiles may have as many stages.

In any case, it appears that there are several different ways in which drivers attempt to stop their vehicles. And a theory of stopping behaviour needs to try to account for all of them—not just select one strategy that happens to fit a theorist's preconceived notions. It is unfortunate that Lee chose to ignore the data from 14 of Spurr's 15 subjects, and did not attempt to provide a more comprehensive description of the different stopping behaviours exhibited by drivers. While—as Robert Frost (1916) suggests in his poem *The Road Not Taken*—it may be admirable in life, when encountering two roads that diverge, to take "the one less traveled," it is not the best direction for a scientist theorizing about human behaviour.

Acknowledgement
We would like to thank Mary Kaiser, of the NASA Ames Research Center Moffett Field, California, for providing us with a copy of Spurr's (1969) original paper.

References

Bloomfield, J.R., Harder, K.A. and Chihak, B.J., 2003, Characteristics of emergency and routine stopping behaviour in simulation studies. Presented at *Vision in Vehicles X*, Granada, Spain, September 7-10, 2003.

Boer, E.R., Yamamura, T., Kuge, N. and Girshick, A. 2000, Experiencing the same road twice: a driving centered comparison between simulation and reality. Presented at *Driving Simulator Conference (DSC2000)*, Paris, France, September 6-8, 2000.

Cavallo, V., Berthelon, C., Mestre, D. and Pottier, A. 1988, Visual information and perceptual style in time-to-collision estimation. In: Gale, A.G., *et al* (editors), *Vision in Vehicles VI*, (Elservier Science Publishers B.V.: Oxford, U.K.), 81-89.

Groeger, J.A., 2000, *Understanding Driving: Applying cognitive psychology to a complex everyday task.* (Taylor & Francis Psychology Press, East Sussex).

Harder, K.A., Bloomfield, J.R. and Chihak, B.J. In press, The Effect of Increasing the Saliency of a Rural Intersection on Driving Behavior. In: Gale, A.G., et al. editors *Vision in Vehicles X*, (Vision in Vehicles Press: Derby, U.K.).

Kaiser, M.K. and Phatak, A.V. 1993, Things that go bump in the light: on the optical specification of contact severity. *Journal of Experimental Psychology: Human Perception and Performance*, **19**, 194-202.

Kim, N-G, Turvey, M.T. and Carello, C. 1993, Optical information about the severity of upcoming contacts. *Journal of Experimental Psychology: Human Perception and Performance*, **19**, 179-193.

Lee, D.N. 1976, A theory of visual control of braking based on information about time to collision. *Perception*, **5**, 437-459.

Spurr, R.T. 1969, Subjective aspects of braking. *Automobile Engineer*, **59**, 58-61.

GENTLE PUSH ...
AN INVESTIGATION INTO SUBTLE SIGNALS

Tom Coombs[1], Martin Bontoft[2] and Diane Gyi[1]

[1] *Department of Human Sciences, Loughborough University, LE11 3TU, UK*
[2] *IDEO, White Bear Yard, 144a Clerkenwell Road, London, EC1R 5DF, UK*

This paper explores signals that exist at the lower limit of perceptibility, such that they are ignorable, and missable. Several benefits are put forward, deriving from various features: the opportunity to signal continuously; the missable nature; the reduced disturbance of nearby others; and the improved aesthetic. Also discussed are the possibilities for altering how likely a signal is to be missed, and how this can be altered for a specific individual. Trials were performed that explored various emotional and behavioural responses to such signals, and their results are presented. It is concluded that despite their obviously reduced ability to deliver information, such signals offer an overall benefit in some specific scenarios.

Introduction

"g e n t l e   p u s h ..." is a term coined to describe an approach to signal design. A signal designed by gentle push principles is one that is delivered very gently. It is perceptible, but only just, and so it is ignorable.

The term refers to the conceptual distinction between information push and pull. In situations where the recipient actively garners the information, such as reading a clock, information is said to be pulled. Where the signalling entity initiates the transfer to a passive recipient, as does a fire alarm, the information is pushed. This paper is concerned with that which is pushed gently: initiated by the signalling device, and delivered subtly.

The prevalent approach to signal design aims to ensure that signals are not missed or ignored. This is rational given that a signal's reason for being is to transfer a piece of information. However it fails to take proper account of the recipient and the environment: an environment that may contain many signals, that may be populated by others to whom the signal represents a disturbance, or a recipient who may be focussed on more important tasks unrelated to the signal.

Issues addressed by g e n t l e   p u s h ...

Interruption is one issue addressed by gentle push. Interruptions can reduce task performance, and cause people to make mistakes (Gillie & Broadbent, 1989), and they can also cause annoyance (Bailey *et al*, 2001). Unpredictable and uncontrollable interruptions cause stress, which can itself have a negative effect on performance. (Cohen, 1980).

These negative effects can be important. On August 16 1987, a Northwest Airlines flight took off from Detroit and crashed almost immediately, killing all but one of the 155 people on board. An interruption while the crew were running through their pre-flight checklist was cited as a causative factor.

Not all interruptions cause error or reduced performance. Some have positive effects. Low complexity, cognitively undemanding task performance increases with interruption (Speier, 1997). Interruptions can positively affect people's ability to remember details of the interrupted task, a phenomenon known as the 'Zeigarnik effect' (e.g. van Bergen, 1968).

Another issue addressed by gentle push is that of social disturbance, caused by a signal that attracts the attention of people other than the intended recipient. This may lead to a loss of productivity similar to that already described, but may be more likely to cause irritation. Social forces may lead to the exclusion of disturbing or irritating signals. This exclusion limits the utility of a signalling device: once it has been excluded or switched off, not even the most important signals can get through.

The third issue addressed by gentle push is the need for aesthetic design. Design of products has moved beyond practicality and function, and the language of the field has come to reflect this with the inclusion of words such as 'pleasure' and 'delight' (e.g. Jordan, 2000). This trend is as relevant to the signals generated by devices as it is to their physical styling.

There are cases of very gentle signals producing benefits even where no obvious issue exists. One example is of the safety-critical environment in which a very gentle bell sounds constantly to indicate that all is well, stopping to indicate a problem. This is only possible with a gentle signal, and it avoids negative consequences of problems with the signalling bell. Other cases cite the ability of gentle signals to indicate a situation continuously such that a signal recipient can more or less subconsciously monitor that signal, and even detect trend information in the signal (Patterson *et al*, 1986). A third case, that of the Arkola bottling plant, shows that by introducing co-workers to signals relating to each other's work, collaboration is encouraged (Gaver *et al*, 1991). Additional signals can be more readily introduced without negative consequences if they are gentle.

Mechanisms of variable awareness

One human ability that is particularly relevant to the current proposition is that of selective attention. This is the ability to take in almost limitless stimuli concurrently, and to filter so as to be able to focus on just one. The effectiveness of the filtering mechanism is demonstrated by the cocktail party phenomenon (e.g. Norman, 1969). A person at a cocktail party is able to filter out the noise of the many nearby conversations, and focus on the words of his/her interlocutor. Despite this focus, the person's ears prick up if his/her name is mentioned nearby. This demonstrates that, rather than blindly filtering those conversations, in fact, each word is processed. This is an example of a stimulus that goes unfiltered, and does so because of its information content, in contrast to the characteristics such as amplitude that we more readily associate with determining how ignorable a stimulus is.

Two types of stimuli are seen here: the easily-filtered words of other conversations, and the person's name, which is difficult for that individual to filter. Other stimuli, such as fire alarms, are universally difficult to filter. This project is concerned with gentle, ignorable signals. They may offer a mechanism by which people are empowered to decide between response and non-response. Other requirements, such as avoidance of disturbing others, may be met by signals that are more noticeable to the target individual, because the individual has become attuned to the signal, as people are to their own name.

Aims and Methods

The project was in two phases. The first phase aimed to sketch out the variety in signalling scenarios. "Signalling scenario" is used to refer to whole situation in which a signal is generated and transmitted, so the signal itself is relevant, as is the generating entity, the recipient, the environment and so on. The aim of investigating this variety was to uncover all the essential characteristics of this type of signal and of the scenarios in which their use would be effective. The phase used both brainstorming and contextual enquiry. Brainstorming served to gather large amounts of data, and contextual enquiry was chosen as a complementary method capable of capturing examples that wouldn't lend themselves to recall in a brainstorm setting.

The second phase aimed to evaluate gentle push concepts. Three concepts were developed from ideas that came up in the brainstorming sessions of phase I. Two of them were investigated using experience prototypes (Buchenau *et al*, 2000). An experience prototype is concerned with responses to an experience, which may be simulated using resources unrelated to the product or concept under development.

In the current case, mobile phones were used, as they were readily available, configurable and remotely controllable. In most cases the participants were happy to use their own devices, which brought the benefit that the signal held real meaning to the participants, a situation difficult to produce experimentally.

The first trial involved asking the participants to set their phones to the lowest, most subtle ringtone. The aim was to investigate the degree to which the participants could attune to their signal, and also explore the behavioural changes and emotional effects of knowing the signal may be missed.

The second trial used an extremely subtle ringtone on a mobile phone, and involved late night signals. The aim was to assess the effect, on both calling and receiving parties, of having a communication channel with no disturbance potential.

The experience prototype trials used interview and focus groups to gather the data.

The third concept was tested by focus group alone, using a basic prototype and scenarios to promote discussion. The concept itself was that of a communication device that delivered an low-disturbance signal to the recipient. That is to say, the signal's physical qualities would be such that it would not be noticed by others, and would be subtle enough not to unduly disturb the recipient. The investigation sought to evaluate the behavioural and emotional responses of individuals to an always-on communication channel.

Results

The first phase of the project produced a set of characteristics of signalling scenarios, and from these was developed a schema for determining the situations in which gentle push signals would be appropriate. The schema is represented in figure 1.

The large oval represents gentle push, and the arrows are the motivators and demotivators for using gentle push signals. Larger arrows are stronger motivators/demotivators. Some of the motivator-demotivator pairs are unequal. For example, "not acceptable if message fails" is a very strong demotivator for using gentle push, because it would itself cause the gentle push option to be dismissed, whereas its opposite, "failure acceptable", would need supporting motivators to make a strong case for the adoption of gentle push.

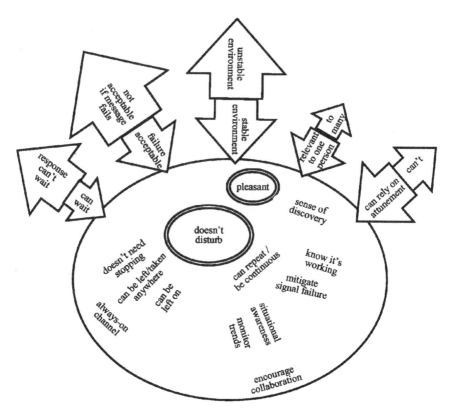

Figure 1 - Motivators, Demotivators and Benefits of using g e n t l e p u s h ...

The following are the results of the investigation of gentle push concepts using prototypes, experience prototypes and scenarios.

In the opinion of the participants, there are many environments that could be made more pleasant without negative consequences. One participant, a doctor, describing the role of a junior doctor, and the huge impact of pager signals on their lives, suggested that even slight improvements to the signal would be very welcome.

Very gentle signals were found to have the potential to create a sense of discovery in the recipient. Receiving the signal was described as less like "being told", and more like having a "hunch". It was thought to be a pleasant experience.

Some participants described becoming attuned to their gentle signal, and that they found it easy to pick the signal out, where others appeared not to notice it.

The opportunity to be able to receive signals in any environment was one that held general appeal. In most cases, participants stated that if the signal were a communication channel, it would have to be accessible by a very select group, in most cases the participants would have it limited to their partner.

Several participants noted that the low-disturbance aspect afforded them greater detachment from the device, for example by not having to consider it when leaving it unattended in the office. The opposite was true when an important signal was expected, the person then had to be more attentive.

Main Findings

The study investigated the potential benefits of designing signals at the very lower limit of perceptibility. A schema describing the effective application of gentle push signals was developed from the data gathered. Various benefits of gentle signals were identified.

Existing evidence showed benefits of such signals in that they can run constantly. A constant signal can be used to contain trend information, or to represent the non-alarm state in safety critical environments, so as to avoid negative consequences of signal failure.

Evidence also suggests that attunement is a powerful tool for targeting signal recipients, the person's name being an example of a highly attuned stimulus. Attunement could allow subtle signals to function effectively.

Gentle signals were seen to have an impact both emotionally and behaviourally. The aesthetic improvement brought a positive emotional response to signalling devices, both from the owner and nearby others. The owner's attachment to a device was also seen to be altered when the device used gentle signals, which had both positive and negative consequences.

Gentle push signals have obvious limitations, but this paper has set out some cases that may benefit from them, and the types of benefits that could be achieved.

References

Bailey, B.P., Konstan J.A., and Carlis., J.V. (2001) The Effects of Interruptions on Task Performance, Annoyance, and Anxiety in the User Interface. In Michitaka Hirose, M. (Ed) *Proceedings of INTERACT'2001*. Oxford: IOS Press

Buchenau, M., Fulton Suri, J. (2000) Experience Prototyping. In: Boyarski, D., Kellogg, W.A. (Eds.) *Proceedings of the Conference on Designing Interactive Systems: Processes, Practices, Methods, Techniques*. New York: ACM Press

Cohen, S. (1980) After-effects of stress on human performance and social behavior: A review of research and theory. *Psychological Bulletin,* **88**(1), 82–108

Gaver, W., Smith, R., O'Shea, T. (1991) Effective Sounds in Complex Systems: The Arkola Simulation. In *Proceedings of the SIGCHI conference on Human factors in computing systems: Reaching through technology*. New York: ACM Press

Gillie, T. & Broadbent, D. (1989), "What Makes Interruptions Disruptive? A Study of Length, Similarity, and Complexity", *Psychological Research* 50(4), 243--50.

Jordan, Patrick W. (2000) *Designing Pleasurable Products*. Taylor & Francis. London

Norman, D.A. (1969) *Memory and Attention*. New York: John Wiley

Patterson, R.D., Edworthy, J., Shailer, M.J., Lower, M.C. and Wheeler, P.D. (1986). Alarm sounds for medical equipment in intensive care areas and operating theatres. *Research Report No. AC598*. Institute of Sound and Vibration

Speier C., Valacich J.S. and Vessey I. (1997) The effects of task interruption and information presentation on individual decision making. In: Proceedings of the XVIII International Conference on Information Systems. Atlanta: Association for Information Systems

Van Bergen, A. (1968) Task Interruption. Amsterdam: North-Holland Publishers.

INTERNATIONAL COLLABORATIVE DESIGN PROJECTS: PROBLEMS, SOLUTIONS AND LESSON PLANS

Andrée Woodcock[1] and Lai-Chung Lee[2]

[1]*The Design Institute, School of Art and Design, Coventry University, Gosford Street, Coventry, UK. email: A.Woodcock@coventry.ac.uk*
[2]*Department of Industrial Design, National Taipei University of Technology.*

Industrial designers often develop concepts for overseas clients, who they may not meet face to face. Such projects are short in duration and exploit the designer's innate knowledge of an indigenous, niche market. They rely on electronic forms of communication (phone, fax, email and videoconferencing). Through analyzing project documents, interviewing designers and case studies we identified the problems of such collaborations, and proposed and tested solutions. Obviously, such research is of practical value, and we have used the research to form a training scheme for those wishing to use ICT to work flexibly and/or with international partners.

Introduction

International Collaborative Design Projects (ICDPs) are one instance of co-operative working, frequently mediated by the use of IT. With a rise in globalisation and multinational working such projects are becoming more frequent. However, not all designers and clients have the support of large corporations. In such cases the designers and clients, often working to strict financial and time constraints may quickly run into problems which, if ignored, impact on the success and quality of the collaboration and the project outcome. This research investigated the problems exhibited in projects between UK designers and Taiwanese clients and manufacturers, with the aim of providing low cost, practical solutions that could be implemented by the project teams themselves.

Discovering problems in ICDPs

Three principle methods were used to uncover problems. As these have been previously documented in Lee *et al.* (1999) and Woodcock *et al.* (2000), they are only briefly summarised here.

Retrospective Documentation Analysis

A systematic analysis of the written documents of 21 international design projects funded between 1993 and 1997, by the Taipei Design Centre revealed 143 problems that could be classified into six main groups relating to time, information flow and communication, brief specification, project management, participation and language. Most of these occurred in the early design stages (i.e. concept design). No relationship was found between the number of problems, project length, number of actors, designers' nationality, or project type. However 70 % of the problems were interpreted as having a major impact on the project and 50% of them remained unresolved.

Obviously there are major shortcomings in this approach. We were relying on documents that had been submitted by project members, some were more comprehensive than others, they had not been written specifically to reveal problems and we did not have access to some types of records (e.g. emails and phone calls). However, with confidentiality agreements and sensitivity surrounding new product development and launch, using project material retrospectively may, in some cases, be the only way of gaining an understanding of the manner in which projects are conducted and the effects of problems on their ultimate success. To mitigate some of these effects a series of follow up interviews was conducted with project representatives.

Follow up Interviews

Interviews were conducted in Taiwan with five representatives of the original projects, with the aims of validating the results, further identifying the nature of the problems and understanding the use of communication technologies in international projects. From the analysis of the transcripts the problems were classified into six main areas: time and project management, information flow/communication, brief specification, project and participation. Linguistic and cultural issues were not seen as being problematic by this sample.

Case Study

The two methods outlined above provided insight into the problems experienced by ICDPs, but we had no real feel for the impact of these problems or how they arose or were resolved. Therefore, a case study approach was used to gather detailed information to assist in the development of low cost intervention strategies. A real 6 week design project was used in which a Taiwanese client and manufacturer commissioned a UK designer to develop a two-in-one machine for thermal binding and lamination. Communication was mediated by phone, fax, email, WWW and videoconferencing. All documents were collected and the videoconference session transcripts subjected to breakdown analysis (Woodcock and Scrivener, 2003) using the categories previously identified.

Development of Intervention Strategies

Table 1 provides examples of the most frequent problems, their probable causes and likely remedies. By supplementing the transcripts and project documents with follow up interviews, it was possible to trace the onset of problems, their resolution (or otherwise) and impact on the design and the relationship between participants. For example, a failure to provide a specification on time meant that the designer started work without crucial information in order to keep to time, when more information was provided he found that he had developed inappropriate concepts.

Table 1. Problems and their intervention strategies

Problems	Possible Causes	Strategy Proposed
Lack of feedback and response	Redefinition of design specification by the client	• Define rules for feedback and response • Remind about time differences
Lack of communication	Unexpected change of personnel during project	Contact participants
Additional work required	Redefinition of scope of design work	• Elaborate the design brief • Clearly define scope of design work
Insufficient marketing information	Redefinition of scope of design work	As above
Ambiguous product brief	Incomplete planning	As above
Lack of understanding of corporate policy /strategy	Lack of understanding	Request the client's information and offer to the designer
Confused terminology	Cultural diversity	Provide a shared vocabulary for the project

Some problems were design specific, such as underestimating the amount of work required to achieve the final design and failure to provide sufficient marketing information; culturally specific problems included linguistic issues and the role of the designer (the client assumed that the designer would undertake market research) and managerial problems included failure to prepare agenda, minutes and action points. Overall the classification of problems established in the earlier work was supported. However two other categories were needed relating to cultural/linguistic issues and technological problems. Linguistic issues were not likely to have found their way into the documents used in the first study (they are low level problems and transitory in nature). Also, few of the earlier projects were supported by videoconferencing. The technological problems experienced by the designer in the case study could be attributed to lack of training, familiarity and technical support.

Table 2. Summary of intervention strategies

Stages	Actions
Project start up	• Train the participants to use the communication media appropriately • Assure the elaboration of the product design brief • Introduce the intervention strategies • Request the client's information and offer to the designer • Provide a shared vocabulary for the project • Define the rules of feedback and response
Pre-meeting	• Develop and gather agenda items • Inform the participants • Monitor emails and contact with the participants
During the meeting	• Facilitate and resolve communication and translation issues • Overcome technological breakdowns • Make notes of any agreements • Remind the participants of agreements and actions
Post-meeting	• Record and distribute the minutes • Trace the agreements and progress on their completion • Monitor emails and contact with the participants

Candidate solutions were generated and discussed with participants in terms of their practicality and whether the design team could implement the changes themselves. Solutions were not 'high tech' or project specific, so would be applicable to other projects. Examples included requiring an agenda and list of action points to be drawn up prior to meetings, spending longer elaborating the design brief, clearly specifying the design work required, providing diagrams and a working project vocabulary to reduce miscommunication.

To test the applicability and efficacy of the proposed strategies a second case study was undertaken on a design project of similar complexity, duration and membership. In this study, the researcher acted as project facilitator – monitoring project communications for potential problems and implementing intervention strategies as the need arose. This led to a reduction in all types of problems and their speedy resolution when they did arise.

Knowledge transfer

The research showed that small ICDPs are subject to a number of easily resolvable problems that may arise due to the inexperience of the participants (e.g. in using computer technology, making and subsequently forgetting action points) or their different priorities (all project participants worked on a number of projects concurrently). The importance of the problems lies in their effects in the overall success of the project. This can be measured by the satisfaction of the partners with the experience and the end product, whether the project was completed on time etc. In a project of one months duration, the failure to adequately specify the design parameters, and provide additional information quickly means that the project runs over time, which in turn impacts on windows of opportunity being lost and other projects falling behind schedule,

Having shown the problems and their potential impact on ICDPs, Lee et al (1999) called for additional training for designers that would enable them to work effectively and efficiently in the global market. On the basis of this research it was considered that additional training was required in 'time management and communication skills for international design projects, cultural and work ethics, universal design issues and international law. This would enable designers and design managers to gain the necessary technical, social and administrative skills required to successfully complete design projects on time and to the satisfaction of clients and end users.'

To this end, in 2002 Woodcock secured a European Social Fund (ESF) grant to provide training for middle managers in SMEs, sole traders and business start-ups in high growth sectors in the Coventry area (i.e. Objective 2, West Midlands). The course was marketed towards those who for whatever reason wanted to work as part of remote teams (either with international partners or as location independent workers).

The training needs uncovered in this research were mapped on to two modules relating to core ICT skills (especially in relation to the effective use of synchronous and asynchronous communication) and the managerial skills required for collaborative working. These were supplemented by a further module looking at the implications of flexible, or location independent working for individuals, work groups, organisations and the local economy. This third module was developed to meet the needs of the local community and also drew on the experiences and evaluation strategies employed in the EU- funded UNITE (Ubiquitous and Integrated Network Environment) project (see Benz et al., this volume).

Conclusions

Design research is concerned with studying designers and their requirements, with a view to better understanding individual and group practice. This research arose out of Lee's experiences in the Taipei Design Centre. It provides an example of the transition from analysis of the design situation, to the iterative development of achievable solutions to the problems designers face, in this instance, in their work with overseas partners, through to knowledge transfer to the wider community. The framework for the knowledge transfer will provide SMEs with opportunities to discuss, share and reflect on their own practice and experience, which will feedback in to research activity.

References

Lee, L.-C., Woodcock, A., and Scrivener, S. A. R., 1999, Breakdowns in the management of international collaborative design project, 6th International Product Development Management Conference, 689-699. Cambridge, U.K.

Lee, L.-C., Woodcock, A., and Scrivener, S. A. R. 2002, Intervention strategies for alleviating problems in international co-operative design projects, Common Ground Conference, September 2002, 38

Woodcock, A., Lee, L.-C., and Scrivener, S.A.R. 2000, Communication and management issues in international collaborative design. In S.A.R. Scrivener, L. J. Ball and A. Woodcock (Eds.), *CoDesigning 2000*, 369-377. Coventry, UK: Springer.

Woodcock, A. and Scrivener, S.A.R. 2003, Breakdown Analysis. In P.T. McCabe (ed.) *Contemporary Ergonomics* (Taylor and Francis, London), 271-276

EXPERIENCES OF ERGONOMICS PRACTITIONERS ON VACATION

Martin Maguire, Colette Nicolle, Suzanne Lockyer
Magdalen Galley, Edward Elton and Zaheer Osman

Ergonomics and Safety Research Institute (ESRI)
Loughborough University, Holywell Building, Holywell Way
Loughborough, Leics, LE11 3UZ, UK.

A group of practitioners collaborated in reporting ergonomic issues experienced during their holidays. While poor ergonomics made a big impression on the vacationers, good well-designed aspects of their holidays demonstrate the benefits that ergonomics can bring.

Introduction

When on holiday, ergonomists sometimes find it difficult to "switch off". Since their discipline is concerned with designing tools, products, equipment and systems to be fit for human use, this can apply as much to being on vacation as being at home or at work. What is distinctive about ergonomics applied to vacations? Firstly, hopes and expectations when going on holiday are usually high so encountering difficulties can cause disappointment. Also people try to relax on holiday so may be more easily caught out by poorly designed equipment. Furthermore, holidaymakers are often novice users of equipment, unaware that a balcony balustrade is too low to provide adequate protection, a deckchair is hazardous or that a beach umbrella will easily blow inside out. This paper describes recent experiences, both positive and negative, for a group of ergonomics practitioners at ESRI, Loughborough University.

Holiday booking and documents

Booking hotels and travel over the Internet is now popular since it allows users to explore different holidays and prices at their leisure. Yet while travel web-sites provide the basic details of the holiday resort and hotel, they often give insufficient background details about the country or local area, or relevant information such as currency and local transport information. It is also difficult to plan a touring holiday on a single website, or to track the itinerary or total cost of the holiday across several sites. Websites should also perhaps provide information about dangers when travelling to certain areas such as civil conflicts which would help to avoid situations such as the kidnapping of tourists and backpackers. Other risks that holiday makers might be unaware of should also be stated such as the dangerous traffic on South African roads every December when vast numbers take to the roads. On a positive note, holiday companies now streamline the documentation with flight tickets and other holiday documents coming in one booklet, with pages that are torn out when needed. Budget airlines streamline documentation still further by providing no ticket and just an email confirmation of a flight booking. Whether this is satisfactory may need further investigation.

Airline travel

Budget airlines offer travel at low cost with no frills. One author tried three UK-based budget airlines in one round trip from Luton to Zurich to Dublin and back to East Midlands Airport. On two of the airlines, seats were not allocated so passengers rushed to get to the front of the queue to get a good seat. For the third airline, which did allocate seats, boarding was a much calmer experience. Another author, travelling to Majorca, found that airport announcements were unclear and they also had to keep getting up to look at the display for the gate number (this was less of a problem at Heathrow where there were more displays placed strategically).

Boarding assistance for people who are heavily laden or who have young children should be provided. One airline recently told a passenger who was struggling with two young children, one in a pushchair, to ask other passengers if she needed help. One author found that they could not check in very early to get rid of cases. So at the restaurant she had to struggle with a food tray, hand luggage, a handbag and a suitcase on wheels. The Ladies toilet was found to have only one hand dryer working and this was blocked by a person washing her hands carefully, so that a queue of people formed. The toilet seat was also found to be very low which could be difficult for an older person with restricted movement.

Passengers often ignore the safety routine presented by cabin crew before take-off. Before presenting the procedures, the pilot or often crew remind the passengers to pay particular attention as some details (e.g. use of a lifejacket) could vary between aircraft. Safety films are commonly used on larger aircraft. A study at Loughborough (Page et al, 1996) of two different safety videos found that neither was wholly successful in increasing the risk perception of those who viewed them due to fatigue, intoxication or information overload. All of these factors need to be taken into account in developing safety videos.

Poor legroom on flights is a common problem and the cramped position of passengers has been cited as a reason for serious conditions such as deep vein thrombosis (DVT). The limited space experienced by one author was exacerbated by the passenger in-front reclining their seat and the neighbouring passengers 'stealing' the armrest. Recent reports of a device available in the USA to prevent other passengers from reclining their seats and therefore impinging on other's personal space could be a potential cause of 'air-rage'. In such circumstances, tasks such as eating a meal become very difficult. Watching the in-flight film was also hampered by glare on the screen from the cabin lights. One author was also surprised to find that when placing her reading glasses in the seat pocket, they dropped through. This was because the pocket was made up of a set of crossed straps, which is easier to clean than a simple, fully enclosed, pouch. The same author also finds when travelling from the United States, that landing cards are handed out early, allowing plenty of time for completion. However for short hops within Europe, she only gets a card just before coming in to land and so there is little time to complete it comfortably and legibly. Holiday company questionnaires handed out in-flight also seem to generate a poor response rate as they are often long and detailed, and come without a pen.

Local transport

Using local transport can be confusing for travellers from other countries. One author arrived in Zurich city centre on a Sunday and was faced with an unmanned ticket machine. It was not clear what ticket type or bus number was needed, and which direction bus they should take. As a result she had to walk to the local train station to get the information and a ticket, and then return to the stop. Having ticket machines in an unexpected place is also problematic. With the extension of the Eurostar rail link from Waterloo to St. Pancras/Kings Cross in London, the ticket machines have been moved out from the Underground station to outside the Kings Cross mainline station, which is confusing for unsuspecting travellers. However, one author arriving at

Dublin airport found the bus service very traveller-friendly. A person was available near the bus stop area to direct passengers to the right bus and to take their luggage. On the bus itself, there was a display and audio announcements of the names of the bus stops and hotels nearby. These were presented well before the stop and were clear and personal.

Announcements of stops for all local transport are generally a good thing for all travellers and in particular for passengers who are blind. London Underground publishes a leaflet and a web-based interactive map that describes the accessibility of the different stations (see http://map.tfl.gov.uk/accessible/). Trams, a popular means of transport in much of Europe are also good for strangers to use for the same reason and have the added advantage of the view.

Driving abroad

Driving abroad is difficult for many people especially when they need to adapt to driving on a different side of the road. One Dutch project partner commented that drivers are more at risk of an accident when they return home and as they relax and forget which side to drive on. Unfamiliar road signs are also a problem. One visiting lecturer to Loughborough reported that, on some highways, lit-up traffic information signs in Rio de Janeiro had to compete with exotic signs advertising nightclubs and bars. An Irish colleague reported that during the national change over from miles to kilometres, two styles of speed limit sign were in operation (one in mph and the other in kph) so that drivers needed to distinguish between the two. Hiring a vehicle abroad poses many problems, i.e. unfamiliar car, unfamiliar road network or jet-lagged driver. One conference speaker reported being impressed when arriving after dark at the rented car lot to find a small light integrated into the key fob, helping to identify the position of the lock.

Hotels and other buildings

Several authors commented about hotel rooms. In Crete it was found that the apartment's air conditioning was noisy at night and only worked when the balcony doors and windows were closed. Also, as the doors would not stay open, the author used a pair of shoes to hold them. Two authors experienced problems with keys and locks. One in Tunisia found that to lock his door when going out, he had to push in the central part of the lock mechanism, inside the room, and then pull the door closed behind him and there was nothing that indicated this procedure. To assist guests, one hotel in Interlaken had a demonstration lock at the reception desk. The author in Crete found that when trying to open their room door that they had opened their neighbour's apartment by mistake. Another author stated that she always judges the comfort of her room by the cleanliness and operability of the shower and the ability to move the showerhead up and down to avoid wet hair. She also needs a place to sit with plenty of light to make up. Often the only suitable place is in the bathroom sitting on the toilet seat or standing at the bathroom sink, holding a magnifying glass to aid her older eyes. To open the 'zesty lemon' soap packet on her recent trip, the author had to use her teeth. The packet urged the guest to 'restore your spirit' – although a taste of soap in the mouth was not the best start.

In a Majorcan hotel, the dinner trolleys in the hotel restaurant were big and cumbersome. This took two waitresses to push it when full and they struggled to push it between the tables, as there was insufficient space and guests had to continually move out of the way. Although intended for relaxation, sun beds can be uncomfortable and unsafe. One author found the plastic ridges on the sun bed dug into his back, so he was forced to buy an airbed to place on top of it. Another found that when lying on his front on the sun bed, and pushing himself up with his arms, this caused the front part of the bed to flip up suddenly and bump him on the head.

Tourist facilities

Facilities such as bank machines, information leaflets, and street signs are an important part of a successful holiday. Bank machines abroad now offer the convenience of access to local currency on the spot without having to bring a large amount into the country. However travellers will not know the exchange rate or how much the ATM commission charge will be until they return home. Bank machines usually present instructions in the customer's own language and the use of colour coded keys (red=cancel, yellow=clear, green=confirm) allow ease of cash withdrawal. One traveller in Turkey however had a problem due to lack of local experience. They tried to enter the amount required in millions by pressing the '0' key six times rather than "000" twice.

One author enjoyed a local road/train trip in Crete. However the lack of a commentary meant that limited information was conveyed. On a day long bus tour with British and German speaking passengers, the guide had so much information to convey that they were speaking in either one language or the other for most of the time which was tiring for all.

Another author, who visited a public building in Zurich, had trouble finding and accessing the ladies room as the chrome door matched the walls, making it look invisible. It was also so heavy to pull open that it appeared locked, while the door sign was engraved with black text so that it could hardly be seen. In Tunisia another author experienced a telephone, which was mounted so high up that it was hard to insert coins into (certainly inaccessible for a person in a wheelchair or someone of shorter stature). During the same holiday, on a quad bike safari across the desert, one member of the group had come with his young son of about 8. The boy rode on the same bike as the expedition leader and was wearing his helmet. However on occasions the driver went on two wheels putting his young passenger in a dangerous position.

Facilities for people who are disabled

Everyone is entitled to an enjoyable holiday so providing facilities for people with disabilities is important. In Venice, stair lifts are provided so that people in wheelchairs can go over some of the numerous bridges in the city. Aircraft boarding arrangements for disabled passengers need careful thought and staff training. A report on the BBC Radio 4's "In Touch" programme (November 2003) told how one blind airline passenger asked for assistance to be led to the plane, only to be forced to sit in a wheelchair and be pushed across to the aircraft.

Many hotels now have rooms that are designed to be accessible by wheelchair. But this may simply mean that the guest can get through door and can turn around in the bedroom and bathroom. The bathroom should also have bars allowing the person to lift him or herself out of the wheelchair and onto the toilet seat. Wheelchair ramps to gain access to the building should not be too steep. A project partner, who is disabled, asked for a double room with a double bed for himself and his wife. Unfortunately, to create sufficient room for a wheelchair, the hotel provides a standard sized room and ¾ sized bed which was very uncomfortable for two people. He was able to get his portable hoist under the bed although this would not have been possible with a Divan bed. In another hotel, he was allocated the 'accessible room', which was located on the 6th floor. In the middle of the night there was a fire alarm. As guests are expected to evacuate the building without using the lift, this caused worry and concern until the alarm stopped without further incident. In contrast, his experience in an accessible room on the ground floor was much more positive because he knew that he could exit in an emergency.

Dealing with an accident or illness

Falling ill or suffering an accident on holiday can pose considerable problems. The husband of one of the authors broke his heal on holiday in Los Angeles. The company required the insured traveller to contact them before contacting a local doctor – not easy in an emergency. Unfortunately the emergency contact number printed in the insurance booklet was incorrect. Making arrangements required many telephone calls, but the author could not find any credit card phones that worked so had to use expensive hotel phones. Another problem was the time difference; the first emergency contact number (the wrong one) was in England, while the correct company was in Quebec. The next problem was to find the locations of hospitals and doctors, and knowing the procedure to follow when visiting them. (Language difficulties can be another problem although this was not so in this case.)

Part of the trip included a conference, which the author's partner was attending. Easy wheelchair hire meant that he could get around. They phoned the hire company (found in the local Yellow Pages), who took all the details and delivered the wheelchair directly to the hotel. It was difficult to get cash because the person in the wheelchair couldn't easily get out of the hotel to change traveller's cheques. They also required extra cash for taxis, and because they needed help, they had to give lots of tips. Taxi drivers were found to be helpful, but car boots had poor space, making it difficult to fit the wheelchair in (although the cars were large). The author also tried coaches, including the shuttle supplied to and from the conference, but the big steps proved to be a problem. The conference was located on three floors of a building with escalators, and only one small lift. The thick carpets in the building made wheeling the chair difficult. The disabled toilet was within the main toilet, so the wheelchair user had to negotiate two doors (one heavy and spring loaded). The cubicle was difficult to get into because of the narrow corridor with a tight turn. The soap, sink, and towels were inaccessible. Refreshment tables and tea urns were also difficult to get to.

On returning home, there was a difficulty in getting a luggage cart at the airport, which required money to be put in to release it. Although the need for rigorous procedures was appreciated, the security checks were found intimidating. The wheelchair user was asked if they could 'walk through' the x-ray machine. They were also subjected to a much closer search than normal (as the wheelchair and plaster could be used to hide objects) including a body search. The author was told to stand out of the way and not to assist her partner. The high point was the helpfulness of the car park staff on arrival who got their car ready and waiting although they returned 2 weeks early and only made contact on arrival at Heathrow.

Conclusions and recommendations

Despite the problems reported, the authors enjoyed their holidays although ergonomic issues did affect their perceptions of them. Designers and planners should appreciate that when a holidaymaker uses equipment or a service they are often a novice or casual user so intuitive design is important. Providing better information before travelling will help people on vacation to adjust to local conditions while obtaining accurate feedback from tourists will help identify ergonomic issues. Finally an unexpected good innovation or design feature will stay in a person's mind and help to create a positive impression of the whole holiday. Based on the findings of this paper, a list of recommendations is being prepared.

Reference

Page M., Southall, D., Bird, R. 1996, *Critique of the Thomson holiday safety video on behalf of the Consumer Safety International*, 16 July 1996, The Research Institute for Consumer Ergonomics (now part of ESRI), Loughborough University, Memo 1411.

Acknowledgement
The authors would like to thank Victoria Haines and Val Mitchell, of ESRI, for their inputs to this paper.

EVALUATING SYSTEMS TO SUPPORT CO-OPERATIVE WORKING – THE UNITE EXPERIENCE

Harald Benz[2], Maryliza Mazijoglou[1], Christoph Meier[2], Stephen Scrivener[1] and Andrée Woodcock[1]

[1]*The Design Institute, Coventry University, Gosford Street, Coventry, UK.*
[2]*Fraunhofer Institute für Arbeitswirtschaft und Organisation, Nobelstrasse 12, Stuttgart, Germany*
email: *A.Woodcock@coventry.ac.uk*

More and more often, people are working in virtual organisations and teams, only infrequently meeting their colleagues face-to-face. The development of computer supported co-operative working (CSCW) systems to support these new ways of collaborating at a distance gives rise to a range of socio-technical problems that need to be considered during system design. This paper considers the issues and evaluation strategies employed during the development of one such CSCW system, the UNITE platform.

Introduction

The UNITE (Ubiquitous and Integrated Teamwork Environment) project was a multidisciplinary, multinational EU project, to develop a web based platform to support collaboration in distributed (i.e., "virtual") teams, either within or across organisations. The aim of the project was to offer project teams and virtual organisations user-friendly, highly efficient co-operative workplaces enabling pervasive, dynamic, secure and transparent binding of project context information with team members, their physical workplace and own information tools – i.e. the UNITE platform. Current commercially available web based, collaboration platforms do not adequately support users who work on several projects simultaneously and collaborate with partners and colleagues using different computing platforms. The UNITE approach was unique in offering the user a team-working platform integrating communication, coordination and collaboration for each active project within a single interface, using the systems resident on their own personal and local computers. This means a team member works within a single and uniform interactive environment, unaware that the system resources vary between members of a project team and between the different teams to which he or she belongs. This removes the need for team members to learn how to use new application and work organisation systems every time they start a new project, thus significantly reducing the overhead involved in distance collaboration. The goals of UNITE were pursued through an iterative development cycle involving:

- elicitation of the requirements and paradigm of co-operative workplaces,
- definition of a suitable platform architecture,
- validation of concept and architecture using prototypes operated by real project teams.

The resultant prototype platform provides a web-based solution that enables teams to set up virtual project workspaces, composed of project documents, bookmarks, shared calendars and individual contact details. The workspace contains all the information a team member needs to find out what is happening on the project and to work with colleagues (see Woodcock et al, 2002 for more details). Compared to single user applications, groupware of which the UNITE platform is an example, is more difficult to evaluate (Sohlenkamp et al. 2000, p. 40-41) because:

1. Usability testing presupposes a critical mass of co-operative functions. A prototype needs to be fairly advanced before it can be meaningfully tested.
2. The complex social and motivational dynamics of co-operative work that CSCW applications need to support are difficult to replicate in a lab setting.
3. Usability testing requires the involvement of a larger number of users.
4. Field tests are problematic in that they imply an intrusion into the workplace of users.

This paper reflects on the evaluation methodology adopted in the project, focussing on how the above limitations informed its design and how the results of its application provide insights for enhancing groupware evaluation methods.

Evaluation strategy

The evaluation had three main goals:

1) To support the iterative development process, progressing from the early interface mock-ups through to operational prototypes.
2) To provide a summative evaluation of the final system - considering the extent to which it met the initial user requirements and whether further work needed to advance the platform into a marketable product.
3) To evaluate UNITE-supported co-operative teamwork platform from the individual user, team and organisational perspectives.

Given the complexity of the intended system, the UNITE team designed a three-phase development plan: mock-up, Basic Platform, Enhanced Platform. A mock-up interface was developed as means of communicating and testing the UNITE concept with the developers and users (i.e., the project partner). The Basic Platform was designed to prove the concept, which was augmented with additional functionality in the Enhanced Platform. This plan, negotiated by all project partners, reflected the technical logic of needing to implement the possible first before dealing with more complicated functionality, and also acknowledged the evaluation teams desire to evaluate prototypes possessing a critical mass of co-operative functions (Sohlenkamp et al., 2000: Bullet Point 1 above).

Real life usability testing places strong demands on trial users, their time and motivation to use such a system especially during the early stages of development. Laboratory based mock trials may be easier to fit in to busy schedules, but are of limited ecological validity because they fail to accommodate the complex social and motivational dynamics of co-operative work (Sohlenkamp et al., 2000: Bullet Point 2 above). However, field tests can be problematic (Sohlenkamp et al., 2000: Bullet Point 4 above). A good compromise is an "in the zoo" approach, where user trials are half way between an exercise and real life. Users conduct a real life activity, limited both in its scope and reality, being an episode after which users return to familiar ways of doing. With the user partner's assistance, a group task, accomplishable by a small number of people, within a limited time frame, was developed.

Given the above, an evaluation plan was devised, using a range of approaches to test usability, utility and work system performance at the individual, team and organisational levels. Also taken into account, were the limitations and possibilities offered by increasingly mature prototypes. Table 1 (which omits the mock-up stage) shows that different methods were required for different aspects. As the Basic Platform provided limited co-operative working functionality, evaluation focussed on the individual and to a lesser extent the team level (indicated by parenthesis in the table). The Enhanced Platform was evaluated at the individual, team and organisational levels.

Table 1: The UNITE evaluation framework

ASPECT	METHOD	LEVEL OF ANALYSIS			
Usability	Usability inspection, diaries, observation, recording, interviews, questionnaires, inventories, breakdown analysis and screen annotations	Basic Platform	• Individual • (Team)	Enhanced Platform	• Individual • (Team)
Utility	Diaries, observations, questionnaire and workshop		• Individual • Team		• Individual • Team
Work System	Project questionnaire		• (Individual) • (Team)		• Individual • Team • Organisational

Application of the framework

Usability

The aim of the UNITE usability evaluation was to inform the continuous improvement of the platform and user interface through the early identification of problems in terms of the LEMES categorization (Neilsen, 1993): Learnability, Efficiency, Memorability, Errors and Satisfaction (LEMES). In the Basic Platform, LEMES feedback was gained through laboratory based usability testing and system walkthroughs with standard inventories and

SUMI (software usability measurement inventory). This enabled all LEMES dimensions to be accessed, while guaranteeing some feedback to designers in the event of it not being possible to trial the Basic Platform as planned. When trialed in the user organisation, the platform did not achieve sufficiently high levels of stability and was evaluated more as a tool to aid work activity, rather than as a piece of usable software. The lack of platform stability probably accounts for the poorer results at the user's organisation, compared to those obtained in the laboratory setting, Table 2. The usability results were fed into the next design cycle to enhance the efficiency, helpfulness and control of the platform. This is reflected in the results from the Enhanced Platform trials, conducted by the UNITE evaluators in the user organization, which yielded results much closer to the laboratory tests, Table 2.

Table 2: SUMI results for Basic and Enhanced Platforms

	Basic		Enhanced
Scale	*Laboratory*	*User Org.*	*User Org.*
Global	52	35	51
Efficiency	44	30	45
Affect	60	32	57
Helpfulness	45	33	50
Control	49	32	45
Learnability	56	49	62

In summary, the approach to usability evaluation was formative – including usability inspection, walkthroughs and testing. The results, accompanied with suggestions for improvement were presented to the developers under items such screen descriptions and structure, graphics, navigation, visibility of system status, compatibility, recognition and recall, language, informative feedback, etc. Where trade-offs needed to be made, user priorities were taken into account.

Utility

A questionnaire covered general issues, such as the overall platform performance in terms of stability and connectivity, etc. Opinion was then sought on the overall range of functions provided and whether users would be willing to pay for a service of this kind. While the Basic Platform did not satisfy the trial users, for the reasons described above, the Enhanced Platform was well received and, in particular, the synchronous collaboration services were much appreciated. To complement the functional tests at the developers' sites, user trial diaries, observations/interviews and workshops were used to investigate the utility of the functions provided. This was performed in close relationship with the usability inspection. Results for each functional area were reported together.

Work System Valuation / Nutzwertanalyse

The introduction into an organisation of advanced IT to support team processes involves complex and interdependent economic, technical, organisational and social change. Pure efficiency calculations, such as profit comparisons, do not to capture the full picture. Hence, we adopted the standard work system valuation method / Nutzwertanalyse (e.g. Fröschle and Niemeier, 1988) as this is highly flexible, both with regard to the criteria employed and the process of assigning values against criteria. Using questionnaires focusing on individual, team and organisation levels, users rated the positive or negative change expected should the UNITE platform were to be implemented in their organisation.

The Challenges of Evaluating Co-operative Systems

Reflecting Sohlenkamp et al.'s (2000) findings, the conditions prevailing in the UNITE user organisation constrained the design of the evaluation process for the UNITE platform to that of an "in the zoo" evaluation strategy. Nevertheless, problems in the system development cycle and in the user organization meant that the evaluation plan was constantly under review such that goals had to be either simplified or abandoned.

Nevertheless, the UNITE experience suggests that the more realistic the trial, the more useful user feedback is for the system designers. Therefore, our recommendation for development projects is not only to plan for broader "in the zoo" or "in the wild" evaluation and trial procedures, but also to plan the development process such that these ambitious goals are not compromised, as is often the case when technical problems are allowed to dominate.

Acknowledgements

This project was co-funded by the European Union under the Information Society Technologies (IST) Programme (Project No. IST-2000-25436). It represents the view of the authors only.

References

Fröschle, H-P. and Niemeier, J. 1988, Assessment of benefits of information technology in the office: A concept for economic valuation and case study. In: Eurinfo '88. *Proceedings of the First European Conference on Information Technology for Organisational Systems*, (ed.) Protonatorios, E.N., Bouwhuis, D., Reim, F. and Bullinger, H.-J. (Amsterdam North-Holland Publishing), 190-197

Nielsen, J.1993, *Usability Engineering* (Morgan Kaufmann, London)

Sohlenkamp, M., Prinz, W. and Fuchs, L. 2000, PoliAwac: Design and evaluation of an awareness enhancing groupware client. *AI and Society Journal*, 14, 31-47

Woodcock, A., Meier C. and Reinema, R. 2002, An integrated approach to support teamworking, in P.T.McCabe (ed.) *Contemporary Ergonomics*, (Springer-Verlag, London), 429-436

EFFECTS OF TIME-OF-DAY
ON INTERPRETING SIGNS AND ICONS

Siné McDougall, Victoria Tyrer & Simon Folkard

Department of Psychology
University of Wales Swansea
Swansea SA2 8PP

The effect of time-of-day on performance in a wide range of cognitive tasks has been well documented. Typically performance improves over the morning followed by a dip in performance early in the afternoon. The practical implications of time-of-day effects are apparent in studies which point out the link between post-lunch performance and the increase in car accidents in the early afternoon. One reason for this may be that drivers are slower to interpret and respond to road signs at these times of day, however, almost nothing is known about the effects of time-of-day on interpreting signs and icons. The evidence to date suggests that variations in performance across the day depend upon a combination of factors including the type of icons presented and task demands. A framework is proposed for understanding these effects and the practical implications of these findings are discussed.

Introduction

There is growing evidence that the frequency of driving accidents coincides with diurnal variations in sleepiness. In the UK, Horne & Reyner (2001) recently reported that the frequency of sleep-related vehicle accidents peak between 0200 and 0600 hours and between 1400 and 1600 hours when sleepiness tends to be highest. Similarly, Lenne, Triggs & Redman (1997) report peaks in road accidents in Australia between 0300 and 0500 with a smaller secondary peak occurring between 1400 and 1500. Lenne et al tested the link between diurnal variations directly by measuring performance at different times of day on a driving simulator. They found that performance was more impaired at 0200 and 0600 hours with an additional drop in performance in the early afternoon. This provides strong evidence for the idea that driving performance is subject to diurnal variations. One of the key issues in driving well is that we are able to interpret, and act upon, road signs quickly and effectively. If there are diurnal changes in sign interpretation, then this has important implications for road safety and also for other areas where icon and sign interpretation needs to be effective (e.g. air traffic control, chemical and nuclear control systems).

Evidence of time-of-day effects

Until recently, only one study had examined changes in the efficacy with which we deal with signs or icons. In this study McFadden & Tepas (1997) asked their two participants to perform a road sign recognition task four times a day (at 0900, 1200, 1500 and 1800) for 18 consecutive days. Their participants were shown a road sign and, when this disappeared from the screen, they were asked to search an array and indicate whether or not the target sign was present. Figure 1(a) provides examples of the two types of signs they used. McFadden and Tepas did find time-of-day effects in participants' ability to respond to traffic signs. However, there were considerable difficulties in interpreting their findings not least because each of the two participants showed somewhat different patterns of diurnal variation. We have since explored these findings in a series of experiments using larger participant groups and have examined the effects of the type of icon presented in more detail.

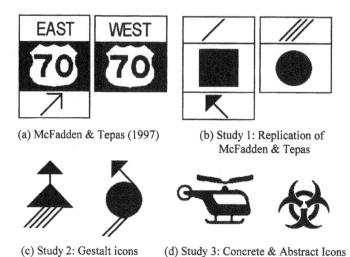

(a) McFadden & Tepas (1997)   (b) Study 1: Replication of
                                    McFadden & Tepas

(c) Study 2: Gestalt icons   (d) Study 3: Concrete & Abstract Icons

Figure 1. Examples of icons and signs used in time-of-day research

Icon Characteristics

Research has shown that there is a great deal of variability in the ease with which signs, symbols and icons can be understood (e.g. McDougall, de Bruijn & Curry, 2000; Rogers, 1986; Stammers & Hoffman, 1991). Ease of interpretation depends, not only on the individual's experience, but also upon the characteristics of the icons. It is generally believed, for example, that simple icons are responded to more quickly than complex icons because they have fewer visual features which can be processed more quickly (Byrne, 1993). Icon concreteness is also important in determining ease of interpretation (e.g. McDougall et al, 2000). Icons which are more concrete, or pictorial, are thought to be easier to recognise than abstract icons because they represent things which we already know about and we can use this knowledge to deduce what they might mean (see Figure 1(d)). Although pictorial icons often cannot represent abstract concepts well, they nevertheless can often be used to convey simpler meanings and provide useful clues for users who may not have encountered these icons before.

The Role of Visual Complexity

In the first experiment of the series, designed to replicate McFadden and Tepas' study, participants searched through displays of icons for a specified target icon and indicated whether or not the target icon was present at 4 times of day (0900, 1200, 1500, 1800). Complex icons consisted of 3 pieces of information: distance (400, 300, 200, 100 yards), route (square, triangle, circle, diamond, star) and direction (straight on, left, right; see Figure 1(b)). Simple icons consisted of two pieces of information: distance and route. In line with previous research, responses were slower for complex icons ($\underline{M}$ = 1675 msecs), which required more visual processing, than for simple icons ($\underline{M}$ = 1321 msecs). Importantly, time-of-day effects were evident in the time it took participants to respond in the icon task. Response times were significantly faster in early/mid-morning (0900/1200) and were slowest at 1500. These findings are summarised in Table 1. They provide support for McFadden & Tepas' original assertion that time-of-day effects are important in determining the ease with which icons can be interpreted.

Table 1. Summary of findings in Studies 1-4

	Type of Icon	Time-of-Day Effects AM (0900/1200)	PM (1500/1800)	Response Type
Study 1	Complex (3 parts)	Faster RTs	Slower RTs	Yes/No Keys
	Simple (2 parts)	Faster RTs	Slower RTs	
Study 2	Complex (3 parts)	Faster RTs	Slower at 1500	Yes/No Keys
	Complex (1 part Gestalt)	Very fast RTs	Slower at 1500	
Study 3·	Concrete	-	-	Mouse click
	Abstract	Slower at 0900	Slower at 1500	
Study 4	Concrete	-	-	Yes/No Keys
	Abstract	Faster at 0900	Slower at 1500	

The Effects of 'Visual Chunking' of Information

Given the nature of the icons used in both of these experiments, it remains an open question whether the differences in response times between complex and simple icons were the result of the number of pieces of information, or objects, to be processed in each (i.e. 3 vs 2) or whether it was the result of the general visual complexity of the icons (since those with 3 pieces of information also contained more visual features, see Figure 1(b)). This is of interest because feature-based theories of visual search suggest that it is the number of features in the icon which are critical in determining response time while object-based theories argue that it is the number of objects to be processed which determine response time (see Goldsmith, 1998, for a review). This was examined in a subsequent experiment in which the visual complexity of the icons to be processed was held constant while the number of objects to be processed was varied (i.e. 3 vs 1). One set was identical to the complex icons shown in Figure 1(b) and examples of the others are shown in Figure 1(c). The latter icons maintain the visual complexity of the icons but reduce the visual processing load associated with them by 'visually chunking' items into a single item. When participants were asked to search icon arrays containing these types of icons responses were much slower for the multi-object complex icons shown in Figure 1(b) ($\underline{M}$ = 1767 msecs) than for the single object icons ($\underline{M}$ = 1215 msecs). This suggests that it is the number of objects, or pieces of visual information, rather than the number of visual features, which is determining participants' response times to the icons. This suggests that object-based, rather than space-based, visual processing is taking place. Practically, this means that icons can be relatively complex providing they can be perceived as a single object. As in the previous study, time-of-day effects were evident: response times were faster in the morning than later in the day with response times being slowest after lunch (at 1500).

The Role of Icon Concreteness

Given the importance attached to the concreteness, or pictorialness, of icons in previous research Study 3 examined whether the changes in performance across the day observed in the previous studies would change when the concreteness of the icons was varied (see Figure 1(d)). In this study, all icons could be processing as a single object and response times for these icons were comparable with the Gestalt icons presented in Study 2 ($\underline{M}$ = 937 msecs). The procedure was identical to Studies 1 and 2 except for the way in which participants were asked to respond. In this study participants were asked to select the target icon using the mouse, rather than pressing Y/N keys to indicate the presence or absence of the target icon.

Although diurnal trends in performance were apparent, the pattern of these trends differed in comparison to previous studies. No time-of-day effects were apparent for concrete icons. This suggests that time-of-day effects in icon interpretation might be reduced by using concrete icons. For abstract icons, response times were slowest at 0900 with a slight post-lunch increase in response times at 1500. The slow responses noted at 0900 contrasted with the findings of Studies 1 and 2 when participants were actually able to deal more quickly with icons in the morning. One possible reason for the change in the pattern of findings is that participants were asked to respond in a different way to the icons (by selecting icons using a mouse). This possibility was explored in Study 4.

The Role of Response Type

Study 4 used the same stimuli as in Study 3 (see Figure 1(d)) and the only difference procedurally was that participants were asked to indicate whether an icon was present in the array using yes/no keys. In this study very few time-of-day effects were apparent. Response times only differed slightly for abstract-simple icons where times at 0900 were faster than at 1500. Comparison of the pattern of findings in Studies 3 and 4 suggest that response type is important (see Table 1). Where mouse click responses are required, responses tend to be slower earlier in the day (0900).

Conclusions

On the basis of the summary of findings presented in Table 1, it is possible to formulate a tentative framework for the way in which time-of-day effects appear to operate when we are dealing with icons. In general, we respond more effectively to icons earlier in the day, particularly at 0900 and become slower as the day goes on. The effects of time-of-day can be considerably ameliorated, however, by appropriate icon design. Icons which consist of fewer visual parts are likely to reduce the visual processing load and lower the time it takes individuals to respond. Wherever possible, concrete icons should be used since these appear to be least affected by time-of-day. Care should be taken to ensure that concrete icons are appropriate for the functions they are supposed to represent since research suggests that pictorial icons may not always be appropriate for more abstract concepts. What is clear from our research is that time-of-day effects do occur in icon interpretation but further research is required to look at how these effects operate in context, outside the laboratory, particularly where the responses required may be complex.

References

Byrne, M.D. 1993, Using icons to find documents: Simplicity is critical, *Proceedings of INTERCHI'93*, 446-453

Goldsmith, M. 1998, What's in a location? Comparing object-based and space-based models of feature integration in visual search, *Journal of Experimental Psychology: General*, **127**, 189-219

Horne, J. & Rayner, L. 2001, Sleep-related vehicle accidents: Some guides for road safety policies, *Transportation Research Part F: Traffic Psychology and Behavior*, **4**, 63-74

Lenne, M.G., Triggs, T.J. & Redman, J.R. 1998, Interactive effects of sleep deprivation, time of day, and driving experience on a driving task, *Journal of Sleep Research and Sleep Medicine*, **21**, 38-44

McDougall, S.J.P, de Bruijn, O. & Curry, M.B. 2000, Exploring the effects of icon characteristics on user performance: The role of icon concreteness, complexity and distinctiveness, *Journal of Experimental Psychology: Applied*, **6**, 291-306

McFadden, E. & Tepas, D.I. 1997, Effects of time of day and task demands on simulated sign recognition, *Proceedings of the Human Factors and Ergonomics Society 41st Annual Meeting*, (Human Factors Society, Santa Monica), 1392-1396

Rogers, Y. 1986, Evaluating the meaningfulness of icon sets to represent command operations. In M.D. Harrison & A.F. Monks (Eds.), *People and Computers: Designing for Usability*, (Cambridge University Press, Cambridge), 586-603

Stammers, R.B. & Hoffman, J. 1991, Transfer between icon sets and ratings of icon concreteness and appropriateness, *Proceedings of the Human Factors Society 35th Annual Meeting*, (Human Factors Society, Santa Monica), 354-358

THE THEORY OF PLANNED BEHAVIOUR:
A USEFUL FRAMEWORK FOR PREDICTING
RULE VIOLATION?

Joyce Lindsay[1] Elaine Ridsdale[1] Gary Munley[2]

[1]Serco Assurance
Warrington, WA3 6AT

[2]Department of Psychology & Speech Pathology
Manchester Metropolitan University, M13 0JA

The Theory of Planned Behaviour (TPB) states that three key elements, namely (i) attitude, (ii) social norm and (iii) perceived behavioural control, contribute to the formulation of behavioural intention, which in turn influences the execution of behaviour. The research reported in this paper describes the application of the TPB as a model for rule violations. Current rule violation assessment techniques concentrate on identifying individual predictors of violating behaviour, or Violating Producing Conditions (VPCs), but do not support the formulation of effective violation reduction strategies. The TPB models VPCs into a coherent structure, which is based on the psychological origins of the rule violation, the three components of the TPB. The TPB model is proposed as the basis for the future development of an alternative violation assessment approach to identify, quantify and reduce rule violations by targeting the psychological origins of the behaviour.

Introduction

Currently accepted techniques for Human Reliability Assessment (HRA) focus primarily on the identification, quantification and reduction of human error. Considerably less effort has been invested in the assessment of rule violations, therefore equivalent assessment approaches are comparatively immature. Rule violations can be defined as "any deliberate deviation from the rules, procedures, instructions and regulations drawn up for the safe operation and maintenance of plant or equipment" (HFRG, 1995). Rule violations can result in a range of consequences, from negligible, to small scale accidents (HFRG, 1995) or incidents of significant gravity, such as the Chernobyl incident. The development of a comprehensive violation assessment technique is therefore equally important to human error assessment in considering the human contribution to system risk. Although human error and rule violation can result in the same consequences, they have different psychological origins (Reason, 1990); human error arises when actions are unintentionally performed outside the prescribed safe operating parameters, whereas violations involve a conscious decision to circumvent or flout rules and procedures. As such, it is unreasonable to expect that remedial measures designed to reduce the likelihood of human error will also target violations effectively.

Rule violation, however, reflects a more complex interaction of motivational, social and control factors, which originate from the attitudes and beliefs held individually, within groups and also those conveyed throughout organisations by management. Rule violation requires systematic consideration, through suitable modelling of these psychological variables that are implicated in the decision making process, in order to understand and predict violation potential and ultimately to reduce these at their source (Reason, 1990). Current techniques available for the assessment of rule violations concentrate on influential contextual factors, VPCs (Violating Producing Conditions) but take little account of their psychological origins. The Theory of Planned behaviour (TPB) is proposed here as a framework to link the contextual influences of VPCs and psychological origins of rule violation.

Methodology

Review Existing Violation Assessment Techniques
There are three key approaches currently adopted in the (qualitative and quantitative) assessment of rule violations: SURVIVE (Survey of Rule Violation Incentives and Effects, Holloway, 1989), Improving Compliance with Safety Procedures (HFRG, 1995) and HEART (Williams, 1996). A qualitative review of these techniques was conducted to determine their strengths and weaknesses, which in turn provided guidance in the formulation of alternative approaches.

Modelling Rule Violations
A number of literature sources were reviewed to identify the fundamental components of the TPB and how this generic theory of human behaviour could be used to link VPCs with the underpinning psychological origins of rule violation. The output of this analysis was a qualitative discussion of the TPB and its potential for modeling the link between the psychological origins and contextual VPCs associated with rule violation.

Populating the TPB Model
In order to populate the TPB model of rule violation, an extensive literature review was conducted to identify a range of rule violations and associated VPCs that could be modelled by the TPB model. A search was conducted on three literature databases using keywords related to rule violation (e.g. obedience, detection and non-compliance): (i) Ergonomic Abstracts, (ii) Science Direct and (iii) Psycinfo. Other sources of information consulted included existing review articles, reference sections of eligible studies and bibliographic reference volumes, producing 87844 potentially relevant papers. The abstracts were reviewed to filter out those papers that contained little useful qualitative data, for example papers which concerned the treatment of violation behaviour of mentally ill patients. A total of 49 papers was selected for detailed qualitative review, whereby the full paper was reviewed to identify rule violations and associated VPCs from different areas of research (e.g. driving violations or violation of PPE regulations in the workplace). For each rule violation the VPCs were recorded in terms of their relationship with the psychological origins of the behaviour. For example violation of the speed limit may be associated with the VPC of low likelihood of detection, which can be linked to driver attitude (the driver believes that they will be unlikely to be detected).

Results

Review Existing Violation Assessment Techniques
The most widely accepted model of rule violations is a contextual classification system by
Lawton (HFRG, 1995). Here violations are assigned to four discreet categories depending
on the context in which they may be committed:
(i)   routine violations conducted on a regular basis,
(ii)   situational violations conducted in response to a specific system status,
(iii)   exceptional violations occur due to problem solving in unusual situations and
(iv)   optimising violations involve changing the environment to improve conditions.
Classification of rule violations by this model suggests that rule violation can be completely
attributed to the operational context (VPCs) and that all operators will respond identically to
a given VPC each time it is present. While this approach provides useful insight into the
effect of VPCs, it does not consider the influence of social, motivational and perceived
control factors. Neither does it consider that the influence of VPCs and associated
psychological origins are not static, (Hsu and Kuo, 2003) and are subject to a dynamic
interaction where the strength of each variable as a mediator varies according to the
situational challenges. For example, violation of the speed limit would be affected by the
VPC of low likelihood of detection, which would in turn affect the attitude of the driver who
may believe that it is acceptable to speed because they will not be caught. However if there
is voluminous traffic their perceived control is diminished, thus over-riding the other VPCs
and attitude.

Assessment of the three main rule violation assessment strategies revealed the following
strengths and weakness that could be used in the development of an alternative assessment
strategy.
(i)   The assessment of rule violation by SURVIVE, HFRG and HEART does not identify
the full range of potential VPCs. Thus an alternative assessment approach should
expand on the list of VPCs.
(ii)   Some of these assessment approaches try to combine the quantitative results of
violation assessment with HEP values obtained from human error assessment, even
though the psychological origins and assessment strategies of human error and rule
violation are not compatible. This should be addressed by considering human error
and rule violation independently before considering the relationship between the two.
(iii)   Little information is provided regarding the origins of the strength of effect assigned
to VPCs included in the current assessment approaches. This was investigated by a
recent meta-analysis (Munley*et al*, 2003) and the findings did not match the values
presented by existing approaches. There is a clear need for further investigation of
the strength of effect of VPCs.

Modelling Rule Violations
According to the TPB, behaviour is determined by intention to engage, which is a product
of: the attitude towards that behaviour, the subjective norm and the perceived level of control
over the planned behaviour (Ajzen, 2002a and 2002b). Thus, there are three elements
involved in the formulation of behavioural intention and execution, namely (i) attitude, (ii)

social norm and (iii) perceived behavioural control (Ajzen, 2002a and 2002b). Just as behavioural beliefs affect attitude and normative beliefs affect subjective norm, so salient control beliefs affect the perceived level of control.

Attitude is a function of behavioural beliefs: personal beliefs and expected outcome (e.g. I believe that it is acceptable to speed because I do not expect an unfavourable outcome). Social norm is a function of normative belief: i.e. the opinions/behaviour of significant others and an individual's motivation to comply with this (e.g. most other drivers violate the speed limit and I feel under pressure to conform). Perceived control is a function of salient control beliefs: internal factors (e.g. driving ability) and external factors (e.g. volume of traffic or power of car). All three factors directly influence the intention to perform a behaviour, but only perceived behavioural control has a direct effect on the execution of the behaviour. Thus intention to behave is a function of perceived internal control (i.e. confidence in skills and abilities) and behaviour is a function of external control (i.e. opportunity and resources available). In terms of rule violation a person may intend to conform to a rule because they believe they have the knowledge and skills necessary to do so (perceived internal control), but they may violate the rule because they don't have the resources available to conform. Thus rule violation could be effectively addressed by consideration of the psychological origins, the contextual VPCs and the relationship between the two factors.

Populating the TPB Model
The literature review identified a range of rule violations and associated VPCs with which to populate the TPB model. These were recorded in a database that reflected the psychological origins of rule violation, as depicted by the TPB. Table 1 shows an example of how rule violations and the associated VPCs were modelled in a TPB database.

Table 1: Application of TPB to Violation of the Speed Limit

Attitude	Social Norm	Perceived Control
Beliefs – it is acceptable to violate the speed limit.	**Opinion of significant others** – most other drivers violate the speed limit.	**Internal** –confidence in driving ability.
Outcomes - the driver believes that the potential severity of the outcome (crash and/or injury) is low.	**Motivation to comply** – the need to comply with the other speeding drivers may vary depending on the situation, personality type and previous driving experience.	**External** – intention to speed is only executed if external conditions allow (traffic levels, weather, etc).

Speeding during driving is influenced by a range of VPCs and psychological origins: (i) a belief that speeding will not lead to an unfavourable outcome (attitude); (ii) a belief that most other drivers speed and an associated motivation to comply (social norm); (iii) a competent driver (internal control) driving a powerful car in light traffic and good weather (external control). According to Reason the most effective way to address rule violation is through the psychological origins therefore the VPCs should be tackled according to their categorisation in the TPB model.

Discussion and Conclusions

In line with his observation that violations are psychologically underpinned by motivational and social factors, Reason suggests that they should be tackled by addressing shortcomings at an organisational level to impact on attitudes and safety culture (Reason, 1990). The TPB model facilitates systematic consideration of precursors to rule violation by categorisation of VPCs according to their psychological origins. The effect of a given VPC depends on the psychological factors associated with the operator at that point in time. For example driver attitude towards speeding may change with increased experience and result in reduced frequency of speed limit violation, even in the presence of VPCs that would previously have encouraged violation. The use of TPB is the first step towards understanding the dynamic relationship between VPCs and psychological origins and identification of how these impinge on rule violation. However, further research would specify this relationship more accurately and provide more comprehensive description of the route to rule violation.

Modelling the precursors to rule violation with the TPB could also be extended in order to develop an alternative methodology for the assessment of rule violation, which tackles the issue from a range of perspectives:
(i)   Identification of the psychological origins of rule violations to focus the development of reduction and/or mitigation strategies. For example if unrealistic beliefs about an outcome leads to rule violation then a possible solution would be to provide education about the potential consequences of rule violation.
(ii)   Prioritise reduction strategies by combining the TPB model of rule violation with the strength of effect data from a recent meta-analysis (Munley *et al*, 2003).
(iii)   Identify the relationship between the psychological origins of behaviour and VPCs.
(iv)   Assess rule violation and human error separately before considering the relationship between human error and rule violation to formulate a combined Human Reliability Assessment (HRA) process.

References
Ajzen, I. 2002, *Behavioural Interventions Based on the Theory of Planned Behaviour*, World Wide Web published paper.
Ajzen, I. 2002, *Constructing a TPB Questionnaire: Conceptual and Methodological Considerations*, World Wide Web published paper.
Holloway, N.J. 1989, *SURVIVE A Safety Analysis Method for a Survey of Rule Violation Incentives and Effects*, Unpublished Report UKAEA.
Hsu, M.H. and Kuo, F.Y. 2003, An Investigation of Volitional Control in Information Ethics, *Behaviour and Information Technology*, 2 (1), 53-62.
Human Factors in Reliability Group. 1995, *Improving Compliance with Safety Procedure: Reducing Industrial Violations*, (HSE Books, London).
Munley, G., Lindsay, J. and Ridsdale, E. 2003, An Application of Meta-Analysis to Studies of Rule Violation, *Contemporary Ergonomics 2003*, (Taylor and Francis, London).

EMBEDDING ERGONOMICS IN IN-CAR INTERFACE DESIGN: THE DEVELOPMENT OF A SIMULATION BASED EVALUATION METHOD

Chen, C.C.[a], Woodcock, A.[a], Porter, S.[b] and Scrivener, S.A.R.[a]

[a] The Design Institute, School of Art and Design, Coventry University, Priory Street, Coventry CV1 5FB, UK
[b] Department of Design and Technology, Loughborough University, Leicestershire, LE11 3TU

This paper provides an overview of the development and validation of a low-cost evaluation method for in-car display and control designers. The method comprises three elements: a computer mock up of the concept design to be evaluated, a driving rig (to approximate the vehicle and the simulated driving task) and a usability evaluation toolset. Having established the effectiveness of the method in providing useful information for iterative design, it was formalised and subsequently tested as a set of guidelines for in-car interface designers.

Introduction

With the development of electronic interactive systems in the car market there has been greater emphasis on in-car comfort and convenience provided by, for example, in-car entertainment, electronic climate control and navigation systems. Such systems are intended to make the driving experience more enjoyable and comfortable. However, research (e.g. Brookhuis and Waard, 1994; Wierwille and Tijerina, 1998) has shown that a driver's workload may increase with the use of such equipment (especially if poorly designed) thereby distracting from the primary driving task. Crash records and studies indicate that complicated controls, together with an ergonomically poor interface, are one of the major causes of road accidents (Wierwille and Tijerina, 1996). Therefore, producing in-car devices that can be operated within acceptable safety limits is an ergonomics issue (Galer Flyte, 1995). A lot of tools and methods have been developed by researchers to evaluate in-car interfaces and support ergonomics in the design process (e.g., Green, 1990). Such methods, often involving user centred evaluation, identify usability problems arising from the use of in-car interfaces and produce data for solution generation. However, the methods are limited in ways that may inhibit their application in design practice. For example, they may often necessitate complicated data analysis, or expensive experimental apparatus, or even involve road tests. Practicing designers rarely have the time, inclination or the ability to use methods requiring complicated procedures and data analysis, which they find difficult to integrate into the concept design process. Additionally, current HCI methods have not been designed to uncover usability problems arising from the use of a device being operated within a broader task context, such as driving.   Hence their validity for evaluating secondary in-car controls and displays is highly questionable.

Notwithstanding ergonomic and design advances, particularly in relation to human-computer interaction, automobile design research (Woodcock and Galer Flyte, 1995; Eost, 1999) has revealed that the design process is still poorly supported by ergonomics information and knowledge; the factors accounting for this situation include poor communication and the failure to integrate ergonomics information and evaluation into the early stages of design (Woodcock and Galer Flyte, 1998; Porter and Porter, 2000).   Thus, it was concluded that there is still a need for methods to bridge the gap between designers and relevant ergonomics data (Woodcock and Galer Flyte, 1995), especially methods that are acceptable to and usable by designers working in the early stages of design.

Consequently, it is argued that there is a lack of useful and usable ergonomically informed methods for in-car interface design in the concept stages of the current design process.   With this in mind, the research described in this paper aimed to develop such a method to resolve these problems.

The Development of the Method

In order to enable designers to evaluate their design concepts, a user centred approach was advocated that would uncover usability problems and generate ergonomics information in a form that designers can use.   Additionally, the method was to be developed in a format that required little time and cost, or the need for complicated experimental apparatus and analytical processes, so as to be really applicable to current design.   One of the weaknesses identified in current HCI methods is the neglect of the primary driving task. Since early concept designs exist only as non-operational prototypes, they cannot actually be used in a real driving situation.   Furthermore, the risk to persons of using experimental devices whilst driving is potentially high, thus legislating against evaluation in real driving situations on ethical grounds. Hence the method incorporated driving simulation to reveal problems arising from the use of the concept interface in a driving situation.

Also as today's designers have high levels of computer literacy, the method would also feature simulation of the interface – thereby ensuring usability testing could take place in the concept design stages.   A usability toolset was developed to uncover problems and generate design information concurrently use of this simulator.

Thus a three-part method was developed to inform the concept design and iterative development of in-car control and display design, featuring
- A low cost driving simulator
- A usability toolset
- An interactive simulation of the object to be evaluated (in this case a touch screen climate control system)

The Usability Toolset
The requirements of the usability toolset were that it should be easy to use (not requiring detailed data analysis), low cost and manageable (not requiring complicated experimental apparatus) and should provide timely, comprehensible results that could inform iterative design. The toolset included objective measures (task completion time,

errors and glances) of user performance to determine the location of usability problems and subjective measures (an ergonomics audit form and usability questionnaire) to collect detailed redesign information from users. The adequacy of the toolset was assessed in terms of the extent to which usability problems emerged and the measures reinforced and complemented each other. The results indicated that usability problems were uncovered in a form likely to support in-car interface ergonomic design.

The Simulated In-Car Interface

Computerised, simulated interfaces have already been demonstrated as effective and efficient ways to evaluate the usability of concept designs. In this research a simulated interface, representing the in-car device concept was produced with fully interactive simulated graphic control buttons and feedback.

This was compared to the performance of a real device through user testing to establish the validity of the simulation. The results indicated that working prototypes (*i.e.*, employing buttons, switches, LCD displays and sculpted surfaces) are not likely to perform significantly differently in terms of usability than screen-based computer simulations.

The Evaluation Rig

Having examined the validity of simulation, the next task was to construct an environment that would allow evaluation to be conducted in the dual task context (i.e., driving while using an in-car device). Therefore a 'low-cost' and 'easy to build' driving rig was constructed in a studio-like space to approximate the driving situation (see Figure 1). The driving rig, comprises four parts, a simulated driving environment; a "computer game" steering wheel and pedals for controlling the 'vehicle'; an adjustable car seat; and a touch screen on which a simulated in-car interface can be displayed. This enables lab-based in-car usability evaluations to be conducted using the usability toolset.

Figure 1: A lab-based in-car usability evaluation

Supporting Iterative Design

An important part of developing the method was to ensure that it could effectively support iterative design. Therefore, the method was used to evaluate an initial concept, with the problems uncovered being used to inform the subsequent redesign of the interface. This 'second design' was then simulated and re-evaluated in a similar way. The results showed improved performance thereby demonstrating that the method supported the redesign process and resulted in a design that yielded enhanced user performance.

The Formalisation of the Method

At this stage of the research, although the method had demonstrated its potential usefulness, it was still very much laboratory based. Therefore the next stage entailed formalising the method so that a designer developing new in-car display concepts could use it. The method was supported by ergonomics guidelines to enable informed solutions to be made. The designer developed a concept for an in-car device interface, which he then tested using the toolkit. The designer was able to use the method successfully to uncover usability problems and, with the help of the guidelines, to formulate revised solutions. This indicated that practicing designers could use the method.

The Practicability of the Method

Another important concern was whether the method has a role in current design practice. Therefore, interviews were conducted with representative in-car interface designers to review the proposed method to determine whether (1) similar methods are employed in the current design process; (2) the method could be incorporated into the current design process; (3) the method was considered to be time and cost effective; and (4) the method was perceived as having value. The results of the interviews support the research claim that there is no current user-centred ergonomically informed method to support in-car interface concept design. As most designers possess high levels of computer ability and can produce computer graphic simulations, without the need for additional training, the designers thought that the method could be adapted to current design practice. It was also found that the proposed time and costs could be absorbed within the current design process, with the possibility that the method might shorten the development time, as usability problems would be uncovered in the early design stages. Hence, the results provide evidence of potential for incorporating the proposed method in automotive design practice.

Conclusion

This research addresses the need for a new method to assist designers in the early evaluation of in-car control and display concepts, which in turn will lead them to adopt a more user-centred approach to this critical design stage. The method developed to fill this gap relies on the simulation of the interface and the driving task, augmented by a set of research methods that designers can use for themselves to evaluate early concepts. The research has also addressed the validity, practicability and usefulness of this approach for practicing in-car interface designers. However, although the potential for the method has been demonstrated, it was not actually applied in the automotive industry. Therefore some uncertainty exists as to whether it will actually prove useful in practice. Future work should focus on confirming its applicability and value in industry.

In conclusion, the research has identified and made attempts to bridge the gap in current evaluation methods by providing a method that designers can use to evaluate in-car interfaces.

References

Brookhuis, K.A. and Waard, D.De . 1994, Measuring driver performance by car-following in traffic, *Ergonomics*, **37**, 3, 427-434

Eost, C.L. 1999, *Capturing Ergonomics Requirements in the Global Automotive Industry*, Ph.D. Thesis, Loughborough University

Galer Flyte, M.D. 1995, The safe design of in-vehicle information and support systems: the human factors issues *International Journal of Vehicle Design*, Special Issue on Vehicle Safety, 158-169

Green, P., Boreczky, J. and Kim, S-Y. 1990, Applications of rapid prototyping to control and display design, *SAE* paper 900470, Warrendale, PA: Society of Automotive Engineers, Inc

Porter, C.S. and Porter, J.M. 2000, Co-designing: designers and ergonomics. In Scrivener, S.A.R., Ball, L.J. and Woodcock, A. (eds.) *Collaborative Design*, (Springer, London), 27-35

Wierwille, W.W. and Tijerina, L. 1996, An analysis of driving accident narratives as a means of determining problems caused by in-vehicle visual allocation and visual workload,   In Gale, A. G. *et al.* (eds.) *Vision in Vehicles II*, Amsterdam: North-Holland, 79-86

Wierwille, W.W. and Tijerina, L. 1998, Modelling the relationship between driver in-vehicle visual demands and accident occurrence" In Gale, A. G. *et al.* (eds.) *Vision in Vehicles VI*, Amsterdam: North-Holland, 233-243

Woodcock, A. and Galer Flyte, M.D. 1995, The opaque interface – The development of an on-line database of ergonomics information for automotive designers" In Allen, G. (ed.) *Adjunct Proceedings of HCI '95: People and Computers*, University of Huddersfield, 96-104

Woodcock, A. and Galer Flyte, M.D. 1998, Supporting the integration of ergonomics in an engineering design environment, In Horvath, I (ed) *TMCE 98. 2nd International Symposium on Tools and Methods for Concurrent Engineering*, 152-166

Analysing marker gaps during work-related activities of daily living using an optometric system

Nancy L. Black[1], Mathieu A. Landry[2], Martha Ross[3] and Edmund N. Biden[3]

[1]Faculté d'ingénierie (industrial engineering sector), Université de Moncton, Moncton, NB, Canada E1A 3E9, blackn@umoncton.ca
[2]Department of Mechanical Engineering, McGill University, Montréal, QC, Canada
[3]Institute of Biomedical Engineering, University of New Brunswick, Fredericton, NB, Canada

Six work-related, goal oriented activities of daily living (ADLs) were recorded for five men and five women with a six-camera VICON 512™ optometric system. Three non-collinear markers were placed on each arm segment and the upper trunk to track movements, while minimising skin-movement artefact. Despite careful camera placement, 57% of recordings had trajectory gaps over 0.167 seconds long, although this was better than the 90% found when tracking one arm at a time with a three-camera system. Results show gap prevalence was highest during changing screwdriver bit and lowest during cutting steak (2.48 and 1.27 gaps / recording, respectively). Distal markers on the wrist and hand accounted for 96% of the recorded gaps. Large calibration movements prior to ADL recordings reduced gaps. Subject motions and anthropometry played roles.

Introduction

Optometric systems are used widely for motion capture and analysis beyond that which is possible with the naked eye. While a single camera is sufficient for general movement interpretation, multiple views of an object are needed to define location and orientation in three-dimensional space. The VICON 512™ and 140™ systems record using cameras of known lens characteristics with infrared strobe lights to distinguish markers covered in retro-reflective material. Digitised marker centres defining these points are recorded for further analysis (Oxford Metrics, 1996). The AUTOLABEL function of the VICON 512™ system allows automatic labelling of markers based on a calibration recording. Unfortunately, with such systems markers may become hidden during activities of daily living (ADLs) which are of interest in ergonomic research (biomch-l 1996). This research quantifies the challenges associated with a six-camera VICON 512™ system recording both arms simultaneously and compares these with a three-camera VICON 140™ system recording one arm's movement at a time. Since the software capabilities differ between the systems, results will be presented by recording system, with greater emphasis on the six-camera set-up.

Recording methodology

Equal numbers of men and women (coded M and F respectively) were studied, five each with the six-camera system, and ten each with the three-camera system. All were of working age (23, 20, 31 and 35.7, 18, 62 average, minimum and maximum, respectively) with no upper limb disability. In advance of the recordings, camera locations were chosen to maximise marker visibility during six work-related ADLs. Tasks were chosen to reflect different levels of dexterity, force and joint involvement while involving both hands (see Table 1). Standardised instructions were used during two training repetitions and at least two recordings of each arm. Typical working environment tools were used.

Table 1. Work-related activities of daily living studied

Title	Instructions
Zipping	Unzip and zip the main U- shaped zipper on the backpack.
Changing	With the screwdriver centred in front of you, remove the bit, and replace it with the loose bit, placing the screwdriver and first bit in the same locations at the end.
Cutting	Using the knife and fork as if you were cutting a tough steak, cut a slice from the plasticine, picking it up and placing it on the other plate. Repeat for a second slice. Then put the knife and fork back.
Hammering	Hammer the nail into the block of wood until it is inserted ½ inch (to the tape line), putting the hammer back where it was initially.
Folding	Fold the sheet of paper into thirds folding on line 1 then on line 2; then insert the sheet into the envelope, tucking the flap in. Place the filled envelope in front of you with the flap down.
Tying	Tie a knot and then a bow with the string on the platform.

Marker placement

Three skin-mounted spherical markers covered in retroreflective tape were placed on each of the arm segments, as well as the upper part of the torso to allow tracking of all relevant segments during a seated task. Marker placements were chosen for minimal skin movement (Williams, 1996) and to ensure maximal marker visibility with both a 3- and 6-camera system during table-top ADLs. Chosen locations ensured that markers on each segment were distinctly non-collinear to determine the segment's three dimensional location and orientation, and to ensure repeatable placements. A three-letter code was used to identify each marker (see Figure 1), preceded by 'R' or 'L' where identical marker locations existed on both right and left sides.

Camera Placement with the three-camera VICON 140^{TM} system

With the three-camera VICON 140 ^{TM} system, cameras were located relative to the subjects' working volume (see Table 2). Since each camera best defines locations in a plane perpendicular to its line of view, locating error is minimised with a 90° camera separation. However if a point is not visible to at least two cameras, three dimensional reconstruction is impossible. Indeed, with two cameras, separation angles of 50° for ADLs with trunk restraint (Romilly et al., 1994) and 40° during feeding movements (Safaee-Rad et al., 1990a) provided a satisfactory compromise between accuracy and maintaining marker visibility. With more than two cameras, redundancy reduces marker locating error. When using a six-camera VICON 370^{TM} system for simple arm reach

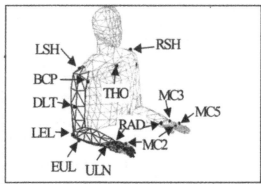

Figure 1. Skin mounted marker locations on the upper limbs and trunk

motion recordings, Williams (1996) found errors less than +/-1 mm for static locations. In the present case, with 35° separation, error levels were small (3 mm in static postures) (Black 2001). Smoothing minimised marker variability without eliminating mean errors. Pilot work found that markers on both arms were not continuously tracked during the ADLs, so the motion of only one arm and the upper trunk were tracked at a time. This required reorientation of the subject, placing camera 2 45° from the mid-sagittal plane toward the right or left depending on the arm of interest.

Table 2. Three-camera system locations

	From centre desk surface		Intercamera separation angles	
Camera	Distance	Vertical angle	Camera 2	Camera 3
Camera 1	2.8 m	30°	35°	60°
Camera 2	2.2 m	18°	-	35°
Camera 3	2.8 m	30°		-

Camera Placement with the six-camera VICON 512™ system
Camera locations with the six-camera VICON 512™ system (see Table 3) allowed simultaneous recording of both upper limbs. This was important given the intra-subject variations in movement, motion and material between repetitions (Black et al., 1999). With the VICON 512™ system, a calibration trial was recorded for each subject to allow use of the AUTOLABEL function to automatically label marker trajectories.

Table 3. Six-camera system location

	From centre desk surface		Intercamera separation angles				
Camera	Distance	Vertical angle	Camera 2	Camera 3	Camera 4	Camera 5	Camera 6
Camera 1	2.9 m	49°	80°	111°	35°	63°	33°
Camera 2	3.8 m	43°	-	36°	100°	20°	49°
Camera 3	2.4 m	7°		-	116°	49°	79°
Camera 4	1.6 m	15°			-	81°	53°
Camera 5	2.8 m	51°				-	31°
Camera 6	2.6 m	60°					-

Results

Gaps in marker trajectories over 10 samples long (0.167 seconds duration) could not be filled by simple interpolation without loss of information. During recording with the three-camera system considering only one arm at a time, such long gaps were present in 90% of the recordings. With the six-camera system, considering both arms simultaneously, this dropped to 57%. Given the superiority of the six-camera system, its detailed recording results are presented here, by variable of interest.

By subject
The number and duration of gaps varied widely across the recorded population. F1 and F7 had the greatest number of gaps in total, while M8 had none at all (see Table 4). F1 did not move prior to ADL movements, and had the highest number of long gaps per recorded ADL. F7 used visibly more extreme and faster motions than other subjects. Notably, despite fixed camera locations and with the exception of F7, as recording sessions progressed, the number of gaps decreased, indicating possible refinement of recording methods. Recordings of women had more gaps than men.

Table 4. Number of long gaps by subject and activity

Activity	F1	M2	F3	F4	F5	M6	F7	M8	M9	M10	Total	Gaps/ Recording
Zipping	13	1	7	18	2	3	14	0	0	1	59	2.19
Changing	14	5	0	5	10	10	15	0	0	3	62	2.48
Cutting	16	0	1	4	2	0	4	0	1	0	28	1.27
Hammering	7	0	15	4	6	4	16	0	0	0	52	2.08
Folding	13	6	7	4	4	12	8	0	4	0	58	2.07
Tying	13	5	1	6	4	0	6	0	0	0	35	1.52
Total	76	17	31	41	28	29	63	0	5	4	294	1.96

By ADL
The number of important marker gaps varied across the ADLs studied (see Table 4). Changing the screwdriver bit was worst, and was most associated with longitudinal rotation of the hand and forearm. Zipping followed, manipulating the closure on a large opaque object. Least problematic was the cutting task, which involved mostly variations in simple shoulder and elbow flexion with little longitudinal rotation. Interestingly, the tying task was next best; here longitudinal rotation was more distal, and the laces manipulated did not cover the markers to be tracked.

By anatomical marker location
Marker gaps were more common distally, with 96% of all gaps occurring at the wrist or on the hand (see Table 5). Gaps varied little between right and left sides. Across all subjects and activities, no gaps were recorded for the LSH, RSH, THO or DLT markers. The MC3 markers had the most frequent gaps, which were nearest to wrist markers, and could be covered by the hand or other markers during flexion or extension. Markers at the wrist were next most prone to gaps. Longest gaps occurred in the markers with the most frequently recorded.

Table 5. Number of gaps by marker across 10 subjects

	SH	DLT	BCP	LEL	EUL	ULN	RAD	MC2	MC3	MC5	THO	Total
Left	0	0	0	6	3	27	44	7	48	11	0	146
Right	0	0	1	1	1	37	37	9	50	12		148

Conclusions

While possible camera placements are infinite, when recording bi-manual ADLs that do not necessarily involve different movements between the upper limbs, both arms should be viewed equally well. By doubling the number of cameras from three, two rather than one arm were recorded simultaneously and gaps were reduced by 36%. During recordings with a six-camera VICON 512^{TM}, markers were most reliably recognised when, prior to the ADL, subjects made large random movements (minimum 40° amplitude variation in elbow and shoulder angles) for at least one second. Distal markers were most susceptible to gaps, especially during longitudinal rotation of the forearm and hand. Large wrist flexion and extension made the smaller MC3 marker near to ULN and RAD difficult to distinguish. The larger inter-marker distances on men may account for their fewer gaps. Marker gaps may be lessened by mounting markers away from the skin, but this would hinder natural movement patterns.

Acknowledgements

The Faculté des études supérieures et de la recherché, Université de Moncton, and the Natural Sciences and Engineering Research Council of Canada funded this work.

References

biomch-l. 1996. Summary – Missing Markers. Log 9609/00031.html. posted Sept 10 by Joshua Swamidas, accessible at http://www.bme.ccf.org/isb/biomch-l/archives

Black, N.L. 2001. *Mechanical energy measurement applied to upper limb powered prostheses*, Unpublished Doctoral Thesis, (University of New Burnswick, Fredericton, NB, Canada)

Black, N.L., Biden, E.N. et Rickards, J. 1999. The importance of qualitative analysis in quantitative activity comparisons, *Proceedings of the Annual Conference of the Association of Canadian Ergonomists*, Hull, QC, 13-15 oct., 5 pp.

Oxford Metrics. 1996. *VICON 140 Version 1.0 User's Manual*, (Oxford Metrics Ltd., Oxford, UK)

Romilly, D.P., Anglin, C., Gosine, R.G., Hershler, C., and Raschke, S.U. 1994. A functional task analysis and motion simulation for the development of a powered upper-limb orthosis. *IEEE Transactions on Rehabilitation Engineering*, **2**, 119-129

Safaee-Rad, R., Shwedyk, E., and Quanbury, A.O. 1990, Three-dimensional measurement system for functional arm motion study. *Medical and Biological Engineering and Computing*, **28**, 569-573

Williams, J. R. 1996, *Some Aspects of the Biomechanics of the Elbow Joint: Relation to prosthetic design*. Doctor of Medicine Thesis. (University of Oxford, Oxford)

THE APPLICATION OF STANDARD HCI EVALUATION METHODS TO WEB SITE EVALUATION

Yang, C.Y., Woodcock, A. and Scrivener, S.A.R.

The Design Institute, Coventry School of Art and Design, Priory Street, Coventry, UK Email: A.Woodcock@coventry.ac.uk

Web design is commonly associated with HCI and as such has borrowed and adapted standard evaluation methods to address web usability. However, websites differ from conventional interfaces in a number of ways and many evaluation methods fail to consider the needs of the designer. Evaluation can be seen as an end in itself, not as a means of providing designers with useful information for redesign. Hence the designer may know that their web site is considered unattractive by a number of visitors, but may not have detailed information to inform redesign. Motivated by a need to provide a useful, easy to use web site evaluation methodology we have investigated the adequacy of existing methods, the requirements of web site designers and web site users before developing a formative evaluation method for web design.

Introduction

With millions of websites, web designers compete for users' time and attention (Nielsen, 2000). Historically, computer software and interfaces developed before web sites, and the community has built up a body of evaluation methods and procedures to address HCI usability. These methods have been applied to websites with some success (Brinck et al, 2002). However such methods have their inherent weaknesses when so applied, for example web sites are used and accessed differently, their interfaces are different from conventional software, they contain different features (Spool, 1999; Winckler et al, 2001), they evolve faster than normal software releases (Winckler et al, 2001; Brinck et al, 2002) with rapidly changing content and updates; and likeability has been shown to be more important than ease of use (Spool, 1999). These inherent differences may mean that standard HCI methods may not be appropriate.

One issue similar to both domains is the need to provide designers with timely and appropriate information that will lead to effective and efficient redesign. One of the most common failures of current methods is their inability to correctly and specifically define user problems (Berkun, 2001) and to provide useful redesign data.

The aim of this research was firstly to consider the appropriateness of standard HCI methods for website evaluation, secondly to determine differences in web site designers and users' expectations from websites, and thirdly to develop a website evaluation method based on user's perceptions that can be used for formative evaluation. Such a method should concentrate on users' perceptions of usability rather than designers' interpretations of them. It is believed that such an approach will lead to a method that will focus designers' attention on the features and functions that users want and need.

Evaluation of existing methods

With these aims in mind, a set of representative HCI evaluation methods - observation, Meaning in Mediated Action (MIMA) (Bourges, 1998), and Website Analysis and Measurement Inventory (WAMMI) (Kirakowski and Cierlik, 1998) - was used in the formative evaluation of a typical, research web site (the UNITE project web site).

Ten culturally heterogeneous students having at least middle level English evaluated the site. The participants were guided through the evaluation, conducted in the context of a set of 14 standard tasks (such as to find out about the project deliverables) developed with the help of the site developer. For the MIMA evaluation the site was broken into its component elements to determine cultural differences in the interpretation of words and images. Task completions were observed and recorded, with special attention being placed on routes to completion and errors. These were later subjected to breakdown analysis (Woodcock and Scrivener, 2003). The evaluation was concluded with the completion of a WAMMI questionnaire that measures attractiveness, controllability, efficiency, helpfulness, learnability, and provides a global usability score.

The site designer evaluated the results derived from the four methods in terms of their usefulness for redesign and concluded that together they provided sufficient information for redesign. For example, the overview provided by WAMMI was supplemented by the detailed information provided from MIMA and the observational studies. However, from the interview it emerged that the issues that the designer was most interested in – keeping the user on the site, likeability and ensuring return visits - were not addressed. An Internet survey was conducted in order to gain further insight into these limitations, both in terms of the formative value of evaluation outcomes and those aspects of usability particular to web sites.

Internet Survey

An international survey of 65 web designers and 69 experienced web users, ranging from 16 to 34 years, was conducted to identify differences in perceptions of usability and user requirements between the two groups, to examine what web designers required of an evaluation method and how user needs might be mapped on to the different components of the web site considered by users and designers when responding to the questionnaire.

The results showed, for example, that users returned to sites that provided helpful and high quality information. They noticed the ease of using of the site and quality of the features they first noticed, though interestingly were not aware of more general features. This clearly has implications for designers who may spend valuable time adding superfluous, unnecessary features to their site. Quality and usability affected whether users stayed on the site and made return visits. Likeability is obviously more idiosyncratic, but may be related to the topic, information, or products.

Through an analysis of the results, it was possible to map designers' goals on to generic web site features (e.g. navigation and visual elements). Then, taking the web users concerns it was also possible to map these on to web site features. These relationships are shown in Table 1

Table 1. Relationship of designers' goals and user issues to web site features

Designers goals	Web site features	Examples of user issues
1. High quality HCI	**Navigation**	LEMES – learnability, efficiency, memorability, effectiveness and satisfaction
2. User retention	**Navigation, visual,** information	Download speed, ease of use, attractiveness, layout, colour scheme, helpful information, site updates
3. Providing appropriate user services (content and functionality)	**Navigation**, information	Easy to use and understand structure, helpful information, layout and updates
4. Likeability	**Visual,** functionality, information	Image, layout, useful information and functions
5. Attraction of more users	Only occurs if 1 to 4 are satisfied	Only if the above are satisfied

Obviously the relationships suggested in Table 1 need further validation. However, such a mapping provides designers with an understanding of which features of a web site might be used to control specific aspects of usability. So if the evaluation showed that users were not satisfied with download speed, this would imply that web site was unlikely to retain users, and that designer should consider enhancing visual and navigational features. This provides designers with an understanding of which aspects of their web site relates to which of their goals.

From the survey we found that users and designers had different expectations in terms of web site usability. For example, users expected a web site to provide diverse functions and information, whereas the designers sought to design a website around a specific theme. Similarly, the users put greater emphasis on information content than visual design, whereas these priorities were reversed for the designers.

Regarding the usefulness of current methods, about 50% of the web designers suggested that seeing users interact with their site was the best way of understanding user problems. They felt that problem statements obtained from responses to usability questionnaires needed a higher degree of specificity to help them in their redesign and they also required information on users' routes to task completion and whether the site reached its target audience.

From the requirements emerging from the questionnaire, our knowledge of web site design, and the strengths and weaknesses of existing methods, a new website evaluation method has been developed.

Development of the new website method

The research outlined above indicated that current single methods do not provide designers with sufficient information for redesign purposes. Individual methods have their strengths e.g., questionnaires provide information about users preferences for contents and visual design, but do not provide rich problem descriptions; observations may provide too detailed information to be easily assimilated into redesign issues. However, just providing designers with a toolbox of methods may not lead to their optimum use. Therefore, a web site method was developed which connected web site goals (as identified by the web designer), usability issues (important to the users) and web site features (information, navigation, graphic components). It is assumed that prior to user testing the designers will have addressed browser interoperability issue and access for those with different visual abilities (e.g. colour blindness, dyslexia).

Table 2. Elements of the proposed method

Designers goals	Examples of user issues	Proposed method(s)
1.High quality HCI	LEMES – learnability, efficiency, memorability, effectiveness and satisfaction	Observation (verbal protocol, route taken, time and errors)
2.User retention	Download speed, ease of use, attractiveness, layout, colour scheme, helpful information, site updates	MIMA, card sorting, tailorable questionnaire, observation
3.Providing appropriate user services	Easy to use and understand structure, helpful information, layout and updates	MIMA, card sorting, tailorable questionnaire, observation
4.Likeability	Image, layout, useful information and functions	Tailorable questionnaire
5.Attraction of more users	Only if the above are satisfied	Tailorable questionnaire

A first stage in conducting the method is for the evaluator to work with the designer to:

- Determine specific usability goals and acceptance criteria
- Decompose the site, and provide definitions (e.g. of navigation elements)
- Provide a set of typical user tasks
- Define target users

This will enable the proposed methods to be customized to the site and the designer's requirements. A representative sample of users (including those with visual impairments) will then engage in in-depth testing based around their use of the site, understanding of its components and perceptions of usability. The evaluation report to the designers will re-present the results of the analysis in terms of the initial usability goals.

A pilot test of the method has been conducted on a commercial site. This showed that the structured evaluation process was efficient. The results from MIMA, observation, and card sorting triangulated to identify problems with task design and information arrangement. It was also possible to tailor a questionnaire to provide useful and specific information on redesign.

The next stages of the research will involve further exploration of the relationship between designers and users goals and web site features. The usefulness of the method will also be further tested on different types of web sites and applications. In addition further attention will also be given to the best ways of communicating usability issues to designers.

Conclusion

The starting place for this research was the need to provide web designers with usability information that could lead to effective redesign and also enable them to uncover the types of information especially related to web usability. Through the Internet survey we found that issues such as retaining and attracting new users were important to designers but not necessarily addressed by current methods. By mapping these issues on to web features and the usability requirements of end users we can develop a tailorable method to support formative web development.

References

Berkun S. 2001, The role of flow in web design, Microsoft Corporation January/February 2001, http://msdn.microsoft.com/library/en-us/dnhfact.html/hfactor10_1.asp?frame=true, accessed on 24/11/2001

Bourges P. 1998, *Handling Cultural Factors in Human-Computer Interaction*, unpublished PhD, University of Derby, U.K.

Brinck T., Gergle D. and Wood S. D. 2002, *Usability for the Web*, Morgan Kaufmann Publishers, USA: 16, 439-440

Chisholm, W., Vanderheiden, G. and Jacobs, I. 1999, *Checklist of Checkpoints for Web Content Accessibility Guidelines 1.0*, http://www.w3.org/TR/WAI-WEBCONTENT/

Kirakowski, J. and Cierlik, B. 1998, Human centered measures of success in web site design, Paper presented at the Fifth Human Factors and the Web Meeting, June 1998.

Ivory, M. Y. and Hearst, M. A. 2002, Improving web site design, *Journal of IEEE Internet Computing*, March-April, 52-63

Newman, M. and Landay, J. 2000, Sitemaps, storyboards, and specifications: A sketch of web site design practice, *DIS'00*, ACM, New York, 263-274

Nielsen J. 2000, *Designing Web Usability: The Practice of Simplicity*, (New Riders Publishing, Indiana USA)

Spool, J. M. 1999, *Web Site Usability, A Designer's Guide*, (Academic Press, USA)

Winckler, M., Pimenta, M., Palanque, P. and Farenc, C. 2001, Usability evaluation methods: What is still missing for the web?, Conference of Human-Computer Interaction; *Proceedings of the HCI International 2001 Usability Evaluation and Interface Design*, 883-887

Woodcock, A. and Scrivener, S.A.R. 2003, Breakdown analysis, P.T.McCabe (ed.) *Contemporary Ergonomics*, (Taylor & Francis, London), 271-276

THE DESIGN OF POLYSENSORY ENVIRONMENTS FOR CHILDREN WITH AUTISTIC SPECTRUM DISORDERS

Darryl Georgiou, Jacqueline Jackson,
Andrée Woodcock and Alex Woolner

*The Design Institute, Coventry University, Gosford Street, Coventry, UK.
email: A.Woodcock@coventry.ac.uk*

It has been estimated that autistic spectrum disorders (ASDs) may affect 50000 families in the UK. Those having ASDs exhibit a wide range of behavioural problems, with limited opportunities to break free and engage with the world. The most effective time for interventions is during childhood. Recently, computers and digital technology are being used to provide stimulating and interactive environments where children may engage with polysensory environments. However, given the wide range of ASDs, if such environments are to be enjoyable, non-threatening and effective, they need to be tailored to meet the sensory needs of the children. This paper introduces the first stage of a project to develop such an environment.

Introduction

Estimates of the incidence of autism in the UK vary from 1:2500 to 1:1000 with most authorities noting a steady increase in incidence over the last decade and an equal distribution amongst social and racial groups. Children with ASD have varying levels of difficulty in processing perceptual, social and cognitive information. This in turn leads to behavioural problems such as short attention spans, lack of curiosity, limited patterns of play and communication leading ultimately to social isolation. Intervention during childhood may be a way of breaking some of this cycle (McEachin et al 1993). For children with ASD, computers and digital technology can provide highly interactive and stimulating environments. Computers may be especially attractive because:

- They present consistent and controllable feedback,
- Children with ASD are often much more willing to engage with them for longer periods (Chen, 1993)
- There is a need to enhance the provision of educational and entertainment facilities for those with ASD – computers would seem an ideal option

- Computers can provide hands-free, exciting multi sensory environments potentially tailorable to meet the needs of differently abled children.

This research adopts a user centred approach to the design, development and evaluation of a polysensory environment by considering the needs and requirements of the children first and foremost, but also their parents, and those involved in their care. These requirements will be embedded in a modular system, tailorable to meet a wide spectrum of needs. We will then develop a set of evaluation metrics designed to measure the performance/acceptability and benefits of the system for the users. In this, the first paper from our research, we will introduce the nature of autism, our approach to capturing user requirements, review current systems and their features, and discuss the challenges autism poses for designers.

The Nature of Autism

Whilst debate and research continues into the cause of autism with biological, neurochemical, neurological and genetic explanations being advocated, no clear-cut answers are on the horizon and there is no cure. All children with an autistic spectrum disorder have the 'triad of impairments' (Wing and Gould, 1979) in the areas of social interaction, communication and imagination. However, what makes ASDs so problematic is that levels of severity and manifestations of the condition are different in each child. This leads to different classifications, for example, where children are considered to be 'higher functioning' they may be diagnosed with Asperger Syndrome (AS) or High Functioning Autism (HFA).

Accompanying ASD is a high incidence of co morbid conditions such as dyspraxia (development coordination disorder) and Attention Deficit Hyperactivity Disorder (AD/HD). These may also be considered to be part of the 'autistic spectrum'. Furthermore Sensory Integration Dysfunction (SID) has been recognised for many years (Ayres, 1972) as a specific condition due to a dysfunction whereby the brain has difficulties processing sensory input accurately. Additionally, children with an ASD have difficulties with central coherence, executive functioning (Baron-Cohen and Swettenham, 1997) and auditory processing (Edelson, 1999). Therefore visual cues are imperative to aid understanding and processing of information and interactions in the physical and social environment.

Regardless of classification, evidence (e.g. Grandin, 1996) suggests that individuals with autism have difficulty integrating some or all sensory experiences (smell, taste, touch, movement, body awareness, sight, sound and the pull of gravity). Unfortunately it is the integration of these experiences that provides the foundation for productive contact with others and the environment. In younger children, these sensory difficulties often seem more pronounced because the child has not yet been taught coping mechanisms, nor have they been 'desensitised' to the sensory stimuli that cause distress. Movement disorders (Teitelbaum, 1998) in children with an ASD also contribute to the overall difficulties in sensory integration.

Digital environments based around visual cues can provide a tailorable environment that may assist in the integration of different experiences and help in co-ordination. However, given the wide range of sensory dysfunction the first requirement of such a system is that it has to be tailorable, or built around the needs of a specific child. For example, a child that is highly sensitive to sound may be extremely distressed by a system based around music.

Our approach to user requirements capture

Therefore an essential part of the design process for the production of a polysensory digital system for children with an ASD, is the need to establish user requirements and provide these in a manner useful to the designer.   An obvious difficulty when establishing user requirements for this particular user group is the fact that the children are unlikely to be able to articulate their needs and preferences directly. As we would like to design a system capable of delivering different sensory experiences that is potentially adaptable to all children within the spectrum, and which would provide different sensory experiences, we need to find out as much as possible about our end users, in non-invasive ways.

The obvious group to provide us with information about children's needs is their parents. It is also important that those who are able to articulate their requirements, namely individuals with High Functioning Autism (HFA) and Asperger Syndrome (AS) be given the opportunity to do so. To this end, in the first year of our research, we are conducting four main requirements gathering exercises:

- Questionnaire – distributed to online mailing lists, by mail-shot (to parents of Blackpool autistic children) and placed on researcher's website for online completion to gain an understanding of the range of ASDs our system will have to accommodate, sensory and interaction preferences and experience with current systems.
- Observations of children using existing sensory environments and at collaborating institutions.
- Reflection/participatory design where appropriate.
- In depth interviews with parents, carers, teachers, psychologists and individuals with AS and HFA.

Review of current systems

A range of sensory products has been developed to support children with disabilities, ranging for stand alone sensory products through to educational software, to highly interactive, novel multi media systems. Although evidence (e.g. (Bogdashina; 2003) suggests that those with an ASD can benefit from sensory input, few sensory products have been specifically designed for this group.

Companies such as ROMPA (Snoezelen), Mike Ayres and The Sensory Company, have designed and manufactured sensory products with the aim of enhancing sensory experiences and increasing sensory awareness in those with varying disabilities. 'Sensory rooms' are now installed in many areas of the country accessed by groups and individuals with a wide range of abilities. Informal observations and experiences show that users derive benefits from their sensory experiences (e.g. in terms of pleasure, engagement, and tranquility) and we would hope to contribute to research in this area.

Until recently, products provided few opportunities for interaction and feedback other than through standard HCI. One notable exception is the Aurora project (http://www.aurora-project.com/) which looks at the therapeutic and educational role of computer robots in developing communication and social interaction skills. Also, virtual learning environments (VLEs) have achieved some degree of success in facilitating environmental interaction and awareness in children with an ASD (Strickland 1998). Systems relying on traditional interfaces, such as mouse and keyboard, provide little opportunity for feedback and interaction. This is problematic for those with ASD who have problems in co-ordination and integrating experiences.

In terms of education, computers/digital technology present consistent and controllable feedback, and as such can be a successful tool to engage children with an ASD (Jordan, 1995). This has led to the development of software for children with special educational needs. Examples of such software for children with ASD include Boardmaker (Mayer, 1980) which focuses on the use of symbols and pictures and 'Mind reading' (Baron-Cohen, 2003) which addresses the recognition of gestures and facial expressions with the aim of helping children make sense of the world around them.

The challenge of autism for system designers

Much of the research and development in human computer interaction centres concerns developing prototype systems meet users requirements. With a broad range of users across the autistic spectrum, these requirements can be many and varied. The challenge for the designer is therefore to develop an HCI system with sufficient flexibility to meet all these needs. Obviously our requirements gathering exercises will provide detailed information for the designer. However, to develop an initial prototype as a testing ground for our ideas we have proceeded from the following assumptions:

1. Those with an ASD have varying degrees of hyper/hyposensitivity possibly in each of their senses. This may be exhibited as extreme reactions to certain colours, tastes or sound in general. This will affect their response to stimulation, how much stimulation is required and the overall effect of the system. In this case user requirements may be seen as polarized in terms of stimulation and sedation. Whilst some users will require a moderate level of sensory input in order to stimulate them, others will demand high levels of input to even get their attention. Therefore, the system needs to be tailorable by the operator to the sensory needs of the individual

2. Certain materials/objects should be avoided as they are dangerous (e.g. those with moving, edible, destructible part) or are known to trigger adverse reactions (e.g. colorants, certain plastics)

3. A distinction has to be made between 'higher' and 'lower' functioning individuals on the spectrum. This will determine the sensory content provided by the system, as it must provide the level of complexity expected by each user in order to make the experience fulfilling, and neither too demanding nor too simple.

4. People on the autistic spectrum are said to enjoy experiences with a degree of repetition. Any changes in an experience or sequence of events may upset a child who relies on, and is comforted by repetition. Whilst appreciating this need, there is also a developmental need to open up new avenues of behaviour. In terms of the system design, it is proposed that an 'aware' system is developed that monitors the behaviour of the user and feeds back this information so the system can adapt itself in real time in a nonthreatening manner, for example, through incremental changes in color, shape and movement.

5. In terms of content, in our earlier modules users have interacted with abstract sounds and shapes through movement. For some users it may be more appropriate to use personalized or favourite imagery. For example abstract shapes could be replaced by favourite cartoon characters. The design must therefore allow for the easy inclusion of information that can be displayed in a meaningful and stimulating way.

Conclusions

Through working co-operatively the research team (social scientists and new media designers) are taking a highly user centred approach to the development of tailorable, interactive polysensory environments for children with ASD. Although system requirements will be specified formally, the designer is part of the data gathering team and as such is not remote from the end user population. This allows him to fully appreciate the range of tailorability that has to be embodied in the final system and explore ways in which this might be accommodated. The need to uncover detailed user requirements for system design will in turn provide more insights into ASD and the design of future systems.

References

Ayres, A.J. 1972, Types of sensory integration dysfunction among disabled learners, *American Journal of Occupational Therapy*, 26

Baron-Cohen, S. and Swettenham, J. 1997, The theory of mind hypothesis of autism: relationship to executive function and central coherence. In D. Cohen and F. Volkmar (eds) *Handbook of Autism and Developmental Disorders*. Wiley

Bogdashina, O. 2003, *Sensory Perceptual Issues in Autism and Asperger Syndrome*. (Jessica Kingsley Publishers, London)

Chen, S.H. 1993, Comparison of personal and computer assisted instruction for children with autism, *Mental Retardation*, **31**, 6, 368-376

Edelson, S.M., Arin, D., Bauman, M., Lukas, S.E., Rudy, J.H., Sholar, M., and Rimland, B. 1999, Auditory integration training: A double-blind study of behavioral, electrophysiological, and audiometric effects in autistic subjects. *Focus on Autism and Other Developmental Disabilities*, **14**, 73-81

Grandin, T. 1996, *Emergence: Labeled Autistic* (Warner Books, London)

Jordan, R. 1995, Computer assisted education for individuals with autism. *Autism and Computer Applications*. Autism-France 3rd International Conference

Mayer, J. 1980, *Augmentive Communication* (Solana Beach, USA)

McEachin, J.J., Smith, T. and Lovaas, O.I. 1993, Long term outcome for children with autism who received early intensive behavioural treatment. *American Journal on Mental Retardation*, **97**, 359-372

Teitelbaum, P., Teitelbaum, O., Nye, J., Fryman, J. and Maurer, R.G. 1998, Movement analysis in infancy may be useful for early diagnosis of autism, *Proceedings of the National Academy of Sciences of the USA*, **95**, 13982-13987

Strickland, B. 1998, Virtual reality for the treatment of autism. in G.Riva. (ed.) *Virtual Reality in Neuro-Psycho-Physiology* (Ios Press.Amsterdam)

Wing, L. and Gould, J. 1979, Severe impairments of Social Interaction and Associated Abnormalities in Children: Epidemiology and Classification

BREAKING THE INFORMATION AND COMMUNICATION BARRIER

Elizabeth Ball

Ergonomics and Safety Research Institute (ESRI)
Loughborough University, Holywell Building
Holywell Way, Loughborough, Leics, LE11 3UZ, UK.

The design and application of methodologies for research involving participants and/or researchers with sensory losses must address their information and communication requirements. This paper demonstrates this need by exploring some of the issues that arose whilst designing a research project on orientation and mobility training for blind and partially sighted adults. The participants' blindness/partial sight and the researcher's deafblindness influenced the choice of research methodologies and how these were applied.

Introduction

Practices of orientation and mobility training in the UK have been driven largely by political and historical factors and have largely ignored the growing body of academic research (Dodds & Howarth, 1995; Social Services Inspectorate, 1988). Little attention has been paid to the experiences of blind and partially sighted people learning the skills of independent travel or to the effectiveness of different approaches to orientation and mobility training.

The research project described here seeks to:

- Identify factors which best enable blind and partially sighted people to learn independent travel skills;
- Examine the experiences of blind and partially sighted orientation and mobility students and their instructors; and
- Investigate the use of blindfolds in the training of students with low-vision.

The Information and Communication Barriers

Vision and hearing losses affect access to information and communication. The design of this research, therefore, had to pay particular attention to information and communication.

Participants include rehabilitation workers and blind and partially sighted people, many of whom may have additional impairments. All participants, irrespective of their degree of vision, must have access to information. Information, therefore, must be provided in a range of formats and methods, to ensure its accessibility to all.

The researcher is deafblind and uses tactile communication. Most participants do not, and therefore communication between the researcher and participants must be enabled through other means. The following sections describe the carefully chosen research solutions and techniques that are being used in this research project to facilitate access to information and communication for both participants and researchers with sensory losses.

Information Solutions

The information barrier is being overcome by providing information in a range of reading formats, including standard print, large print, audiotape, braille, computer disk and email. To ensure that a consistent high quality of information is sent to participants, prior to dispatch, a third party checks:

- Visual appearance of print/large print information;
- Sound quality of audio information; and
- Compatibility of electronic information with screen magnification and screen reader software.

A person who can read both braille and print and who regularly uses information on audiotape as well as both screen magnification and screen reader software performs this checking. This ensures consistency across all formats. Braille information is produced manually or computerised transcription is manually corrected and proof read.

When participants prefer, information is read to them over the telephone. To avoid unease at using BT TextDirect/RNID Typetalk (the UK's telephone relay service that relays calls between deaf textphone users and hearing telephone users), initial telephone contact is made by a hearing person. Subsequent telephone contact is made by the researcher via the relay service.

When participants must sign consent forms, copies of the form are provided in the participants' preferred reading format and/or read to the participant. However, they are required to sign a print form, with assistance if appropriate.

Participants may respond in their preferred format and are advised that a third party will transcribe print and audio correspondence.

Communication Solutions

The communication barrier proved to be more challenging. The research methodologies needed for this research are mostly qualitative. Interviews are essential to explore people's views on what is important in enabling blind and partially sighted people to learn independent travel skills and to investigate the experiences of those involved in orientation and mobility training. Interviews require effective and efficient communication between interviewer and interviewee. Analysis of interview data requires detailed notes and/or an accurate record of the interview.

The chosen solution is to interview the majority of participants by email. Email interviews have a number of advantages:

- Email is a method of communication that is accessible to the researcher and to many potential participants, eliminating the need for a human aid to communication;
- It ensures that participants can take part at a time and location convenient for them;
- It eliminates the need for participants to travel to be interviewed in person or for the researcher to visit them; and
- It is a format that allows participants time to reflect on their responses, should they wish to do so, before returning them.

However, email interviews also have a number of disadvantages:
- It reduces interaction between researcher and participant. In an attempt to minimise this effect, participants are encouraged to contact the researcher to clarify any questions and the researcher asks if she may contact them with further questions or to seek clarification;
- Only verbal content is transmitted by email and, therefore, nonverbal content is inevitably lacking. However, its asynchronous nature would keep this potential problem to a minimum, as, for example, nonverbal communication to determine turn taking is not necessary in email but is necessary in face-to-face conversations; and
- Not all potential participants have access to email. For this reason a small number of interviews are held face-to-face.

One unexpected issue arose with email interviews. In face-to-face interviews it is obvious to participants that the researcher is deafblind. In email interviews it is not. However, some people participating in email interviews realise that the researcher is blind, usually because they have had previous contact with or read about the researcher. In one case an interviewee asked to change some of his responses on the grounds that he had realised the researcher is blind. Why he wished to do this is unclear. Neither is it clear whether other participants respond differently depending on whether or not they know of the researcher's sensory losses nor how significant an impact this effect may have on the validity of the research results.

Face-to-face interviews, in this research, require a human aid to communication. The chosen solution is to use speech to text reporting. In speech to text reporting a trained speech to text reporter types everything that is said. The transcript is displayed in real-time on a computer screen and, in this case, on a braille display.

There are a number of advantages to interviews facilitated by a speech to text reporter:
- It provides the researcher with rapid access to the participants' responses;
- The transcript can be saved for later use; and
- Its synchronous nature maximises interaction between participant and researcher.

However, there are also a number of disadvantages:
- Speech to text reporters must be paid for their time, including idle time between interviews, and for travel expenses for every interview. As only a small number of interviews are held face-to-face the cost is not prohibitive;
- Text-based communication lacks the social presence of face-to-face interaction and does not convey nonverbal information such as tone of voice; and

- Participants may feel uneasy about the presence of the speech to text reporter and this may be exaggerated for those participants who are unable to see the transcript as it is produced. To minimise any unease, it is ensured that participants understand the role of the reporter and the confidentiality policy. Participants also have the option to read the transcript, in real-time, using screen magnification and/or synthesised speech output. However, synthesised speech can cause a distraction for participant and reporter.

Woodcock (2001) describes the success of using remote speech to text reporting for research interviews. In remote speech to text reporting, telecommunications are used to enable a reporter in a remote location to transcribe an interview and the transcript is displayed at the interview location. This approach has a number of advantages because it:
- Eliminates travel expenses and dead-time for the reporter;
- Increases flexibility in the scheduling of interviews; and
- Reduces unease at the physical presence of a human aid to communication.

However, technical failure of any part of the equipment used for remote speech to text reporting could lead to a total breakdown in communication between researcher and participant. If the participant knew neither the deafblind manual alphabet nor braille and had insufficient vision to use a braille alphabet card, there would be no means of communication. To avoid this, a communicator-guide, who could facilitate communication through the use of the deafblind manual alphabet, would have been needed at the interview location. However, this would cancel out any advantage of remote speech to text reporting. Remote speech to text reporting, therefore, was not a suitable option.

Interviews facilitated by a deafblind manual interpreter were also considered as an alternative to speech to text reporting. The deafblind manual alphabet is a form of tactile fingerspelling. Words are spelled out by touching the deafblind person's hand in different ways for each letter of the alphabet. Advantages of this approach include:
- An increased sense of social presence and human interaction; and
- An increased (though still limited) amount of nonverbal information.

However, it was rejected because:
- Deafblind manual occupies the hands. Note taking is, therefore impossible and no record of the interview would have been available for later use; and
- Deafblind manual is a slow method of communication, especially for this researcher.

Using observation of orientation and mobility lessons to augment interview data was also considered. This would mean:
- Making and recording observations on the move;
- Not interfering with the lesson; and
- Interviewing student and/or instructor after the lesson to discuss observations.

The research needed both auditory and visual information to be conveyed. Possible solutions included:
- Having the information relayed by a communicator-guide;
- Having the observations made by a third party according to the interviewer's instructions; or
- Using audio or video recording and subsequently having these transcribed.

The first option would have been slow and would, therefore, interfere with the lesson. It would also prevent notes from being taken. The second and third would have resulted in the data not being available to the researcher until some time afterwards and, therefore, would have precluded the possibility of discussing observations immediately after the lesson. In this research, the subjective views of students and instructors are of greater relevance than data on what actually happens during orientation and mobility lessons. Observations, therefore, would have had only limited value and would have entailed significant practical difficulties. It was therefore rejected.

The research also involves experiments on how training under blindfold affects the use of sensory information. This involves the recording of brief, quantitative responses from participants. Though using a deafblind manual interpreter was rejected in this research for qualitative interviews, the chosen solution for these experiments is to use a communicator-guide to describe visual responses and to facilitate communication. This is because:

- Both visual and auditory information must be conveyed to the researcher. Communicator-guides are specifically trained to do this. Other human aids to communication are not;
- The researcher must be able to communicate with the participant anywhere in the experiment room. Deafblind manual can be used almost anywhere. Speech to text reporting, when the output must be read from a braille display, is not portable;
- As only concise, quantitative measures need to be recorded, the conflict between needing hands for both note taking and communication is minimal; and
- The amount of communication in experiments is less than in qualitative interviews. Therefore, a slow pace to communication is less significant. However, it is important to ensure that participants understand and do not feel uneasy about delays in communication.

Conclusion

Sensory losses of both participants and researcher created unique requirements for information and communication in this research project. Many potential difficulties were identified during the research design but these were overcome and accessible methodologies devised, whilst still ensuring useful, valid and reliable data is collected. The research is ongoing and the success of the chosen methodologies will become clearer as the research progresses. It demonstrates the need for researchers designing research projects involving people with sensory losses to consider these, and how information and communication barriers may be avoided throughout the research process.

References

Dodds, A.G. & Howarth, C.I. 1995, The Blind Mobility Research Unit: 1965-1995. *British Journal of Visual Impairment*, 13(3).
Social Services Inspectorate. 1988, *A Wider Vision: The Management and Organisation of Services to People who are Blind or Visually Handicapped*. DoH, London.
Woodcock, K. 2001, Real-time Remote Reporting for communication access. Http://www.deafened.org/rrr.htm

CORPORATE RISK: AN ERGONOMICS PERSPECTIVE

Nigel Heaton

Human Applications
The Elms
Elms Grove
Loughborough
Leicestershire
LE11 1RG

Changes in the way that organisations are required to assess, reduce and monitor risk, combined with the need to provide appropriate controls assurance presents an opportunity for the ergonomics community to place ergonomics on to the Board agenda. This paper examines the impact of the Turnbull report on internal control and the HM Treasury requirement for Government bodies to manage risk. Our experience is that many organisations have both weak models for comparing disparate risk and are unaware of the nature and extent of ergonomics risks. This paper reports some ideas for helping organisations to manage risk and suggests how ergonomists could use some of the concerns of corporate governance to raise the profile of ergonomics and improve the effectiveness of ergonomics controls.

Introduction

In 1999 the Institute of Chartered Accountants published a document: "Internal Control: Guidance for Directors on the Combined Code". The aim of the document was to provide London Stock Exchange listed companies with a "sound system of internal control to safeguard shareholders' investment and the company's assets".

The guidance is more commonly known by the name of the Chairman of the internal control working party who produced it - the 'Turnbull Report' (after Nigel Turnbull, an executive director of Rank Group Plc).

The themes of the Turnbull report are based around the concept of risk management. Specifically, the board of any listed company is
"required to consider:
- the nature and extent of the risks facing the company;
- the extent and categories of risk which it regards as acceptable for the company to bear;
- the likelihood of the risks concerned materialising'
- the company's ability to reduce the incidence and impact on the business of risks that do materialise; and
- the costs of operating particular controls relative to the benefit thereby obtained in managing the related risks."

Turbull's guidance has more recently been adopted by central Government, under the auspices of the Treasury's "Orange book"[1]. Government bodies are now required to produce

---

[1] Management of risk: A strategic overview (with supplementary guidance for smaller bodies - HM Treasury 2001)

Turnbull reports, demonstrating that they understand how to manage risk and that they have appropriate controls in place.

Over the last two years, Human Applications has worked with a number of large organisations, both Plcs and Government Departments. One of the most surprising risks (to boards) that has emerged from proper consideration of organisational risk is the risk associated with poor ergonomics. For some companies this might be the risk of multiple claims for Musculo-Skeletal Disorders, for others it is simply the huge cost of sickness and absence caused by poorly designed jobs.

The aim of this paper is to present an example of how a number of large companies are now considering ergonomics risks. We have worked with senior managers and boards allowing them to put ergonomics risks in a wider context. This paper will examine how a technique known as the "Top X" principle has been applied and how organisations are learning to manage ergonomics better.

Managing Risk

The starting point in putting ergonomics into a management framework is to ensure that organisations understand what risk is. The term risk is used by a plethora of people to mean exactly what they want it to mean. The cornerstone of effective risk management is to agree a common grammar for describing risk and a set of rules for producing risk assessments. We recommend that the simplest risk models employ two independent variables - the **hazard** (that which has the potential to cause harm), - also know as impact (in Corporate Governance parlance) and **likelihood** - a measure of exposure to the hazard or the chance of the impact occurring. These two factors combine to produce **risk**. However, it is meaningless to talk about risk in an abstract form. Risk requires a context, typically a task or activity, i.e. the risk of X occurring resulting in Y when undertaking activity Z. This simple model can be applied to any significant risk faced by organisations.

The model needs to be extended if an organisation is to produce a statement of controls assurance to meet the needs of Turnbull or the Treasury. There must be a framework for risk that ensures risks are:
- Valid
- Reliable
- Comparable

This requires the description of the risks to be based on auditable data.

One of the challenges in managing risk is that many organisations have no common ground for comparing disparate risks. The net effect of this is that risks that are notionally "well understood" such as financial risks, receive more attention and resource than so called "soft risks", e.g. those risks relating to human resource (HR). As a consequence, organisations that face significant soft risks may only become aware of the risk after it has been realised. For example, when notified of an intention to sue from an employee who is claiming damages due to organisational negligence.

Our experience when working with organisations that are attempting to get to grips with corporate risk management is that they have no real experience of quantifying and comparing risks. This means that at best, risk management is opaque - "the risk is high because I say it is high" and at worst is completely inappropriate. "I missed that risk because I never realised that it was a problem and I didn't even know that we did that".

Many organisations are desperately seeking a simple and appropriate methodology for describing and managing risk.

Understanding Activities

The starting point for assessing risk is to understand what the organisation actually does. The disparate range of activities is what gives rise to risk and our experience is that many senior managers simply do not understand what happens at the coal-face. This was epitomised by a meeting with a Board to discuss corporate risk whilst outside the window an employee was using a compactor in a very dangerous manner. We were assured that a) that what we were seeing wasn't actually happening and b) employees didn't use the compactor.

In essence, companies need to have either a formal or an informal method for determining what they do. They also need to understand how different stakeholders come with their own risks and issues. We find that teaching organisations basic ergonomics techniques such as stakeholder analysis and task analysis significantly increases the quality of their risk assessments.

Risk Prioritisation

The Orange book identifies more than 20 different risk types. If each manager attempts to find multiple examples of each risk for every activity undertaken, organisations are presented with a mountain of paperwork. The problem is knowing what risks matter. We recommend that organisations undertake a walkthrough process. For any significant risk, the challenge is to make a justifiable assessment of the impact and the consequence of that impact in shorthand. Essentially, the walkthrough should take the task analysis and for each activity provide an estimate of the range and scale of impacts, coupled with an estimate of each likelihood (again supported by auditable data).

At a walkthrough level, the assessment should tell a brief story - what is the activity, what is the impact, why will the impact happen and how bad will it be, plus 3-5 "golden nuggets" of information to explain why the impact is more or less likely to occur.

Using this simple model, it becomes apparent that in many production environments, the risk associated with running a line is more than just the line breaking or the quality of the output deteriorating. There are significant risks associated with workers on the line who may, for example, be exposed to MSD risks. The likelihood of these risks being realised might be significantly higher than the likelihood of the line braking down, as issues of preventative maintenance and robust engineering are much better understood and actioned.

As no organisation has an infinite fund to control risk, the walkthrough process is the start of getting issues onto the agenda. A good risk management system should have a range of comparable "people" risks at a walkthrough level that can be initially ranked and compared with engineering, financial, environmental and other risks. It is almost inconceivable that an organisation that employs people will not have some people risks.

The Top 'X' Principle

Even at a walkthrough level, organisations can end up with individual managers having to deal with 10's of risks. We know that people struggle to deal with half a dozen problems simultaneously, how can any manager be expected to manage 30, 40 or 50 disparate risks? Our knowledge and experience of people leads us to take a different approach. Firstly, for any risk, no matter how trivial, is there a quick fix? If so, do it now. Then, with those risks that remain, identify the most significant, based on a comparative ranking process and determine which as the top X risks, where X is a number between 5 and 10.

The top X risks need to be assessed further. A more detailed explanation needs to be given about the activity, the impact and the consequence and considerably more detail needs to be provided about the likelihood. It is often at this stage that a considerable number of mitigating controls are uncovered, such as training, information, safe working practices, etc. The only reason to perform the detailed assessment is to allow the manager to understand what additional controls are required and to ensure that there is sufficient detail to provide an audit trail in the event of anything happening. Typically, it is the top X risks that form the basis of the risk register. Note two important details, firstly the top X risks will change on a very regular basis, as one risk is reduced to a tolerable level, a new one will be added. Secondly, within an organisational hierarchy, the top X risks of a senior manager will not be the amalgamation of the entire top X risks of more junior managers. Indeed, we would expect the only risks to flow up the management chain to be those that require large budgets or have strategic implications for the organisation.

Ergonomics Risks

The simplest method for getting ergonomics risks on to the management agenda is thus to demonstrate that:
1)   There is a reasonably foreseeable ergonomics risk
2)   The risk is significant when put into the context of the operation of the business and thus should be managed
3)   It is possible to manage the ergonomics risk by putting in more effective controls

Our experience of working with large organisations is that these risks become self evident provided that the organisation can agree on the common framework for risk reporting. As there is now an imperative for all Government organisations to have such a framework and as many Plcs are also introducing risk reporting frameworks, ergonomics becomes a much easier "sell".

The challenge for the ergonomist is to explain the benefits of good ergonomics using the terminology and frameworks that relate to risk management. Thus, describing the impact of poor ergonomics in terms of the effect on the business and emphasising the value of ergonomics controls against a background of controls assurance become important tools for the ergonomist to deploy.

Our experience is that if risk management teams are given the correct tools to describe and manage risks, many of the risks that they come up with of their own accord are ergonomics risks. For example, when working with a company that had a huge call centre handling all of its business, the notion that an "RSI epidemic" could cause the call centre to close came from the attendees on the workshop not from the ergonomist. This lead to a vigorous debate on what RSI was, what caused it and why the "epidemic" might happen. It was interesting that there was no debate on the catastrophic impact of only 5% of the operators going off sick, nor of the long term reputational damage to the company, the managers accepted that as given. The result of the discussion saw ergonomics firmly on the management agenda as just another item that they needed to manage to ensure that the business remained profitable (and hopefully a good place to work).

Conclusions

The requirement for organisations to have a better understanding of the risks that they face offers an opportunity for the ergonomics community. In describing corporate risk management, it is clear that the risks associated with the activities of people can be very significant. The challenge is to persuade organisations that the key to managing those risks lies in ergonomics and that the people to use to help manage the risks are ergonomists.

For ergonomists to engage in an organisation's risk management structure they must understand the terminology used in corporate governance. They must be able to demonstrate that ergonomics risks are both knowable in advance and comparable with any other category of risk.   Finally, they must be able to show that there are definite benefits in improving ergonomics controls, that will yield dividends to both those exposed to the risks and also those who are required to manage them.

ANALYSIS OF COGNITIVE ACTIVITIES OF NUCLEAR POWER PLANT CONTROL ROOM OPERATORS IN CASE OF ABNORMAL CONDITIONS

Kamran Sepanloo, Reza Jafarian

Atomic Energy Organization of Iran, North Karegar Ave., Tehran, Iran

Supervisory and control tasks performed in the control room of nuclear power plants depend to a large extent, on the operators' cognitive activities. Depending on the complexity of the task, cognitive functions can be categorized in skill-based, Rule-based, or Knowledge-based levels. In this paper, a model of operator cognitive performance is developed and used for identification and calculation of average basic human error probabilities for a number of selected operators of Bushehr Nuclear Power Plant (BNPP-1) control room assuming the occurrence of two accidents: 1) Small Loss of Coolant Accident (SLOCA), and 2) Steam Generator Tube leakage (SGTL), with consideration of two affecting performance shaping factors (PSFs): Time Pressure and Task Complexity.

Introduction

SLOCA and SGTL are time consuming events. The average cognitive processing time of 1.76 and 2.42 hours are reported for small LOCA and steam generator tube leak respectively, with the assumption that there are no delays due to operators' confusion with other events [Sheridan, 1983]. A small LOCA is assumed to be caused by a small break in a portion of letdown piping of primary circuit of a nuclear power plant. It causes the flow of the primary coolant into the containment building, resulting in a loss of coolant inventory. The role of human factors in the design of control rooms in proper handling of the emergencies by the operators is quite significant. As an example, in one advanced control room simulator, in case of a simulated loss of coolant accident, 500 emergency lights go on or off within the first minute and 800 within the second minute after the initial alarm. These in turn cause severe stress and complexity to the crew and move their cognition function level into knowledge-based level [Sheridan, 1983]. In this analysis, it is assumed that in case of occurrence of the SLOCA and SGTL the operator cognition functions are in Knowledge-based level.

The nominal HEP is the probability of a given human error when the effects of plant specific PSFs have not yet been considered. The basic HEP (BHEP) is the probability of human error without considering the conditional influence of other tasks. The conditional HEP (CHEP) is a modification of the BHEP to account for influenced of other tasks or events.

THERP (Technique for Human Error Rate Prediction) is a linear model which is commonly used as the reliability model to describe the expected error rate as a function of the PSFs. The reliability model and the PSFs are represented as the following equation [Hollnagel, 1998], [Swain, Guttmann, 1983]:

$$P_{EA} = HEP_{EA} * \sum_{1}^{N} PSF_{K} * W_{K} + C$$

Where:
P_{EA}= the probability of a specific erroneous action,
HEP_{EA} = the corresponding human error probability,
C= numerical constant,
PSF_{K} = numerical value of performance shaping factor,
W_{K} = the weight of PSF_{K}, and
N=2 (the number of PSFs; Time Pressure & Task Complexity)

The actual probability that a specific type of action goes wrong is thus expressed as the basic HEP for that action type modified by the influence of the PSFs. The value of the P_{EA} is derived simply by multiplying the HEP_{EA} by the weighted sum of the PSFs, i.e., by combining them in a linear fashion. In this analysis, THERP method has been used to calculate the P_{EA}s.

Rasmussen's Stepladder Model

In this paper, a variation of the Rasmussen's stepladder model is used as the basis for error modeling [Gertman *et al*, 1993], [Hollnagel, 1998]. The objective is to develop a context-free model of erroneous actions. Distinguish is made among three types of performance that correspond to separate levels of cognitive functioning. This is commonly known as the skill-based, Rule-based and Knowledge-based (SRK) framework. A representation of the Rasmussen model is shown in Figure 1. Knowledge-based mistakes occur when people do not have ready procedure to apply to the situation they face. In this analysis, the two steps of problem solving and procedure selection are combined into one-steps the decision-making (D.M.). Thus, in contrast to failures at the skill-based and rule-based levels, knowledge-based failures are supposed to reflect the inherent qualities of novice performance.

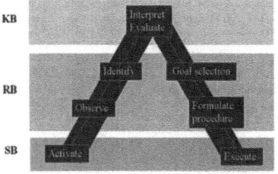

Figure 1. General representation of Rasmussen model [Hollnagel, 1998]

Analysis of BNPP-1 Operator's Cognitive Functions

The Attribute Rating Survey

Since characteristics of performance are different for each accident, two specific accidents are considered for operators' response. Operators were selected carefully to give a good estimate of real situations. Each attribute is scaled from one to seven, with one and seven indicating lowest and highest impacts on the performance, respectively. 26 selected operators of BNPP-1 responded to the survey by filling the checklists. None of the operators had any experience in nuclear power plant operation but had passed their training career.

The Model and Analysis

The model used for the analysis is shown in Figure 2. The model is variation of Rasmussen's stepladder model. The proper way of analysis of these two shaping factors is by using a full-scope simulator. Since the development of Bushehr NPP-1 full-scope simulator is not finished yet, the analysis of effects of the factors are somehow simulated by a replacing method, i.e. by use of computerized questioners (checklists). In one set of questioners; the time available to the operators was short, while they were strongly requested to answer to all questions. In other set of questioners, the number and ambiguity of the correct answers were increased, while there was no limitation in time.

For simulating the recovery, we asked some senior operators to help the examined operators to rethink over the impervious wrong answers.

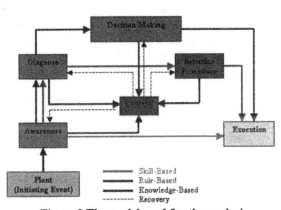

Figure 2.The model used for the analysis

- **Awareness**

To quantify the awareness failure, the curve fit approach (using optimal extrapolation) has been used. The nominal human error probability equation obtained by the experimental data [Azadeh, 1993], using the CURVE-FIT software is as the following:

$Y_{awareness (LOCA)} = 1.01907 * exp. (-1.889093 E-02 * t)$

$Y_{awareness (SGTL)} = 1.246984 * exp. (-0.2207275 * t)$

Where Y is the nominal amount of cognitive failure probability

- **Diagnosis**

The nominal human error probability equation obtained by the experimental data [Azadeh, 1993], using the CURVE-FIT software is as the following:

$Y_{Diagnosis\,(LOCA)} = 1.001472 * exp. (-1.470159\ E\text{-}03 * t)$
$Y_{Diagnosis\,(SGTL)} = 1.022996 * exp. (-2.273592\ E\text{-}02 * t)$

- **Decision-making**

The operator decision-making including problem solving and procedure selection is considered as shown in figure 2.

The nominal human error probability equation obtained by the experimental data [Azadeh, 1993], using the CURVE-FIT software is as the following:

$Y_{D.M.\,(LOCA)} = 1.001395 * exp. (-1.394067\ E\text{-}03 * t)$
$Y_{D.M.\,(SGTL)\,Recovery} = 1.00033 * exp. (-3.302184\ E\text{-}04 * t)$
$Y_{D.M.\,(SGTL)\,Non\text{-}Recovery} = 1.000533 * exp. (-5.322146\ E\text{-}04 * t)$

Calculation of Probability of Erroneous Action

The Basic Human Error Probability is calculated using a software, developed in Visual Basic6 language. The average values of PSFs coefficients for Bushehr NPP-1 control room is considered as:

$(\alpha_{T.P.}, \alpha_{T.Com.}) = 1.25, \quad C = 0$

Where $\alpha_{T.P.}, \alpha_{T.Com.}$ are coefficients that modify the values of PSFs assumed for the two attributes (i.e., task complexity and time pressure) based on the results of questioners.

Thus, the probability of the erroneous (BHEP) operator's decision making in the LOCA and SGTL is calculated as the following:

$$P_{EA_{K.B.}}^{SmallLoca} = \left[1 - HEP_{EA_{Awareness}}^{Nominal} \right] * \left[1 - HEP_{EA_{Diagnosis}}^{Nominal} \right] * HEP_{EA_{D.M.}}^{Nominal} * \sum_{1}^{n} PSF_k W_k$$

The results of calculations are presented in the tables 1 and 2.

Table 1. Probability of erroneous action in D.M. process for KB –behavior – SLOCA

t (min.)	$\alpha_{(T.P.)}$, $\alpha_{(T..Com.)}$	P_{EA}
35	Ideal Operator: 1	0.192
	BNPP Operators: 1.25	0.24
38	1	0.151
	1.25	0.189
65	1	0.016
	1.25	0.020
68	1	0.0127
	1.25	0.0159

Table 2.Probability of erroneous action in D.M. process for KB –behavior - SGTL

t (min.)	$\alpha_{(T.P.)}$, $\alpha_{T.(Com.)}$	P_{EA} (with recovery)	P_{EA} (Non recovery)
120	Ideal Operator: 1	0.45948	0.45957
	BNPP Operators: 1.25	0.57435	0.57447
130	1	0.37689	0.37697
	1.25	0.47112	0.47121
190	1	0.11479	0.11482
	1.25	0.14349	0.14352
200	1	0.09416	0.09418
	1.25	0.11770	0.11773

Conclusion

A cognitive model was developed based on Rasmussen stepladder model and used to determine the average basic human error probability of BNPP-1 control room operators. It was found that in the case of small LOCA, "misdiagnosis" is the main effective factor in the cognitive functions of selected operators of Bushehr NPP-1, but in the Steam Generator Tube leak accident, "decision-making" is the main effective factor.

By inclusion of recovery in small LOCA, it was observed that the magnitude of human errors in the levels of diagnosis and decision-making decreases, but in SGTL accident the decreases due to recovery was identified only in decision-making step.

Refrences

Azadeh, M., A., Modeling and Simulations of Integrated Information Processing and Operators' Decision Styles to Enhance the Reliability of Nuclear Power Plants, A Dissertation Presented to the Faculty of the Graduate School University of Southern California, In Partial Fulfillment of the Requirements for the Degree DOCTOR OF PHILOSOPHY, January 1993.

Gertman, David I. ,Blackman, Harold S. , *Human Reliability and Safety Analysis Data Handbook*, ISBN 0-471-59110-6 (alk. Paper), 1993.

Hollnagel, E. , *Cognitive Reliability and Error Analysis Method*, Free specimen copy gladly sent on request: Elsevier Science Ltd, The Boulevard, Langford Lane, Kidlington, Oxford, OX5 IGB, UK, ISBN 0-08-0428487, 1998.

Sheridan, T.B. (1983). Measuring, Modeling and augmenting reliability of man-machine system automatica, 19 (6), 637-645.

Swain, A.D. ,Guttmann, H.H. ,NUREG/CR-1278 *Handbook of Human Reliability Analysis with Emphasis on Nuclear Power Plant Applications* Final Report, SandiaNational Laboratories, Albuquerque, NM 87185, Manuscript Completed: June 1983.

CONCEPTUALIZING *EMOTIONAL ERGONOMICS* AND EXPLORING WAYS TO EMPOWER WORKPLACE DYNAMICS

Zawahir Siddique

PhD Student
Department of Engineering Design and Manufacture
Faculty of Engineering, University Malaya
Kuala Lumpur, Malaysia

The increasing intense competitive demands of the 21st century workplace challenge everyone, everywhere; to adapt in order to prosper under the fast paced trends in technological advancement. Hierarchies are morphing into networks; labour and management are uniting into teams; wages are coming in new mixtures of options, incentives and ownership. All these transitions owe immensely to the giant catalyst role played by *Emotional Intelligence*. On the other hand, Ergonomics as a science continues to grow and ensure the well being of employees at work place. *Emotional Intelligence* may just turn out to be the inevitable ingredient lacking in Ergonomics to empower workplace dynamics. Integrating these two giant fields to ensure the standards of the emotional as well as the physiological well being of employees at the work place; and hence to launch a revolutionary concept of *Emotional Ergonomics*, is the core essence of this paper.

Introduction

The evolution, application and popularity of the revolutionary concept, *Emotional Intelligence* has, no doubt, infused dramatic changes into the work place dynamics, in recent years. Ergonomics, on the other hand, as a multi-disciplinary science provides a standardized approach to the analysis of work systems with the emphasis on the interactions between human beings and machines. The guiding philosophy of Ergonomics is *Fitting the Job to the Man* (FJM). Ergonomics, therefore, "fits the job to the man" at the physiological, behavioural and cognitive levels.

Ergonomics, as a science, has had phenomenal influence on improving the work environment in ensuring the physical well being of the human body. Ergonomics has taken the human factor, and pushed it to the forefront of industry and corporations as a major consideration in work place dynamics. Though Ergonomics, as a science, has played a tremendous role in ensuring the well being of the human being at the physiological level and to a significant extend, at cognitive levels, its contribution to the emotional well being of the human being requires thorough introspection.

As a science, Ergonomics continues to grow, and becomes increasingly important to the work place. *Emotional Ergonomics*, which could be hypothetically defined as *"the science that deals with the influence of Emotional Intelligence of the employees towards the empowerment of work place dynamics"*, must, therefore become the natural extension of this growing field of Ergonomics.

The influence of *Emotional Ergonomics* would be, ideally, to prepare the work environment to be aware of the impact of emotion on performance; and make it sensitive to what it can do to enhance the process, as well as to help employees develop their own emotional intelligence in an effective manner. Increasing employee retention starts with an assessment to find out what may or may not be meeting employee needs, and creating undue emotional pressure. *Emotional Ergonomics* would contribute to help the employees obtain the skills that they would need to improve their emotional stability in the work place, leading to an increase in productivity and overall workplace empowerment.

The *Humanistic* Approach of Ergonomics

Measuring the 'humanness' of either work or people is being done in bits and pieces for several years. There is a need to create some relatively simple, but valid and effective means of assessing the issues so that the positions of both humans and work can be reliably gauged. Awareness of the full breadth of the human component in ergonomics has continued to climb for the last decade with the emphasis on more macro ergonomic approaches by Hendrick (1995). Nagamachi and Imada (1994) have demonstrated the powerful effects of a safety program that features high levels of attention being paid to what they call *human ware* in addition to software and hardware.

Hart (1998) emphasized on *Fitting the person to the person* and shed more light on his argument: "this possibility involves fitting both the person and the task(s) to the context of human being. Bringing the human and the work closer to the evolutionary reality of humanness will be helpful".

Participatory Ergonomics, another evolutionary concept, is a hybrid of other organizational effects designed to accomplish more than just ergonomics. Employees from all levels, functions and organizations work and communicate collectively in functional or natural groups or teams, using ergonomics as a forum. Through the participatory ergonomics process, a commitment is made to agree upon and attain desirable outcomes for the organization and work system (or macro ergonomics); and Workstations and specific tasks and jobs associated with them (or micro ergonomics) (Hendrick 1987; Imad and Nagamachi 1995; Wilson 1995).

The traditional view of ergonomics argues that the individual and the working system must operate in close harmony for efficient operation of the system. More recently, however, this view of the fundamental relationship between man and machine has been questioned by an increasing number of authors as being too simplistic. For instance, discussing the human-computer interaction, Eason (1991) argues that this viewpoint "as a form of conversation between different kinds of participants" misses the rich complexity of the interaction. Eason points out that we interact with machines not merely to exchange messages but to engage in complex tasks in the real world. The man-machine interaction, then, has a meaning, which is over and above that which can be expressed by simple, straightforward analyses of the component parts. Both the individual operator and the nature of the task and its outcomes inject this meaning into the system.

Wisner (1989) takes the argument beyond just interacting with computers to encompass the full domain of ergonomics. He argues that what is specific to ergonomics, as well as psychology, is that it should not try just to investigate specific 'properties of man'. Rather it should attempt to "understand how man uses his own properties in terms of a story, his own story and that of humanity, the part of humanity to which he belongs".

Paul Branton (1916-90) was perhaps the first ever ergonomist to infuse the person-centred philosophy into the science of ergonomics. Brantonian view of *person-centred ergonomics* emphasized a major shift in the way in which ergonomists should view the role of the operator within the working situation. The strength of Brantonian view, within this person-centred approach, lies in the need to consider not just the person's actions at work, but the purposes that underlie the actions. In doing so, Branton took a 'humanistic' view of behaviour. Thus, he viewed the person as central to the ergonomic system, who brings to the situation needs, wishes and desires, as well as the physical and cognitive abilities, which are necessary to carry out the task. Branton's person-centred view thus emphasized the importance of exploring factors beyond the mere physiological function. The person-centred approach viewed the human operator as purposive, information seeking, uncertainty reducing and responsible. Branton argued that by identifying the core nature of human input to the system, it is possible to devise novel ways of measuring it. A cybernetic model was also presented of behaviour taking a cyclical, servo-based, form in adaptation to internal and external pressures.

The shift in ergonomics thinking that has emerged in recent years has, therefore, been towards understanding more the nature of the individual within the system. The original concept of person and machine being almost equal partners is gradually giving way to a view which argues that the partners are not equal; that the person should play a more central role within a working system.

Impact of Emotion at Workplace

The past two decades have witnessed substantial advances in understanding the structure and role of emotions in human behaviour. Industrial/Organizational (I/O) psychologists and other applied researchers have recognized the relevance of such advances for understanding workplace behaviour, producing a number of recent articles, special issues on emotions and emotions at work.

Scherer (1994) provides a compelling reason for concern with emotions. He maintains that emotions are an interface that mediates between environmental input and behavioural output. This interface has strong ties to motivational-implementation systems and helps ensure that the central needs of a workplace are met.

Emotions are central components of human reactions to many types of stimuli and can influence organizational behaviour in a number of ways. Some of the ways are direct, such as the triggering of behaviour by emotions; where as other ways are indirect, such as emotions influencing behaviour through mediating mechanisms like motivation or cognition. Emotions are both bodily states and mental states, and they are also part of a larger information processing system (Clore, 1994; Isen, 2000)

Depending on their orientation, some researchers view emotions as primarily biological (Panksepp, 1998) while others view them as primarily psychological (Lazarus, 1991). The challenge that lies ahead is to convert emotionally suppressive workplaces into emotionally expressive workplaces.

Empowering Workplaces through Emotional Ergonomics

Conceptually, *Emotional Ergonomics* is a 'yet to be explored' topic. Apart from Dr. William Howatt's short and inadequate paper *Emotional Ergonomics in the Work Place*, there is hardly any source of information available on this topic.

The concept of *Emotional Ergonomics* should be floated into challenging work place environments. To introduce this new concept into the work place, the employer should be motivated to make the commitment to the necessary research by conducting extensive assessments to evaluate what is adding and taking away from the employees. Poor peer review systems, lack of communications and meaningful feedback, fear and intimidation are but a few that would be true contenders. To explore the information in the assessment phase, the employer would be prompted to conduct detailed surveys to determine what factors would add emotional strain to employees. This information could be gathered through workflow studies, focus groups, 360-degree surveys, and applying the various Emotional Intelligence assessment tools like *Multifactor Emotional Intelligence Scale, Emotional Quotient Map Self Scoring Version, Bar-On Emotional Quotient Inventory* and a variety of other tools that are being developed and explored on the work force.

The top three reasons employees leave work place are that they feel they are not respected and valued; they perceive they are not being paid to their potential; and there is no clear path for career growth and/or advancement within the organization (Howatt, 2002). In the workplace, as in all aspects, emotions may serve to motivate, organize, direct, and activate behaviour, but also may be disruptive to the other appropriate wok related and social behaviours. Converting the emotional weaknesses into strengths is the key to an empowered workplace. This is where Emotional Intelligence could provide the cutting edge. Cooper and Sawaf (1996) defined "Emotional Intelligence" as "the ability to sense, understand and effectively apply the power and acumen of emotions as a source of human energy, information, connection and influence".

Launching Ergonomics into a Resurgent Future

Caring about the employees' emotional condition will impact decisions about the architectural considerations and general environmental considerations (Ryerson, 1999). For instance, greater understanding about the effects of colour and lighting on the emotions of the workers would be vital. This resurgent approach would compliment the contributions of Ergonomics in merely attributing colour and lighting conditions to the mere physiological needs of the ears and eyes of the workers. This new direction will transform sterile workplace environments into warm, pleasant and productive surroundings. The integration of indoor waterfalls, easy go work layouts, aromatherapy and other ambience-oriented enhancements into the workplace will inevitably lead to increased productivity and health within the employees and consequently boost their emotional resilience.

The value of ergonomics is perceived to be measurable, such as the reduction in lost time due to back problems after new ergonomic chairs are supplied. *Emotional Ergonomics* aspires to develop formulas and models that could be used to measure the impact of *Emotional Intelligence* on work place dynamics. What would be the costs involved for not developing environments that are emotionally ergonomic? One only needs to look at the costs involved in replacing staff, hiring, training and sick/stress time off! *Emotional Ergonomics* may not be a quick fix solution initially, but once it is implemented it would be a mainstay and would eventually boost morale, productivity and employee retention. The impact of *Emotional Intelligence* is growing at a magnificent pace, but there is one vital domain where it needs to penetrate: Ergonomics. Therefore, quite inevitably, *Emotional Ergonomics* is a giant catalyst concept in the making!

References

Cherniss, C. and Goleman, D. 2001, *The Emotionally Intelligent Workplace*, First Edition, (Jossey-Bass, San Francisco)

Cooper, R.K. and Sawaf, A. 1996, *Executive EQ: Emotional Intelligence in Leadership and Organizations*, First Edition, (Grosset/ Putnam, New York)

Goleman, D. 1995, *Emotional Intelligence: Why It Can Matter More Than IQ*, First Edition, (Bantam Books, New York)

Goleman, D. 1998, *Working with Emotional Intelligence*, First Edition, (Bantam Books, New York)

Kroemer, K., Kroemer, H. and Elbert, K. 2001, *Ergonomics: How To Design For Ease And Efficiency*, Second Edition, (Prentice Hall, New Jersey)

Lord, R.G., Klimoski, R.J and Kanfer, R. 2002, *Emotions in the Workplace*, First Edition, (Jossy-Bass, San Francisco)

Oborne, D.J, Branton, R., Leal, F., Shipley, P. and Stewart, T. 1993, *Person Centred. Ergonomics: A Brantonian View of Human Factors*, First Edition, (Taylor and Francis, London)

Payne, R.L. and Cooper, C.L. 2001, *Emotions At Work: Theory, Research And Applications For Management*, First Edition, (John Wiley and Sons LTD, Chichester)

Schultz, D. and Schultz, S.E. 1998, *Psychology And Work Today*, Seventh Edition, (Prentice Hall, New Jersey)

Painting pictures in words: the development of an information web site for audio transcription

Debbie Taylor and Caroline Parker

Glasgow Caledonian University, Glasgow G4 0BA

It is said that a picture is worth a thousand words. However this is unlikely to be the case for people who cannot see the picture well, or cannot see it at all. Audio description (AD) is used to provide additional narrative to television, film or theatre to literally describe to visually impaired people the events that are important to the story line but which are not conveyed by the dialog or soundtrack. This paper describes the development of a prototype web-site which was designed to test the potential for use of web technology to support the dissemination and use of audio description materials in the UK. The results suggest that although there are usability issues to address there is a place for well designed AD websites. It is likely that they will support the experience of the visually impaired cinema and theatre goer and could bring entertainment more in line with disability discrimination acts.

Introduction

Going to the cinema or the theatre is a common and enjoyable social activity and one that most readers of this paper will have experienced on many occasions. For people with low or no vision however the dramatic arts use of visual forms of communication can greatly reduce its accessibility. It is said that a picture is worth a thousand words, and sighted people are unlikely to notice the extent to which actions and expressions are used to communicate the storyline of a film, play or operatic piece: the visually impaired audience however may be left with a mix of music, footfalls and noise, in between islands of dialogue. Audio description (AD) attempts to bridge the gaps between these islands.

What is Audio description?
Audio description provides an additional narrative to theatre, film and television (and also any other visual images) for the benefit of visually impaired people. Ideally, it verbally fills in details of a scene that carry meaning for the storyline but are normally communicated purely visually e.g. clothing styles, body language, colours and landscapes. These details are inserted during pauses in the actual dialogue or narration of the relevant media. For example, here is a section of proposed audio description for Steven Spielberg's "Close Encounters of the Third Kind":

> "The little boy has slipped out of his bed and is padding down the stairs in his Boston University t-shirt over his pyjama bottoms towards the open porch door and the bright light outside…he turns his head towards another sound…he toddles into the kitchen and stares wide eyed at the mess on the floor. He raises his head

towards the noise, his big round eyes fascinated.   His mouth opens in calm
surprise."   (ITC Guidelines for Audio Description, p.12)
Without this audio description and some visual cues, this scene from the movie would
be largely meaningless.

As a means of formalised description of entertainment, the concept was first
developed in the 1970s by Gregory Frazier at San Francisco State University and the first
audio description for television was created in 1987 by the Boston broadcasting company,
WGBH (Tanton et al., 2000). Its usefulness has been demonstrated in several research
projects. For example in Schmeidler & Kirchner's 2001 study those people who used AD
said that it made the programs more enjoyable, interesting and informative and that they
were more comfortable talking with sighted people about the content of the programs that
had AD on them. However, despite the passing of nearly 30 years since its emergence
Audio Description is not universally available or widely used.

Cost and universal access
The additional cost of providing this service has obviously been a barrier to the
widespread provision of AD in the past but this may now be changing. Under the UK's
Disability Discrimination Act (DDA) and the Americans with Disabilities Act (ADA), it
is against the law for service providers to treat people with disabilities less favourably
because of their disability. Since October 1999 in the United Kingdom, service providers
have had to make "reasonable adjustments" to the way they provide their service and as
of October 2004 other reasonable adjustments may have to have been made to their
premises so there are no physical barriers stopping or making it unreasonably difficult to
use the service. Ergo, it can be said that theatres, cinemas and TV will have to provide a
"reasonable" amount of audio descriptions to make them accessible. They may also have
to provide audio information e.g. about events/programmes. There is likely therefore to
be pressure on companies to provide AD as part of their accessibility measures.

Lack of awareness
It is also the case that visually impaired people may not be aware of the AD services
provided or find out about them too late for them to be used. Information about audio
described events can be circulated around local and national organisation e.g, Glasgow
and West of Scotland Society for the Blind (GWSSB) and the Royal National Institute for
the Blind (RNIB) or a tape or Braille programme of events sent out on request. However,
this approach is only useful for those who are members or users of such organisations,
and not all visually impaired people are. They also operate fairly slowly and as film and
theatre productions tend to be available in a location for relatively short periods of time
the information can be out of date before it arrives.

Audio Description Information Website

As the pressures on cinema organisations and theatres increase there is likely to be a rise
in the availability of AD material, e.g. by 2008 all digital channels should have 10% of
content audio described (ITC, 2003), and the remaining problem will be access to it. A
potential solution to this problem might be the development of an AD specific website.
According to the RNIB (2003) thousands of visually impaired people in the UK use the
Internet. With around one million visually impaired people recorded in the UK alone,
(the RNIB believe this figure to be closer to 2 million) and this figure set to rise over the

next few years, it is fair to assume that the amount of visually impaired computer users will also rise. If web accessibility guidelines are adhered to visually impaired people can access the web fairly easily with devices such as screen readers and Braille readers, large text, magnifiers and talking browsers.

To test the feasibility of this idea a prototype web-site was developed as part of a project at Glasgow Caledonian University. The prototype web site, which uses a single film as an example, demonstrated the kind of features that could be included in an online information service for audio described media. Currently theatres and cinemas in the UK provide this sort of information in an analogue tape format or not at all. According to research so far a lot of the information below is not available without specifically requesting it. The web site included information such as:

- The cinema's programme of audio described and non-audio described films.
- Details of how to access the cinema: everything from transport information on how to reach it to what facilities the cinema has for visually impaired people for example Braille or large print information about films.
- Spoken word information describing the kind of aspects that visually impaired people would miss out on when watching a film.
- An audio described trailer or clip of a film.

The web site could also be a marketing tool and links from the cinema in question's web site, information in the standard programme as well as using the standard route of national and regional organisations for the blind to make this a well-visited web site.

Design

Web page design
Cascading Style Sheets (CSS) were chosen as a method of tidy, accessible design quite early on in the process. A design that was as uncluttered and easy to follow as possible was necessary for the target audience. Furthermore, from a design level, it makes the web site far easier to maintain.

The other key design choice was the use of XHTML 1.0 transitional. It is the latest standard for marking up web pages and requires a stricter mark-up approach. This is particularly useful because, if validated correctly, there should be no mistakes in the mark-up and it should make for a smoother read for a screen reader or refreshable Braille display. XHTML transitional rather than strict was used was because of the need to cater to users with older browsers. The main design of the site is quite linear with a "Quicklinks" page for those who are familiar with the web site. Because screenreaders are linear in their output, making the links as descriptive as possible was essential e.g. Introduction instead of 'click here'!

It proved initially quite difficult to cater for the needs of both blind *and* partially sighted users simultaneously. Blind users may utilise Braille and screen readers whereas partially sighted users often use specialised magnifying equipment. The first group need an easy to follow format, and instructions only need to be placed early on in the page to be accessible, however issues like font size and placement have also to be considered if text enlarging systems are used. The solution here was to have a fairly large font to start with and, again, because style sheets rather than inline settings have been used, they are easy for the user to change should they choose to do so. Minimalist design was employed and extensive use of graphics was avoided. These would have made the site less accessible because of download times and would also clutter the screen and make it harder for those using magnifiers to read.

A concept called 'breadcrumbs' was applied. These are cues to make it as unlikely as possible for users to get lost. The breadcrumbs in this site consisted of "You are here: Home>>Technical explanation." The "home" part of the breadcrumb is always a link to the home page to provide an extra method of finding a way around the site.

As this was a prototype site and would be inviting people to comment on it a technical explanation and introduction page were also provided. Access to previous reports was also given. The thinking behind all this information was to give testers and insight into exactly what they were testing should they choose to find out more about it.

An accessibility section was also created for partially sighted testers who may want to change the style of the web page. Instructions were given on how to turn off style sheets in Internet explorer and, for the rest of the instructions and for other browsers, an outside link to VIVID - Virtual Information for the Visually Impaired and Dyslexic – was provided. A link on creating one's own style sheet was also provided.

Audio material

One of the problems with existing analogue provision is that the user cannot skip parts or go back without fast forwarding the tape, a fairly time consuming and hit and miss operation. After analysis the analogue material was broken down into meaningful chunks and each one made available on the web-page. The background material files were recorded using a simple computer microphone and saved as a WAV format. They were recorded in 16 bit stereo using Cool Edit Pro™ so as to have as little sound degradation as possible. Helix Producer Basic 9™ was used to to further prepare them for fast streaming. This approach produces little loss of sound quality.

Video material

In order to accommodate users with slow Internet access and yet still provide some video footage for trailers the DivX™ format was chosen. To synchronise the audio and visual components both Macromedia™ and Quicktime Pro were explored but rejected in favour of SMIL™ (Synchronised Multimedia Integration Language) as this had a more flexible approach to timing. Unfortunately this flexibility came at a price, the need for the user to have the latest version of a multimedia player (e.g. RealPlayer™) resident on their machines. Unfortunately creating a highly compressed format for low bandwidth users had serious consequences for the video quality. In order to get over this problem users were offered two choices, one to view a streamed and highly compressed version of the trailers and two, to download a better quality format for viewing off line. The video file was taken from a DivX™ version of Monsters, Inc™. It was edited down using Virtual Dub™ version 1.5.1. This is an extremely basic editing tool that allows easy, graphic lead editing of trailers. The trailer was first shortened to 20 seconds and then halved in size. By decreasing the size, this makes for a more compressed trailer to stream online.

Content

An existing trailer from the Pixar™ & Disney™ film Monsters Inc™ was used to create an example of AD on the web site. The background material consisted of information about two of the main characters from Monsters, Inc: Mike Wazowski and James P Sullivan. A short description was recorded on what they looked like. A background information description was also created for the main set that features in the trailer: the scare factory. One of the criticisms of other AD exercises is that they talk over the dialogue (Chong, 1998). The trailer was developed to ensure that this did not happen.

User testing

Invitations to test the site were posted on Yahoo forums pertaining to visual impairments. The questionnaire is part of the web site at http://www12.brinkster.com/webdeb/audiodescribed

A questionnaire was devised to accompany the web-site and this was used to find out the following information: what hardware/software the user was utilising to access the Internet; what files the user listened to; and whether they thought the information provided would be useful to them in going to see a movie.

Results

Six people responded to the request for feedback, a much lower number than was hoped for. However this was very much a preliminary investigation of the approach and, according to Neilsen & Landauer's research (1993) the response should have been sufficient to find 80% of the problems with the site. Males and females were equally represented in the sample, one was in the 16-30 age bracket, three were aged between 31-45 and two between 46-60. Five of the respondents used a screen reader to access the site and the other used magnifying equipment. Four out of the six respondents had broadband access. Four out of the six respondents stated that they had not had access to this type of materials before. Four people also stated that they would prefer to use Internet to access this type of material. Most of the respondents (4/6) found the navigation of the site easy but the linear nature of the navigation was felt to be limiting e.g. "navigation should be circular, not linear." All respondents rated navigation as very easy using their preferred access method (reader and magnifier). None of the respondents made use of the user definable functions and so no data was gathered on the utility of these.

Four out of the six said they had no problems with streaming, one low bandwidth user reported problems and one did not reply to the question. None of the users bothered to download the files and comments suggest that this is because the activity is a responsive rather than a reflective one e.g. "Even though my connection is fast, I thought it is not really necessary to download something that I will probably not use and listen to again so I thought it would be easier and quicker to stream them."

All of the four who answered the question said that they thought the description material was helpful and all felt that more would have been useful. Most respondents (4/6) said that they thought the trailer with the description was enjoyable. Five out of the six respondents said that the trailer would make it more likely that they would go and see the film. Asked about the usefulness of AD generally the comments were also positive e.g. "the more AD the better"; "I would just have a better sense of the film. Also, I would know that the film makers took the time to have it [audio] described"; "It would give the film a three dimensional quality that is often lacking in films for people who are blind or low vision."

Not all feedback came through the official feedback form. Some more informal correspondence was received by way of e-mail and this very strongly indicated that people were unwilling to test the site because they didn't want to install the RealOne™

player, some objecting to this in very strong terms. Another issue raised by an email correspondent was the asynchronous nature of the communication, users he felt, wanted the information while they were watching and not before the event.

Discussion

The results from this preliminary investigation suggest that there is a place for well designed AD websites: that they are likely to provide a means of improving the experience of the visually impaired cinema and theatre goer and would bring entertainment more in line with disability discrimination acts.

The exercise highlighted many technical and usability issues which will be of great value for future web-site developments. The results suggest that the quality needed for smooth viewing of video requires a broadband connection and this suggests that this technology will not be able to be fully utilised in the UK until broadband is more universally available. The need to reduce reliance on specific multimedia playback software may also need to wait until such players have evolved a little further.

The extent of the site content and the low number of users however mean that further usability exercises will need to be carried out, with larger numbers and with a wider range of access methods. It would also be very useful to test usability issues in situations where the material is downloaded live during performance i.e. via a WAP phone or PDA, or where the material is downloaded onto digital playback equipment.

Acknowledgements

This paper would not have been possible without the input and contributions of those who participated in the study and took the time to email with their views. Thanks also to Scottish Opera for their support.

References

Chong, P (1998) Audio Description: Accessory or Accessibility? *The Braille Monitor*, vol **41** (No 7) pp472

Neilsen,J. & Landauer, T.K. (1993) A Mathematical model of the finding of usability problems. *Proceedings of ACM INTERCHI'93 Conference* (Amsterdam, The Netherlands, 24-29 April 1993), pp 206-213

ITC Guidelines for Audio Description (2003) http://www.itc.org.uk/itc_publications /codes_guidance/audio_description/index.asp

Royal National Institute for the Blind (2003). http://www.rnib.org.uk/xpedio/groups /public/documents/PublicWebsite/public_PressCentreStatistics.hcsp)

Schmeidler, E & Kirchner, C (2001), Adding Audio Description: Does It Make a Difference? *Journal of Visual Impairment and Blindness*. Vol. 95 (Number 4) pp197-p212

Tanton, N & Ware, T & Armstrong, M (2000) Access Services for Digital Television: Matching the Means to the Requirement for Audio Description and Signing. *International Broadcasting Convention (IBC 2000), Amsterdam, 8-12 September, Conference Publication.*

Workshops as a cost-effective way to gather requirements and test user interface usability? Findings from WMSS, a decision support development project.

Caroline Park[a]

Caroline Parker[a]

Denise Ginsberg[b]

[a]*Centre for Research in Systems and People*
Computing Division, Glasgow Caledonian University,
Glasgow, G4 0BA

[b]*ADAS Boxworth, Boxworth, Cambridge, CB3 8NN*

The Weed Management Decision Support system (WMSS) project is a 3.5 year long, collaborative LINK funded project. Its aim is to provide support for decision making in the control of weeds in winter wheat crops. One of the cornerstones of user-centered design is iteration and it was therefore essential that a project like WMSS, which prides itself on its user-centred approach, continued to consult users at every stage. However, like many research-based developments the WMSS budget for all activities was extremely limited and cost-effective mechanisms for user-inclusion had to be identified.

This paper describes the use of one such method, workshops, within the context of this project. It details the structure and planning of these sessions and in particular examines their usefulness as mechanisms for identifying user-interface and functionality issues at an early stage of decision support system design. The positive and the negative aspects of the approach are examined and the paper concludes that workshops, while limited in scope in comparison with face to face interviews are a very valuable tool.

Introduction

The UK devoted 1.902 million hectares of land to wheat production in 2003 and produced an estimated 14.83 million tones of wheat: a return of roughly 8 tonnes per hectare (Defra, 2003). This high level of productivity would not be possible without the use of chemicals to reduce the numbers of highly competitive weeds. The down side is that herbicides are expensive both in economic and environmental terms. Arable farm incomes have reduced considerably in recent years and the cost of herbicides has been significant (£95 million in 1999, Defra 2001). There is also considerable concern among environmentalists that that herbicide use may have a detrimental impact on the environment e.g. the presence of herbicides in water. The UK government is committed to supporting activities that will minimise pesticide use and has funded a considerable amount of research and development in this area. WMSS, is one of the projects funded by Defra under the Arable LINK programme[1]

The Weed Management Decision Support system (WMSS) project is a 3.5 year long, user-centred collaborative project which started in September 2000. Its aim is to provide support for decision making in the control of weeds in winter wheat crops. The project will encapsulate the most recent research and information on weed management and biology in a computer based decision support system: its aim being to enhance crop profitability at the same time as addressing environmental and biodiversity issues.

One of the cornerstones of user-centered design is iteration and it was therefore essential that a project like WMSS, which prides itself on its user-centred approach, consults users at every stage.   However, like many research-based developments the WMSS budget for all activities was extremely limited and cost-effective mechanisms for user-inclusion had to be identified.   WMSS needed cost-effective mechanisms for gathering user requirements and for testing the initial design ideas with users.

Approaches to user involvement

There are a number approaches available to those who wish to pursue the user-centered approach to software development: approaches that involve the user very closely in the development process i.e. DSDM (Howard, 1997) and Agile Programming Methods (Boehm, 2002) and tools for eliciting user opinion within these, and less intensive approaches i.e. interviewing, questionnaire, focus group and workshops.

Value of the workshop approach

Recruiting users into the design team (e.g. in DSDM) is possibly the most expensive option, unless they can be persuaded to give their time for nothing.   The 'hostage' problem is also something which has to be considered (Damodaran, 1996). It has been noted that users can become so involved in the development that they become developers rather than users and no longer provide useful opinions on usability issues.

Postal surveys are cheap ways of obtaining the views of a large number of people but are only really useful when answers to specific, simple, concrete questions are required.   Postal surveys are not so useful for identifying fuzzier issues such as the range of questions users ask or for feedback on a particular user interface design.

Face-to-face interviews are more appropriate than postal surveys or user team members but are also time consuming and costly.   The user population in agriculture is, by its very nature, distributed widely across the country and scheduling more than two or three people into a day is difficult.   The act of trying to fit people into a timetable to reduce the travelling effort also absorbs a considerable amount of time.   Workshops are much faster and easier to organise, requiring only that a sufficient quantity of people are invited to a pre-specified place to ensure a good discussion.

Workshops are also cheaper to run than interviews. There are minimal travelling and staff costs. For example, obtaining the views of eight people may take three days of interviewing, plus another couple of days to organise. A workshop, even if it takes two days to organise will only take half a day to run. Two workshops can be run 'back to back' to involve even more people for a similar outlay. As those invited to attend these workshops are usually very interested in participating they rarely request expenses and the travelling costs are therefore minor. The cost of providing refreshments and a buffet lunch are trivial in comparison.

---

[1] Defra (LK0916)

Use of workshops within WMSS

Workshops have been used within the WMSS project over the past two years for two main purposes: user requirements analysis and design development. The format and purpose of the two styles of workshop are outlined in this section.

Table 1: outline of requirements focused workshops

Topic	Minutes
Introduction to WMSS and to the focus group	5
Focus on issues related to aspects of decision making about weeds.	80
Participants were taken through four stages of weed decision making to refresh their minds about the issues they might examine when making decisions. At each stage they were asked to think about the questions they asked and the information they used to make decisions at this point. Discussion was encouraged.	
Coffee break	5
Areas in which additional support is needed	30
While the issues were fresh in their minds they were asked to split up into groups and list areas in which they felt more scientific support would be useful. They then came back together to create a group list to rank areas in order of importance to them.	
Most important weeds	10
Participants were asked to list and rank the weeds they dealt with in the order of their perceived importance.	
Feedback on the types of approach suggested by WMSS	20
Some of the mechanisms WMSS might use to deliver the information were outlined and participants asked to comment on them	
Availability of weather and observation data	20
As this information is essential to the running of the models within a weeds DSS it was important to find out what access participants already had to it.	
Completion of questionnaire	10

The session was designed to provide answers to as many of the questions raised by WMSS collaborators as possible. Where questions did not fit naturally into the discussion format they were collated into a questionnaire for participants to complete individually at the end of the session.

User requirements workshop
Four workshops involving 20 people took place shortly after the start of the LINK project. The workshops were designed to fulfil three main functions:
     a) To identify end-users main concerns associated with weed control decisions.
     b) To prioritise the areas in which users felt more scientific support was required.

c) To find out specific answers to questions the technical partners felt important to pose to users prior to engaging in the first stage of design and development.

All groups followed the same general format and were scheduled to take 3 hours each. Participants gave their time on a voluntary basis. Morning groups started at 9.30am and afternoon groups started at 1.30pm. A buffet lunch was provided. The general outline of the session is shown in Table 1.

Design development workshops

The next phase of the project required a different style of workshop; one based more on reaction than reflection. This phase started about 6 months after the beginning of the project and has lasted until recently when the field based usability trials began. Apart from the period of the foot and mouth crisis the project has run between 4 and 8 workshop sessions a year. The aim of these workshops was to involve a new set of users and to:

1. Validate the design decisions based on user requirements
2. Identify usability issues with the current design
3. Expand and add detail to the task requirements
4. Answer specific design questions posed by the technical partners

Although each of the workshops held in this period varied in the content and form of the material presented to and discussed by the participants they all followed a general pattern. In the first part of the session the participants were given an overview of the project, the current status of development and an overview of the main screens and their function as well as some indication of the format of the day. The second part of the workshop was devoted to the elicitation of feedback on the current state of design. In the earliest workshops quite a lot of this feedback was given during the walkthrough of the main screen ideas as these only existed as drawings or mock-ups in PowerPoint and there was no interactivity to speak of. However as soon as any part of the system allowed user interaction this became the focus of the workshop activity. Participants were split into groups of two or three to carry out this exercise. .

In order to structure the users interaction they were asked to carry out tasks, based on the original user requirements, and which reflected the expected use of the system. The tasks were listed on a task sheet which also held questions pertinent to each part of the system the users viewed. These varied from session to session and were a mix of general usability questions (e.g. How easy did you find creating a weed list?) and specific design queries (e.g. 'Do you want the value for margin or for yield to be the default on this screen'). A coffee break usually followed or was part of this section of the day and the session ended with a group discussion of the main issues.

The third part of the workshop was generally devoted to issues that the developers needed specific input on e.g. identifying the way in which users categorized weed species in order to inform the structure of the weed encyclopaedia and was conducted either in small groups or as a full session depending on the nature of the questions to be answered.

In the final part of the workshop the facilitator summarized the main points arising during the day and listed them on flip charts, seeking the participant's agreement on them and some indication of the importance and priority of the changes proposed.

Pro's and con's of the workshop approach
The workshop approach has been a useful one for the WMSS project in that it facilitated the gathering of user requirements and a forum for testing the emerging design with a

sample of end users. Workshops are also relatively cheap to run and less time consuming to organize than face to face interviews. They however have a down side.

Sample: as those who participate in the workshops do so in their own time and for no remuneration they are inevitably a biased sample, often polarized in their views about the technology. Most participants tended to be more technology aware than the average.

Unpredictability: again as a result of their voluntary nature it is hard to predict how many of those who state they will attend workshops will actually turn up. Given the nature of the industry however you can almost predict a low turn out if the weather is good and a full house if the land is too frozen or wet to get out on to.

Effort to organise. When one of the authors first started running workshops some 8 years ago there was always a strong response and little in the way of persuasion was required. The industry is getting smaller however and each person in it seems to be working increasingly long hours. It is therefore becoming a more time consuming activity to organize events and ensure adequate turn out.

Need for strong facilitation. While running a workshop is not a difficult task people management skills are essential. It is not always easy to get a view from all participants in the face of one or more strong voices and sessions have to run to time if they are to succeed.

Level of detail. Although the workshop approach does allow for a greater level of discussion and for putting flesh on issues it does not compare to the unstructured or semi-structured interview in this respect. In an interview it is possible to spend time exploring issues to some depth and to collect case material to support them.

Conclusion

In conclusion we feel that the workshop approach is an extremely valuable one, and despite the small numbers and sampling profile involved appears to generate a reasonably representative response to requirements and design, if the responses of new groups of users and to questionnaires are any judge. It is a pragmatic solution to the problem of involving people in the development of small scale and tightly funded software system. As an answer to the problem of falling turn out we have discovered that running workshops in collaboration with existing social or industrial societies and using pre-existing meeting dates is a fruitful way forward.

Acknowledgements

The authors would like to acknowledge the support of Defra who fund the Arable LINK scheme and the WMSS project, and also that of the projects industrial and research collaborators: HGCA, Syngenta, Bayer CropScience, Du Pont, BASF, Dow Agrosciences, ADAS, SAC, Rothamsted Research, Silsoe Research Institute.

References

Boehm. B, (2002) Get ready for agile methods, with care. *IEEE Computing,* **35** (1) Jan 2002

Damodaran. L, (1996) User involvement in the system design process – a practical guide for users. *Behavior and Information Technology,* **15** (6) p363-377

Defra (2001) Arable crop botany and weed science (AR04) – Policy rationale and scientific objectives. *aims.defra.gov.uk/homepage/documents/30.doc,* January 2001.

Defra (2003) http://www.defra.gov.uk/farm/arable/pdf/Market%20and%20 Production%20_web_.PDF.

Howard. A, (1997) A new RAD based approach to commercial information systems development: the dynamic system development method. *Industrial Management and Data Systems,* **97** (5-6) p175(3)

Parker. C.G., Clarke. J.H., (2001) Weed Management: Supporting better Decisions. *BCPC conference, Brighton, November 2001.*

TOWARDS A STANDARD METHOD TO TEST ANTI-SLIP DEVICES

Gunvor Gard and Glenn Berggård

Dept of Health Sciences, Luleå University of Technology , Hedenbrovägen, SE-961 36 Boden, Sweden.
Dept of Environmental Planning and Design, Luleå University of Technology, SE-971 87 Luleå, Sweden

The aim of this study was to describe the development towards a standard method to test anti-slip devices and main results from test of anti slip devices on a larger group of healthy individuals. Method development have been on-going and methods for evaluations of perceived walking safety and balance, videorecordings of walking postures and movements, measures of time to take on and off each anti-slip device, advantages/disadvantages with each anti-slip device as well as evaluation of priority for own use according to three criteria; safety, balance and appearance have been developed. In a test on healthy individuals three different designs of anti-slip devices were evaluated; heel device, foot blade device and whole-foot device on different slippery surfaces, gravel, sand, salt, snow and ice. The results showed that a heel device was perceived to be the safest. The heel device was perceived to fit the shoe and to be stable at heel-strike and stable on ice. The heel device had the highest priority according to walking safety, walking balance and choice for own use.

Introduction

Accidental falls are a common health problem and one third of persons aged 65 or more fall at least once yearly (Tinetti et.al.,1994). Injuries occur in approximately half of the falls and 10% of them lead to serious injuries (van Weel et al,1995). To develop strategies to prevent accidents and injuries due to slips and falls is important (Grönqvist, 1999, Gard och Lundborg 2000,2001). The traffic environment can be a risky environment where pedestrians can be injured due to slippery pavements and roads particularly in winter time. The total risk for slip and fall accidents depends on the interaction between individual behaviour, the task or activity to be performed and the external environment . Internal factors as individual factors and external factors as environmental factors interact with each other. Methods to reduce the risks for slip and fall accidents need to deal with internal as well as external factors and the opportunity to handle risk situations, the coping ability (Grönqvist, 1995).

Lord et al (2001) have summarised and classified the risk factors for accidents as psychosocial and demographic factors, postural stability factors, sensory and neuromuscular factors, medical factors, medication factors and environmental factors. There are also individual factors such as age and gender. Walking ability and walking patterns can be seen as risk factors for slips and falls among both men and women. Risky walking patterns and situations have been identified: these include a tripping walking pattern, rough and uneven surfaces and short length of the footsteps. In analyses of walking patterns and walking problems, different aspects can be observed: heel strike, toe off, length of steps, height of steps, symmetry of steps, rhythm of steps, rotation of different parts of the body and deviations in walking patterns (Grönqvist et al, 2001; Gard,2000, Gard & Lundborg, 2000,2001).). The primary risk factor for slipping accidents is according to Grönqvist, poor grip and low friction between the footwear (foot) and the underfoot surface (pavement) (Grönqvist,1995).

In Europe the work concerning EC type-examination and certification of Personal Protective Equipment for anti-slip protection (also called anti-slip devices) is mainly the responsibility of CEN Technical Committee TC 161 (foot and leg protectors). The essential health and safety requirements for the design, construction and manifacturing of personal protective equipment including anti-slip devices are set forth in the directive 89/686/EEC and its amendments, whereas the European standards have been worked out to devise practical solutions as how to harmonize those essential requirements. Currently there are no standards for anti-slip devices, wherefore these have been assessed directly against the essential health and safety requirements stipulated in the Council Directive 89/686/EEC (Grönqvist and Mäkinen, 1997). There is an urgent need to develop European harmonized standards for anti-slip devices in use by workers at the workplace as well as for the elderly pedestrians and for other private users. The focus of the standard should be on assessing the preventive effect of the device, including ergonomic and functional aspects. There is a need for subjective and functional methods for testing and evaluating safety aspects of anti-slip devices. Using an appropriate anti-slip device as a pedestrian may prevent the risk of slips and falls on ice and snow. When trying to improve the function of anti-slip devices and the possibilities of safe pedestrian movement during winter, methods to test anti-slip devices need to be developed.

The aim of this paper is to present the development of a standard method to test anti-slip devices as well as the main results of test of anti slip devices on healthy individuals in Sweden.

The development process

Method development

The development towards a standard method to test anti slip devices have been made by method development followed by experimental tests. The methods that have been developed are rating scales for perceived walking safety and balance, observations from videorecordings of walking postures and movements, measures of time to take on and off each anti-slip device, measures of advantages/disadvantages with each anti-slip device and listing of priority according to three criteria; safety, balance and appearance/own use. (Gard and Lundborg, 2000,2001). These methods were developed to describe functional problems in walking with different anti-slip devices and have been found to be reliable (Gard and Lundborg, 2000,2001). The percentage of agreement of the walking safety scale was 86% and the corresponding agreement of the walking balance scale was 88% (Gard and Lundborg, 2000,2001). Walking safety was defined according to Holbein and Chaffin (1997) as a stable posture is one where the body's centre of mass is within the base of support. Walking balance is defined according to Winter (1995) as a term to describe the dynamics of the body posture to prevent falling.

Four rating scales for evaluation of observed walking movements have also been developed (Gard and Lundborg, 2000,2001). The dimensions evaluated were 1) walking posture and movements including normal muscle function in the hip and knee 2) walking posture and movements in the rest of the body (head, shoulders and arms), 3) heel-strike and 4) toe-off. All four dimensions were evaluated by observation scales ranging from 0 to 3. Interreliability test of these scales has shown the percentage of agreement between two observers to be 85%, 80%, 86% and 85% respectively (Gard and Lundborg, 2000,2001).

Experimental set-up

For the experimentall tests five different walking areas with different slippery surfaces, each 10 meters long, have been used:

Ice with sand ($180g/m^2$)
Ice with gravel (4-8 mm ca 150 g/m^2)
Ice with 3-5 mm snow
Ice
Ice with salt ($9g/m^2$)

To evaluate realistic walking situations in the traffic environment, such as a pedestrian crossing, the walking cycle for each walking area was divided into six parts:

1. Walk "normally" across the whole area
2. Turn around
3. Walk rapidly 4-5 steps
4. Stop
5. Walk backwards 4-5 steps
6. Walk "rapidly" across the whole area

Before the test each subject had the opportunity to walk without anti-slip device on these six parts of the experimental walking cycle without stress so they were familiar with the situation. In the test situation each subject walked with each anti-slip device on the five different slippery surfaces.

Tests of 107 healthy subjects in different ages

In the development of a norm material 107 subjects, from the general population in the north of Sweden participated in the test of anti-slip devices. The anti-slip devices were new on the Swedish market and represented three different designs of anti-slip devices; heel device, foot blade device and whole-foot device (figure 1). The sequence of the devices was randomized for each subject. The subjects had no earlier experience from walking with the anti-slip devices.

Figure 1 in here

Results

What methods could be recommended in a European standard for anti-slip devices? In our opinion both subjective and objective methods must be used. The methods used in our study were all practical, functional and user-friendly and can be used in a Nordic and later a European standard. Video-recording of the walking cycle with a focus on heel-strike and toe off is important with analyses of similarities and differences in walking movements and postures between anti-slip devices and surfaces. We used two video cameras for registration of body movements. For the future more advanced video equipment can be used, focusing directly on the feet during the walking cycle, particularly focusing on heel-strike and toe-off, as they are the most relevant parts to study. Reliability test of a video-registration method focusing on heel-strike and toe off can be recommended. The results showed that the movements in the rest of the body were not of any significant importance when walking with anti-slip devices. Standing balance can today be objectively studied by measuring static postural sway in the antero-posterior and medio-lateral planes and by recovery of posture on a moving platform. However safe walking on slippery surfaces are more complicated, as it is influenced by each individual's momentary limits of stability, which may vary over time and type of slippery surfaces. For the future walking balance, in itself, could be studied on a moving platform. For attaining dynamic stability of locomotion the step length and the walking cycle time control appear to be important In a European standard for anti-slip devices step length and walking cycle time ought to be recorded and compared between anti-slip devices and surfaces. Also studies of the walking pattern in a frontal plane concerning abduction of the legs on different slippery surfaces could be done and compared. The rating scales for perceived safety and balance, the list of subjective priorities for own use and the listing of perceived advantages and disadvantages were functional and practical methods and could be recommended for further use. It is important to include both objective and subjective methods in a standard. This methods can be included in a Nordic and later in a CEN-standard for type-examination of anti-slip devices.

The subjects perceived that walking with an anti-slip device could improve the walking safety on snow, ice and salt. The heel device was perceived to be the safest one on all five surfaces, followed by the toe-device and the whole-foot device. The subjects perceived that an anti-slip device can improve the walking balance on snow and ice, but not on the other surfaces. The heel device was perceived to be the one with the best walking balance on ice and snow.
Most subjects walked with a normal muscle function in the hip and knee and with normal movements in the rest of the body when walking with or without an anti-slip device on all surfaces. The heel device was perceived as the most rapid one to take on and the toe device as the most rapid one to take off. No significant differences were noted in time to take on or off between the devices. Advantages mentioned were that all three devices were perceived as having good foothold. The heel device was perceived to fit the shoe and to be stable at heel-strike. The toe device was easy portable and stable on ice. The whole-foot device was comfortable to walk with and safe on snow. The heel device had the highest priority according to walking safety, walking balance and choice for own use

Conclusion

In a Nordic standard for testing anti-slip devices both subjective and objective methods should be recommended.

References

Gard,G., Lundborg, G., 2000. Pedestrians on slippery surfaces during winter – methods to describe the problems and practical tests of antiskid devices. *Journal of Accident Analysis and Prevention* **32**: 455-460.

Gard,G., Lundborg, G.,2001. Test of anti-skid devices on five different slippery surfaces. *Journal of Accident Analysis and Prevention,* **33**:1-8.

Gard G., 2000. Prevention of slip and fall accidents.Risk factors – methods – suggestions for prevention. *Physical Therapy Review* **5**:177-184.

Grönqvist, R.,1995. *A dynamic method for assessing pedestrian slip resistance.* Thesis. Finnish Institute of Occupational Health, Helsinki.

Grönqvist, R.,1999. *Slips and Falls.* In: Biomechanics in Ergonomics, (Ed. S Kumar), Chapter 19, Taylor and Francis, London.

Grönqvist, R., Mäkinen, H.,1997.*Development of standards for anti-slip devices for pedestrians.* In Fifth Scandinavian Symposium on Protective Clothing,may 5-8, Elsinore, Denmark.

Holbein,M.A., Chaffin,D.B.,1997. Stability limits in extreme postures:effects of load positioning, foot placement and strength.*Human Factors* **39**,456-468.

Lord, S.R., Sherrington, C., Menz, H.B., 2001. *Falls in older people.* Cambridge Universityu Press.

Tinetti, M.E., Baker,D.I., McAvay,G., Claus,E.B., Garrett,P., Gottschalk,M.,K.,Trainor K., Horwitz, R.I.,1994.A multifactorial intervention to reduce the risk of falling among elderly people in the community *N Engl J Med* **331**:821-7

Van Weel, C., Vermeulen, H., van der Bosch, W.,1995. Falls, a community care perspective.*Lancet* **345**:1549-51.

Winter,D.A.,1995.*Anatomy,biomechanics and Control of Balance during Standing and Walking,* Ontario,Canada:University of Waterloo.

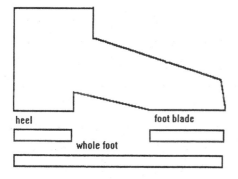

heel   foot blade

whole foot

Figure 1. Principal design of the anti-slip devices; No. 1 is a heel device, No 2 a foot blade or toe device, No 3 a whole foot device.

AUTHOR INDEX

SUBJECT INDEX

For Product Safety Concerns and Information please contact our EU representative GPSR@taylorandfrancis.com Taylor & Francis Verlag GmbH, Kaufingerstraße 24, 80331 München, Germany

Printed and bound by CPI Group (UK) Ltd, Croydon, CR0 4YY

01/05/2025

01858496-0001